EXCITATION–CONTRACTION COUPLING IN SKELETAL, CARDIAC, AND SMOOTH MUSCLE

ADVANCES IN EXPERIMENTAL MEDICINE AND BIOLOGY

Recent Volumes in this Series

Volume 304
REGULATION OF SMOOTH MUSCLE CONTRACTION
Edited by Robert S. Moreland

Volume 305
CHEMOTACTIC CYTOKINES: Biology of the Inflammatory Peptide Supergene Family
Edited by J. Westwick, I. J. D. Lindley, and S. L. Kunkel

Volume 306
STRUCTURE AND FUNCTION OF THE ASPARTIC PROTEINASES: Genetics, Structures, and Mechanisms
Edited by Ben M. Dunn

Volume 307
RED BLOOD CELL AGING
Edited by Mauro Magnani and Antonio De Flora

Volume 308
CELLULAR AND MOLECULAR MECHANISMS IN HYPERTENSION
Edited by Robert H. Cox

Volume 309A
PURINE AND PYRIMIDINE METABOLISM IN MAN VII, Part A: Chemotherapy, ATP Depletion, and Gout
Edited by R. Angus Harkness, Gertrude B. Elion, and Nepomuk Zöllner

Volume 309B
PURINE AND PYRIMIDINE METABOLISM IN MAN VII, Part B: Structural Biochemistry, Pathogenesis, and Metabolism
Edited by R. Angus Harkness, Gertrude B. Elion, and Nepomuk Zöllner

Volume 310
IMMUNOLOGY OF MILK AND THE NEONATE
Edited by Jiri Mestecky, Claudia Blair, and Pearay L. Ogra

Volume 311
EXCITATION-CONTRACTION COUPLING IN SKELETAL, CARDIAC, AND SMOOTH MUSCLE
Edited by George B. Frank, C. Paul Bianchi, and Henk E. D. J. ter Keurs

---

EXCITATION–CONTRACTION COUPLING IN SKELETAL, CARDIAC, AND SMOOTH MUSCLE

Edited by

George B. Frank
University of Alberta
Edmonton, Alberta, Canada

C. Paul Bianchi
Thomas Jefferson University
Philadelphia, Pennsylvania

and

Henk E. D. J. ter Keurs
University of Calgary
Calgary, Alberta, Canada

PLENUM PRESS • NEW YORK AND LONDON

Library of Congress Cataloging-in-Publication Data

Excitation-contraction coupling in skeletal, cardiac, and smooth
muscle / edited by George B. Frank, C. Paul Bianchi, and Henk E.D.J.
ter Keurs.
p. cm. -- (Advances in experimental medicine and biology ; v.
311)
"Proceedings of the Third International Symposium on Excitation
-Contraction Coupling in Skeletal, Cardiac, and Smooth Muscle, held
June 26-30, 1991, in Banff, Alberta, Canada"--T.p. verso.
Includes bibliographical references and index.
ISBN 0-306-44194-2
1. Muscle contraction--Congresses. 2. Calcium channels-
-Congresses. 3. Potassium channels--Congresses. 4. Excitation
(Physiology)--Congresses. I. Frank, George B. II. Bianchi, C.
Paul (Carmine Paul), 1927- . III. ter Keurs, H. E. D. J.
IV. International Symposium on Excitation-Contraction Coupling in
Skeletal, Cardiac, and Smooth Muscle (3rd : 1991 : Banff, Alta.)
V. Series.
[DNLM: 1. Calcium Channels--physiology--congresses. 2. Muscle
Contraction--drug effects--congresses. 3. Muscle, Smooth--drug
effects--congresses. 4. Muscles--drug effects--congresses. W1
AD559 v. 311 1992 / WE 500 E96 1991]
QP321.E93 1992
591.1'852--dc20
DNLM/DLC
for Library of Congress 92-3573
CIP

Proceedings of the Third International Symposium on
Excitation-Contraction Coupling in Skeletal, Cardiac, and Smooth Muscle,
held June 26-30, 1991, in Banff, Alberta, Canada

ISBN 0-306-44194-2

Printed in the United States of America

PREFACE

The Third International Symposium on Excitation-Contraction Coupling in Skeletal, Cardiac, and Smooth Muscle, organized by George Frank, C. Paul Bianchi, and Henk E.D.J. ter Keurs, was held in Banff Centre, Banff, Alberta, Canada during June 26 to June 30, 1991. The theme of these symposia has been to recognize the similarities and dissimilarities of excitation-contraction coupling in skeletal, cardiac, and smooth muscle. Cross fertilization of concepts of excitation-contraction coupling in these three types of muscle has occurred since the early studies in the late fifties and early sixties on skeletal muscle. Investigators in each field meet only at specialized symposia which exclude investigators in the other fields. The purpose of the symposia has been to bring together international investigators studying excitation-contraction coupling in skeletal, cardiac, and smooth muscle so that we may learn from each other and hence provide a more global concept of excitation-contraction.

The Third International Symposia has accomplished its objective as we recognize that calcium channels of the sarcolemma and the sarcoplasmic reticulum play key essential roles in excitation-contraction coupling in all three types of muscles. In skeletal muscle the recognition that E-C coupling consists of two parallel mechanisms, one dependent upon a dihydropyridine voltage-sensitive sensors coupled to calcium release from the terminal cisternae via the ryanodine sensitive channel in the foot structure of the triad. The foot structure is dependent upon a membrane bound calcium and binding site associated with the voltage sensor which upon release is involved in the charge transfer observed during excitation-contraction coupling. The single twitch and rapid discharge of calcium is dependent upon the feet-operated channels. However, when frequency of stimulation is increased, a second coupling mechanism which involves calcium influx through L-calcium channels and is related to the positive staircase and tetanus tension development. The concept of two parallel systems one involved in the single twitch and both involved in regulating force development as frequency of stimulation is increased.

Two predominant types of calcium release channels from the sarcoplasmic reticulum of striated, cardiac, and smooth muscles were discussed. The first type consists of a calcium release of calcium channel involved in rapid calcium release and the second, the IP_3-operated channel which possesses much slower kinetics of calcium release and may involve depression of the calcium release of calcium channels allowing calcium-dependent functions of the cells to take place without necessarily involving contraction.

The ground work for advancing to a new level of understanding of excitation-contraction coupling in skeletal, cardiac, and smooth muscle was laid by the 3rd International Symposium on excitation-contraction coupling.

The Editors

CONTENTS

SPEAKERS

PRESENTERS

Na/Ca EXCHANGE AND FIRST MESSENGER Ca IN SKELETAL MUSCLE EXCITATION - CONTRACTION COUPLING

Brian A. Curtis

University of Illinois College of Medicine at Peoria
Peoria, IL 61656 USA

Depolarization initiates a process in the wall of the transverse tubular system (t system). This process gives rise to a transient electrical signal - charge movement - which is closely linked to the release of Ca from the sarcoplasmic reticulum (SR) via the foot process. Two types of linkage have been discussed at this Conference: 1) structural links spanning the t-SR gap, such as the rigid rod model (Schneider and Chandler, 1973), and 2) chemical links including trigger Ca (Bianchi, 1969 and Frank, 1980).

A Ca LINK The rapid activation of Ca-induced Ca release from the SR (Fabiato, 1984 and Smith *et al.*, 1986) and the regenerative nature of contraction induced by Ca application to skinned muscle fibers (Podolsky, 1964) suggest a first messenger Ca linking t system with the foot process of the SR and a second messenger Ca released by the SR activating the contractile proteins. Ca effectively stimulates Ca release from isolated SR vesicles (Meissner, 1984) and from foot processes in bilayers (Hymel *et al.*, 1988). When SR release of second messenger Ca is blocked, intracellular dyes show a small increase in intracellular Ca (first messenger Ca?) soon after t system activation (Baylor *et al.*, 1989).

The source of first messenger Ca has remained elusive because skeletal muscle will contract in Ca-free solutions (Curtis, 1963) and when voltage gated Ca channels are blocked (Gonzalez-Serratos *et al.*, 1982). A protected source of first messenger Ca is the rapidly exchanging Ca store on the wall of the t system (Winegrad, 1965). 3Na/Ca exchange is a depolarization activated mechanism of Ca entry (Sheu & Blaustein, 1986), which could produce both the charge movement signal and Ca influx. Both whole skeletal muscle (Cosmos & Harris, 1961)

Excitation-Contraction Coupling in Skeletal, Cardiac, and Smooth Muscle
Edited by G.B. Frank *et al.*, Plenum Press, New York, 1992

and sarcolemmal vesicles (Gilbert & Meissner, 1982) show Ca movement directly linked to reciprocal movement of Na. The missing link is to demonstrate the existence of first messenger Ca, which I will do in this paper.

CALCIUM INFLUX A contraction related Ca influx has been demonstrated in whole sartorius muscle (Bianchi & Shanes, 1959) and in single muscle fibers (0.02 pmolCa/twitch/fiber, Curtis, 1966). Ca entry into contracting and D600 paralyzed single fibers can be parsed between: 1) a resting influx (0.5 pmol Ca/min); 2) a concurrent, voltage sensitive, contraction unrelated influx (1.1 pmol Ca/min depolarized); and 3) a contraction related influx (1.3 pmol Ca/contracture) which occurs *after* repolarization (Curtis & Eisenberg, 1985). The later influx is refilling one or more Ca stores emptied during the preceding contracture.

t-WALL Ca STORE The source of first messenger Ca has been a major concern because fibers will contract once in Ca-free solutions if the membrane voltage is maintained by current passage (Curtis, 1963), or the addition of Mg (Luttgau, 1963). These fibers must be returned to Ca containing solutions before they will contract again (Luttgau & Spieker, 1979). Fibers in 1 mM EGTA show a diminution of twitch tension with a time constant of some 60 min before tension drops in half at 25 min of exposure. Even after 40 min of exposure, these fibers show some twitch tension (Armstrong *et al.*, 1972). Hence, t lumen Ca is an unlikely source of first messenger Ca.

A Ca store can be localized to the transverse tubular wall by radioautography and quantitated by Ca efflux studies. Winegrad (1965) localized a rapidly exchanging Ca fraction to the Z line and a slowly exchanging Ca fraction to the SR. After 5 min of ^{45}Ca influx, the Z line source contained 10 times more Ca than expected in the t lumen. "Most of the exchange of intracellular calcium with extracellular calcium takes place between transverse tubules and terminal cisternae" (Winegrad, 1970).

Study of the rate of ^{45}Ca efflux cannot localize the sources of Ca, but can characterize these sources with greater accuracy. ^{45}Ca efflux (Curtis, 1970) is described by two exponentials with time constants of 18 and 300 min---strongly suggestive of efflux from two compartments; these are very similar to the two anatomical compartments shown by Winegrad, both in exchange rate and size. The Ca content of the slower time constant compartment is very similar to the Ca content of the SR as determined by electron probe analysis (Somylo *et al.*, 1981). Using the morphometric parameters of Mobley & Eisenberg (1975) and a 140 um fiber 0.7 cm long, the SR contains 95-100 pmol Ca, the walls of the transverse tubule 8-10 pmol Ca, and the t lumen 0.7-0.9 pmol Ca. The Ca efflux rate from the t wall is unaffected by low Ca solutions.

The t wall contains 3-5 pmol/fiber dihydropyridine receptors (Schwartz *et al.*, 1985). At this conference Graham Lamb pointed out that most voltage gated Ca channels are open only 5-10% of the time when activated. If these Ca channels are typical, all of the dihydropyridine receptors could be Ca channels which bind 2 Ca/channel or a total of 6-10 pmol Ca. Most of the Ca in the t wall may well be bound to Ca channels.

The amount of charge moved in a frog fast fiber in response to depolarization may be as large as 50 nC/ uF (Rios & Brum, 1987). This is the equivalent of 0.25 pmol of Ca entering a fiber via a 3Na/Ca exchanger. For charge movement to occur, a cation site, rapidly accessible to the extracellular space, must be occupied. This site binds Ca, Mg, and Na in the ratio 100:10:1 (Brum *et al.*, 1988). Both dihydropyridines (Brum & Rios, 1987) and tetracaine (Almers & Best, 1976) block charge movement and subsequent contraction.

I conclude that Ca bound in the transverse tubular wall is in a strategic position to supply first messenger Ca and will do so in Ca-free conditions. Ca binding to the t wall is necessary before charge movement, and, hence, contraction can occur.

3Na/Ca EXCHANGE 3Na/Ca exchange has been demonstrated in a wide variety of tissues (Philipson, 1985). In whole skeletal muscle, coupled Ca and Na fluxes have been demonstrated by Cosmos & Harris (1961), Watson & Winegrad (1973), and Caputo & Bolanos, (1978). Gilbert & Meissner (1982) and Donoso & Hidalgo (1989) showed similar fluxes in sarcolemminal vesicles.

Evidence for 3Na/Ca exchange in contracting skeletal muscle comes from studies of the rate of rise of tension, which should be controlled by the rate of presentation of first messenger Ca to the release site on the foot process. The Na/Ca exchange blocker dichlorobenzamil, elevated Na_0, and low pH (an inhibitor of Na/Ca exchange) reduce both the rate of rise of tension in K contractures and the depolarization induced inward driving force on the Ca side of the Na/Ca exchanger (Curtis, 1988).

Elevated Na_i increases both the driving force for depolarization induced Ca influx and the rate of rise of contracture tension. Dichlorobenzamil, low Na_o, and low pH each delay the beginning of the repolarization-repriming sequence in a manner consistent with a reduced outward driving force on the Ca depleted site. They do not alter the rate constant of repriming (Ca reuptake by the outward facing but depleted site). All of these observations are consistent with a depolarization induced translation of a 3Na/Ca exchanger moving first messenger Ca from a position in exchange with Ca_o into the fiber to activate the foot process of the SR *and* to generate the charge movement signal.

I have reviewed many threads of evidence consistent with first messenger Ca entering the muscle fiber via a voltage activated 3Na/Ca exchanger. First messenger Ca links transverse tubular depolarization and release of second messenger Ca from the SR. These threads of evidence have yet to be woven into a coherent pattern largely because legitimate skepticism lingers about the existence of first messenger Ca. In the experimental section of this report, I will demonstrate a small (0.3-0.6 pmol) Ca influx from the transverse tubular Ca store into the muscle directly related to contraction, which I believe is the long sought first messenger Ca.

METHODS

The experimental apparatus consisted of interconnected influx and efflux chambers quite similar to those previously described (Keynes, 1951 and Curtis & Eisenberg, 1985) except for the replacement of the G-M tube with a thin scintillation crystal and photomultiplier covered with a thin plastic sheet. The counting efficiency of this combination is 3%, in contrast to 0.2% for a parafilm covered G-M tube. This order of magnitude increase in sensitivity makes measurement of 0.01 pmol Ca possible. Bundles of 3-5 fibers were dissected from the tibialias anterior muscle of recently collected Mexican frogs, *Rana berlinderia*. All bundles gave at least 80% of the initial twitch tension at the end of the experiment for the data to be described here. Group data are given as the average +/- standard error of the mean with the number of observations in parentheses.

RESULTS

^{45}Ca ADDED DURING RECOVERY After tension was induced with 190 mM potassium methanesulfonate and the bundle relaxed spontaneously (top left, Fig 1), a 0.4 min rapid flush of Ringer washed the elevated K out of the influx chamber. 0.3 ml of very concentrated (2mCi/ml) ^{45}Ca Ringer was added to the 0.3-0.4 ml of Ringer surrounding the muscle bundle. 1 min later (at the end of the repriming period) the ^{45}Ca was flushed away. The bundle was placed over the scintillation crystal and counted while Ringer flowed briskly past, carrying off the ^{45}Ca as it left the bundle. Panel A of Fig 1 shows the loss of Ca* (cpm in the bundle multiplied by the specific activity of the influx solution) calculated per fiber in the bundle. The efflux can be described by two exponentials, which are highly suggestive of two compartments. I will designate them Ca_{fast} (t_c=4min) and Ca_{slow} (t_c=162 min). They represent Ca exchange with the t wall and the SR, respectively.

THE Ca FAST COMPARTMENT After 1 min of exposure to ^{45}Ca (Fig 1A), some 4 pmol Ca entered or exchanged with the Ca_{fast} compartment. There is excellent reason to believe the Ca_{fast} compartment represents the 8-10 pmol Ca bound to the wall or the t system. With a time constant of 4 min, only 2 pmol Ca would be expected to exchange. ^{45}Ca refilling an emptied compartment is a much more likely explanation. The amount of refilling of the Ca_{fast} compartment depends upon the duration of exposure to ^{45}Ca. At 0.5 min, 4.7 pmol Ca enters the Ca_{fast} compartment, at 1 min, 10.6 pmol Ca (Curtis, 1991) and at 2 min, 16.7 pmol Ca enters the Ca_{fast} compartment. Essentially all of the Ca expected in the Ca_{fast} compartment is released during a contracture and regained during the repriming period following repolarization. This uptake rate is very similar to the rate of mechanical repriming. Refilling the Ca_{fast} compartment may be the rate limiting step of repriming. Suffice it to say that the Ca_{fast} compartment can be readily filled with ^{45}Ca in 1-2 min immediately after repolarization rather than the 10-20 min of exposure to ^{45}Ca needed for exchange in a resting fiber.

The right side of Fig 1 shows the effects of following the same influx protocol with a second contracture. Fig 1B shows the efflux is described by a single exponential (t_c=128 min), from the Ca_{slow} compartment. During the second contracture, the Ca_{fast} compartment lost all the ^{45}Ca gained during the preceding influx period (Curtis, 1990). This observation is further evidence of the complete turnover of the Ca_{fast} comparment during each contracture-repriming cycle; more details await a future paper.

THE Ca SLOW COMPARTMENT The second component of ^{45}Ca efflux shown in Fig 1A has a long (162 min) time constant and is from the Ca_{slow} compartment which represents the 80-100

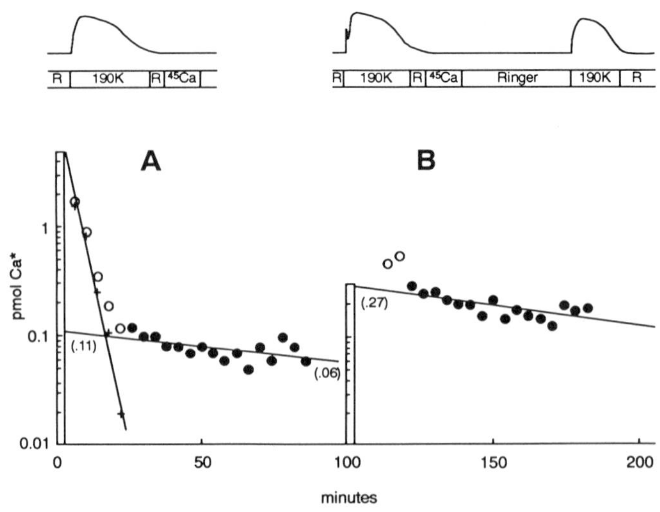

Figure 1 Extra Ca enters the muscle interior if the ^{45}Ca influx period is followed by a second contracture. A muscle bundle underwent a potassium contracture and was then exposed to ^{45}Ca for 1 min (upper left). Subsequent Ca* of the bundle is shown in A and consists of two exponentials: Ca* efflux from two compartments, Ca_{fast} and Ca_{slow}. Then the bundle contracted, exposed again to ^{45}Ca, but now contracted a second time (upper right). Ca* (B) can be described by a single exponential, Ca* efflux from a single compartment, Ca_{slow}. The second depolarization emptied the Ca_{fast} compartment *and* transfered an extra 0.1 pmol Ca to Ca_{slow}. This additional Ca uptake into the Ca_{slow} compartment, $Ca_{transfer}$, which averaged 0.53 pmol Ca, is related to the final depolarization-contraction cycle.

pmol Ca in the SR. The bundle took up 0.11 pmol Ca/fiber into the Ca_{slow} compartment during the 1 min exposure to ^{45}Ca; this is the resting influx. During 1 min of exposure, 0.62+/-0.12 pmol Ca (14) entered the Ca_{slow} compartment (1990 data). In 1991, the rate of flushing was increased, and the average resting influx into the Ca_{slow} compartment fell to 0.30+/-0.10 pmol Ca/min (8), significantly ($P<.05$) lower than the 1990 data. It is quite possible that the slower flush (done to minimize splashing of ^{45}Ca) left an elevated K, and part of the resting influx was a depolarization stimulated influx. Resting Ca influxes from other studies are 0.5 pmol Ca/fiber/min (Curtis, 1966 and Curtis & Eisenberg, 1985). The value 0.30 is significantly ($P<.05$) different. When measured previously, resting influx included exchange with both the Ca_{fast} and Ca_{slow} compartments. Resting Ca influx is probably entering the fiber via nonspecific pathways and is then actively transported into the SR.

The average time constant for the Ca_{slow} compartment, 238+/-45 min (8) is not significantly (P<.05) different than the 300+/-40 min previously reported (Curtis, 1970) from data on collected efflux samples taken during 3-4 hr efflux experiments. Such long efflux periods were impractical in this counting configuration. At 90 min of efflux (Fig 1A), the bundle was counting just 6 cpm above a 20 cpm background.

When the ^{45}Ca influx period is followed by a contracture (top right) a different pattern of efflux follows, Fig 1B. For practical purposes, only the Ca_{slow} compartment contains ^{45}Ca and loses ^{45}Ca with a t_c = 128 min. Just prior to the influx the Ca_{slow} compartment contained 0.06 pmol Ca, and just after the second contracture it contained 0.27 pmol Ca, a gain of 0.21 pmol Ca. This bundle took up 0.11 pmol Ca into the Ca_{slow} compartment during the ^{45}Ca influx period itself (Fig 1A). The remaining 0.10 pmol Ca must be related to the second depolarization-contracture cycle. Because the Ca_{fast} compartment was the only possible source of ^{45}Ca during the second contracture, this 0.10 pmol Ca, $Ca_{transfer}$, must have transfered from the Ca_{fast} compartment to Ca_{slow} compartment during the second contracture. The specific activity of the Ca_{fast} compartment is probably less than the specific activity of the influx solution, so the value of $Ca_{transfer}$ from the Ca_{fast} compartment to the Ca_{slow} compartment is a low estimate.

The paired difference between resting ^{45}Ca uptake into the Ca_{slow} compartment and ^{45}Ca content of the Ca_{slow} compartment following a second contracture, $Ca_{transfer}$, was 0.53+/-0.28 (8) pmol Ca, which is significantly (P<.05) different than 0. This 0.5 pmol represents a transfer from the Ca_{fast} compartment to the Ca_{slow} compartment related to the depolarization - contraction cycle of the second contracture. Is $Ca_{transfer}$ related to the initiation of contraction or is it flowing through late opening, voltage gated Ca channels?

BLOCKING VOLTAGE GATED Ca CHANNELS There are many ways to block voltage gated Ca channels. Most of them shorten the contracture or interfere with the dihydropyridine binding sites which I believe are the binding sites for the Ca_{fast} compartment. Ni was chosen because it does not interfere with the dihydropyridine binding sites and contracture time-tension curves are slightly prolonged. Voltage gated Ca channels in the t system are blocked by 2-4 mM Ni (Almers *et al.*, 1981 and

Stefani & Chiarandini, 1982). Ni blocks the 1.1 pmol Ca/min voltage sensitive, contraction independent Ca influx (Curtis & Eisenberg, 1985).

Fig 2A shows the effect of Ni applied before, during and after 2 min of ^{45}Ca influx just after relaxation. The Ca* efflux is once again described by two exponentials suggestive of two components, Ca_{fast} and Ca_{slow}. Both these compartments are much the same as in Ringer. The Ca_{fast} compartment (t_c= 3 min) took up 5 pmol Ca. 2 mM Ni did not interfere with the turnover of the Ca_{fast} compartment in the 5 bundles tested. The Ca_{slow} compartment (t_c=558 min) took up 0.18 pmol Ca. The average resting uptake in Ni Ringer into the Ca_{slow} compartment, 0.45+/-.07/2min, was no different than in Na Ringer.

When a second contracture followed the ^{45}Ca influx period (Fig 1B), once again only the $_{slow}$ compartment contained ^{45}Ca, taking up (0.94 - 0.15)= 0.79 pmol Ca.

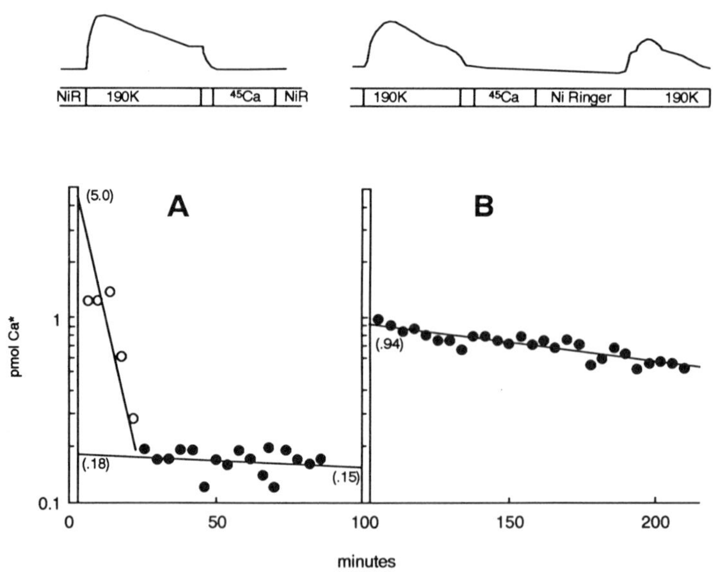

Figure 2 Blockage of depolarization activated Ca channels by Ni does not alter the extra Ca influx into the Ca_{slow} compartment during a subsequent contracture. The muscle bundle was in 2 mM Ni Ringer for the entire experiment. ^{45}Ca was again added during the recovery period and Ca* efflux can be described by two exponentials (A). Following the second ^{45}Ca influx period (upper right) a second potassium contracture was induced and Ca* efflux (B) can be described by a single exponential, Ca* efflux from the Ca_{slow} compartment. The extra Ca influx into the Ca_{slow} compartment was 0.61 pmol Ca.

Of this, 0.18 entered during the influx period, so (0.79 - 0.18) = 0.61 pmol transferred from the Ca_{fast} to the Ca_{slow} compartment during the second contracture. The average $Ca_{transfer}$ during contractures in Ni Ringer was 0.59+/-0.32 (5) pmol Ca, no different than in Na Ringer.

A second type of experiment in Ni Ringer exposed the bundle to 4 mM Ni Ringer only during the 3 min between the ^{45}Ca influx and the second contracture, Fig 3B. Here $Ca_{transfer}$ was 0.24 pmol Ca.

I conclude that $Ca_{transfer}$ is not entering the Ca_{slow} compartment via a voltage gated Ca channel.

Ca-FREE SOLUTIONS Experiments very similar to that shown in Fig 3 were done exposing the bundle to 0 Ca, 4Mg, 1 EGTA (Luttgau & Spiecker, 1979) during the 3 min wash between the ^{45}Ca influx and the second contracture. $Ca_{transfer}$ was 0.52+/-0.1 (3) pmol Ca. One bundle failed to give a second contracture, but twitched at the end of the experiment; $Ca_{transfer}$ was -0.16 pmol Ca. Clearly when the bundle contracts after 3 min in 0 Ca solution, Ca is transferred from the Ca_{fast} to the Ca_{slow} compartment and in the same amount as in normal Ca Ringer.

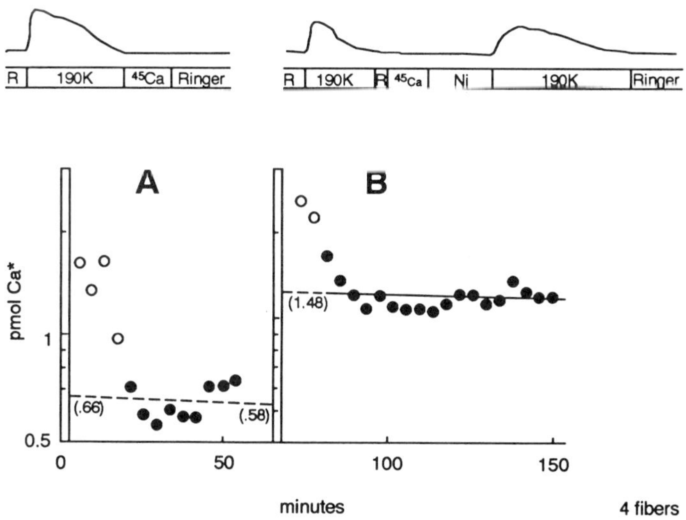

Figure 3 Short application of Ni does not alter $Ca_{transfer}$. 4 mM Ni was added only during the 3 min period between the second ^{45}Ca influx period (upper right) and the second contracture, which was prolonged by the Ni. Here $Ca_{transfer}$ was 0.24 pmol Ca. $Ca_{transfer}$ clearly does not enter via voltage gated channels.

$Ca_{transfer}$ is observed in the absence of ionized Ca in the t lumen. I conclude the Ca_{fast} compartment must be the source of $Ca_{transfer}$ and $Ca_{transfer}$ does not pass through an ionized form in the t lumen on its way to the Ca_{slow} compartment. Voltage gated Ca channels are twice as unlikely as the mechanism for the extra Ca entry into the Ca_{slow} compartment associated with contraction.

BLOCKING CHARGE MOVEMENT The possibility remains for a depolarization linked, contraction independent mechanism of $Ca_{transfer}$ during the second contracture. Tetracaine blocks charge movement, and hence, contraction. Fig 5A shows the by-now familiar pattern of refilling of the Ca_{fast} compartment and resting influx into the Ca_{slow} compartment. In Fig 5B, ^{45}Ca influx was followed by 3 min of 0.25 mM tetracaine, the second contracture was very small. The resulting Ca* efflux pattern shows contributions from both Ca_{fast} and

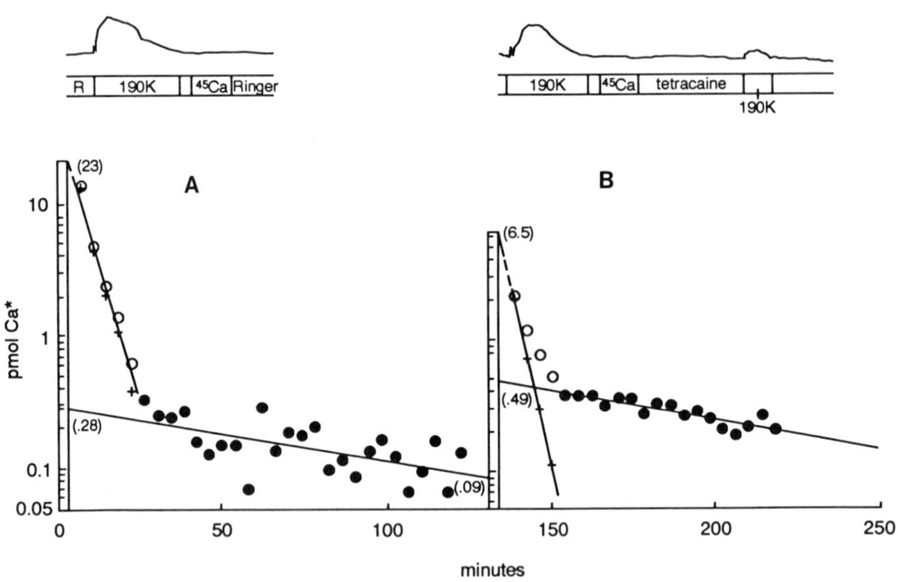

Figure 4 Blockage of depolarization induced charge movement, and consequently contraction, abol ishes $Ca_{transfer}$. When 0.25 mM tetracaine was added in the 3 min between the second ^{45}Ca influx period and the second depolarization (upper right), a very small contracture insued. $Ca_{transfer}$ was 0.12 pmol. The average for 3 non-contracting bundles was -0.09 pmol Ca. $Ca_{transfer}$ is indeed the link, first messenger Ca, between transverse tubular depolarization and release by the sarcoplasmic reticulum of second messenger Ca.

Ca_{slow} compartments. Refilling of the Ca_{fast} compartment is a bit reduced. I had thought that the Ca in the Ca_{fast} compartment was released by depolarization and would be lost in an experiment like this one. Perhaps fibers need to contract in order to squeeze the contents of the t lumen, including ^{45}Ca released from the Ca_{fast} compartment, out into the extracellular space. $Ca_{transfer}$ was 0.12 pmol Ca, very small, and there was a small contracture which should have been caused by a small movement of Ca from the Ca_{fast} to the Ca_{slow} compartments. In the other bundle tested, $Ca_{transfer}$ was -0.22 pmol with an even smaller contracture.

I conclude that $Ca_{transfer}$ is zero when the fibers do not contract and that $Ca_{transfer}$ is related to the initiation of contraction. $Ca_{transfer}$ is indeed first messenger Ca.

CAFFEINE Direct release of Ca from the SR can be induced by caffeine and bypasses much of the E-C coupling chain including first messenger Ca. When a second contracture is induced by caffeine, the efflux pattern is essentially unchanged, and influx into the Ca_{slow} compartment is at resting levels.

DISCUSSION

I have demonstrated an 0.5 pmol Ca influx, $Ca_{transfer}$, from the Ca_{fast} compartment (transverse tubular wall Ca) into the Ca_{slow} compartment (sarcoplasmic reticulum Ca) which occurs whenever the fiber contracts. $Ca_{transfer}$ is observed in contracting fibers when voltage gated Ca channels are blocked by Ni and when the t lumen contains no Ca. The average $Ca_{transfer}$ for all contracting bundles was 0.54+/-0.24 (12), significantly ($P<.05$) different from 0. When contraction is blocked by tetracaine or incomplete recovery in 0 Ca, the average $Ca_{transfer}$ was -0.09+/-0.10 pmol Ca (3). I conclude $Ca_{transfer}$ is first messenger Ca linking t system depolarization with release of Ca from the SR via the foot process. We would expect the following properties of first messenger Ca:

1. Ca store in t tubular wall
2. Released by/with charge movement
3. Release is graded, continous and reversible
4. Receptor site on SR isolated from myoplasm
5. Store depleted during contracture
6. Store replenished during recovery

A model incorporating these criteria and a 3Na/Ca exchange

mechanism is shown in Fig 5. This model is described much more extensivly elsewhere (Curtis, 1988).

At rest (Fig 5A), both Ca on the exchanger (0.5 pmol) and Ca bound to the dihydropyridine sites (10 pmol) are in exchange with the t lumen Ca (0.7-0.9 pmol) and in turn with the extracellular space surrounding the muscle fiber. The first two constitute the Ca_{fast} compartment described in the Results section. Given the Na and Ca concentration gradients and the -95 mV resting membrane voltage, the 3Na/Ca exchanger is in equilibrium.

Upon depolarization (Fig 5, A > B), the driving force on the 3Na/Ca exchanger suddenly changes to move Ca into the cell. The inward Ca gradient is so large that the exchanger translates only once. As the exchanger translates, a charge movement type signal is generated. Maximum charge movement, 50 nC/μF, converts to 0.25 pmol Ca moving into the fiber--good agreement with the 0.5 pmol Ca observed for $Ca_{transfer}$. As Ca enters the delimited space in or between the foot processes, it stimulates the foot process to release Ca from the SR. Ca is the only known stimulus which will release SR Ca in intact and skinned fibers as well as isolated SR vesicles and feet in bilayers. Chemical transmission may well explain the lack of a 1:1 relationship between the particles in the t wall and the feet of the SR (Block *et al.*, 1988).

The rate of translation of the Na/Ca exchanger controls the amount of Ca entering the delimited space (Ca_{t-sr}) around the receptor (R), which in turn controls SR Ca release. Increasing the rate of translation by increasing the net Ca inward driving force on the 3Na/Ca exchanger should increase the rate of rise of tension. When contractures are induced with 95 mM K and either 95 mM Na or TRIS, the rate of rise of tension is faster with TRIS because the inward Ca driving force is unopposed by an inward Na driving force. Likewise, when Na_i is elevated by poisoning Na/K active transport with oubain, the rate of rise of tension increases (Curtis, 1988).

Slowing the rate of translation of the Na/Ca exchanger should slow the rate of rise of tension. Dichlorobenzamil, a Na/Ca exchange blocker (Siegl *et al.*,1984) slows the rate of rise of tension. Low pH inhibits Na/Ca exchange in many tissues and slows the rate of rise of tension (Curtis, 1988).

First Messenger Ca, Ca_{t-sr}, slowly diffuses out of the delimited space until Ca_{t-Sr} falls below threshold and t-SR transmission fails--a constant observation in all prolonged depolarizations.

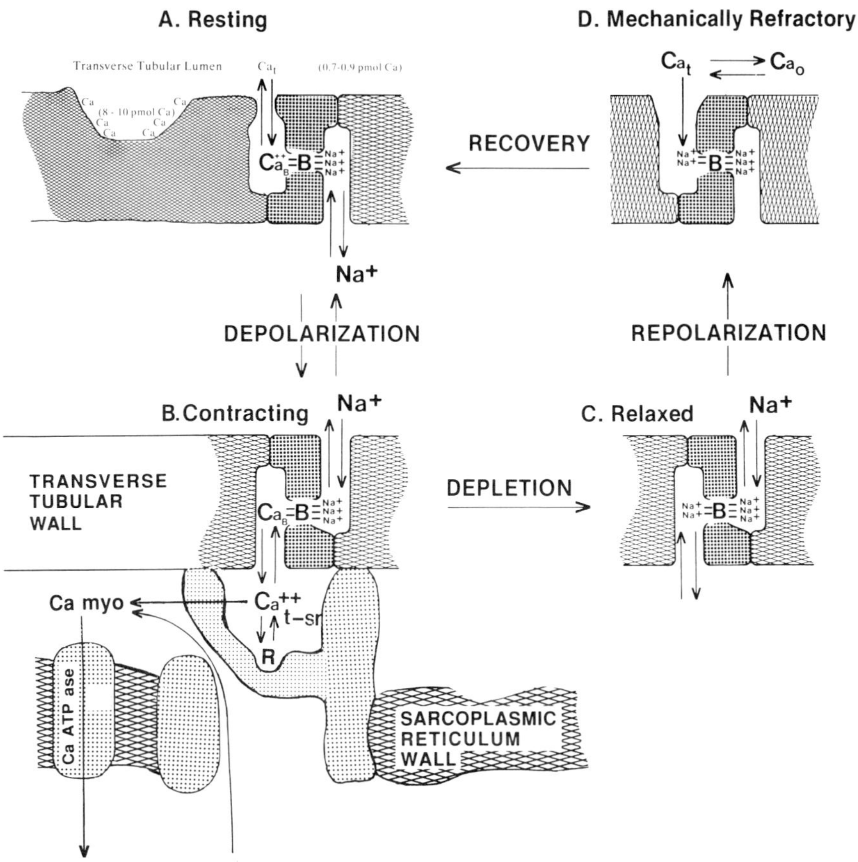

Figure 5 A 3Na/Ca exchange model of excitation-contraction in skeletal muscle. At rest (A) the Ca site faces the t lumen and exchanges with Ca in the t lumen as does the much larger Ca_{fast} store. Upon depolarization (B), the difference in chemical potentials of Na and Ca gives a net inward force on the Ca side of the exchanger resulting in translation so the Ca site now faces a small space in the foot process where it stimulates Ca-activated Ca release from the SR and subsequent contraction. As depolarization continues, the small space is depleated of Ca and contraction relaxes spontaneously (C). Repolarization reorients the exchanger (D); the depleted Ca site faces the t lumen and slowly refills with Ca to return to the primed state (A).

The exchanger in the relaxed muscle (Fig 5C) is depleted of Ca.

To regain the ability to contract, the fiber must first repolarize (C > D) moving the 3Na/Ca exchanger to its original configuration. Second, even after repolarization, a Ca and time dependent process, repriming (D > A), must occur before full contracture tension can once again be produced. Recovery is blocked in 0 Ca (Luttgau, 1963) and is slow in 0.2 mM Ca (Curtis, 1964).

I have shown that reuptake into the Ca_{fast} compartment is time dependent with much the same t_c as repriming. The beginning of the repriming reaction (refilling the Na/Ca exchanger with Ca) depends on the rate of translation of the Na/Ca exchanger back to its resting configuration. Dichlorobenzamil, low pH, and low Na delay the start of repriming, but none alters the rate.

I have developed a kinetic model which adequately reproduces both twitch and contracture tension, as well as the experimental findings described in the last few paragraphs (Curtis, 1988).

All of this evidence points to a 3Na/Ca exchanger operating in the E-C coupling pathway to move first messenger Ca into the fiber. Simultaneous measurement of both Ca and Na fluxes is impractical, and no specific extracellular Na/Ca exchange blocker is available. There is hope, a highly specific cytoplasmic Na/Ca blocker has recently been isolated (Li, 1991).

I conclude that a first messenger Ca exists in skeletal muscle and its depolarization induced translation from a binding site on the transverse tubular wall to the release site on the foot process of the sarcoplasmic reticulum is by a 3Na/Ca exchanger.

REFERENCES

Almers, W., and Best, P. M., 1976, Effects of tetracaine on displacement currents and contraction of frog skeletal muscle, *J. Physiol.*, 262:583-611.

Almers, W., Fink, R., and Palade, P. T., 1981, Calcium depletion in frog muscle tubules: the decline of calcium current under maintained depolarization, *J. Physiol.*, 312:177-207.

Armstrong, C. M., Bezanilla, F. M., and Horowicz, P., 1972, Twitches in the presence of ethylene glycol bis (B-aminoethyl ether), *N*, *N*-tetraacetic acid, *Biochim. Biophys. Acta*, 267:605-608.
Baylor, S. M., Hollingworth, S., and Marshall, M. W., 1989, Effects of intracellular ruthenium red on excitation-contraction coupling in intact frog skeletal muscle fibres, *J. Physiol.*, 408:617-635.
Bianchi, C. P., 1969, Pharmacology of excitation-contraction coupling in muscle, *Fed. Proc.*, 28(5):1624-1628.
Bianchi, C. P., and Shanes, A. M., 1959, Calcium influx in skeletal muscle at rest, during activity, and during potassium contracture, *J. of Gen. Physiol.*, 42(4):803-815.
Block, B. H., Imagawa, T., Campbell, K. P., and Franzini-Arm strong, C., 1988, Structural evidence for direct interaction between the molecular components of the transverse tubule/sarcoplasmic junction in skeletal muscle, *J. Cell Biol.*, 107: 2587-600.
Brum, G., and Rios, E., 1987, Intramembrane charge movement in frog skeletal muscle fibres: properties of charge2, *J. Physiol.*, 387:489-517.
Brum, G., Fitts, R., Pizarro, G., and Ríos, E., 1988, Voltage sensors of the frog skeletal muscle membrane require calcium to function in excitation-contraction coupling, *J. Physiol.*, 398:475-505.
Cosmos, E., and Harris, E. J., 1961, *In vitro* studies of the gain and exchange of calcium in frog skeletal muscle, *J. Gen. Physiol.*, 44:1121-30.
Curtis, B. A., 1963, Some effects of Ca-free choline Ringer solution on frog skeletal muscle, *J. Physiol.*, 166:75-86.
Curtis, B. A., 1964, The recovery of contractile ability following a contracture in skeletal muscle, *J. Gen. Physiol.*, 47(5): 953-964.
Curtis, B. A., 1966, Ca fluxes in single twitch muscle fibers, *J. Gen. Physiol.*, 50(2):255-267.
Curtis, B. A., 1970, Calcium efflux from frog twitch muscle fibers, *J. Gen. Physiol.*, 55:243-53.
Curtis, B. A., 1988, Na/Ca exchange and excitation-contraction coupling in frog fast fibres, *J. Mus. Res. and Cell Motil.*, 9: 415-427.
Curtis, B. A., and Eisenberg, R. S., 1985, Calcium influx in contracting and paralyzed frog twitch muscle fibers, *J. Gen. Physiol.*, 85:383-408.
Donoso, P., and Hidalgo, C., 1989, Sodium-calcium exchange in transverse tubules isolated from frog skeletal muscle, *Biochim. Biophys. Acta*, 978:8-16.
Fabiato, A., 1984, Dependence of the Ca^{2+}-induced release from the sarcoplasmic reticulum of skinned skeletal muscle fibres from the frog semitendinosus on the rate of change of free

Ca^{2+} concentration at the outer surface of the sarcoplasmic reticulum, *J. Physiol.*, 353:56P.
Frank, G. B., 1980, The current view of the source of trigger calcium in excitation-contraction coupling in vertebrate skeletal muscle, *Biochem. Pharmacol.*, 29:2399-406.
Gilbert, J. R., and Meissner, G., 1982, Sodium-calcium ion exchange in skeletal muscle sarcolemmal vesicles, *J. Mem. Biol.*, 69:77-84.
Gonzälez, S., Brum, G., and Pizarro, G., 1991, Effects of procaine on calcium release in skeletal muscle fibers, *Biophys. J.* 59:62a.
Gonzälez-Serratos, H., Valle-Aquilera, R., Lathrop, D. A., and Del Carmen Garcia, M., 1982, Slow inward calcium currents have no obvious role in muscle excitation-contraction coupling, *N.ature*, 298:292-4.
Hymel, L., Inui, M., Fleischer, S., and Schindler, H., 1988, Purified ryanodine receptor of skeletal muscle sarcoplasmic reticulum forms Ca^{2+}-activated oligomeric Ca^{2+} channels in planar bilayers, *Proc. Nat. Acad. Sci.* , 85:441-445.
Keynes, R. D., 1951, The ionic movements during nervous activity, *J. Physiol.*, 114:119-150.
Li, Z., Nicoll, D. A., Collins, A., Hilgemann, D. W., Filoteo, A. G., Penniston, J. T., Tomich, J. M., and Philipson, K. D., 1991, Identification of a peptide inhibitor of the cardiac sarcolemmal Na^{+}-Ca^{2+} exchanger, *Biophys. J.*, 59:138a.
Luttgau, H. C., 1963, The action of calcium ions on potassium contractures of single muscle fibres, *J. Physiol.*, 168:679-697.
Luttgau, H. C., and Spiecker, W., 1979, The effects of calcium deprivation upon mechanical and electrophysiological parameters in skeletal muscle fibres of the frog, *J. Physiol.*, 296: 411-429.
Meissner, G., 1984, Adenine nucleotide stimulation of Ca^{2+}-induced Ca^{2+}release in sarcoplasmic reticulum, *J. Biol. Chem.*, 259:2365-2374.
Mobley, B. A., and Eisenberg, B. R., 1975, Sizes of components in frog skeletal muscle measured by methods of stereology, *J. Gen. Physiol.*, 66:31-45.
Philipson, K. D., 1985, Sodium-calcium exchange in plasma membrane vesicles, *Ann. Rev. Physiol*, 47:561-71.
Podolsky, R. J., 1964, The maximum sarcomere length for contraction of isolated myofibrils, *J. Physiol.*, 170:110-123.
Reuter, H., 1991, Ins and outs of Ca^{2+} transport, *Nature*, 349: 567-568.
Schneider, M. F., and Chandler, W. K., 1973, Voltage dependent charge movement in skeletal muscle: a possible step in excitation-contraction coupling, *Nature*, 242:244-6.

Schwartz, L. M., McCleskey, E. W., and Almers, W., Dihydropyridine receptors in muscle are voltage-dependent but most are not functional calcium channels, *Nature*, 314:747-750.
Sheu, S-S., and Blaustein, M. P., 1986, Sodium/calcium exchange and regulation of cell calcium and contractility in cardiac muscle, with a note about vascular smooth muscle, in: H. A. Fozzard et al., eds., *The Heart and Cardiovascular System*, New York: Raven Press, Ch. 26.
Siegl, P. K. S., Cragoe, E. J., Trumble, M. J., and Kaczorowski, G. J., 1984, Inhibition of Na^{+}/Ca^{2+} exchange in membrane vesicle and papillary muscle preparations from guinea pig heart by analogues of amiloride, *Proc. Nat. Acad. Sci.*, 81:3238-42.
Smith, J., Coronado, R., and Meissner, G., 1986, Single channel measurement of the calcium release channel from skeletal muscle sarcoplasmic reticulum, *J. Gen. Physiol.*, 8:573-88.
Somlyo, A. V., Gonzalez-Serrato, H., Shuman, H., McClellan, G., and Somlyo, A. P., 1981, Calcium release and ionic changes in the sarcoplasmic reticulum of tetanized muscle: an electron probe study, *J. Cell. Biol.*, 90:577-94.
Stefani, E., and Chiarandini, D., 1982, Ionic channels in skeletal muscle, *Ann. Rev. Physiol.*, 44:357-372.
Watson, P. T., and Winegrad, S., 1973, A possible sodium-calcium exchange in skeletal muscle, *Fed. Proc.*, 32:374, Abst.
Winegrad, S., 1965, The location of muscle calcium with respect to the myofibrils, *J. Gen. Physiol.*, 48:997-1002.
Winegrad, S., 1970, The intracellular site of calcium activation of contraction in frog skeletal muscle, *J. Gen. Physiol.*, 55:77-88.

ACKNOWLEDGEMENT

Supported by a Grant in Aid, Illinois Affiliate, American Heart Association.

THE SIGNIFICANCE OF Na^+ IN E-C COUPLING IN MUSCLE

Vladimir Nesterov

Sechenov Institute of Evolutionary Physiology and Biochemistry
Leningrad, USSR

A great diversity of forms of multicellular organisms having come into being during animal evolution has determined the exclusively structural and functional variety of their muscles differing, particularly, in their mode of activation. In vertebrate twitch muscles a crucial step in excitation- contraction (E-C) coupling concerns the mechanisms whereby the electrical activity of the T-tubular membrane (TTM) induces the release of Ca^{2+} from the sarcoplasmic reticulum (SR). The investigation of this key question of the E-C coupling has resulted in two principal hypotheses about the nature of the T-SR signal transmission: (i) Ca^{2+} release from SR triggered by a change in tubular membrane polarization and (ii) Ca^{2+} release induced by specific chemical transmitter from the T-system.

Electro-mechanical hypothesis (i) postulates that voltage dependent "charge movements" within the TTM gate the Ca^{2+} channels in the feet of the T-SR junction by altering long-connecting molecules (Schneider and Chandler, 1973; Schneider et al., 1981). Correlations between charge movement, intracellular Ca^{2+} transients and contraction under different conditions support the involvement of this mechanism in E-C coupling (Adrian et al., 1976; Kovacs et al., 1979). However, a molecular structure of a messenger from TTM to SR has not determined; there are experimental results inconsistent with this hypothesis - some organic drugs greatly depress charge movement but at the same time potentiate the twitches (Frank, 1984; Hui et al.,1984). Therefore we can consider charge movement as a possible initial step in the intracellular voltage-dependent Ca^{2+} mobilization, but the molecular mechanism of its involvement in twitch triggering is yet remained to be discovered.

According to the chemical conception (ii) during excitation some quantity of trigger mediator diffuses through T-SR junction and induces Ca^{2+} release from SR. The mechanism with inositol 1,4,5-trisphosphate ($InsP_3$) has been proposed as a messenger in E-C coupling (Vergara et al., 1985; Volpe et al., 1985, 1986). However it was found that $InsP_3$ could be the chemical transmitter in smooth muscles but in skeletal and even in cardiac muscles the response to $InsP_3$ is too slow and the $InsP_3$ concentration formed during a single twitch (about 2 nM) is not enough to trigger the contraction (Fabiato, 1985; Lea et al., 1986; Mikos and Snow, 1987; Walker et al., 1987; Varsanyi et al., 1989).

Several authors have demonstrated that calcium ions entering the muscle fiber from the extracellular space via activated Ca^{2+} channels of the TTM or having been displaced by the incoming action potential (AP) from binding sites at the TTM play a role in a

Excitation-Contraction Coupling in Skeletal, Cardiac, and Smooth Muscle
Edited by G.B. Frank *et al.*, Plenum Press, New York, 1992

signal transmission involved in E-C coupling ("Ca^{2+}-induced Ca^{2+} release") (Frank, 1980, 1982; Luttgau et al., 1986). The obtained results have revealed that in skeletal muscle fibers L-type Ca^{2+} channels in the TTM are opened and provide only the long lasting depolarization contractions but seem to play no essential role in the twitch response (Luttgau and Spiecker, 1979; Gonzalez-Serrates et al., 1982; Stanfield and Ashcroft, 1982; Frank, 1984, 1990).

Difficulties in the explanation of available results in terms of the hypotheses considered make it urgent to envisage the significance of the potential-dependent inward Na current across the TTM in E-C coupling in skeletal muscle fibers. Our general proposal about the possible involvement of Na ions in processes of Ca^{2+} release from SR appeared initially as a result of a comparative investigation of Na^+ and K^+ distribution in functionally different muscles (Table 1). It was brought out that muscle specialization directed towards acceleration of contraction is accompanied by an increased extrusion of Na^+ from muscle fibers leading, as a result, to an enlarged $[Na^+]_o/[Na^+]_i$ ratio across the plasmic membrane (Fig. 1).

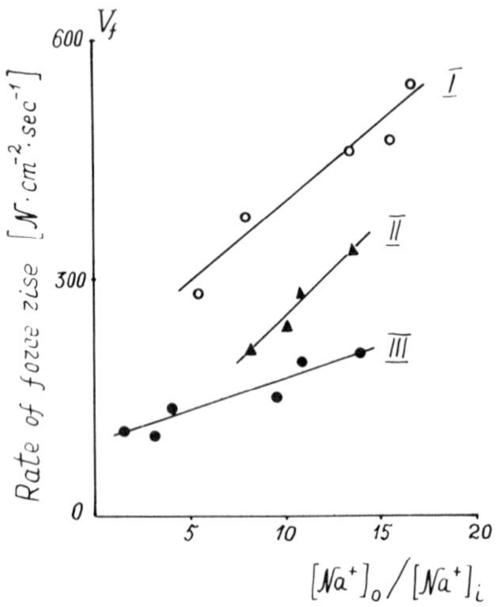

Figure 1. Relationship between maximum rates of isometric force rise (V_f) during a twitch and transsarcolemmal Na^+ distribution ($[Na^+]_o/[Na^+]_i$) ratio of different skeletal muscles of white rat (I, 37°C), starling (II, 38°C) and grass frog (III, 20°C). $V_f = \triangle F_n / \triangle T$, $[N \cdot cm^{-2} \cdot sec^{-1}]$, where F_n is a force developed by the muscles (F_m) normalized per unit of cross-sectional area of the muscle ($F_n = F_m/S_f$); $\triangle T$ - the time-interval from the start of the response to the beginning of tension decrease. Each point is the mean of several determinations of the functionally specific skeletal muscles (Nesterov, 1984).

Taking into account these data we suggested that a Na^+ current flowing through TTM during the AP could transmit the trigger information to an intracellular calcium signalling system, perhaps by a "Na^+-induced Ca^{2+} release" mechanism (Nesterov and Fedorov, 1971; Nesterov, 1975). In this case it was reasonable to assume that the rate of a twitch tension rise should correlate with the amount of Na^+ (or Li^+) carried by the current into fibers along the transmembrane electrochemical gradient. Our data (e.g. shown in Figure 1) and the results of French authors obtained as well on skeletal muscle fibers corroborated this assumption.

Concentrations are expressed in meq/l intracellular ($[M^+]_i$) and extracellular ($[M^+]_o$) water. The number of animal species studied is in the numerator; another numbers designate a quantity of individuals studied.

Table 1. Distribution of Na^+ and K^+ in resting intact muscles of some animals.

Muscles	$[Na^+]_i$	$[Na^+]_o$ / $[Na^+]_i$	$[K^+]_i$
SEA INVERTEBRATES			
Smooth muscles (19/375)	83 ± 5	5,5	92 ± 7
Cardiac muscles (2/23)	45 ± 6	10,2	116 ± 7
Striated muscles (7/170)	33 ± 4	14,0	126 ± 10
SEA BONY FISHES			
Smooth muscles (7/74)	35 ± 3	4,9	137 ± 4
Cardiac muscles (7/107)	18 ± 3	9,4	143 ± 3
Skeletal muscles (7/208)	8 ± 1	21,2	157 ± 7
FROG (Rana temporaria)			
Smooth muscles (28)	41 ± 4	2,6	124 ± 4
Cardiac muscles (35)	35 ± 3	3,1	121 ± 3
Skeletal muscles:			
m.rectus abdominis (17)	32 ± 2	3,3	116 ± 2
m.iliofibularis (67)	11 ± 2	9,7	142 ± 3
WHITE RAT			
Smooth muscles (14)	28 ± 3	5,1	123 ± 5
Cardiac muscles (38)	17 ± 3	8,4	118 ± 5
Skeletal muscles:			
m.soleus (88)	20 ± 2	7,1	137 ± 4
m.plantaris (88)	8 ± 1	19,1	156 ± 3

In current clamp and voltage clamp conditions Caille et al. (1978, 1979, 1981) observed that at 20°C a part of the mechanism responsible for the Ca^{2+} release from the SR as well as contraction directly depended on the influx of sodium ions through the tubule membrane. Different procedures that modified the pulse amplitude and duration of the late inward sodium current altered the corresponding contraction in the same way. A direct relation between Na^+ entry and contraction was also reported by Potreau and Raymond (1982). In special experimental conditions (Ca^{2+}-free solutions containing TTX) they established a close relationship between a slow inward Na^+ current through L-type Ca^{2+} channel in TTM and contraction amplitude. The relationship between a phasic component of tension and the slow sodium current developed in a OCa^{2+}-EGTA solution was observed in cardiac muscle (Bonvallet et al., 1984). Recently the significance of Na^+ in E-C coupling has been confirmed by Leblank and Hume (1990) who provided evidence for sodium current-induced release of Ca^{2+} from cardiac SR.

For the elucidation of possible mechanisms of Na^+ involvement in E-C coupling it was reasonable to study muscles when sodium ions in the external solution were replaced by isomorphic lithium ions. Unlike a radioactive isotopes the lithium ions possesses only a restricted physical-chemical similarity to Na^+ that permits to differentiate the specific for these two ions processes in the functioning muscle fibers. This method of approach was applied by many authors (Jonemura and Sato, 1967; Nesterov and Senchenkova, 1975; Caille et al., 1979; Potreau and Raymond, 1982; Nesterov, 1985). Nevertheless, an application of the Li^+-sound for the E-C coupling mechanism investigation was not widely spread because of discrepancy between the results of different researches. According to some authors (Caille et al., 1979) the substitution of Na^+ in Ringer solution by Li^+ leads to the decrease of the twitch amplitude; according to others (Nesterov and Senchenkova, 1975; Frank, 1980) - the replacing of Na^+ by Li^+ potentiates the twitches until enough Li^+ enters the muscle fibers to depolarize them.

Taking into account these data we continued studying that question. Activation of a contraction was used as the intracellular $[Ca^{2+}]$ indicator (Nesterov, 1985). Figure 2 represents the time course of changes in tissue cations contents before and after immersion of frog sartorius muscles in Li^+ Ringer's solution (Li^+-R). It can be seen that from the beginning the sodium ions are exchanging for Li^+, firstly more rapidly in extracellular space. In some minutes as the transmembrane electrochemical Li^+ gradient is getting higher the cellular accumulation of Li^+ starts. By 20 min of the immersion in Li^+ -R solution a substantial part of intracellular Na^+ (>50%; $p<0,05$) is substituted by Li^+ and from about 2 h of the immersion the intracellular potassium ions begin exchanging for Li^+. In these conditions during the first 10 min of exposure in Li^+-R the maximum twitch tension increases to about 120% of the control and after that the amplitude has maintained for up to 20-25 min before its decreasing (Fig. 3). Hence the potentiation of the twitches is determined by the augmentation of these cations gradients on the

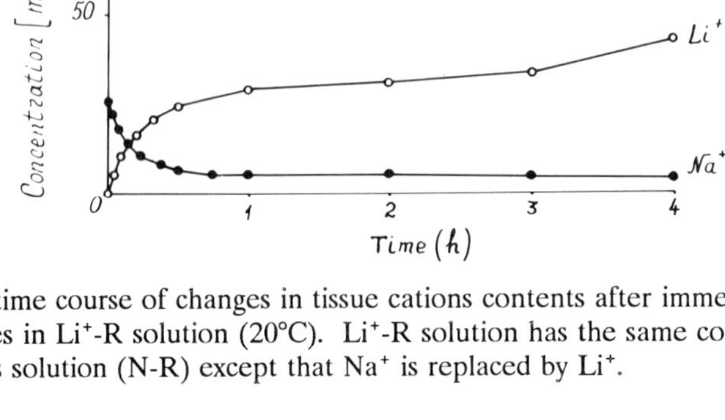

Figure 2. The time course of changes in tissue cations contents after immersion of frog sartorius muscles in Li^+-R solution (20°C). Li^+-R solution has the same composition as normal Ringer's solution (N-R) except that Na^+ is replaced by Li^+.

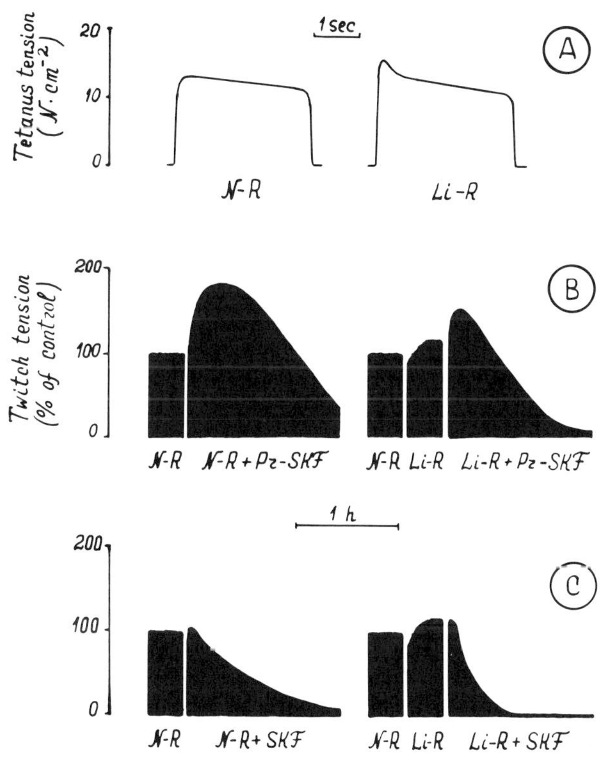

Figure 3. The effect of Li^+-R on tension during tetanus (A; 50 Hz) and twitches (B and C; 0,1 Hz) of frog skeletal muscles (20°C). The solutions containing 4.10^{-5} M of SKF or Pr-SKF were made by adding the drugs to the N- or Li^+- Ringer's solutions.

plasmic membrane - up to this time they have been getting the most and directed: Li^+- into the fibers and Na^+ - vice versa. The time of the peak potentiation reaching of the twitches depends on a muscle size: in frog's small extensor longus digiti 1V muscle the duration of replacement of extracellular Na^+ by Li^+ is shorter (1-2 min) as compared with the greater sartorius muscle (about 10 min). In experiments with single muscle fibers the cations substitution in surroundings and correspondingly the twitch potentiation had occurred, we suppose, before the recording was made. The application of depolarizing current for compensation of the fiber hyperpolarization (in Li^+-R) used by Caille et al. (1979) could also intensify and accelerate the depressing effect of Li^+-R. Therefore uncontrolled spikes in Li^+-R can potentiate the twitches of skeletal muscle fibers as compared with the twitches in Na^+-R and the main reason of this potentiation is the enlarged current of lithium ions through TTM along the higher, as compared with Na^+ current, electrochemical gradient.

The increase in the size of contractile response produced by the substitution of Na^+ in the Ringer solution by Li^+ can be seen during the tetanus developed by electrical stimulation (Fig. 3, A). The twitch responses in Li^+-R can be modulated by the addition of some chemicals to the Ringer's solution (Fig.3, B and C). Well-known oxidase inhibitor SKF-525A (N,N-diethylaminoethyl-2,2-diphenyl-valerate) has exerted negative inotropic effects in Na^+ and Li^+-R; its acetylenic derivate (N,N-diethylaminoethyl-2,2-diphenyl-2- propargylacetate; Pr-SKF) which displays veratrine-like action on membrane Na^+-channels by preventing sodium current from inactivating (G.Frank and V.Nesterov, unpublished results), has potentiated the twitches in Na^+- and Li^+-R solutions, supporting the idea about the Na^+ (Li^+) current-induced release of Ca^{2+} from the SR.

Another moment to be investigated yet is the mechanism of Ca^{2+} release from the SR under the influence of Na^+ (Li^+) influx through the TTM. Several models could be used for an explanation how activation of Na^+ currents is coupled to the release of Ca^{2+} from the SR.

It was shown in vitro that Ca^{2+} release from the SR vesicles can be evoked by a drastic change in the membrane surface charges distribution during a rapid replacement of K^+ in the surroundings by less permeant Na^+, Li^+, $Tris^+$ or $Cholin^+$ (Endo, 1977; Caswell and Brandt, 1981). In this case on the SR membrane an electric field is generating (with inside negative membrane potential) which opens voltage-gated Ca^{2+} channels. We suppose the same mechanism of Ca^{2+} release can be used in muscle fibers in vivo. On excitation some amounts of Na^+ (or Li^+) ions enter through the TTM and produce a transient rise of positive charges quantity near the external surface of the SR terminal cisternae membrane. A diffusion potential formed by the cations on the SRM acts on the voltage sensor of Ca^{2+} channels and opens the latter. So the Na^+ (Li^+) ions may directly and primarily trigger a release of Ca^{2+} from the SR. In turn these Ca^{2+} ions as a secondary trigger interact with the Ca^{2+} receptors of SR Ca^{2+} channels and cause the main Ca^{2+} release from the SR into myofilament space.

The release of Ca^{2+} from the SR may be induced by influx of Ca^{2+} via Na^+/Ca^{2+} exchanger in TTM (Curtis, 1988; Donoso and Hidalgo, 1989; Bridge et al., 1990) in response to a rise of Na^+ in the T-SR space during a single spike. The experiments have confirmed this assumption. An increase of Ca^{2+} in Ringer solution intensify the Pr-SKF-induced potentiation of twitches in frog sartorius muscle. The same model was suggested by Leblank and Hume (1990) for cardiac myocytes. Both models pointed out above and operating as a possible initial E-C coupling stage via the mechanism of Na^+- and/or Ca^{2+}-induced Ca^{2+} release from the SR are schematically presented on the Figure

4. It is to be emphasized that any transient increase of Na^+ influx in T-SR junctional space trough TTM Na^+ channels (by means of veratrine, Pr-SKF, etc.) enhances the Ca^{2+} entry in this space via Na^+/Ca^{2+} exchanger and, hence, increase the contractile force. And vice verse any decrease of inward Na^+ entry into triad junction (reduction of $[Na^+]_o$, pharmacological blockade of Na^+ channels by TTX, SKF, etc.) will reduce the rate of the tension rise.

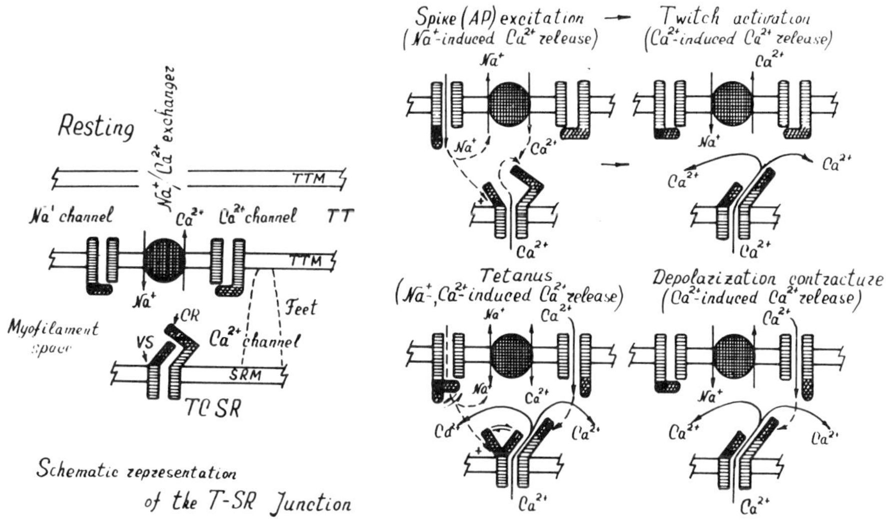

Figure 4. A Na^+- and/or Ca^{2+}-induced Ca^{2+} release schematic model of E-C coupling in skeletal muscle fibers of vertebrates. During a single spike excitation the Na^+ ions enter through the open Na^+ channels into T-SR junctional space. The cations activate a Na^+/Ca^{2+} exchange (reverse mode) and as positive charges cause the voltage-gated Ca^{2+} channels in SRM to open (I open state). Ca^{2+} ions passing through these channels from the SR and entering via Na^+/Ca^{2+} exchanger from outside induce main Ca^{2+} release (II open state) from the SR. During rapid repetitive stimulation (tetanus) twitches are fused and arising depolarization of TTM open the slow Ca^{2+} channels for extracellular Ca^{2+}. The latter induces the main Ca^{2+} release from the SR. During long lasting depolarizations of TTM (high K^+ contractures) there is only extracellular Ca^{2+}-induced main Ca^{2+}-release from the SR. TTM - transverse tubula membrane; VS - voltage sensor of the SR Ca^{2+} channel; CR - Ca^{2+} receptor of the SR Ca^{2+} channel; TC SR - terminal cisternae of the SR; SRM - membrane of the SR.

The release of Ca^{2+} from the SR by means of these mechanisms would take place if the Na^+ triggering signal can produce sufficient rise of Na^+ transient at a local space of T-SR junction. We have estimated the increase of Na^+ in a restricted volume between TTM and SRM of a frog sartorius muscle fiber during a single twitch. Taking into account the available data about a density and properties of the Na^+ channels in TTM (Venosa, 1974; Jaimovich et al., 1976; Franzini-Armstrong and Nunzi, 1983; Luttgau and Stephenson, 1986) we have found the elevation of $[Na^+]_{T\text{-}SR}$ above the normal $[Na^+]_i$

average by 25 mM (Nesterov, 1988). In this case the rate of $[Na^+]$ rise in T-SR junctional restricted volume during single spike activation is about 25 M sec^{-1} that is much more sufficient to trigger Ca^{2+} release from the SR as compared with model experiments in vitro (Kim et al.,1983).

SUMMARY

Recent studies have revealed arguments in favour of the possible triggering role of Na^+ ions in E-C coupling in skeletal muscle fibres of vertebrates: (i) Na^+ is one of the four major inorganic cations widespreaded in the biosphere ununiformly (with gradient) distributed across plasmic membranes of all muslce fibres; (ii) there is correlation between contractile parameters and a pattern of transsarcolemmal Na^+ distribution in skeletal muscles; (iii) "Na^+ current - induced Ca^{2+} release" mechanism is corresponded to the criterions for intracellular mediators: a) excitation of plasmic membrane increases $[Na^+]$ in junctional space; b) increase of $[Na^+]$ in surroundings in vitro induces efflux of Ca^{2+} from SR; c) estimated rate of $[Na^+]$ increase in junctional space in vivo is exceeded the threshold that induces Ca^{2+} from the SR in vitro; d) there is endogenic system (Na^+, K^+-ATPase) of quick removal of Na^+ from junctional gap of triads; e) pharmacological modification of Na^+ current through T-tubule membrane leads to correlated changing in twitch response.

A definite order of Na^+- and Ca^{2+} transmembrane triggering fluxes involved in E-C coupling in fast skeletal muscle fibers provide a very protective intracellular functional system of Ca^{2+} regulation, coordinated in time and space, and garantees the most complete dependence of voluntary muscle contractions on the CNS control.

References

Adrian, R.H., Chandler, W.K., and Rakowski, R.F., 1976, Charge movement and mechanical repriming in skeletal muscle, J. Physiol. (L.), 254: 351.

Bezamilla, F., and White, M.M., 1986, Properties of Ionic Channels in Excitable Membranes, in: "Physiology of Membrane Disorders", II ed., T. Andreoli et al., eds. Plenum Med. Book Company, N.Y. and L.

Bridge, J.H.B., Smolley, J.R., Spitzer, K.W., 1990, The Relationship Between Charge Movements Associated with I_{Ca} and $I_{Na\text{-}Ca}$ in Cardiac Myocytes, Science, 148: 376.

Caillé, J., Ildefonse, M., and Rougier, O., 1978, Existence of a sodium current in the tubular membrane of frog twitch muscle fibre; its possible role in the activation of contraction, Pflügers Arch., 374: 167.

Caillé, J., Ildefonse, M., and Rougier, O., 1979, Evidence for an action of sodium ions in the activation of sodium ions in the activation of contraction of twitch muscle fibre, Pflügers Arch., 379: 117.

Caswell, A.H., and Bzandt, N.R., 1981. Ion induced release of calcium from isolated sarco-plasmic reticulum, J. Membrane Biol., 58: 21.

Curtis, B.A., 1988, Na/Ca exchange and excitation-contraction coupling in frog fast fibres, J. Muscle Res. and Cell Motil., 9: 415.

Donoso, P., and Hidalgo, C., 1989, Sodium-calcium exchange in transverse tubules isolated from frog skeletal muscle, Biochem. Biophys. Acta, 978: 8.

Endo, M., 1977, Calcium release from the sarcoplasmic reticulum, Physiol. Rev., 57: 71.

Fabiato, A., 1985, Time and calcium dependence of activation and inactivation of calcium from the sarcoplasmic reticulum of a skinned canine cardiac Puzkinje cell, J. Gen. Physiol., 85: 247.

Frank, G.B., 1980, The current view of the source of trigger calcium in excitation-contraction coupling in vertebrate skeletal muscle, Biochem. Pharmacol., 29: 2399.

Frank, G.B., 1982, Roles of extracellular and "trigger" calcium ions in excitation-contraction coupling in skeletal muscle, Can. J. Physiol. Pharmacol., 60: 427.

Frank, G.B., 1984, Blockade of Ca^{2+} channels inhibits K^+ contracture but not twitches in skeletal muscle, Can. J. Physiol. Pharmacol., 62: 374.

Frank, G.B., 1990, Dihydropyridine calcium channel antagonists block and agonists potentiate high potassium contracture but not twitches in frog skeletal muscle, J.J. Physiol., 40: 205.

Franzini-Armstrong, C., and Nunzi, G. 1983, Functional feet and particles in the triads of a fast twitch muscle fiber, J. Muscle Res. Cell Motil, 4:233.

Gonzales-Serratos, M., Valle-Aguilera, R., Lathrop, R., del Carmen Garcia, M., 1982, Slow inward calcium currents have no obvious role in muscle excitation-contraction coupling, Nature, 298: 292.

Hui, C.S., Milton, R.L., and Eisenberg, R.S., 1984, Charge movement in skeletal muscle fibres paralysed by the calcium-entry blocker to 600, Proc. Natl. Acad. Sci. USA, 81: 2582.

Jaimovich, E., Venosa, R.A. Shrarger, P., Horowicz, P., 1976, Density and distribution of tetrodotoxin receptors in normal and detubulated frog sartorius muscle, J. Gen. Physiol., 67: 399.

Kim, D.H., Ohnishi, S.T., Ikemoto, N., 1983, Kinetic studies of calcium release from sarcoplasmic reticulum in vitro, J. Biol. Chem., 258: 9662.

Kirsch, G.E., Nichols, R.A., and Nakajima, S., 1977, Delayed rectification in the transverse tubules, J.J. Gen. Physiol., 70: 1.

Kovács, L., Rîos, E., and Schneider, M.R., 1979, Calcium transients and intramembrane charge movement in skeletal muscle, Nature (L.), 179: 391.

Lea, T.J., Griffiths, P.J., Tregear, R.T., and Ashley, C.C., 1986, An examination of the ability of inositol, 1,4,5-trisphosphate to induce calcium release and tension development in skinned skeletal muscle fibres of frog and crustacea, FEBS Lett., 207: 153.

Leblanc, N., and Hume, R., 1990, Sodium Current-Induced Release of Calcium from Cardiac Sarcoplasmic Reticulum, Science, 248: 372.

Lüttgau, H.C., Gottschalk, G., and Berwe, D., 1986, The role of Ca^{2+} in inactivation and paralysis of excitation-contraction coupling in skeletal muscle, Fortschr. Zoologie, 33: 195.

Lüttgau, H.C., and Spiecker, W., 1979, The effects of calcium deprivation upon mechanical and electrophysiological parameters in skeletal muscle fibres of the frog, J. Physiol. (L.), 296:411.

Lüttgau, H.C., and Stephenson, G.D., 1986. Ion movements in skeletal muscle in relation to the Activation of Contraction, in: "Physiology of Membrane Disorders", II ed., T. Andreoli et al, eds, Plenum Med. Book Company, N.Y and L.

Mikos, G.L., and Snow, T.R., 1987, Failure of inositol 1,4,5-trisphosphate to elicit or potentiate Ca^{2+} release from isolated skeletal muscle sarcoplasmic reticulum, Biochem. Biophys. Acta, 927: 256.

Nesterov, V.P., 1975, Membrane translocations of Na^+ in skeletal muscles function, in: "Structure and Functions of Biological Membranes", Nauka, Moscow.

Nesterov, V.P., 1985, On the mechanisms of Na^+ (Li^+) involvement in the skeletal muscle intracellular signal transmission, Sechenov Physiol. J. of USSR, 71: 985.

Nesterov, V.P., 1988, Possible mechanisms of Na^+-induced release of calcium ions from the sarcoplasmic reticulum of skeletal muscle fibres of vertebrates, Physiol. J. (Rus.), 34: 60.

Nesterov, V.P., and Fedorov, V.V., 1971, On the possible role of Na and K ions in electro-mechanical coupling, J. Evol. Boich. Physiol. (Rus.), 7: 303.

Nesterov, V.P., and Senchenkova, A.A., 1975, On the mechanisms of the excitation-contraction coupling in the frog phasic muscle fibres, Cytology, (Rus.), 17: 167.

Potreau, D., and Raymond, G., 1982, Existence of a sodium-induced calcium release mechanisms on frog skeletal muscle fibres, J. Physiol., 333: 463.

Schneider, M.F. and Chandler, W.K., 1973, Voltage dependent charge movement in skeletal muscle: a possible step in excitation-contraction-coupling, Nature (L.), 242: 244.

Stanfield, P.R., and Ashcroft, F.M., 1982, Calcium currents of frog and insect skeletal muscle fibres measured during voltage clamp, Can. J. Physiol. and Pharmacol., 60: 508.

Varsanyi, M., Messer, M., and Brandt, N.R., 1989, Intracellular localization of inositol-phospholipid-metabolizing enzymes in rabbit fast-twitch skeletal muscle, Eur. J. Biochem., 179: 473.

Venosa, R.A., 1974, Inward movement of sodium ions in resting and stimulated frog's sartorius muscle, J. Physiol., 241: 155.

Vergara, J., Tsien, R.Y., and Delay, M., 1985, Inositol 1,4,5-trisphosphate: a possible chemical link in excitation-contraction coupling in muscle, Proc. Natl. Acad. Sci. USA, 82: 6352.

Volpe, P., Di Virgilio, F., Pozzan, T., and Salviati, G., 1986, Role of inositol 1,4,5-trisphosphate in excitation-contraction coupling in skeletal muscle, FEBS Lett., 197: 1.

Volpe, P., Salviati, G., Di Virgilio, F., and Pozzan, T., 1985, Inositol 1,4,5-trisphosphate induces calcium release from sarcoplasmic reticulum of skeletal muscle, Nature, 316: 347.

Walker, J.W., Somlyo, A.V., Goldman, Y.E., Somlyo, A.P., and Trentham, D.R., 1987, Kinetics of smooth and skeletal muscle activation by laser pulse photolysis of caged inositol 1,4,5-trisphosphate, Nature (L.), 327: 249.

Yonemura, K., and Sato, M., 1967, The testing membrane potential and cation movement in frog muscle fibers after exposure to lithium ions, J.J. Physiol., 17: 678.

CELLULAR Mg^{++} ACCUMULATION IS ALTERED BY EXTRACELLULAR Na^{+} AND DIRECTLY AFFECTS AGONIST-INDUCED MOBILIZATION OF Ca^{++} IN VASCULAR SMOOTH MUSCLE

Suraj S. Shetty, John V. Zawadzki, and George B. Weiss

CIBA-GEIGY Corporation, Research Department
556 Morris Avenue
Summit, NJ 07901, U.S.A.

ABSTRACT

Effects of altered Na^{+}, Ca^{++} and Mg^{++} concentrations on ^{45}Ca and ^{28}Mg distribution and binding as well as of changes in cellular Mg^{++} on mobilization of Ca^{++} by added norepinephrine (NE) were examined in the rabbit aortic media-intimal layer. Uptake of ^{45}Ca at cellular high affinity sites was decreased by Mg^{++} much more than ^{28}Mg uptake was altered by Ca^{++}. Substitution of Na^{+} affects ^{45}Ca uptake primarily at extracellular (La^{+++}-accessible) binding sites. Muscles were pre-loaded with Mg^{++} by incubation in a low-Na^{+} solution (75% Na^{+} replaced isosmotically with sucrose) for 30 min followed by a 90 min exposure to a similar solution also containing 15 mM $MgCl_2$. These tissues, upon examination in normal (154 mM) Na^{+}-containing solution, indicated decreased retention of that cellular, high-affinity Ca^{++} fraction important for NE-induced contractile response. Accordingly, release of ^{45}Ca from this site and associated tension responses to added NE were attenuated in these muscles. These results suggest that variations in extracellular Na^{+} concentration modulate binding and subsequent mobilization of activator Ca^{++} by agonists through alterations in cellular Mg^{++} content in vascular smooth muscle.

INTRODUCTION

Numerous studies have suggested a regulatory role for Mg^{++} in the excitation-contraction (E-C) coupling process in excitable cells (Saida, 1982; Ford and Driska, 1986; White and Hartzell, 1988). Although extracellular Mg^{++} is known to modify vascular smooth muscle contractility, few studies have directly examined the effect of altered intracellular Mg^{++} concentrations on Ca^{++} mobilization and associated muscle tension changes (Ford and Driska, 1986). The relative insensitivity of cellular Mg^{++} content to changes in extracellular Mg^{++} concentration (Palaty, 1971, 1974; Shetty and Weiss, 1988) has made measurement of these variables difficult.

We have previously reported that the accumulation of ^{28}Mg in rabbit aorta is increased more than 5-fold when the external Na^{+} concentration is decreased by 75% for 90 min, and most of this accumulated Mg^{++} (more than 86%) is still retained 90 min after restoration of extracellular Na^{+} (Shetty and Weiss, 1988). Thus, incubation in low Na^{+} is a technique that

Excitation-Contraction Coupling in Skeletal, Cardiac, and Smooth Muscle
Edited by G.B. Frank *et al.*, Plenum Press, New York, 1992

accelerates loading and subsequent retention of elevated cellular concentrations of Mg^{++}. This inverse relationship between external Na^{+} concentration and cellular Mg^{++} accumulation can be used to increase the cellular Mg^{++} content of vascular smooth muscle and to characterize subsequent effects of this on the Ca^{++} fraction mobilized by NE.

METHODS

New Zealand White rabbits weighing 2.7 to 3.2 kg were sacrificed by injections of sodium pentobarbital (50 mg/kg) into the ear vein. The thoracic aorta was excised rapidly, placed in a Petri dish containing Tris-buffered bathing solution and aerated with 100% oxygen. The aorta was cut into a spiral strip of 3 to 4 mm width according to the method of Furchgott (1960). The adventitial layer was then separated from the media-intimal layer and muscle strips 6-10 mm long were prepared (Karaki and Urakawa, 1977).

The normal bathing solution contained (mM) NaCl, 154; KCl, 5.4; $CaCl_2$, 1.5; glucose, 11.0; and tris (hydroxymethyl) aminomethane, 6. Solutions containing either no added calcium (0-Ca solution) or no added magnesium (0-Mg solution) were also used; and various concentrations of Mg^{++} (as $MgCl_2.6H_2O$), Ca^{++} (as $CaCl_2$), or Sr^{++} (as $SrCl_2$) were added to different solutions as specified. The other bathing solutions used were identical to the normal solution except that the NaCl was included in reduced amounts (115.5 or 38.5 mM) corresponding to reductions of 25 or 75%. Isosmolarity was maintained in these solutions by inclusion of appropriate amounts of sucrose. All solutions were prepared with demineralized water, adjusted to pH 7.4 with small volumes of 6 N HCl, aerated with 100% oxygen and maintained at a constant temperature of 34.0 ± 0.5°C. Radioactive Ca^{++} (^{45}Ca, 21 mC/mg) was obtained from Du Pont (Wilmington, DE), and radioactive Mg^{++} (^{28}Mg, 37-180 mC/mg) was obtained from Brookhaven National Laboratory (Upton, NY).

Uptake of Ca^{++} was estimated in rabbit aorta media-intimal preparations by measuring ^{45}Ca as tissue spaces (tissue/medium ratios) as described previously (Hudgins and Weiss, 1969; Goodman et al., 1972; Karaki and Weiss, 1979, 1980a,b). To obtain ^{45}Ca tissue spaces (either total or cellular after washout), aortic strips were equilibrated in normal solution and then were placed in solution containing ^{45}Ca (0.4 μC/ml) for 30 to 60 min after a 30-90 min preincubation period in nonradioactive solutions of specified composition. For total ^{45}Ca uptake, the tissue was removed from the radioactive solution after the desired time interval, blotted lightly with Whatman No. 5 filter paper, dipped rapidly four times (2-4 sec) in similar but nonradioactive solutions, blotted again and weighed. Muscles were then placed in scintillation vials and were dialyzed overnight with 2 ml of EDTA (20 mM). The radioactivity was measured by scintillation spectrometry using Ready Safe (Beckman) as the scintillation cocktail. The procedure for determining the cellular (La^{+++}-resistant) ^{45}Ca uptake was the same as above except for a 60-min washout in isosmotic La^{+++} solution at 0.5°C before weighing. The La^{+++} solution contained (mM): $LaCl_3$, 80.8; glucose, 11.0; and tris (hydroxymethyl)aminomethane, 6; it was adjusted to pH 6.8 at 0.5°C with 1 M maleic acid (to an approximate final concentration of 3 mM). After washout in La^{+++} solution, the muscles were blotted, weighed and dialyzed as above. Tissue Ca^{++} content was calculated as the product of the ^{45}Ca tissue/medium ratio (minus the extracellular ^{14}C-sucrose space of 0.41 ml/g) and the Ca^{++} concentration in the incubation medium (Karaki and Weiss, 1979). The sucrose space correction was omitted in La^{+++} washout experiments. To obtain ^{28}Mg uptake (total), a similar experimental protocol was adopted except that the incubation solution contained ^{28}Mg (1-2 μC/ml) instead of ^{45}Ca. All data are expressed as the mean ± S.E. Statistical significance was determined either by Student's t test for paired data or by Neuman-Keul's multiple range test for multiple comparisons with a single control. Differences with $P < 0.05$ were considered significant.

For tension measurements, one end of each aortic media-intimal strip was attached to a glass rod which was then placed into a muscle bath (final volume, 20 ml) containing normal solution and the other end of the muscle strip was attached to a Grass FT-03 force displacement transducer. After a minimum of 90 min equilibration under a resting tension of 2 g, isometric contractions were recorded using a Dynograph recorder (R612).

RESULTS

Previous studies (Karaki and Weiss, 1980b; Karaki et al., 1983) have demonstrated that exposure of rabbit aortic smooth muscle to 1.5 mM Sr^{++} selectively decreases the La^{+++}-resistant, high-affinity Ca^{++} fraction released by NE. A qualitatively similar effect was also observed with 1.5 mM Mg^{++} but only when the ^{45}Ca uptake was performed in the presence of the added Mg^{++} (Table 1). The amount of ^{45}Ca bound to this site was not changed when added Mg^{++} was present prior to but not during the ^{45}Ca uptake period (Karaki et al., 1983), suggesting that the Ca^{++} fraction decreased by added Mg^{++} is replenished on reexposure to a 0-Mg solution.

Table 1. Effects of Mg^{++} and Sr^{++} on La^{+++}-resistant high affinity Ca^{++} uptake

Incubation Conditions[a]	La^{+++}-resistant Ca^{++} Uptake	
	After 1.5 mM Sr^{++}	After 1.5 mM Mg^{++}
	nmol/g muscle ± S.E.	
60 min ^{45}Ca incubation control	83.4 ± 2.5	85.3 ± 3.5
Sr^{++} or Mg^{++} added for final 30 min of ^{45}Ca incubation	24.4 ± 0.7*	61.4 ± 3.5*
Sr^{++} or Mg^{++} added for 60 min ^{45}Ca incubation + 30 min preceding	19.9 ± 0.8*	53.1 ± 2.8*

Data from Karaki et al (1983), tables 4 and 5
[a] In 0.03 mM Ca^{++} solutions, N= 6-14
* Significantly different ($P < 0.05$) from control

Incubation of rabbit aorta in a low-Na^{+} solution containing added Mg^{++} results in increased accumulation of Mg^{++} in tissues (Shetty and Weiss, 1988). However, use of this technique to examine interactions between cellular Mg^{++} and Ca^{++}, is complicated because low-Na^{+} solutions can also affect Ca^{++} binding and movements in vascular smooth muscle (Ozaki and Urakawa, 1979, 1981). To determine the effects of reduced external Na^{+} concentration on different fractions of bound Ca^{++} in the media-intimal layer of rabbit aorta, ^{45}Ca uptake studies were carried out. In these experiments, the amount of ^{45}Ca taken up by the smooth muscle during a 60-min ^{45}Ca incubation period was measured. Decreasing the external Na^{+} concentration by 25% or 75% resulted in significant increases in total ^{45}Ca uptakes at low affinity binding sites (in 1.5 mM Ca^{++} incubation solution) and at high affinity binding sites (in 0.03 mM Ca^{++} incubation solution) without any concomitant changes at cellular binding sites (La^{+++}-resistant, low- or high-affinity) (Table 2). This suggests that reducing external Na^{+} enhanced binding of Ca^{++} primarily at extracellular (La^{+++}-accessible) binding sites.

Table 2. Effect of external Na^{+} concentration on ^{45}Ca accumulation in the media-intimal layer of rabbit aorta

External NaCl[a] Conc.	Reduction	^{45}Ca Uptake Conditions	^{45}Ca Uptake[b]	Change
mM	%		nmol/g	%
154.0	-	1.5mMCa^{++}	3229 ± 82	-
115.5	25	1.5mMCa^{++}	3776 ± 93*	+16.9
38.5	75	1.5mMCa^{++}	5491 ± 151*	+70.1
154.0	-	0.03mMCa^{++}	1039 ± 13	-
115.5	25	0.03mMCa^{++}	1219 ± 40*	+17.3
38.5	75	0.03mMCa^{++}	1282 ± 21*	+23.4
154.0	-	1.5mMCa^{++},La^{+++}	429 ± 31	-
115.5	25	1.5mMCa^{++},La^{+++}	458 ± 29	+ 6.8
38.5	75	1.5mMCa^{++},La^{+++}	543 ± 33	+26.6
154.0	-	0.03mMCa^{++},La^{+++}	90 ± 12	-
115.5	25	0.03mMCa^{++},La^{+++}	67 ± 7	-25.6
38.5	75	0.03mMCa^{++},La^{+++}	72 ± 9	-20.0

[a] Isosmotic sucrose substituted for NaCl
[b] N = 6
* Significantly different ($P < 0.05$) from control

As observed with ^{45}Ca (Weiss, 1977), uptake of ^{28}Mg analyzed by a Scatchard-coordinate plot indicated the presence of at least two distinct uptake sites (Shetty and Weiss, 1988). To determine whether external Ca^{++} alters binding of Mg^{++} at specific sites in rabbit aortic smooth muscle, the uptake of ^{28}Mg in 0-Mg^{++} incubation solution (for high affinity sites) or in 1.5 mM Mg^{++} incubation solution (for low affinity sites) was measured in the presence or absence of added Ca^{++}. Results of these experiments summarized in Table 3 indicate that Mg^{++} binding does not change in the presence of 1.5 mM extracellular Mg^{++} and is only slightly changed in the absence of added Mg^{++}. Similar examination of the effects of Mg^{++} and Ca^{++} accumulation in the presence of either 1.5 mM or 0.03 mM Ca^{++} revealed that 1.5 mM Mg^{++} did not affect either the total ^{45}Ca uptake or the La^{+++}-resistant ^{45}Ca uptake into low-affinity binding sites (Table 4). However, La^{+++}-resistant ^{45}Ca uptake in the presence of 0.03 mM Ca^{++} was decreased by Mg^{++} as described earlier (Table 1). Thus, Mg^{++} significantly decreases only cellular high-affinity Ca^{++} in a manner similar to Sr^{++} (Table 1).

The reversible nature of the changes in Ca^{++} binding, produced by short-term variations in external Mg^{++} or Na^{+} concentrations, permitted measurement of changes in cellular retention of NE-releasable, high-affinity Ca^{++} in muscles acutely loaded with Mg^{++} in a low-Na^{+} solution. This experimental approach preincubated muscle strips in a nonradioactive, low-Na^{+} (38.5 mM, 75% reduction), low-Ca^{++} (0.03 mM) solution for 30 min followed by additional preincubation for 90 min in a similar solution with or without

Table 3. Effect of external Ca^{++} concentration on ^{28}Mg accumulation in the media-intimal layer of rabbit aorta

External Ca^{++}, Mg^{++}	^{28}Mg tissue/medium ratios[a]	Change
mM	ml/g	%
1.5 Ca^{++}, 0 Mg^{++}	2.33 ± 0.07	-
0 Ca^{++}, 0 Mg^{++}	1.99 ± 0.06*	-14.6
1.5 Ca^{++}, 1.5 Mg^{++}	0.84 ± 0.02	-
0 Ca^{++}, 1.5 Mg^{++}	0.93 ± 0.04	-10.7

[a] N = 6
* Significantly different ($P < 0.05$) from control

15 mM added Mg^{++}. Incubation in low-Na^{+} solutions containing added Mg^{++} has been previously shown to promote entry and subsequent prolonged retention of Mg^{++} in rabbit aorta (Shetty and Weiss, 1988). Muscles loaded with Mg^{++} were rinsed rapidly (2-4 sec) in normal Na^{+}, low Ca^{++} (0.03 mM)-containing solution with no added Mg^{++}, and incubated for 60 min in a similar solution containing ^{45}Ca with or without added NE (10^{-6}M). The La^{+++}-resistant high-affinity Ca^{++} retention in muscle strips subjected to the above experimental protocols is shown in Fig. 1. Tissues loaded with Mg^{++} decreased retention (82%) of that high-affinity, cellular fraction released by NE and there was little additional release from this binding site by NE. These Ca^{++} flux data measurements can be correlated with tension experiments in which the contractile response to added NE (10^{-7}M) is reduced by more than 30% in

Table 4. Effect of external Mg^{++} concentration on ^{45}Ca accumulation in the media-intimal layer of rabbit aorta

External Ca^{++}, Mg^{++}	^{45}Ca Uptake[a]	Change
mM	nmol/g	%
Total Ca^{++} uptake:		
0 Mg^{++}, 1.5 Ca^{++}	3229 ± 82	-
1.5 Mg^{++}, 1.5 Ca^{++}	2987 ± 82	-7.5
Total Ca^{++} uptake:		
0 Mg^{++}, 0.03 Ca^{++}	1038 ± 13	-
1.5 Mg^{++}, 0.03 Ca^{++}	992 ± 53	-4.4
After La^{+++} washout:		
0 Mg^{++}, 1.5 Ca^{++}	429 ± 31	-
1.5 Mg^{++}, 1.5 Ca^{++}	383 ± 49	-10.7
After La^{+++} washout:		
0 Mg^{++}, 0.03 Ca^{++}	90 ± 12	-
1.5 Mg^{++}, 0.03 Ca^{++}	43 ± 4*	-52.2

[a] N = 6
* Significantly different ($P < 0.05$) from control

tissues previously loaded with Mg^{++} in low-Na^{+} solution and subsequently examined in normal Na^{+} solution.

DISCUSSION

The response of smooth muscle cells to contractile agents is dependent on the free intracellular Ca^{++} concentration, which, in turn, is directly related to the amounts of Ca^{++} bound to and released from cellular binding sites, as well as influx from the extracellular compartment or associated superficial sites (Weiss, 1982b). Previous studies have shown that the Ca^{++} taken up by rabbit aortic smooth muscle is accumulated at two types of sites (Weiss, 1977, 1978), a high affinity site that can be detected at low Ca^{++} concentrations (up to 0.2 mM extracellular Ca^{++}) and a low affinity site that is occupied at higher Ca^{++} concentrations (above 0.2 mM extracellular Ca^{++}). Use of an isosmotic La^{+++} solution at 0.5°C further delineated each of these sites into superficial (La^{+++}-accessible) and less superficial (cellular or La^{+++}-resistant) Ca^{++} sites. It was further shown that contractile agents like NE mobilized specific Ca^{++} fractions, observed as

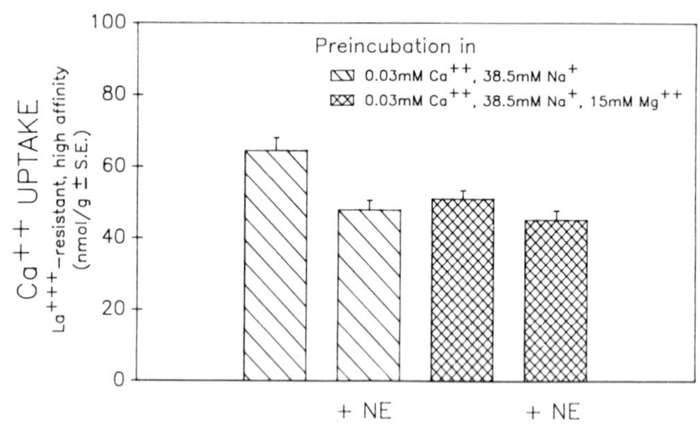

Fig. 1. Effect of Mg^{++} on the NE-induced decrease in ^{45}Ca uptake. After preincubation, muscles were incubated in similar solutions containing ^{45}Ca for 60 min. Each bar represents the average for 4 members of experimental sets of 4 muscles. All three experimental bars are significantly different ($P < 0.05$) from control.

decreased ^{45}Ca retention, from cellular high-affinity sites (Karaki and Weiss, 1979). This decrease in ^{45}Ca retention is believed to correlate with a simultaneous intracellular Ca^{++} release, an increased level of intracellular Ca^{++} and a subsequent contractile response that is rapid in onset and not well maintained (the phasic component of the NE-induced response). Displacement of this Ca^{++} fraction by Sr^{++} or Mg^{++} was accordingly shown to be associated with attenuated contractile responses to added NE (Karaki et al., 1983). The displacement of cellular high-affinity Ca^{++} by external Mg^{++} persisted as long as the muscle was exposed to the cation and was reversed upon exposure to a 0-Mg^{++} solution.

The decreased cellular retention of high-affinity Ca^{++} fraction, observed in muscles pre-loaded with Mg^{++}, indicates an inverse relationship in the cellular binding of Ca^{++} and Mg^{++} in rabbit aortic smooth muscle. A similar relationship was also noted in chicken heart cells where an increase in cellular Ca^{++} was shown to release Mg^{++} from intracellular binding sites (Murphy et al., 1989; Rotevatn et al., 1989). Competition between cellular Ca^{++} and Mg^{++} was thought to modulate free intracellular Mg^{++} in these cells. From our studies, it is not possible to ascertain if the two divalent cations compete for mutual binding sites or if a site-site interaction between discrete ionic binding sites inhibits occupancy at one site when the other site is occupied (negative cooperativity). Our experiments also do not reveal the anatomic locations of these binding sites. Such a determination is rendered difficult by the likelihood that cellular Ca^{++} and Mg^{++} are unevenly distributed within various subcellular compartments (Weiss, 1982a; Cech et al., 1980; Gunther el al., 1984; Maguire, 1984).

By regulating the amount of Ca^{++} bound to cellular high-affinity sites, intracellular Mg^{++} also controls the availability of activator Ca^{++} for tension development. This was clearly demonstrated by the decrease in observed NE-induced Ca^{++} mobilization and associated contractile response in muscles pre-loaded with Mg^{++}. Ford and Driska (1986) also showed that conditions changing cellular Mg^{++} content altered maximum contractility in vascular smooth muscle. With porcine carotid arterial strips pre-loaded with Mg^{++} by incubation in high-K^{+}, Ca^{++}-free solutions containing variable amounts of Mg^{++}, they demonstrated that alterations in contractility are directly related to the cellular Mg^{++} content. Thus, while cellular Ca^{++} plays the specific role of an active regulatory ion in the E-C coupling process, cellular Mg^{++} appears to modulate this process by controlling the set point or gain of the process. The ability of Mg^{++} to modulate this process is apparently a function of its intracellular concentration which, in turn, is inversely related to external Na^{+} concentration (Shetty and Weiss, 1988).

The preceding discussion indicates that the availability of cellular, high-affinity activator Ca^{++} in vascular smooth muscle is regulated by a Na^{+}-Mg^{++}-Ca^{++} interaction. These ionic interactions, schematically represented in Fig. 2, illustrate the following: a) external Na^{+} inhibits active transport of Mg^{++} either by altering the binding of Mg^{++} to the translocator sites or by altering the translocation process itself (Shetty and Weiss, 1988), b) accumulated free intracellular Mg^{++} is retained by the cell and equilibrates with Mg^{++} present at cellular high-affinity sites, and c) binding of Mg^{++} to high-affinity sites alters the cellular retention of the high affinity Ca^{++} important for NE-induced tension response. Alternatively, Mg^{++} and Ca^{++} may compete for the same high-affinity cellular binding sites and, in this manner, alter the availability of bound activator Ca^{++} for tension development.

Reductions in extracellular Na^{+} concentration, therefore, initiate a complex sequence of ionic events including increased Mg^{++} uptake, increased Mg^{++} binding to high-affinity sites, decreased Ca^{++} binding to high-affinity sites, decreased release of cellular high-affinity Ca^{++} after NE exposure, and decreased contractile response (Fig. 3). Although a large reduction in external Na^{+} (75%) during relatively short time intervals was employed to demonstrate this succession of events in the present study, it is nevertheless conceivable that such a Na^{+}-Mg^{++}-Ca^{++} interaction is physiologically relevant and plays an important role in E-C coupling process in vascular smooth muscle when small reductions in extracellular Na^{+} concentrations are maintained over an extended period of time.

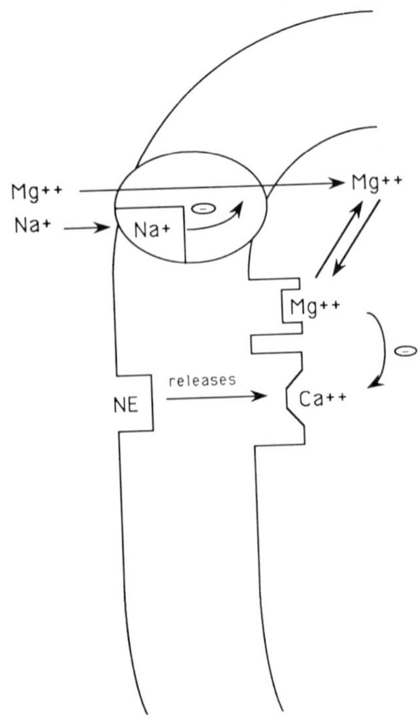

Fig. 2. Coupling of Mg^{++} uptake to extracellular Na^{+} and to Ca^{++} binding and release. For explanation, see text.

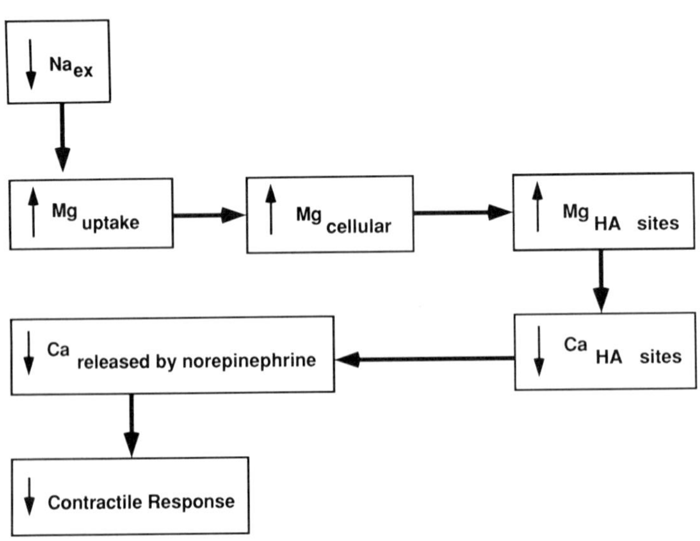

Fig. 3. Diagram of postulated sequence of ionic events coupling changes in extracellular Na^{+} to the contractile response in vascular smooth muscle.

REFERENCES

Cech, S. Y., Broaddus, W. C. and Maguire, M. E., 1980, Adenylate cyclase: The role of magnesium and other divalent cations, Mol. Cell. Biochem., 33:67-92.

Ford, G. D. and Driska, S. P., 1986, Influence of altering cellular magnesium content on vascular smooth muscle contractility, Am. J. Physiol., 251:C687-C695.

Furchgott, R. F., 1960, Spiral-cut strips of rabbit aorta for in vitro studies of responses of arterial smooth muscle, Methods Med. Res., 8:177-186.

Goodman, F. R., Weiss, G. B., Weinberg, M. N. and Pomarantz, S. D., 1972, Effects of added or substituted potassium ion on ^{45}Ca movements in rabbit aortic smooth muscle, Circ. Res., 31:672-681.

Gunther, T., Vormann, J. and Forster, R., 1984, Functional compartmentation of intracellular magnesium, Magnesium Bull., 6:77-81.

Hudgins, P. M. and Weiss, G. B., 1969, Characteristics of ^{45}Ca binding in vascular smooth muscle, Am. J. Physiol., 217:1310-1315.

Karaki, H., Hatano, K. and Weiss, G. B., 1983, Effects of magnesium on ^{45}Ca uptake and release at different sites in rabbit aortic smooth muscle, Pflügers Arch., 398:27-32.

Karaki, H. and Urakawa, N., 1977, Possible role of endogenous catecholamines in the contraction induced in rabbit aorta by ouabain, sodium depletion, and potassium depletion, Eur. J. Pharmacol., 43:65-72.

Karaki, H. and Weiss, G. B., 1979, Alterations in high and low affinity binding of ^{45}Ca in rabbit aortic smooth muscle by norepinephrine and potassium after exposure to lanthanum and low temperature, J. Pharmacol. Exp. Ther., 211:86-92.

Karaki, H. and Weiss, G. B., 1980a, Effects of stimulatory agents on mobilization of high and low affinity site ^{45}Ca in rabbit aortic smooth muscle, J. Pharmacol. Exp. Ther., 213:450-455.

Karaki, H. and Weiss, G. B., 1980b, Dissociation of varied actions of norepinephrine on ^{45}Ca uptake and release at different sites in rabbit aortic smooth muscle, J. Pharmacol. Exp. Ther., 215:363-368.

Maguire, M. E., 1984, Hormone-sensitive magnesium transport and magnesium regulation of adenylate cyclase, Trends Pharmacol. Sci., 5:73-77.

Murphy, E., Freudenrich, C. C., Levy, L. A., London, R. E. and Lieberman, M., 1989, Monitoring cytosolic free magnesium in cultured chicken heart cells by use of the fluorescent indicator furaptra, Proc. Natl. Acad. Sci. (USA), 86:2981-2984.

Ozaki, H. and Urakawa, N., 1979, Na-Ca exchange and tension development in guinea-pig aorta, Naunyn-Schmied. Arch. Pharmacol., 309:171-178.

Ozaki, H. and Urakawa, N., 1981, Involvement of a Na-Ca exchange mechanism in contraction induced by low-Na solution in isolated guinea-pig aorta, Pflügers Arch., 390:107-112.

Palaty, V., 1971, Distribution of magnesium in the arterial wall, J. Physiol. (Lond.) 218:353-368.

Palaty, V., 1974, Regulation of cell magnesium in vascular smooth muscle, J. Physiol. (Lond.), 242:555-569.

Rotevatn, S., Murphy, E., Levy, L. A., Raju, B., Lieberman, M. and London, R. E., 1989, Cytosolic free magnesium concentration in cultured chick heart cells, Am. J. Physiol., 257:C141-C146.

Saida, K., 1982, Intracellular Ca release in skinned smooth muscle. J. Gen. Physiol., 80:191-202.

Shetty, S. S. and Weiss, G. B., 1988, Alterations in ^{28}Mg distribution and movements in rabbit aortic smooth muscle, J. Pharmacol. Exp. Ther., 245:112-119.

Weiss, G. B., 1977, Calcium and contractility in vascular smooth muscle, in: "Advances in General and Cellular Pharmacology," T. Narahashi and C. P. Bianchi, eds., Vol. II, pp. 71-154, Plenum Publishing Corp., New York.

Weiss, G. B., 1978, Quantitative measurement of binding sites and washout components for calcium ion in vascular smooth muscle, in: "Calcium in Drug Action," G. B. Weiss, ed., pp. 57-74, Plenum Publishing Corp., New York.
Weiss, G. B., 1982a, Patterns of intracellular Ca^{++} distribution and mobilization in smooth muscle, in: "Calmodulin and Intracellular Ca^{++} Receptors," ed. by S. Kakiuchi, H. Hidaka, and A. R. Means, eds., pp. 111-123, Plenum Publishing Corp., New York.
Weiss, G. B., 1982b, Delineation of actions of calcium antagonists in vascular smooth muscle, in: "Ca^{++} entry blockers, adenosine and neurohumors," G. F. Merrill and H. R. Weiss, eds., pp. 63-72, Urban and Schwarzenberg.
White, R. E. and Hartzell, H. C., 1988, Effects of intracellular free magnesium on calcium current in isolated cardiac myocytes, Science, 239: 778-780.

CALCIUM AND POTASSIUM CHANNEL REGULATION BY PROTEIN PHOSPHORYLATION IN SMOOTH MUSCLE CELLS OF GUINEA PIG TAENIA COLI

Hideyo Yabu, Kazuo Obara and Toshihiro Usuki

Department of Physiology
Sapporo Medical College
Sapporo 060, Japan

INTRODUCTION

Generally speaking, Ca^{2+} influx through membrane in smooth muscle cells may play a crucial role in Excitation-Contraction Coupling. Mammalian smooth muscle cells are universally endowed with a lot of L-type, dihydropyridine-sensitive voltage-dependent Ca^{2+} channels and various kinds of K^+ channels. Actually, Ca^{2+} conductances are increased by membrane depolarization causing Ca^{2+} influx and subsequent contraction.

Recent electrophysiological studies have suggested that the ion currents that pass through a certain Ca^{2+} channel can be modulated by reversible protein phosphorylation and dephosphorylation reaction. Our study, concerns the modulation mechanism of the L-type Ca^{2+} channel by means of the phosphorylation of the putative protein component of the Ca^{2+} channel by Ca^{2+}/phospholipid-dependent protein kinase (PK-C) and the de-phosphorylation of phosphorylated channel proteins by various kinds of phosphoprotein phosphatases.

It is well known that, in the guinea pig taenia coli, acetylcholine has been found to activate the muscarinic receptor leading to their contraction. The modulation of gating of the Ca^{2+} channel and the activation of the muscarinic receptor are also examined in connection with the phosphorylation of Ca^{2+} channels.

MATERIALS AND METHODS

The single smooth muscle cells were isolated with use of collagenase (Type 1, Sigma CO.) by the method of Obara (1984). The freshly isolated cells were then incubated in modified KB medium (Isenberg and Klockner, 1982).

Whole-cell currents were recorded using a patch clamp technique (Hamill et al., 1981). An EPC 7 patch clamp amplifier was used for recordings. In inward current recording, the patch pipette was filled with 140 mM Cs^+ to block K^+ outward current. To decrease Ca^{2+}-mediated inactivation of the Ca^{2+} channel, we recorded the inward currents carried

Excitation-Contraction Coupling in Skeletal, Cardiac, and Smooth Muscle
Edited by G.B. Frank *et al.*, Plenum Press, New York, 1992

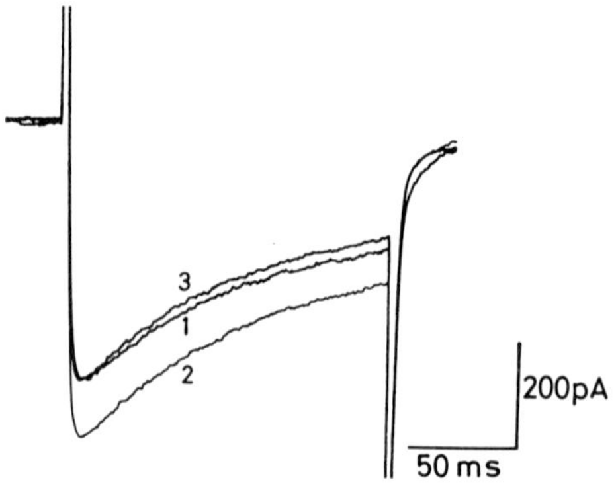

Fig. 1. Effect of carbachol and H-7 on the voltage-dependent inward current.
The drugs were applied for 3 min prior to depolarization. Inward current was elicited by depolarizing steps from -80 to +10 mV in the absence (1) and the presence of 100 nM carbachol (CCh) (2) or of 100 nM CCh and 100 μM H-7 (3).

by Ba^{2+}. Voltage clamp pulses from a holding potential of -80 mV to a test potential of +10 mV were applied at a frequency of 0.067 Hz.

RESULTS AND DISCUSSION

Muscarinic receptor agonist, carbachol (CCh), enhances the inward calcium channel currents

Whole-cell currents were recorded in smooth muscle cells isolated from guinea pig taenia coli. The peak inward current-voltage relations indicated that the maximum current was obtained at +10 mV, the threshold potential for the activation was about -40 mV and estimated reversal potential was about +60 mV, which is expected for a Ca^{2+}-selective channel current. The inward currents in whole-cell recording were very sensitive to the dihydropyridine (DHP) antagonist, nifedipine. These results suggest that the DHP-sensitive Ca^{2+} channel current appears to be the predominant current component in whole-cell recording in guinea pig taenia coli smooth muscle cells.

The effect of CCh was first studied at a test potential of +10 mV, where the inward current is largest (Fig. 1). CCh was applied by bath perfusion. Recording trace No.1 is the control current and No.2 is the current 3 min after the addition of CCh. The peak inward current increased by 25%. The decay time did not change compared to that of the control current. The effect of CCh was to increase the currents at all potentials tested. And the apparent threshold, potential obtained maximum current and reversal potential were not significantly changed. No detectable shift in the current-voltage relationship could not be attributed to a change in the voltage sensitivity of the Ca^{2+} currents.

Figure 2 shows the CCh effect on the magnitude of the peak inward current. The increase in the inward current by CCh was reversed by muscarinic receptor antagonist, atropine. This means that the CCh effect on the inward current was really mediated by muscarinic receptor stimulation. The most direct interpretation of the above results might be that CCh exerts its effect on the voltage-dependent Ca^{2+} channel. However, two alternatives should be considered. The first of these is that CCh activates the Na^+ currents. The second possibility is that CCh acts on Cl^- currents. In this study, CCh had no effect on the threshold of activation and the reversal potential. This evidence indicated that the effect of CCh on the inward current cannot be explained by the effect of CCh on the Na^+ and Cl^- currents.

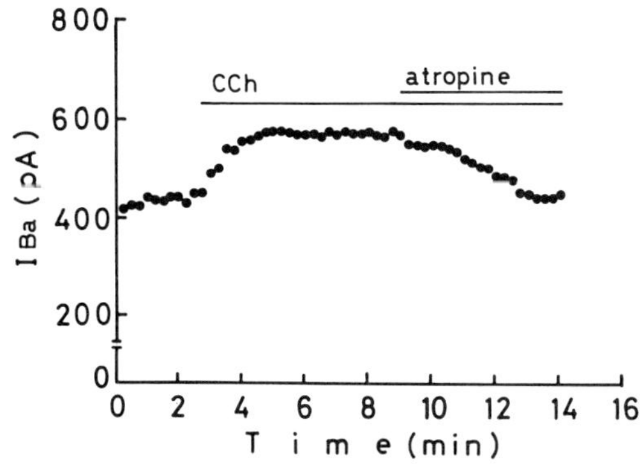

Fig. 2. Time course of carbachol and atropine effects on the voltage-dependent inward current.
Inward currents were elicited by depolarizing steps from -80 to +10 mV. Carbachol (CCh, 100 nM) and atropine (5 μM) were applied for the duration of the horizontal bar.

Benham et al. (1985) reported that, in rabbit ileum, muscarinic agents themselves are able to gate open a channel and therefore produce a net inward current. It is called the operation of receptor operated ion channels. In this study, CCh did not affect the holding current when the membrane potential was held constant at a potential near -80 mV. This means that, in the isolated smooth muscle cells of guinea pig taenia coli, muscarinic agents cannot by themselves gate open Ca^{2+} channels but can alter the gating caused by changes in membrane potentials.

The involvement of phosphorylation in the modulation of calcium channel

Amiloride. Amiloride, a K^+-sparing diuretic, has been shown to inhibit various kinds of protein kinases (Ozaki et al., 1987; Ishihara et al., 1989). Amiloride (100 μM) suppressed the inward current by about 15% of the control current. This suggests that some kinds of endogenous protein kinase activities partially modulate the Ca^{2+} channel current.

Isoproterenol, cyclic AMP and protein kinase A. Recently, the most common phosphorylation hypothesis of the modulation mechanism of Ca^{2+} channels is that they appear in the cardiac myocytes. It was shown that isoprenaline or the intracellular application of cyclic AMP (cAMP) or the catalytic subunit of cAMP-dependent protein kinase (PK-A) increased the amplitude of the Ca^{2+} current in guinea pig ventricular cells (Kameyama et al., 1986).

We examined whether the PK-A contributes to the modulation of the Ca^{2+} channel in smooth muscle cells. A bath application of β-adrenergic agonist, isoproterenol, dibutyryl cAMP (DBcAMP) and H-8, a specific PK-A inhibitor, had no effect on the inward current in our study. These results are consistent with the report that the cAMP mediated modulation of Ca^{2+} channel is absent in the rabbit ileum smooth muscle (Ohya et al., 1987).

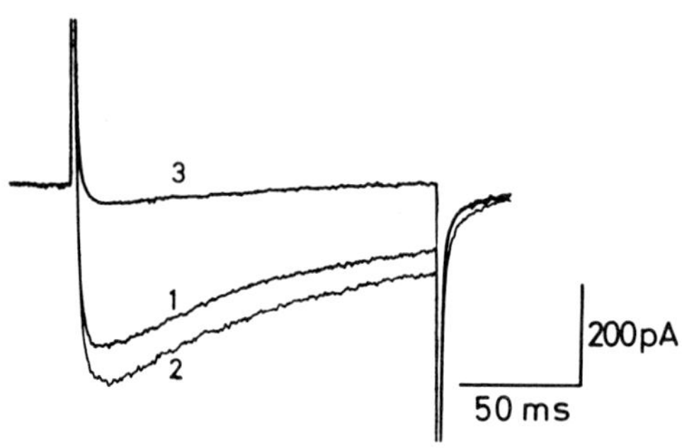

Fig. 3. Effects of phorbol ester and nifedipine on the voltage-dependent inward current.
The drugs were applied for 3 min prior to depolarization. Inward current was elicited by depolarizing steps from -80 to +10 mV in the absence (1) and the presence of 100 nM phorbol 12,13-dibutyrate (PDB) (2) or of 100 nM PDB and 200 nM nifedipine (3).

The involvement of PK-C activity in the modulation of calcium channel currents

In smooth muscle cells, as in many other cell types, muscarinic agents have been reported to increase the breakdown of phosphatidylinositol 4,5-bisphosphate (PIP_2), which indicates that inositol trisphosphate (IP_3) and diacylglycerol (DAG) can be generated. Actually, Salmon and Honeyman (1979) showed, using smooth muscle cells, that muscarinic stimulation resulted in a rapid accumulation of phosphatidic acid and that DAG is also generated. Therefore, it seemed possible that a product of phosphoinositide (PI) breakdown might mediate the muscarinic effect on the voltage-activated channels in our cells. DAG, one product of PI breakdown, increases the activity of PK-C. Remembering the data in Fig. 1, in which the current supported by CCh was suppressed by H-7, a PK-C inhibitor, we tested the involvement of PK-C activity for the modulation of the inward current of guinea pig taenia coli smooth muscle cells with use of tumor promoting phorbol ester, a direct activator of PK-C, phorbol 12,13-dibutyrate (PDB).

The effect of PDB on the inward currents was first studied at a test potential of +10 mV, where the inward current is largest (Fig. 3). When applied externally, PDB increased the peak inward current by 25-50% of the control current at all potentials tested and did not change the decay of the current. To examine the specificity of the PDB effect, the inactive form of phorbol ester, 4α-phorbol 12, 13-didecanoate (4α-PDD) was included in the bathing solution. Even at a high concentration of 500 nM, the inward current was not noticeably changed, therefore, the possible involvement of nonspecific actions of phorbol structure in the PDB-induced response can be ruled out.

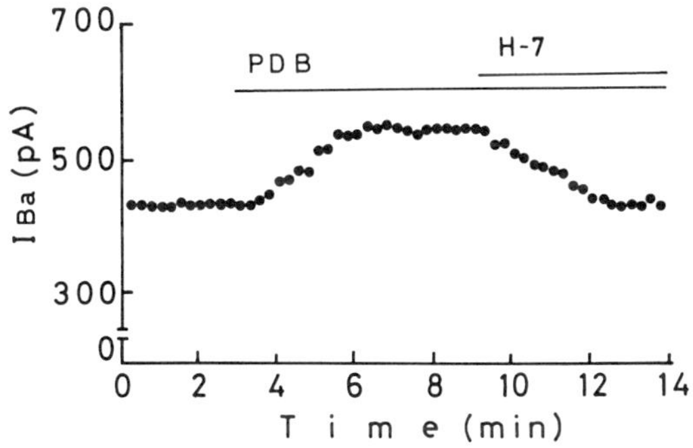

Fig. 4. Time course of phorbol ester and H-7 effects on the voltage-dependent inward current.
Inward currents were elicited by depolarizing steps from -80 to +10 mV. Phorbol 12, 13-dibutyrate (PDB, 100 nM) and H-7 (100 μM) were applied for the duration of the horizontal bar.

The PDB effect on the Ca^{2+} current is shown in Fig. 4. In the presence of PDB, the current is increased and an additional application of 100 μM H-7 caused a decrease in the inward current to the control level. As shown in Fig. 3, recording trace No. 3, the addition of 200 nM nifedipine to the bathing solution attenuated the current in cells previously stimulated by 100 nM PDB, confirming that the PDB-stimulated currents were the Ca^{2+} channel currents passing through the L-type, DHP-sensitive Ca^{2+} channels. This study suggest that the DHP-sensitive, L-type voltage-dependent Ca^{2+} channel was modulated by phorbol activation of PK-C. PK-C activation may, therefore, cause the phosphorylation of the L-type Ca^{2+} channel or regulatory protein(s), resulting in an increase in Ca^{2+} conductance which could occur through either an increase in the number of channels available for the voltage activation or an increase in single channel open probability or conductance, or both. H-7 suppressed partially the control Ca^{2+} channel current by about 15%. The suppression is due to the inhibition of endogenous PK-C activity. Without any agonist stimulation, endogenous PK-C activity contributes to the modulation of Ca^{2+} channel gating. This also could indicate that the phosphorylation by PK-C is not a prerequisite for maintaining a large part of control inward current.

The involvement of protein phosphatase activity in the modulation of the calcium channel

According to the current theory of Ca^{2+} channel activation in heart muscle, okadaic acid, a potent inhibitor of protein phosphatase, has a strong inotropic action on the heart muscle (Hescheler et al., 1988) that is due to an increase in the open-state probability of Ca^{2+} channels, which is thought to be regulated by cAMP-dependent phosphorylation which is determined by the activity balance between protein kinase and protein

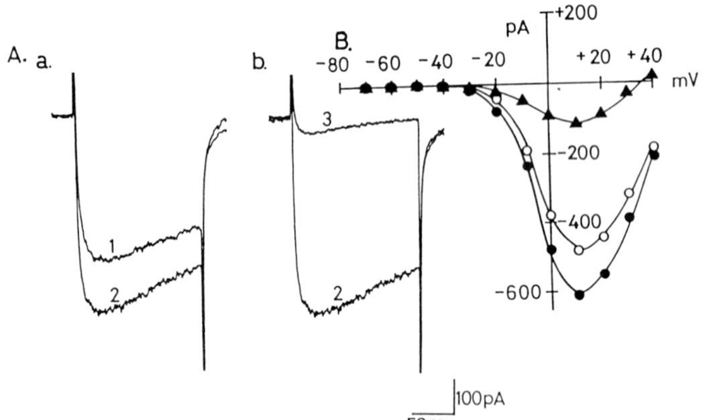

Fig. 5. Effect of CL-A on the voltage-dependent inward currents. The drugs were applied for 3 min prior to depolarization. A: Inward current elicited by depolarizing steps from -80 to +10 mV in the absence (1) and presence of 10 nM CL-A (2) or of 10 nM CL-A and 200 nM nifedipine (3). a: Effect of CL-A. b: Effect of CL-A and nifedipine. B: Peak current-voltage relations for the cell in the absence (O) and the presence of 10 nM CL-A (●) or of 10 nM CL-A and 200 nM nifedipine(▲). Leak currents were subtracted.

phosphatase. In this study, we used calyculin A (CL-A), as a protein phosphatase inhibitor, provided by Dr. Karaki, Tokyo University. CL-A is novel marine toxin isolated from marine sponge Discodermia calyx collected in the Gulf of Sagami, Japan (Kato et al., 1986). In 1989, Dr. Karaki's group reported that CL-A had a potent inhibitory action on phosphoprotein phosphatase type-1 and type-2A. The characteristics of CL-A closely resembles those of okadaic acid, a marine toxin isolated from Halichondria okadai, however, CL-A is a more effective inhibitor of protein phosphatase type-1 than okadaic acid (Ishihara et al., 1989). Furthermore, it was found that CL-A increases in cytosolic free Ca^{2+} concentration ($[Ca^{2+}]_i$) in rat aorta (Ishihara et al., 1989). This increase in $[Ca^{2+}]_i$ was inhibited by verapamil or by removal of extracellular Ca^{2+}, and was potentiated by Bay K 8644. Yabu et al. (1990) reported, preliminarily, that CL-A enhanced the inward current of smooth muscle cells from guinea pig taenia coli.

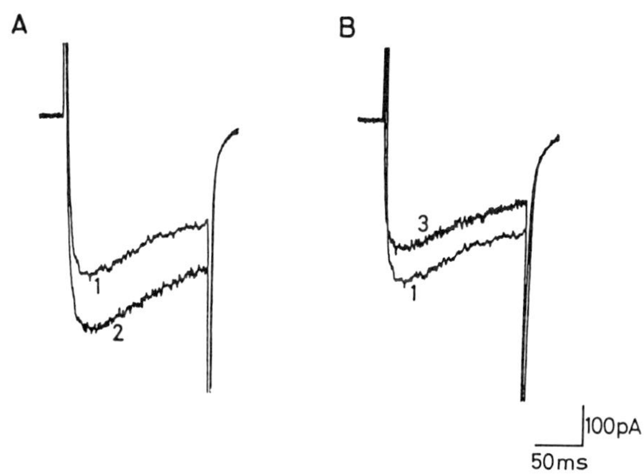

Fig. 6. Effect of CL-A and H-7 on the voltage-dependent inward currents. The drugs were applied for 3 min prior to depolarization. Inward current was elicited by depolarizing steps from -80 to +10 mV in the absence (1) and the presence of 10 nM CL-A (2) or of 10 nM CL-A and 100 μM H-7 (3). A: Effect of CL-A. B: Effect of CL-A and H-7.

As shown in Fig. 5, CL-A (10 nM) increased the inward current by an average of 27.0 ± 6.0% (mean ± S.E., n=6) (Fig. 5Aa). CL-A did not appear to affect the threshold of activation of inward current or the reversal potential obtained by extrapolation (control: 51.9 ± 2.4 mV, plus CL-A: 52.2 ± 2.2 mV; mean ± S.E., n=4). CL-A also had no effect on the voltage at which the maximum inward current was recorded (Fig. 5B). The CL-A effect was dose-dependent and the maximum response was observed at about 10 nM CL-A. Simultaneous application of CL-A (10 nM) and nifedipine (200 nM) decreased the inward current by about 90% of the control amplitude which was almost equal to the amplitude decrease by nifedipine alone (Fig. 5Ab).

In the presence of CL-A, an additional application of 100 μM H-7, a PK-C inhibitor, caused a decrease in the inward current by 15.0 ± 5.0% (mean ± S.E., n=4) of the control amplitude which was almost equal to the amplitude decrease by H-7 alone (Fig. 6B, recording trace No. 3).

Table 1 summarizes the effects of agonist and some other agents which are able to modify the phosphorylation steps on the inward currents. The peak of the inward current in the presence of agonist or other agents was normalized to the peak of the control current. The increasing effect was not over 50% of the control current and the decreasing effect was no more than about 15% of the total control current. This indicates that the modulation of Ca^{2+} channel gating of smooth muscle cells from guinea pig taenia coli is partly dependent on the PK-C-dependent phosphorylation of channel protein.

Table 2 summarizes the effects of agonist and agents affecting the phosphorylation steps on the half-time of decay ($t_{1/2}$) of the inward current. As showed in Table 2, CCh, PDB and CL-A did not affect the $t_{1/2}$. The mechanism whereby they did not affect the rate of decay of the Ca^{2+} inward current remains to be established. It is possible that CCh, PDB and CL-A increase either the availability or open probability of L-type Ca^{2+} channels, or both, but they do not modify the voltage-dependent inactivation process.

Table 1. Effect of drugs on the voltage-dependent inward current.

Drugs		Relative amplitude	
Carbachol	($1x10^{-7}$ M)	1.24 ± 0.06	(n=5)
Carbachol plus atropine	($1x10^{-7}$ M) ($5x10^{-6}$ M)	1.03 ± 0.02	(n=3)
Carbachol plus H-7	($1x10^{-7}$ M) ($1x10^{-4}$ M)	0.99 ± 0.02	(n=3)
Phorbol 12,13-dibutyrate	($5x10^{-7}$ M)	1.50 ± 0.15	(n=7)
Phorbol 12,13-dibutyrate	($1x10^{-7}$ M)	1.23 ± 0.04	(n=3)
Phorbol 12,13-dibutyrate plus H-7	($1x10^{-7}$ M) ($1x10^{-4}$ M)	1.05 ± 0.03	(n=3)
Calyculin A	($1x10^{-7}$ M)	1.23 ± 0.07	(n=4)
Calyculin A	($1x10^{-8}$ M)	1.27 ± 0.06	(n=6)
Calyculin A plus H-7	($1x10^{-8}$ M) ($1x10^{-4}$ M)	0.85 ± 0.05	(n=4)
4α-phorbol 12,13-didecanoate	($5x10^{-7}$ M)	1.03 ± 0.04	(n=4)
Amiloride	($1x10^{-4}$ M)	0.85 ± 0.03	(n=3)
H-7	($1x10^{-4}$ M)	0.85 ± 0.04	(n=4)
Atropine	($5x10^{-6}$ M)	1.01 ± 0.02	(n=3)

Each value represents the mean ± S.E.
The voltage-dependent inward currents were elicited by step depolarization from -80 to +10 mV. The drugs were applied for 3 min prior to depolarization. The peak current in the presence of drugs was normalized to the peak current in the absence of drugs.

Table 2. Effects of drugs on the half-time of the voltage-dependent inward current.

Drugs		Half-time of decay (msec)		
		Control	plus drug	n
Carbachol	($1x10^{-7}$ M)	173.4 ± 17.2	170.9 ± 17.5	4
Phorbol 12,13 dibutyrate	($1x10^{-7}$ M)	166.8 ± 43.3	164.3 ± 38.1	6
Calyculin A	($1x10^{-8}$ M)	160.0 ± 16.0	151.6 ± 28.4	4

Each value represents mean ± S.E.
The voltage-dependent inward currents were elicited by step depolarization from -80 to +10 mV. The drugs were applied for 3 min prior to depolarization.

The involvement of phosphorylation in the modulation of potassium outward current

Isoproterenol increased the Ca^{2+}-independent K^{+} outward current in a concentration dependent manner. The threshold concentration was 1 nM and maximum effect was obtained at over 1 μM. DBcAMP enhanced the Ca^{2+}-independent K^{+} outward current by about 50% of control and then it was suppressed by H-8, a PK-A inhibitor, to the control level 3 min after addition of H-8 (Fig. 7). These results indicate that Ca^{2+}-independent K^{+} outward currents are partly modulated by the phosphorylation of K^{+}-channel protein through PK-A activity. To verify the idea, CL-A, a novel protein phosphatase inhibitor, was introduced. CL-A increased the

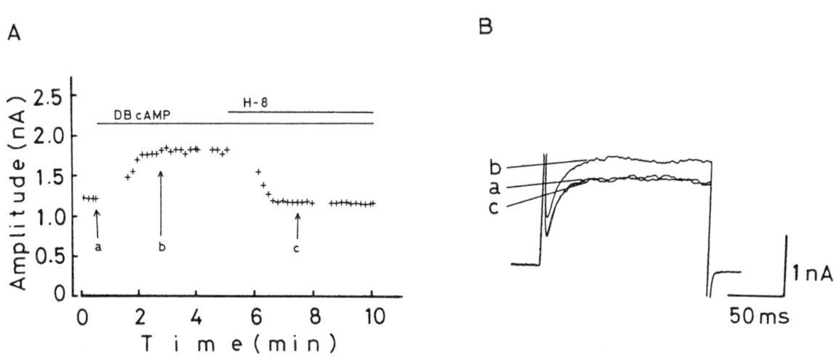

Fig. 7. Effect of dibutyryl cAMP (DBcAMP) and H-8 on the Ca^{2+}-insensitive K^{+} outward current.
A: Time course of the effect of DBcAMP and H-8 on the peak current of K^{+} currents elicited by depolarizing steps from -60 to +40 mV. Horizontal bars indicate duration of application of DBcAMP (200 μM) or H-8 (100 μM). B: Sample recordings of K^{+} current respectively from the points of time labeled in A.

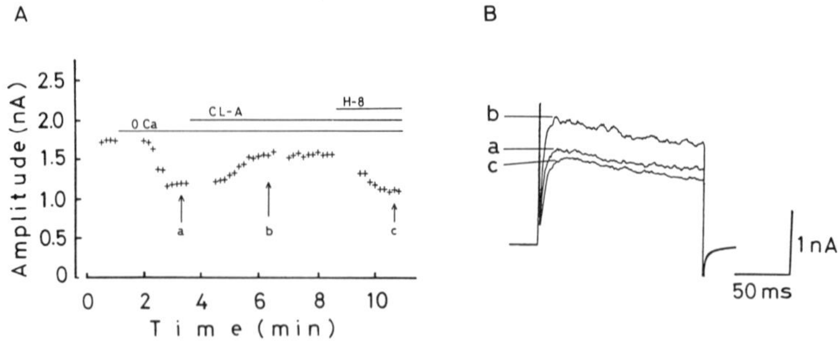

Fig. 8. Effect of calyculin A (CL-A) and H-8 on the Ca^{2+}-insensitive K^+ outward current.
A: Time course of the effect of CL-A and H-8 on the peak current of K^+ currents elicited by depolarizing steps from -60 to +40 mV. Horizontal bars indicate duration of application of CL-A (10 nM) or H-8 (100 μM) B: Sample recordings of K^+ current respectively from the points of time labeled in A.

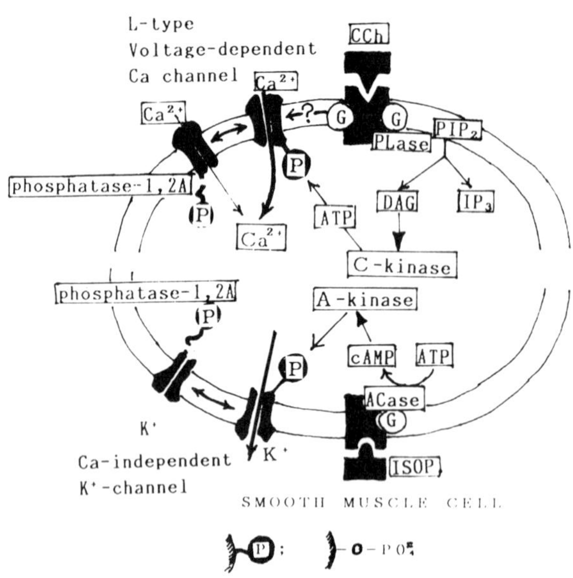

Fig. 9. A simplified scheme for the modulation of voltage-dependent Ca^{2+} channel and K^+ channel.
Carbachol (CCh); Isoproterenol (ISOP); GTP-binding protein (G); Phosphatidylinositol 4,5-bisphosphate (PIP_2); Inositol trisphosphate (IP_3); Diacylglycerol (DAG); Phospholipase (PLase); Adenylate cyclase (ACase); Inorganic phosphate (P).

Ca^{2+}-independent K^+ outward current by over 30% of the control current (Fig. 8). The combination of CL-A and H-8 decreased the increased current supported by CL-A. This indicates that it is the proportion of phosphorylated and dephosphorylated channel protein that balances the open-closed state of the Ca^{2+}-independent K^+ channel.

CONCLUSION

A schematic presentation based on our experimental results is shown in Fig. 9.

Gating of L-type Ca^{2+} channels of smooth muscle cells from guinea pig taenia coli is modulated partially by phosphorylation of the channel or a related protein component through PK-C activity. Muscarinic receptor stimulation supports this PK-C activity by generating DAG. On the other hand, K^+ outward current is also partly regulated by the phosphorylation of channel protein through PK-A activity. Both channels are downward regulated by the dephosphorylation of channel protein by protein phosphatase type-1 or type-2A.

REFERENCES

Benham, C.D., Bolton, T.B. and Lang, R.J., 1985, Acetylcholine activates an inward current in single mammalian smooth muscle cells, Nature, 316: 345-347.

Hamill, O.P., Marty, A., Neher, E., Sakmann, B. and Sigworth, F.J., 1981, Improved patch-clamp techniques for high-resolution current recording from cells and cell-free membrane patches, Pflügers Arch., 391: 85-100.

Hescheler, J., Mieskes, G., Rüegg, J.C., Takai, A. and Trautwein, W., 1988, Effects of a protein phosphatase inhibitor, okadaic acid, on membrane currents of isolated guinea-pig cardiac myocytes, Pflügers Arch., 412: 248-252.

Isenberg, G. and Klockner, U., 1982, Calcium tolerant ventricular myocytes prepared by preincubation in a "KB Medium", Pflügers Arch., 395: 6-18.

Ishihara, H., Martin, B.L., Brautigan, D.L., Karaki, H., Ozaki, H., Kato, Y., Fusetani, N., Watabe, S., Hashimoto, K., Uemura, D. and Hartshorne, D.J., 1989, Calyculin A and okadaic acid: Inhibitors of protein phosphatase activity, Biochem. Biophys. Res. Commun., 159: 871-877.

Ishihara, H., Ozaki, H., Sato, K., Hori, M., Karaki, H., Watabe, S., Kato, Y., Fusetani, N., Hashimoto, K., Uemura, D. and Hartshorne, D.J., 1989, Calcium-independent activation of contractile apparatus in smooth muscle by calyculin-A, J. Pharmacol. Exp. Ther., 250: 388-396.

Kameyama, M., Hescheler, J., Hofmann, F. and Trautwein, W., 1986, Modulation of Ca current during the phosphorylation cycle in the guinea pig heart, Pflügers Arch., 407: 123-128.

Kato, Y., Fusetani, N., Matsunaga, S. and Hashimoto, K., 1986, Calyculin A, a novel antitumor metabolite from the marine sponge *Discodermia calyx*, J. Am. Chem. Soc., 108: 2780-2781.

Obara, K., 1984, Isolation and contractile properties of single smooth muscle cells from guinea pig taenia caeci, Jpn. J. Physiol., 34: 41-54.

Ohya, Y., Kitamura, K. and Kuriyama, H., 1987, Modulation of ionic currents in smooth muscle balls of the rabbit intestine by intracellular perfused ATP and cyclic AMP, Pflügers Arch., 408: 465-473.

Ozaki, H., Kojima, T., Moriyama, T., Karaki, H., Urakawa, N., Kohama, K. and Nonomura, Y., 1987, Inhibition by amiloride of contractile elements in smooth muscle of guinea pig taenia cecum and chicken gizzard, J. Pharmacol. Exp. Ther., 243: 370-377.

Yabu, H., Yoshino, M., Usuki, T., Someya, T., Obara, K., Ozaki, H. and Karaki, H., 1990, Modification by calyculin A of inward Ca currents in smooth muscle cells isolated from guinea pig taenia coli, in: "Frontiers in Smooth Muscle Research", N. Sperelakis and J.D. Wood, eds., pp623-626, Wiley-Liss, New York.

POTASSIUM CHANNEL ACTIVATION IN VASCULAR SMOOTH MUSCLE

G. Siegel[1], J. Emden[1], K. Wenzel[1], J. Mironneau[2], and G. Stock[3]

[1]Institute of Physiology, Biophysical Research Group, The Free University of Berlin, D-1000 Berlin 33, Germany
[2]Laboratoire de Physiologie Cellulaire et Pharmacologie Moléculaire, Université de Bordeaux II, F-33076 Bordeaux, France
[3]Cardiovascular Pharmacology, Research Laboratories of Schering AG, D-1000 Berlin 65, Germany

INTRODUCTION

K^+ channel opening causing membrane hyperpolarization in vascular smooth muscle and thereby vasodilatation, cannot be rated as the 'newly discovered' action principle of certain pharmacological substances but as an ubiquitous physiological mechanism for the relaxation in smooth muscle. Acidification of the serum-pH-value is such an effector influence initiating vasodilatation. Drugs such as pinacidil, nicorandil, minoxidil sulphate and cromakalim effect vasodilatation by membrane hyperpolarization of the vascular smooth muscle cells as well, which in some tissues raises the membrane potential to a value close to the K^+ equilibrium potential [36]. Although the number of K^+ channels observed is steadily but rapidly increasing, they can still be classified. They are usually subdivided according to their mode of activation. Some are activated strictly voltage-dependently, others by a variation in the intracellular Ca^{2+} concentration, and some by the internal concentration of ATP, Na^+, cyclic nucleotides etc. The heterogeneous group of K^+ channel openers may be a potential therapy for hypertension, asthma, peripheral vascular disease, and diseases of the heart and nervous system. The central starting point of their physiological mode of action is the hyperpolarization of the smooth muscle cells which leads to relaxation by closing T- and/or L-type Ca^{2+} channels without, in the classical sense, the participation of cAMP or cGMP [34]. We would like to discuss this problem from a physiological point of view and venture a wider definition of the term 'K^+ channel opener'. This seems to be justified by the fact that cyclic nucleotides also elicit a membrane hyperpolarization in the vascular smooth muscle.

MATERIALS AND METHODS

Experimental Preparations and Solutions

Intracellular potential recordings, measurements of isometric tension and radioisotope efflux were performed on the vascular smooth muscle of common carotid and coronary arteries [26,29]. The carotid arteries were surgically removed from sacrificed dogs or rabbits within 3-4 min. In order to examine the participation of G-proteins in the receptor-mediated K^+ channel opening, rabbits were treated in a test series with pertussis toxin (PTx) for 4-5 days. PTx was first injected intravenously in a single dose of 2.5 μg/kg body weight [10]. Endothelium denuded segments were obtained by gently scraping off the endothelial cell layer. The preparations were subjected to light and electron microscopy in order to verify the effectiveness of the procedure. Human coronary arteries were taken from heart transplant patients who had been suffering

Excitation-Contraction Coupling in Skeletal, Cardiac, and Smooth Muscle
Edited by G.B. Frank *et al.*, Plenum Press, New York, 1992

from a dilative cardiomyopathy or extensive atherosclerosis. There was a clear differentiation between normal and atherosclerotic vessels where the preparations had been examined by light and electron microscopy. All arterial segments were equilibrated in a Krebs solution of the following composition (mmol/l): Na^+ 151.16; K^+ 4.69; Ca^{2+} 2.52; Mg^{2+} 1.1; Cl^- 145.4; HCO_3^- 16.31; $H_2PO_4^-$ 1.38; glucose 7.77 (temp. 37°C; pH 7.35). The solution was usually aerated with a 95% O_2 - 5% CO_2 gas mixture (carbogen). Different pH values were obtained by changing the O_2/CO_2 ratio of the gas mixture with which the Krebs solutions were aerated. Blood substitute solutions were equilibrated with $O_2/CO_2/N_2$ mixtures containing different oxygen concentrations, but a constant concentration of carbon dioxide (5ml/dl) to vary the oxygen tension. Microelectrodes consisting of platinum cathodes were used for continuous oxygen tension measurements in the Krebs solution close to the blood vessel strip [9].

The PGI_2 analogue, iloprost (Schering, ZK 36374; mol. wt. 360.5 g/mol; 16R,S-methyl-18,18,19,19-tetradihydro-6a-carbaprostaglandin I_2), was used in the concentration steps 10^{-9}, 10^{-8}, 10^{-7}, 10^{-6}, and $3 \cdot 10^{-6}$ mol/l. The dihydropyridine derivative, cicletanine (IHB-IPSEN; mol. wt. 298.17 g/mol;D-L-1,3-dihydro-3-(4-chlorophenyl)-7-hydroxy-6-methyl-(3,4-c)-furopyridine hydrochloride), was first dissolved in dimethylsulfoxide (Sigma, Deisenhofen) at a concentration of 0.671 mol/l (stock solution) [3]. From this, the concentration steps 10^{-6}, 10^{-5}, 10^{-4}, and 10^{-3} mol/l were obtained in a Krebs solution. The corresponding DMSO concentrations were 0.021, 0.21, 2.1, and 21 mmol/l. A similar procedure was applied to ajoene, a natural product of garlic. Pure, oily ajoene was first dissolved in DMSO at a concentration of 1 mmol/l (stock solution) and the concentration steps 10^{-6}, 10^{-5}, 10^{-4}, and 10^{-3} mol/l were also obtained in a Krebs solution [24]. The corresponding DMSO concentrations were 0.659, 6.59, 65.9, and 659 μmol/l. The effects of cicletanine or ajoene were measured as the relative change to DMSO control experiments. Indomethacin (Vonum, Econerica) was added at a concentration of 10^{-5} mol/l and 6-hydroxydopamine (Aldrich-Chemie, Steinheim) at a concentration of $1.8 \cdot 10^{-3}$ mol/l. Each solution was freshly prepared.

Mechanical Recording

4-5 mm long cylinders from the carotid or coronary arteries were cut lengthwise. Since the muscle cells of arteries are arranged circularly, the folded-out vessel rings, 3-15 mm long, were attached at the cut ends to an isometric tension measuring device (KWS 522.C, K 52 C; Hottinger-Baldwin, Darmstadt, Germany) [32]. The specimens were superfused with the Krebs solution at 10 ml/min for 30 min at an initial tension of 2 g. After this time, the tension was in equilibrium and amounted from 1.116 ± 0.243 g (n=6) (human coronary arteries) to 2.016 ± 0.097 g (n=8) (canine carotid arteries). The solutions with different iloprost concentrations were then applied consecutively for 30 min each, first at rising, then at falling concentrations (reversibility). Cicletanine and ajoene were also applied in a cumulative manner and the reversibility was tested [24,32]. The incubation time was 15 min for all concentration steps. In the experiments with varying oxygen tension, solutions with first decreasing and then increasing oxygen partial pressures were applied for 15 min each. After these periods the mechanical tension had reached a steady-state value for each concentration step [9,25,32]. The flow rate was 50-60 ml/min in order to bring about quick changes in P_{O_2} and reach low pressures. Voltage and tension values were averaged for identical effector concentrations.

Electrical Recording

Glass capillary microelectrodes filled with 3 mol/l KCl were used to record intracellular membrane potential. These recordings and the mechanical tension registration were performed simultaneously. The electrode resistances ranged from 60-100 MΩ and the tip potentials from -40 to -80 mV. The electrodes were shielded to just below the tips; otherwise conventional recording techniques were used. The microelectrode was inserted into the arterial muscle cell from the intimal surface [25,31]. Arteries in a normal Krebs solution with membrane potentials between -50 and -80 mV were selected for the final averaging. Membrane potentials more positive than -50 mV (20% of all impalements) had to be discarded because, according to our own measurements and to those of other authors, they stem from endothelial cells [17].

Isotope Flux Measurements

Carotid or basilar artery segments from the dog were saturated for at least 3 hs in a radioactive Krebs solution containing $^{24}Na^+$ and $^{42}K^+$. In addition, the test solutions contained iloprost at a concentration of 10^{-6} mol/l. Washout of the radioisotopes was performed in the corresponding non-radioactive solution. The activity in the preparation was measured using VIP series coaxial Ge(Li) detector systems with a peak height to Compton plateau height ratio of 30:1 (Ortec, Oak Ridge, TN, USA). This permitted adequate discrimination of the γ-lines (system resolution 2.2 keV using 1.333 MeV photons) of ^{24}Na at 1.37 MeV and ^{42}K at 1.52 MeV by means of stabilized single-channel analyzers (SCA-N-3; Elscint, Haifa, Israel) [23]. The combined γ-spectra from the preparation were measured continuously over 40 min, accumulated at 10 s intervals, and transferred to a PDP 11/23 Plus (Digital, Maynard, CA, USA). During the measurements, a strictly constant and laminar flow of the washout medium was maintained. The fluid in the measurement chamber was completely exchanged 1.2 times per second (chamber volume 2.3 ml, flow rate 160 ml/min). Experiments with dye indicators confirmed the favourable flow conditions. Thus, the accumulation of tracer ions in the preparation chamber, interstitium, or on the surface of the specimen was avoided. After each experiment the content of Na, K, Mg, Ca and Cl in the preparation as well as in the incubation and washout solution was determined by atomic absorption spectroscopy.

Patch-Clamp Recording

Portal veins from Wistar rats (150 g) were dissected free of connective tissue and single cells were obtained using a dispersal procedure similar to that previously described [15]. Whole-cell membrane currents were measured using a standard patch-clamp technique. Patch pipettes had resistances of 2-3 MΩ. The data were analyzed with a Plessey 6220 microcomputer and the illustrations were drawn with an X-Y plotter. The series resistance, measured at 10 kHz, was 4-6 MΩ and no correction for errors in measurements of potential was applied [28].

The extracellular medium (reference solution) contained: NaCl 130, CsCl or KCl 5.6, $CaCl_2$ or $BaCl_2$ 5, $MgCl_2$ 0.24, glucose 11, HEPES 8 mmol/l (pH 7.4; temp. 30°C). The pipettes were filled with solutions containing: CsCl or KCl 130 mmol/l, Cs- or K-pyruvate, Cs- or K-succinate, Cs- or K-oxaloacetate, HEPES 10 mmol/l (pH 7.3 with CsOH or KOH). The stimulation frequency was 0.05 Hz.

Determination of Cyclic Nucleotide Levels

Mesenteric arteries of the rabbit were cut into segments 5-7 mm in length (6-11 mg wet wt.). The segments were suspended in a Krebs-Ringer bicarbonate buffer containing indomethacin (10^{-5} mol/l) and stored in an ice bath. After one hour, the segments were blotted dry and suspended in a Krebs-Ringer bicarbonate buffer at 37°C, containing indomethacin (10^{-5} mol/l), 3-isobutyl-1-methyl-xanthine(10^{-3} mol/l, Janssen, Neuss) and papaverine ($1.1 \cdot 10^{-5}$ mol/l) [32]. After 2 min, iloprost, dissolved in ethanol, was added in concentrations as given in Fig. 2. The reaction was halted after 5 min with 500 μl trichloroacetic acid. The segments were homogenized and centrifuged at 3,000 r.p.m. for 5 min at room temperature. After the trichloroacetic acid had been extracted with ethyl ether, the cAMP content of the supernatant was assayed by radioimmunoassay (RIA test kits of NEN, Dreieich).

RESULTS

Vasorelaxation Through Membrane Hyperpolarization

The activation of K^+ channels in the membrane of vascular smooth muscle cells implies their hyperpolarization. Hyperpolarization effects relaxation via T- and/or L-type Ca^{2+} channels which are closed voltage-dependently between 0 and -80 mV [16,28,32]. Which Ca^{2+} channels are affected by this closure is dependent on the actual membrane potential before hyperpolarization. The consequence is a fall in intracellular Ca^{2+} activity and thus vasodilatation. Numerous effector influences can evoke such a hyperpolarization. It is of clinical interest whether there is a membrane physiological correlate to the well-known vasodilatation with acidosis and the vasoconstriction with alkalosis (Fig. 1). The reason for these mechanical changes is a hyperpolarization with acidosis and a depolarization with alkalosis [27,31].

Prostacyclin (PGI_2), a main but unstable metabolite of arachidonic acid cascade in vascular tissue, is often considered to be a vasorelaxing autacoid in hypoxic dilatation. In our studies we were interested to see whether the vasodilatation observed has an electrophysiologic correlative. Because of the chemical instability of natural PGI_2, we used iloprost, a stable prostacyclin analogue. As can be seen in Fig. 2, a concentration-dependent hyperpolarization of membrane potential was found under iloprost. Iloprost hyperpolarized the membrane by 7.4 mV maximally, whereas this effect was reduced with very high doses. Part B of the figure shows the dose-dependent relaxation of the vascular strip. The half-maximal effect on membrane potential and tone was observed with a $2 \cdot 10^{-8}$ mol/l concentration [23,28,32].

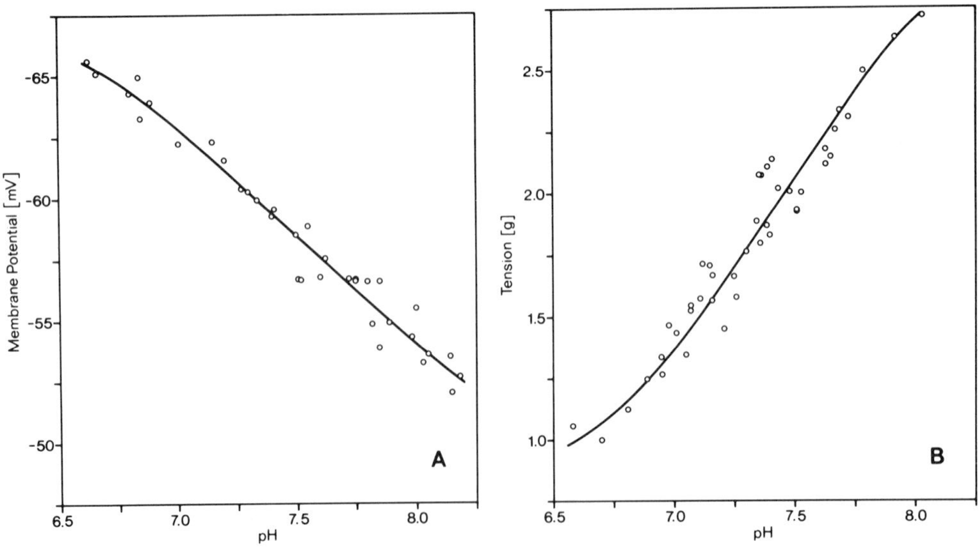

Fig. 1. (A) Membrane potential of vascular smooth muscle in relation to the external pH value of the Krebs solution. (B) Tension developed in isolated carotid segments as a function of the extracellular H^+ concentration.

The compound cicletanine, a dihydropyridine derivative, was also investigated for possible K^+ channel opening properties. This compound lowers blood pressure but, contrary to earlier expectations, does not stimulate prostacyclin synthesis. As cicletanine is virtually insoluble in blood substitute solutions, the influence of the solvent (DMSO) alone was contrasted to the effect of cicletanine plus solvent, illustrated in Fig. 3. While DMSO depolarized and contracted the vascular smooth muscle cells, a concentration-dependent hyperpolarization and relaxation appeared under cicletanine [32].

The net effect of cicletanine is illustrated in Fig. 4. Cicletanine (10^{-3} mol/l) hyperpolarized the membrane from -61.0 ± 0.4 mV (n=5) to -69.8 ± 2.2 mV (n=5; P < 0.005), while the force decreased from 1.194 ± 0.087 g (n=5) to 0.303 ± 0.198 g (n=5; P < 0.005). The strong hyperpolarization by $\Delta V = 8.8$ mV suggests that cicletanine has K^+ channel opening properties. The

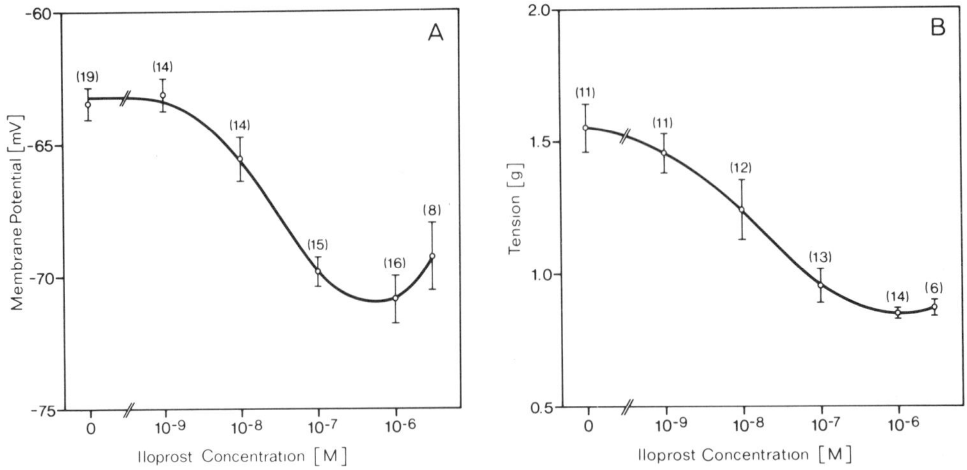

Fig. 2. Membrane potential (A) and tension (B) of isolated carotid segments as a function of the iloprost concentration in the Krebs solution. The preparations were incubated for 30 min at each iloprost concentration step.

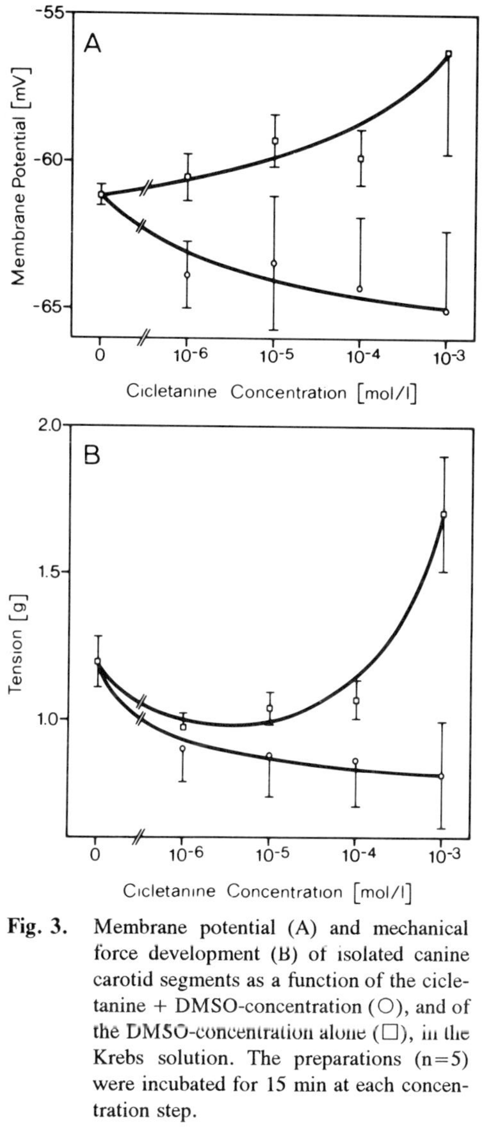

Fig. 3. Membrane potential (A) and mechanical force development (B) of isolated canine carotid segments as a function of the cicletanine + DMSO-concentration (○), and of the DMSO-concentration alone (□), in the Krebs solution. The preparations (n=5) were incubated for 15 min at each concentration step.

increase of the $^{42}K^+$ efflux, discovered by Weston's studies [35], points to a K^+ channel opener as well, although the rise observed in adenylate cyclase activity does not allow its complete integration into this effector group when defining it classically.

When comparing the effects of iloprost with those of ajoene, a natural product of garlic, there was a similar course in membrane potential and vessel tone for concentrations between 10^{-6} and 10^{-3} mol/l (Fig. 5). Cumulative 15 min applications of ajoene hyperpolarized the membrane and reduced the tone [22,24]. A concentration of 10^{-5} mol/l ajoene shifted the membrane potential from -61.1 ± 0.4 mV (n=6) to -67.1 ± 1.3 mV (n=6; $P < 0.002$); at the same time, the tension fell from 1.541 ± 0.052 g (n=6) to 1.376 ± 0.036 g (n=6; $P < 0.03$). Half-maximal effect occurred at a concentration of $2.4 \cdot 10^{-6}$ mol/l ajoene. The concentration step of 10^{-3} mol/l led to depolarization and contraction. Thus, hyperpolarization of the membrane and inhibition of basal tension in function of the ajoene concentration run parallel. The maximum hyperpolarization of the cells amounted to 6.0 mV, the maximum relaxation to 0.165 g. In these experiments, it was striking that with decreasing concentrations, the reversibility in membrane potential and tone was markedly reduced, i.e. a significant after-effect of the substance could be observed.

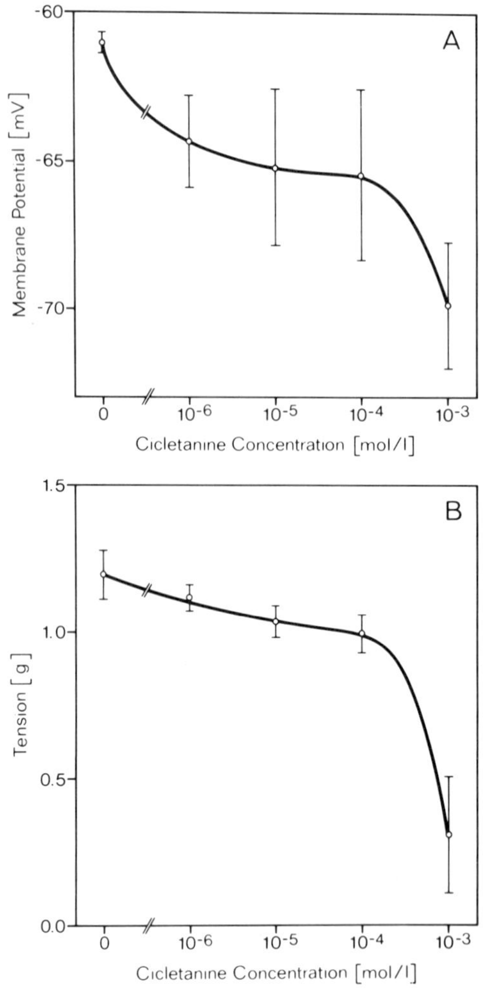

Fig. 4. Membrane potential (A) and tension (B) of isolated strips of the canine carotid artery as a function of the cicletanine concentration in the Krebs solution. The preparations (n=5) were incubated for 15 min at each cicletanine concentration step.

Vasodilatation, effected by a decrease in oxygen partial pressure or by a rise in blood flow, is due to membrane hyperpolarization. The cause of this hyperpolarization is quite different in each case, however, the end result is always elevated K^+ channel opening. The following experiments were designed to elucidate the role of the vasodilators prostacyclin (PGI_2) and endothelium-derived hyperpolarizing factor (EDHF) in oxygen deficiency. By recording membrane potential and tension as a function of O_2 partial pressure, a dose-dependent hyperpolarization of 9.4 mV and relaxation of 0.688 g in the canine carotid artery smooth muscle cells was observed when P_{O_2} was reduced from 550 to 35 mm Hg (Fig. 6). A further decline in P_{O_2} reduced this hyperpolarization and relaxation. Depolarization and contraction may even occur. These effects can be almost completely abolished by an α_1-receptor blockade [25,28,31,32]. The application of 6-hydroxydopamine (6-OHDA), which destroys the terminal sympathetic nerve endings [1], augmented the hypoxic hyperpolarization and relaxation in the entire P_{O_2} range between 550 and 0 mm Hg (cf. Fig. 7d), i.e. below 35 mm Hg, depolarization and contraction fail to appear.

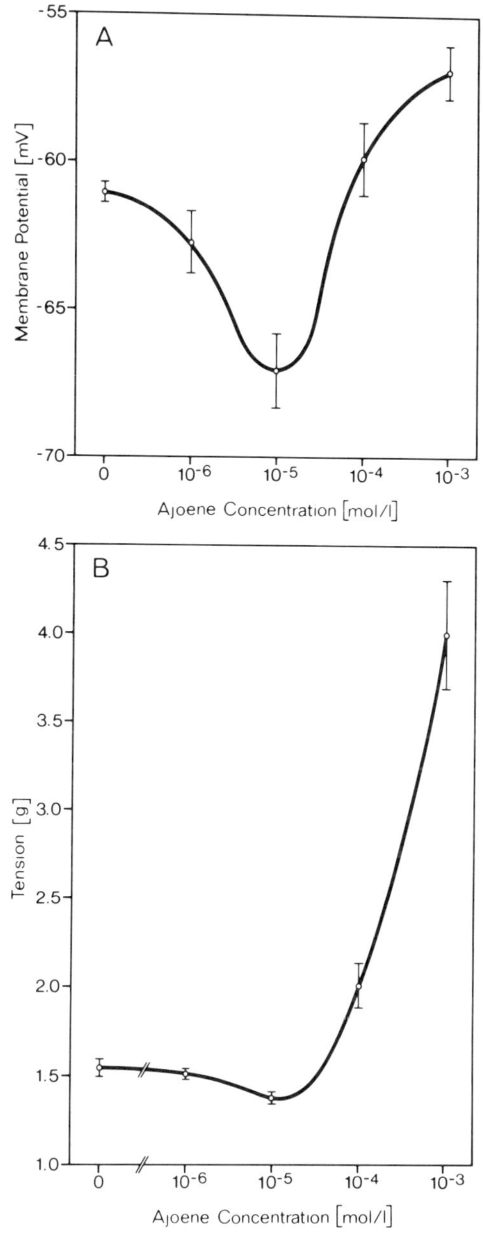

Fig. 5. Membrane potential (A) and tension (B) of isolated carotid segments as a function of the ajoene concentration in the Krebs solution. The preparations (n=6) were incubated for 15 min at each ajoene concentration step.

Fig. 7 illustrates the results under control conditions (a), in indomethacin-treated (b), deendothelialized (c), and in 6-hydroxydopamine-preincubated preparations (d). The following quantitative conclusions can be drawn from the steady state changes in voltage and tone in the carbogen Krebs solution, and from the alterations of the courses of the curves at variable P_{O_2}, about the participation of the different vasodilatory and vasoconstrictory factors: The difference of the curves a-d

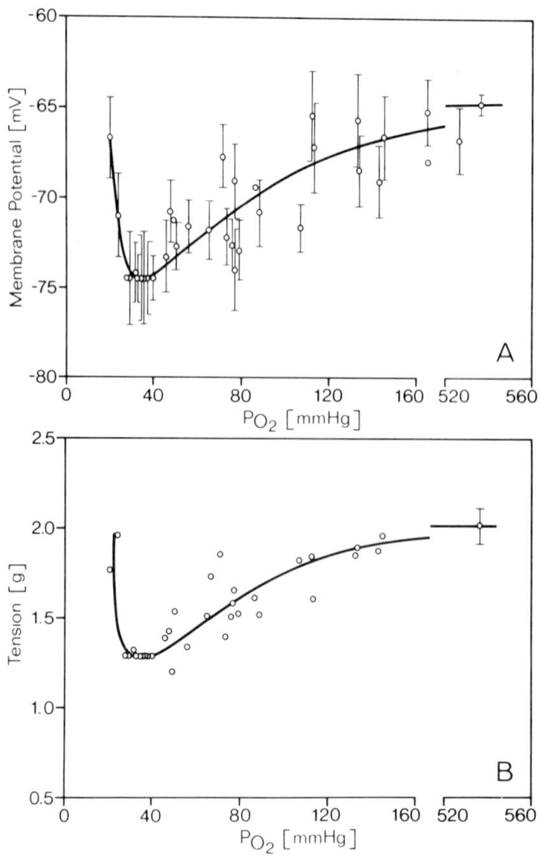

Fig. 6. Membrane potential (A) and mechanical force development (B) of isolated carotid segments in response to changes in the oxygen tension of the Krebs solution.

predominantly reflects the effect of a release of noradrenaline [cf. 5,6,20]. Below 35 mm Hg P_{O_2}, 20-HETE and/or endothelin may also play a role because an α_1-receptor blockade, or the application of 6-OHDA, did not completely abolish the hypoxic vasoconstriction [14]. Curve b shows the relationship with indomethacin, so that the difference of b-a represents the effect of prostacyclin (20-30%). Curve c was measured in preparations without endothelium. There is only a slight hyperpolarization and relaxation with a decrease in P_{O_2}, both of which are due to a release of prostacyclin from the vascular smooth muscle cells. The difference of the curves c-b practically illustrates the effect of the endothelial dilator EDHF (70-80%). In conclusion, the response of vascular smooth musculature to a decline in the oxygen tension presents itself as a balanced interplay between hyperpolarizing-vasodilatory (PGI_2, EDHF) and depolarizing-vasoconstrictory factors (noradrenaline, 20-HETE, endothelin). The same holds true for the adjustment of muscle tone under stationary resting conditions because all these compounds have a basal release [32].

From Fig. 8 one can see how the lowering of O_2 partial pressure influences membrane potential and tension of vascular smooth muscle cells in human coronary arteries taken from heart transplant patients. Circles represent the results of control experiments and squares represent the results of atherosclerotic vessel segments. Voltage and tension as a function of P_{O_2} show similar courses to those which we have already seen in the canine carotis. There is a continuous hyperpolarization and relaxation with a reduction of P_{O_2} from 550 mm Hg to 35 mm Hg, and a depolarization and contraction between 35 and 0 mm Hg. It is striking that in the case of atherosclerosis, vascular smooth muscle cells are more depolarized and have more tone [32]. This is true for all O_2 tensions.

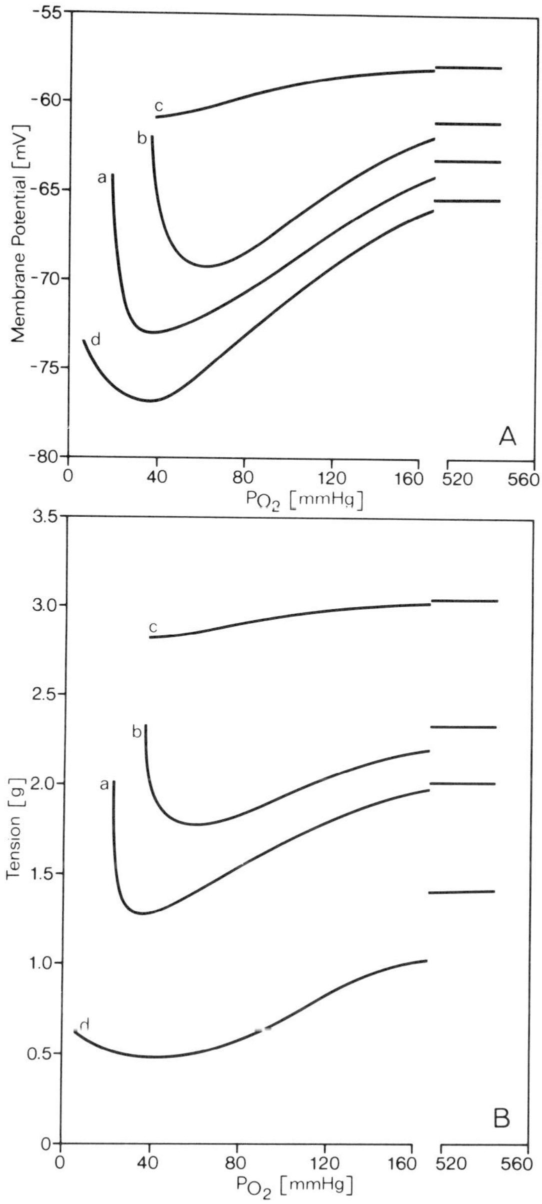

Fig. 7. Membrane potential (A) and tension (B) in isolated vascular strips of the *A. carotis communis* of the dog in dependency on the oxygen partial pressure of the Krebs solution. Mean values from 28 experiments. *a:* Normal Krebs solution; *b:* Krebs solution with indomethacin (10^{-5} mol/l); *c:* normal Krebs solution, endothelial layer of the carotid segments removed; *d:* normal Krebs solution, preparations pretreated by 6-hydroxydopamine ($1.8 \cdot 10^{-3}$ mol/l).

The endothelial function is possibly limited in this tissue. Iloprost, in a concentration of 10^{-7} mol/l, is able to induce additional hyperpolarization and relaxation at all O_2 partial pressures, even with maximal hyperpolarization and dilatation at about 35 mm Hg P_{O_2}. In normal and in atherosclerotic

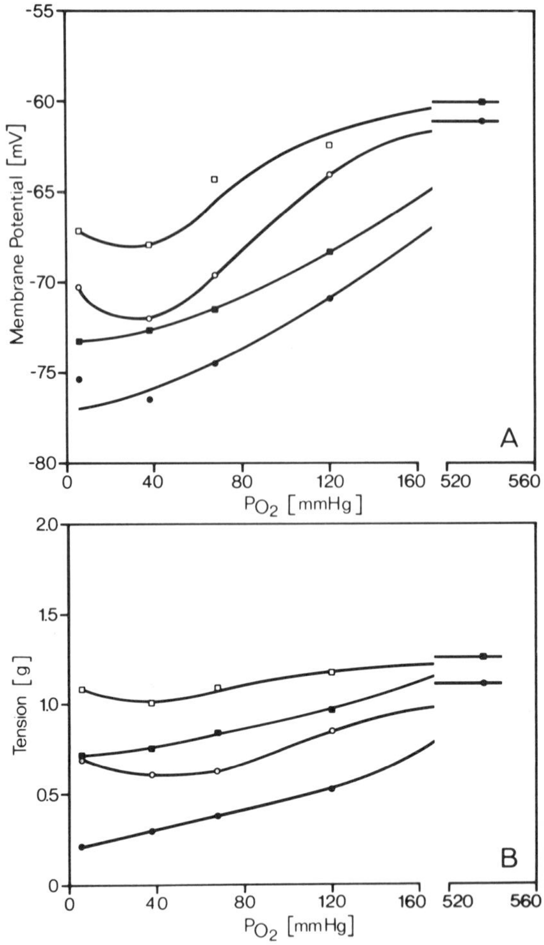

Fig. 8. Membrane potential (A) and tension (B) in isolated vascular strips of human coronary arteries dependent on the oxygen partial pressure of the Krebs solution. Mean values from five experiments each. (○) Normal coronary arteries with intact endothelium, (●) plus iloprost 10^{-7} mol/l; (□) arteriosclerotic coronary arteries, (■) plus iloprost 10^{-7} mol/l.

blood vessels, the hypoxic depolarization and contraction with oxygen pressures below 35 mm Hg is prevented by iloprost. These results emphasize that also with an oxygen deficiency, vasodilatation can be attained by K^+ channel opening.

Mechanisms of Hyperpolarization

As has been shown, various substances, and a whole range of changes in the physical state, lead to vascular relaxation via membrane hyperpolarization. How can the changes in membrane potential be explained? First of all the effects of pH. Fig. 9 shows a semilogarithmic plot of the $^{42}K^+$ efflux of basilar artery segments at pH steps of 6.8, 7.3, and 7.8. The flux curves have a regularly graded, pH-dependent course. They reveal that the slowest exchange (phase 3) occurs with almost an identical time constant at all pH steps, whereas exchange phase 2 has the smallest time constant at pH 6.8, the largest at pH 7.8 (left inset). In addition, the amplitude falls regularly as pH increases, which also indicates a flux reduction with alkalinization. Calculations with the individual compartments and kinetic coefficients show that the membrane hyperpolarization of vascular smooth muscle cells under acidosis is due to a rise in passive K^+ permeability (K^+ channel opening) with a simultaneous fall in passive Na^+ permeability.

These results led to the question of how iloprost effects hyperpolarization of the membrane. When the Ca^{2+} channel current was tested by whole cell patch-clamping in the rat portal vein with Ba^{2+} ions acting as charge carriers, it was increased by merely 10% (Fig. 10). However, measuring the K^+ currents in either Ba^{2+} or Ca^{2+} solutions had drastic effects. In a Ba^{2+} solution, depolarizing voltage pulses produced an initial inward current followed by an outward current. Iloprost (10^{-6} mol/l) strongly enhanced the outward current, thereby totally suppressing the net inward current. In a Ca^{2+} solution, depolarizing pulses only led to an outward current which could be divided into an early 'peak current' and a smaller 'sustained current' [28]. Again, iloprost strongly enhanced the outward current. If we consider the relationship between voltage pulse and peak outward current, iloprost caused the latter to increase over the whole voltage range -60 mV $< V <$ +30 mV, and by about 90% at a test potential of +30 mV (Fig. 11). This was also true when the electrogenic Na^+/K^+ pump was blocked by ouabain. The rate of increase in the outward current was unchanged. Therefore, a direct contribution of the electrogenic ion pump to the membrane hyperpolarization can be excluded. According to these findings, iloprost may be characterized as a K^+ channel opener.

Iloprost did not show an effect on the Ca^{2+} channel current, but augmented both the voltage-dependent and the Ca^{2+}-activated K^+ outward current. This property is again emphasized by combined $^{42}K^+/^{24}Na^+$ efflux experiments (Fig. 12). In the $^{42}K^+$ efflux (Fig. 12A), we detected a rise in K^+ permeability of 340% causing the hyperpolarization under iloprost (10^{-6} mol/l). The much steeper course of the K^+ decay under iloprost is immediately recognizable. After the exponential analysis and computation of a compartment model based on five experiments, we found an average increase of 250% in passive K^+ efflux with iloprost [23,28]. In spite of this enormous increase in K^+ permeability under iloprost, both flux curves in Fig. 12 start from the same K^+ concentration [32]. This is a clear indication that the active K^+ inward transport also has to be raised in order to maintain the internal K^+ concentration. A surprising finding was observed in the $^{24}Na^+$ efflux, which is 98% active transport in this tissue (Fig. 12B). Here too, the efflux curve is significantly steeper

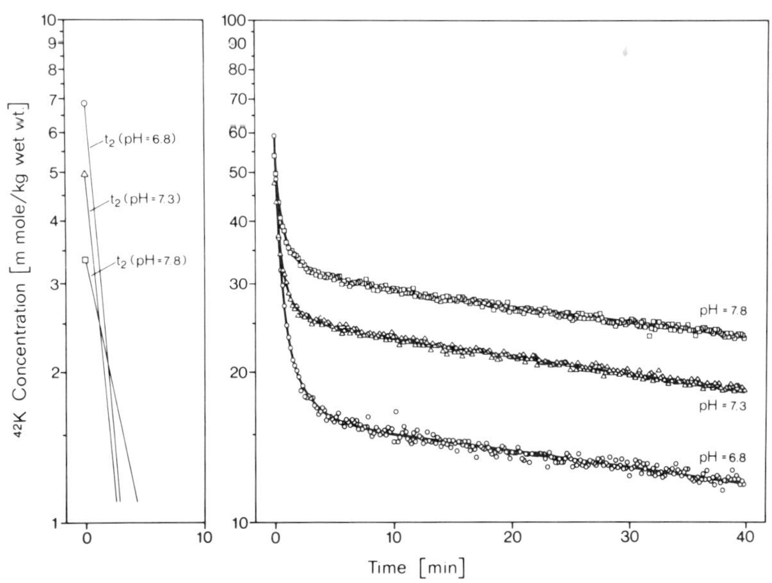

Fig. 9. $^{42}K^+$ efflux in the basilar artery of the dog according to the direct method in Krebs solutions with pH 6.8 (○), 7.3 (Δ), and 7.8 (□). The graph shows the time course of radioactive K^+ decrease within the preparations at intervals of 10 s in single experiments. The superimposed lines represent the optimal triple-exponential functions found by computer fitting. In the left inset the pH-dependent exchange phases 2 are represented separately.

with iloprost than in the control experiment. A flux calculation resulted in a 50% increase in active transport rate [23]. In principle, both the rise of passive K^+ permeability and the electrogenic Na^+ outward transport lead to membrane hyperpolarization.

In order to arrive at a molecular description of the iloprost response, we first considered the PGI_2 receptor to which iloprost also binds. We assumed that the K^+ channel opening is induced via a G-protein. Experiments with pertussis toxin (PTx) resulted in membrane hyperpolarization and vasorelaxation with iloprost, being even more accentuated in the carotid musculature of PTx-pretreated animals than in the untreated control group (Fig. 13). This series of experiments indicates that cAMP may play a part in iloprost-induced vasorelaxation and that proteins of the G_i-type are also involved. After blocking the G-proteins in question with PTx, cAMP may well have increased intracellularly. This figure demonstrates that an additional hyperpolarization is combined with this supposed rise in cAMP [32].

The differing behaviour of the vessels treated with high external K^+ concentrations compared with the classical K^+ channel opener cromakalim implies cAMP participation. While iloprost causes a dose-dependent relaxation of the mesenteric artery even in 70 mmol/l $[K^+]_o$, this can no longer be observed with cromakalim. Fig. 14 shows that iloprost does effect an augmentation of the cAMP concentration. After 5 min incubation, the course of cAMP reflects the concentration-response curves for membrane potential and muscle tone (cf. Fig. 2).

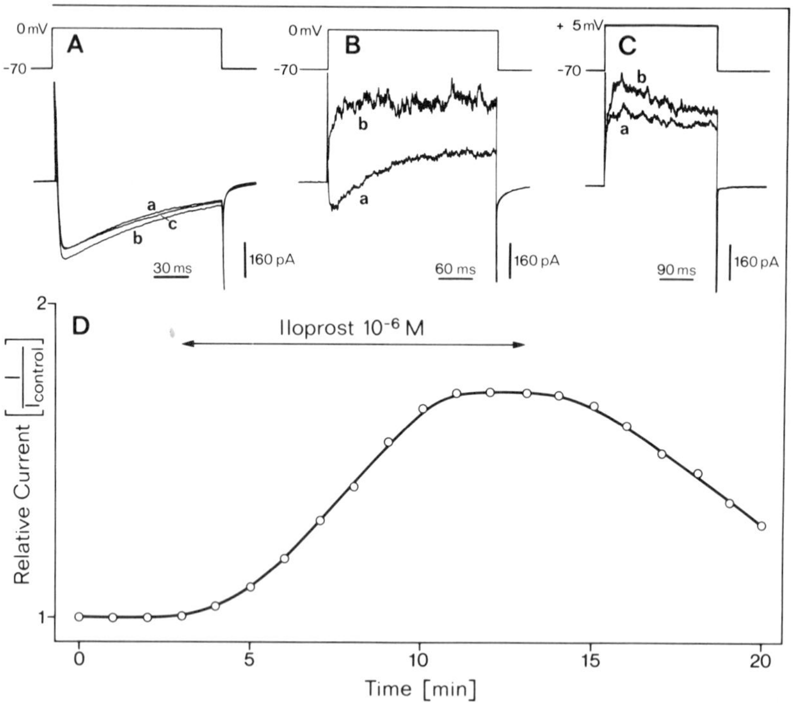

Fig. 10. Effects of iloprost (1 μmol/l) on Ca^{2+} channel and K^+ channel currents recorded in single vascular smooth muscle cells isolated from the rat portal vein. (A) Inward Ba^{2+} currents in response to a depolarization from a holding potential of -70 mV to 0 mV in the reference solution (a), after adding iloprost for 5 min (b), and after 5 min of washing out (c). The pipette solution contained CsCl instead of KCl. (B) Membrane currents obtained for a depolarization to 0 mV in 5 mmol/l Ba^{2+} solution (a), and after the addition of iloprost for 5 min (b). The pipette solution contained KCl. (C) Membrane currents obtained for a depolarization to +5 mV in another cell in 5 mmol/l Ca^{2+} solution (a), and after the addition of iloprost for 5 min (b). The pipette solution contained KCl. (D) Effect of iloprost on maximal K^+ currents recorded in 5 mmol/l Ca^{2+} solution and with KCl in the pipette solution. K^+ currents were obtained in response to depolarizations from a holding potential of -70 mV to +10 mV. A steady stimulatory effect of iloprost was obtained within 7 min.

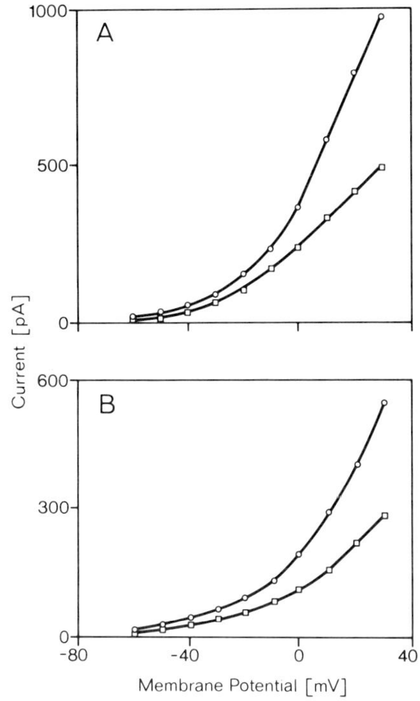

Fig. 11. *A:* Current-voltage relationships established for the peak outward K^+ current, obtained in 5 mmol/l Ca^{2+} solution. The pipette solution contained KCl. Reference solution (□) and after the addition of 1 μmol/l iloprost for 7 min (○). Data shown are from a typical experiment. Similar estimates were obtained from 4 separate experiments. *B:* Current-voltage relationships established for the peak outward K^+ current, obtained in 5 mmol/l Ca^{2+} solution. The pipette solution contained KCl. The cell was incubated for 10 min in 100 μmol/l ouabain before establishing the curve (□), and then iloprost was added for 7 min (○). Data shown are from a typical experiment. Similar estimates were obtained from 4 separate experiments.

When carrying out a more exact time analysis of the K^+ outward current in the whole cell voltage-clamp investigations, a rapid increase of the voltage-dependent K^+ current in a Ba^{2+} solution can be demonstrated (Fig. 15). However, this current together with the Ca^{2+}-activated K^+ current in a Ca^{2+} solution only rises after a delay. Apparently, cAMP plays a part in the increase of the Ca^{2+}-activated K^+ outward current. As suggested by Sadoshima et al. [19], the cAMP-dependent protein kinase seems to increase the Ca^{2+} sensitivity of the Ca^{2+}-activated K^+ channel. Obviously, one or two types of K^+ channels (perhaps the Ca^{2+}-activated K^+ channel and the delayed rectifier K^+ channel) are stimulated by phosphorylation with cA-PK [33]. The hyperpolarizing and relaxing effect of iloprost may be explained by cA-PK interventions at the Ca^{2+}-activated K^+ channel, the sarcolemmal Ca^{2+} pump, and the $I_{Ca(s)}$. In particular, cAMP-induced hyperpolarization due to an

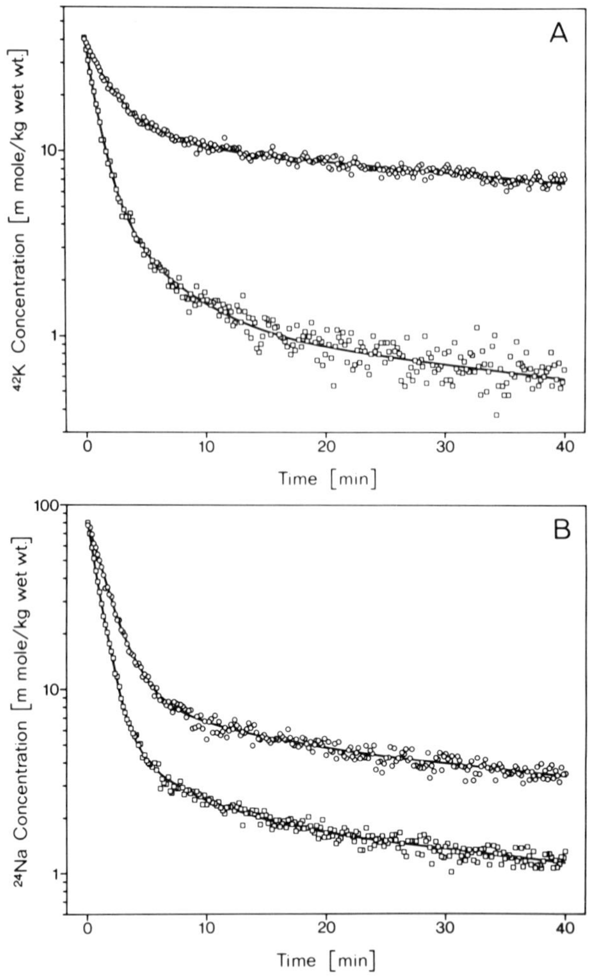

Fig. 12. *A:* $^{42}K^+$ efflux of the carotid artery of the dog in Krebs solutions without (○) or with 10^{-6} mol/l iloprost (□). The graph shows the time course of the radioactive K^+ decay within the preparations at intervals of 10 s in single experiments. The superimposed lines represent the optimal double-exponential functions found by computer fitting. *B:* $^{24}Na^+$ efflux of the carotid artery of the dog in Krebs solutions without (○) or with 10^{-6} mol/l iloprost (□). The graph shows the time course of the radioactive Na^+ decay within the preparations at intervals of 10 s in single experiments. The superimposed lines represent the optimal triple-exponential functions found by computer fitting.

increase in g_K cannot be distinguished from a hyperpolarization possibly elicited by a more direct effect of the drug on K^+ channels.

Fig. 16 shows the stationary activation curve 'tension vs. membrane potential' of the canine carotid artery smooth muscle. A number of different effector parameters are at the root of this curve [31]. Since in the investigations mentioned, the effector influences on membrane potential and vascular tone had been measured simultaneously, the elimination of the former parameters from the dose-response curves was possible [21]. Plotting the developed tension as a function of the

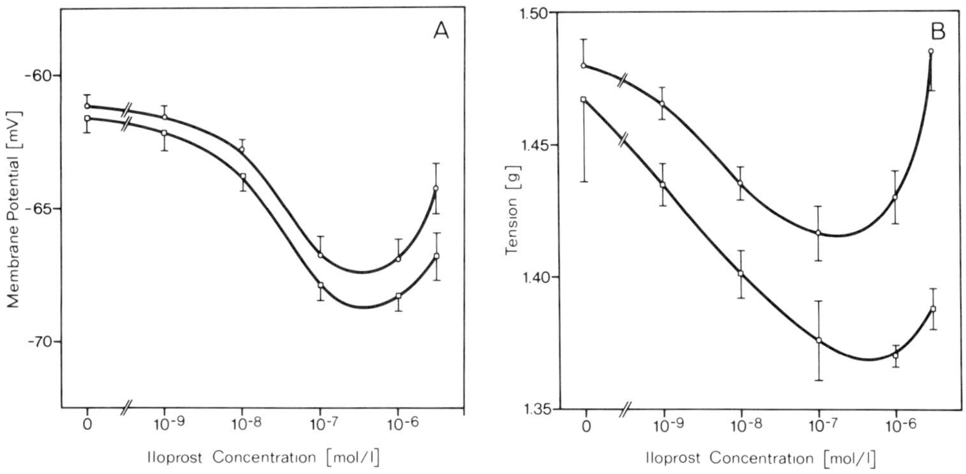

Fig. 13. Membrane potential (A) and tone (B) in vascular strips of the carotid artery of untreated (○) and pertussis toxin-pretreated rabbits (□) as a function of the iloprost concentration in the Krebs solution. The initial pre-tension of the preparations was 2 g. Application of iloprost for 10 min at each concentration step.

membrane potential at each identical effector concentration produced the depolarized and hyperpolarized portions of the stationary activation curve. The point marked on the curve represents the membrane potential and tension under normal conditions. The range of effectivity of a hyperpolarization, caused by so-called K^+ channel opening, is in the lower linear and the nonlinear, hyperpolarized section of the curve. From the sigmoid, stationary activation curve and the values measured, it is clear that the same change in tone occurs with a defined change in membrane potential, no matter from which effector influence the latter originates [21,31]. Thus, this curve provides strong evidence for the existence of electromechanical coupling in vascular smooth muscle, at least in the carotid artery.

DISCUSSION

A membrane hyperpolarization of vascular smooth muscle cells is generally a prerequisite for vasodilatation under physiological conditions. Such a hyperpolarization can simply be brought about through an increase in the open state probability for K^+ channels. A variety of K^+ channels has also been described for smooth muscle which differ through their activability and single channel conductance. Different K^+ channel types, with their specific biophysical properties and pharmacological sensitivities, come about because of genetically coded channel proteins with slightly different amino-acid sequences. G-proteins, Ca^{2+}, and a number of kinases participate in their regulation. The clue to the initiation of vasorelaxation is a tight electromechanical coupling [21]. The strong voltage-sensitivity of the Ca^{2+} channel open probability provides an explanation for this [16].

Various compounds and differing physical states can produce a more or less pronounced membrane hyperpolarization. How much the smooth muscle cells relax depends on, among other things, their membrane potential before the intervention. Acidification of the extracellular pH value [27,31], a fall in the O_2 partial pressure [9,25,28,31,32], the application of iloprost [23,28,30,32], cicletanine [32] or ajoene [22,24] are only a selection from the many possibilities which lead to hyperpolarization.

The hyperpolarization with acidosis can be explained by an increase in K^+ permeability with a simultaneous decrease in Na^+ permeability. The reasons for the hyperpolarization with prostacyclin (iloprost) and with a reduction in O_2 partial pressure are much more complicated. Iloprost induced a drastic increase in both the voltage-dependent and Ca^{2+}-activated K^+ outward current over the

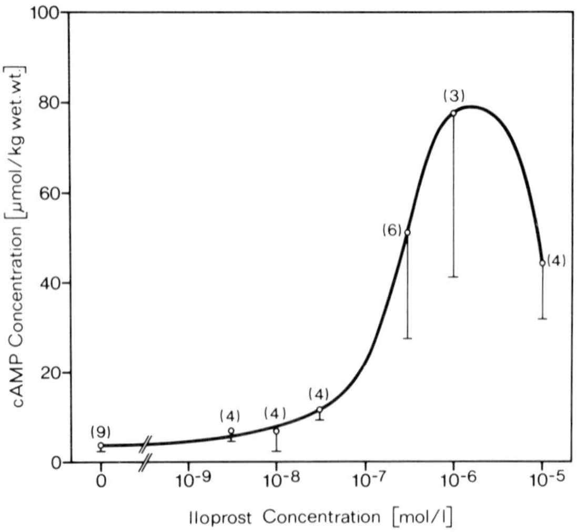

Fig. 14. Concentrations of cAMP in vascular segments of the rabbit mesenteric artery in response to the iloprost concentration in the Krebs solution. Incubation with iloprost for 5 min at each concentration step.

whole potential range in patch-clamp investigations. The rise in Na^+ permeability by 40%, and in K^+ permeability by 340%, as revealed by tracer studies, demonstrates once again that following treatment with this drug, the permeability of the cell membrane increases drastically and almost selectively for K^+ ions. The slight increase in Na^+ and Ca^{2+} permeability, as found in the flux and voltage-clamp studies, may signal the opening of an unspecific channel for small cations. Since this rise in permeabilities occurs to a noticeable extent only with high iloprost concentrations, it may explain the reversal of membrane hyperpolarization and vasorelaxation found in the same concentration range. This reduction in hyperpolarization and relaxation was not seen with cicaprost which does not have the prostaglandin E_1-like side effects of iloprost [32]. Finally, the studies with

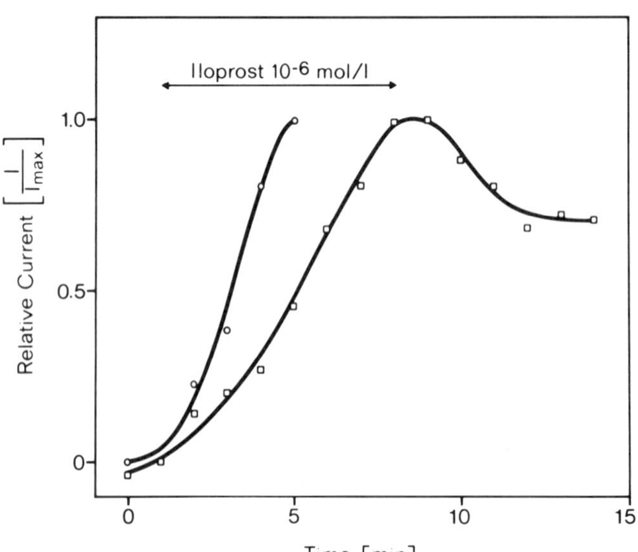

Fig. 15. Time analysis of the K^+ outward current I/I_{max} under iloprost (10^{-6} mol/l) in patch-clamp investigations. (○) Ba^{2+} solution; (□) Ca^{2+} solution.

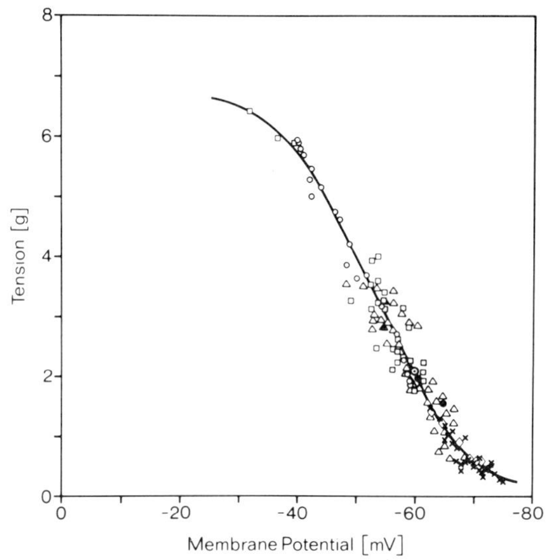

Fig. 16. Dependency of mechanical tension on the membrane potential in isolated carotid arteries (stationary activation curve). The membrane potential was changed by varying the extracellular concentration of H^+ (△), K^+ (□), Ca^{2+} (●), norepinephrine (○) or prostacyclin (◇) or by lowering the oxygen tension (×). The point marked on the curve (⊙) indicates membrane potential and mechanical force under control conditions.

PTx, the studies of the time course of the increasing K^+ outward current, and the determination of cAMP resulted in the following effector chain of PGI_2 and EDHF. The hyperpolarizing and relaxing effect of iloprost may be explained by cA-PK interventions at the Ca^{2+}-activated K^+ channel, the sarcolemmal Ca^{2+} pump and the $I_{Ca(s)}$. In principle, the effect of EDHF in O_2 deficiency seems to be the same, but its effector chain proceeds via cG-PK [33]. This is in contrast to cardiac muscle. We presume that an ATP-sensitive channel, especially with respect to the effect of prostacyclin is not responsible for the K^+ channel opening resulting in hypoxic hyperpolarization, but the Ca^{2+}-activated K^+ channel.

Comparing the effect of hypoxia on membrane potential and tension of normal and arteriosclerotic human coronary arteries which were taken from heart transplant patients reveals a distinct difference. During oxygen deficiency, control preparations showed a maximal hyperpolarization of 10.7 mV and a maximal relaxation of 0.466 g. Arteriosclerotic arteries, however, became hyperpolarized by merely 6.6 mV and relaxed by 0.258 g. The relation was quite similar for a carbogen Krebs solution (resting, control conditions). In comparison to normal coronary arteries ($V = -61.1 \pm 0.4$ mV; $T = 1.115 \pm 0.243$ g; $n=6$) arteriosclerotic vessels were depolarized ($V = -60.0 \pm 0.8$ mV; $n=10$) and more contracted ($T = 1.262 \pm 0.076$ g; $n=10$). Thus, it may be concluded that in arteriosclerotic blood vessels synthesis and release of vasodilators are diminished. The beneficial effect of a compensation of the PGI_2 lacking was demonstrated by the application of iloprost (10^{-7} mol/l). This prostacyclin analogue 'transformed' arteriosclerotic coronaries into normal ones as far as their membrane potential and tone response to a reduction in O_2 partial pressure is concerned. Above all, membrane depolarization and vasoconstriction failed to appear with O_2 pressures below 35 mm Hg even in arteriosclerotic vessels. This is a very important result because such low P_{O_2} values were found during coronary angiography in the large coronary blood vessels of man with extensive arteriosclerosis. Therefore, even lower values may be expected in peripheral arteries and arterioles.

Cicletanine and ajoene lead to vascular relaxation via membrane hyperpolarization as does iloprost. Assuming that patients have a plasma level of $3.1 \cdot 10^{-5}$ mol/l cicletanine for some time, and 90% of this compound is bound to proteins [7], the effective concentration is $3.1 \cdot 10^{-6}$ mol/l. Under normal conditions, this cicletanine concentration hyperpolarized the smooth muscle cell membrane by 3.8 mV; the relaxation of the vascular strip amounted to 0.128 g. This effect can be explained by closure of T-type Ca^{2+} channels with hyperpolarization. We observed a similar behaviour under iloprost [23,28]. These experiments support the idea of including cicletanine in the group of K^+ channel openers. Nevertheless, only detailed binding studies and flux experiments can validate whether K^+ channel opening or Ca^{2+} antagonistic properties exist. In our preparations, prostacyclin did not participate in the effect of cicletanine since unchanged dose-response curves were recorded under indomethacin. We could not confirm the participation of the eicosanoid system in the mechanism of the hypotensive action of cicletanine [4].

Pure, oily ajoene resulted in a concentration-dependent hyperpolarization and relaxation in arterial vasculature [22,24]. Only very high doses, irrelevant for man, attenuated and reversed these effects. As for nicorandil (ΔV = 14.4 mV) [13] and iloprost (ΔV = 7.4 mV) [23,30,32], new vasodilator substances, the strongly hyperpolarizing effect of 6.0 mV of ajoene, could also be attributed to an increase in the K^+ permeability in the smooth muscle cells of the carotid artery. This effect is not surprising because essential constituents of garlic oil can be seen as being structurally related to arachidonic acid derivatives and behave in a similar fashion as these to membrane lipids. Since the ajoene concentration from one single garlic clove is 0.18 μmol/l throughout the extracellular space in man, our studies have proven a relevant membrane hyperpolarization and inhibition of tone.

How the hyperpolarization effects relaxation of the blood vessel is explained by the activation curve [21,31]. Every hyperpolarization leading to vasodilatation closes L- and/or T-type Ca^{2+} channels after all. The deciding factor is that a certain Ca^{2+} channel open probability exists with a resting potential of about -60 mV in the vascular smooth muscle cells [8,16]. In the linear range of the activation curve, a potential change of 5.1 mV corresponds to an alteration in tone of 1 g. This linear part comprises about 20 mV and thus 70% of the force developed. The activation curve can be shifted along the potential axis, for example under the influence of cyclic nucleotides or IP_3, or steepened in its degree of coupling [2,11,18]. A classical example is the effect of noradrenaline, which releases Ca^{2+} ions from intracellular stores, if present, and finally augments the affinity to the contractile filaments. On no account does this involve an abolition of electromechanical coupling. The force development is still voltage-dependent; only the system has switched to a shifted activation curve. The theory is confirmed by earlier observations in which a voltage-dependence of ROC-linked processes has been described [2]. Finally, the tight electromechanical coupling expressed by the stationary activation curve is supported through measurements in the vascular smooth muscle by Hermsmeyer [12], demonstrating a very tight electromechanical coupling, and through investigations of the strong voltage dependency of the Ca^{2+} channel found by Nelson et al. [16], who observed a reduction in tension of 50 % with a 2 mV hyperpolarization. Our coupling curve establishes a 50% vasorelaxation for 2.5 mV hyperpolarization.

SUMMARY

1. Numerous compounds and changes in physical state functions shift the membrane potential of vascular smooth muscle to more negative values. The consequence is a vasodilatation because Ca^{2+} channels are closed. K^+ channel opening frequently causes the hyperpolarization.

2. Acidification of the blood substitute solution and a fall in O_2 partial pressure dilate arterial vessels. Acidosis is associated with a rise in K^+ permeability and a simultaneous fall in Na^+ permeability. Prostacyclin has a 20-30% share, and EDHF a 70-80% share, in hypoxic vasodilatation. Experiments with iloprost (PGI_2 analogue) confirmed the K^+ channel opening properties of this drug. A voltage-dependent K^+ channel and a Ca^{2+}-activated K^+ channel, via the influence of cA-PK or cG-PK, are responsible for the hyperpolarization with iloprost and with oxygen deficiency.

3. Cicletanine and ajoene cause a concentration-dependent membrane hyperpolarization and are potent vasodilators. A cicletanine concentration, which is attained by the dosage given to patients, is sufficient to produce these effects. Ajoene exerts a hyperpolarizing and vasodilating influence even in a concentration which may occur in the extracellular space by the administration of a single garlic clove.

4. The stationary activation curve 'developed force vs. membrane potential' satisfactorily explains the effects of K^+ channel openers. The tight electromechanical coupling expressed by this curve comprises a 50% vasorelaxation for a 2.5 mV hyperpolarization. In the linear part of the curve, the coupling ratio is 5.1 mV/g.

5. In the vascular smooth muscle, vasorelaxation can be evoked by membrane hyperpolarization which is linked to a simultaneous increase in K^+ outward current and $^{42}K^+$ efflux. In the case of substances whose influence is solely or partially receptor-mediated, cyclic nucleotides may be involved in vasorelaxation. Since cyclic nucleotides also hyperpolarize through an increase in K^+ conductance, the resulting dilatation often cannot be divided into its single components. Therefore, it is sensible not to give the term "K^+ channel opener" too fine a definition. The term should be applied to all substances and changes in physical states which predominantly increase the open probability of K^+ channels finally via a conformational change in the cell membrane. For example, giving an acidic blood substitute solution (acidosis) is an intervention opening K^+ channels. Which K^+ channel and which single channel conductance is concerned in a particular case, and which 'mediator' may participate, become secondary questions.

ACKNOWLEDGEMENTS

The authors thank Mrs. Ch. Fuhrmann for her excellent technical assistance and Mr. H. Ewald from the mechanical workshop for his constant help. We are grateful to Mrs. M. Krawczynski for her outstanding work in preparing the illustrations and Mr. P. Holzner for the photographs. We are indebted to Mrs. A. Scheuermann for the translation and editorial elaboration of the manuscript.

REFERENCES

1. Aprigliano, O., and Hermsmeyer, K., 1976, In vitro denervation of the portal vein and caudal artery of the rat, *J. Pharmacol. Exp. Ther.* 198:568-577.
2. Bolton, T. B., 1979, Mechanisms of action of transmitters and other substances on smooth muscle, *Physiol. Rev.* 59:606-718.
3. Braquet, P., Guinot, P., and Tarrade, T., 1988, Cicletanine: biology, pharmacology and clinical sciences, *Drugs Exp. Clin. Res.* 14:71-230.
4. Calder, J. A., Schachter, M., and Sever, P. S., 1991, Acute relaxant effect of cicletanine in human subcutaneous resistance arteries, *Blood Vessels* 28:279.
5. Fidone, S., and González, C., 1982, Catecholamine synthesis in rabbit carotid body *in vitro*, *J. Physiol. (Lond.)* 333:69-79.
6. Fidone, S., González, C., and Yoshizaki, K., 1982, Effects of hypoxia on catecholamine synthesis in rabbit carotid body *in vitro*, *J. Physiol. (Lond.)* 333:81-91.
7. Fredj, G., 1988, Clinical pharmacokinetics of cicletanine hydrochloride, *Drugs Exp. Clin. Res.* 14:181-188.
8. Ganitkevich, V.Ya., and Isenberg, G., 1990, Contribution of two types of calcium channels to membrane conductance of single myocytes from guinea-pig coronary artery, *J. Physiol. (Lond.)* 426:19-42.
9. Grote, J., Siegel, G., Zimmer, K., and Adler, A., 1988, The influence of oxygen tension on membrane potential and tone of canine carotid artery smooth muscle, *Adv. Exp. Med. Biol.* 222:481-487.
10. Haeusler, G., de Peyer, J.-E., and Schultz, G., 1987, Vascular effects of α_1- and α_2-adrenoceptor agonists in vitro and in hypertensive rats, *J. Cardiovasc. Pharmacol.* 10 [Suppl. 4]:15-18.
11. Hashimoto, T., Hirata, M., Itoh, T., Kanmura, Y., and Kuriyama, H., 1986, Inositol 1,4,5-trisphosphate activates pharmacomechanical coupling in smooth muscle of the rabbit mesenteric artery, *J. Physiol. (Lond.)* 370:605-618.

12. Hermsmeyer, K., 1979, High shortening velocity of isolated single arterial muscle cells, *Experientia* 35:1599-1602.
13. Kajiwara, M., Droogmans, G., and Casteels, R., 1984, Effects of 2-nicotinamidoethyl nitrate (nicorandil) on excitation-contraction coupling in the smooth muscle cells of rabbit ear artery, *J. Pharmacol. Exp. Ther.* 230:462-468.
14. Kauser, K., Clark, J. E., Masters, B. S., Ortiz de Montellano, P. R., Ma, Y.-H., Harder, D. R., and Roman, R. J., 1991, Inhibitors of cytochrome P-450 attenuate the myogenic response of dog renal arcuate arteries, *Circ. Res.* 68:1154-1163.
15. Loirand, G., Pacaud, P., Mironneau, C., and Mironneau, J., 1986, Evidence for two distinct calcium channels in rat vascular smooth muscle cells in short-term primary culture, *Pflügers Arch.* 407:566-568.
16. Nelson, M. T., Patlak, J. B., Worley, J. F., and Standen, N. B., 1990, Calcium channels, potassium channels, and voltage dependence of arterial smooth muscle tone, *Am. J. Physiol.* 259:C3-C18.
17. Northover, B. J., 1980, The membrane potential of vascular endothelial cells, *Adv. Microcirc.* 9:135-160.
18. Rüegg, J. C., 1986, "Calcium in muscle activation", Springer-Verlag, Berlin Heidelberg New York London Paris Tokyo.
19. Sadoshima, J.-I., Akaike, N., Kanaide, H., and Nakamura, M., 1988, Cyclic AMP modulates Ca-activated K channel in cultured smooth muscle cells of rat aortas, *Am. J. Physiol.* 255: H754-H759.
20. Shaw, K., Montagne, W., and Pallot, D. J., 1990, Biochemical studies on the release of catecholamines from the rat carotid body *in vitro*, *in:* "Arterial Chemoreception", C. Eyzaguirre, S. J. Fidone, R. S. Fitzgerald, S. Lahiri, D. M. McDonald, eds., Springer-Verlag, New York Berlin Heidelberg London Paris Tokyo Hong Kong, pp 87-91.
21. Siegel, G., 1986, Membranphysiologische Grundlagen der peripheren Gefäßregulation, *Physiol. akt.* 1:31-52.
22. Siegel, G., 1991, Garlic and vasoregulation, *Cardiology in Practice* [Suppl.], p 7.
23. Siegel, G., Carl, A., Adler, A., and Stock, G., 1989, Effect of the prostacyclin analogue iloprost on K^+ permeability in the smooth muscle cells of the canine carotid artery, *Eicosanoids* 2:213-222.
24. Siegel, G., Emden, J., Schnalke, F., Walter, A., Rückborn, K., and Wagner, K. G., 1991, Wirkungen von Knoblauch auf die Gefäßregulation, *Med. Welt* 7a:32-34.
25. Siegel, G., Grote, J., Schnalke, F., and Zimmer, K., 1989, The significance of the endothelium for hypoxic vasodilatation, *Z. Kardiol.* 78 [Suppl. 6]:124-131.
26. Siegel, G., Jäger, R., Nolte, J., Bertsche, O., Roedel, H., and Schröter, R., 1974, Ionic concentrations and membrane potential in cerebral and extracerebral arteries, *in:* "Pathology of Cerebral Microcirculation", J. Cervós-Navarro, ed., Walter de Gruyter, Berlin New York, pp 96-120.
27. Siegel, G., Kämpe, Ch., and Ebeling, B. J., 1981, pH-dependent myogenic control in cerebral vascular smooth muscle, *in:* "Cerebral Microcirculation and Metabolism", J. Cervós-Navarro, E. Fritschka, eds., Raven Press, New York, pp 213-226.
28. Siegel, G., Mironneau, J., Schnalke, F., Schröder, G., Schulz, B.-G., and Grote, J., 1990, Vasodilatation evoked by K^+ channel opening, *Prog. Clin. Biol. Res.* 327:299-306.
29. Siegel, G., Roedel, H., Nolte, J., Hofer, H. W., and Bertsche, O., 1976, Ionic composition and ion exchange in vascular smooth muscle, *in:* "Physiology of Smooth Muscle", E. Bülbring, M. F. Shuba, eds., Raven Press, New York, pp 19-39.
30. Siegel, G., Stock, G., Schnalke, F., and Litza, B., 1987, Electrical and mechanical effects of prostacyclin in the canine carotid artery, *in:* "Prostacyclin and Its Stable Analogue Iloprost", R. J. Gryglewski, G. Stock, eds., Springer-Verlag, Berlin Heidelberg New York London Paris Tokyo, pp 143-149.
31. Siegel, G., Walter, A., Bostanjoglo, M., Jans, A. W. H., Kinne, R., Piculell, L., and Lindman, B., 1989, Ion transport and cation-polyanion interactions in vascular biomembranes, *J. Membrane Sci.* 41:353-375.
32. Siegel, G., Wenzel, K., Schnalke, F., Mironneau, J., Schultz, G., Schröder, G., Schillinger, E., Grauhan, O., and Hetzer, R., 1990, Prostacyclin analogues as K^+ channel openers, *Clin. Pharmacol.* 7:72-96.
33. Sperelakis, N., and Ohya, Y., 1990, Cyclic nucleotide regulation of Ca^{2+} slow channels and neurotransmitter release in vascular muscle, *Prog. Clin. Biol. Res.* 327:277-298.
34. Weston, A. H., 1990, The pharmacology of smooth muscle potassium channels, *Clin. Pharmacol.* 7:1-18.
35. Weston, A. H., 1990, *in:* "Cicletanine and K^+ Channel Opening", Institut Henri Beaufour, Le Plessis-Robinson, France.
36. Weston, A. H., and Abbott, A., 1987, New class of antihypertensive acts by opening K^+ channels, *Trends Pharmacol. Sci.* 8:283-284.

EXCITATION-CONTRACTION COUPLING IN RAT SKELETAL MUSCLE CELLS: EVOLUTION DURING IN VITRO MYOGENESIS

Christian Cognard, Bruno Constantin, Michèle Rivet, Nathalie Imbert, Colette Besse[1], and Guy Raymond

Laboratory of General Physiology, CNRS URA 290 and [1]S.U.M.E.B.; Faculty of Sciences, 40 Av. Recteur Pineau F-86022 POITIERS Cedex France

INTRODUCTION

The fusion of mononucleated myoblasts into multinucleated myotubes is a crucial event during myogenesis since it precedes (without being a prerequisite) the appearance of a lot of molecules and processes vital for the cell: T system formation and sarcotubular organization, contractile proteins arrangements, voltage-dependent calcium channels, sarcoplasmic reticulum Ca-ATPase, junctional nicotinic receptor (see reviews from Schneider and Olson, 1988; Nathanson, 1989). Therefore the study of events taking place before and after fusion will be of great importance for the understanding of mechanisms controlling the main function of muscle cells, that is contraction.

We have recently shown that in rat myoballs, the high threshold calcium current ($I_{Ca,L}$), which could play a role in excitation-contraction coupling regulation (Brandt et al., 1990), is responsible for a part of contraction (Rivet et al., 1989) and the other part being controlled by a nifedipine-sensitive complex (Cognard et al., 1990), likely the voltage-sensor. An other role for $I_{Ca,L}$ during myogenesis might be to fill intracellular stores with calcium (Cognard et al., 1988).

Since in cultured myotubes, the myoplasmic calcium level depends on extracellular calcium (Jacobs et al., 1991) and that the development of T-tubules and junctional SR is poorly coordinated (Schiaffino et al., 1977), one can expect that changes in excitation-contraction occur during myogenesis in relation with the control of contraction by $I_{Ca,L}$, loading of internal calcium stores and ultrastructural organization.

The present experimentation was conducted to follow the appearance of voltage-dependent calcium currents and of excitation-contraction coupling mechanisms in relation with ultrastructural organization and calcium transients before and following the myoblastic fusion.

METHODS

Satellite cells were obtained from hindlimbs of 1-3 days newborn rats as already described (Rivet et al., 1989) plated (2ml/dish) in 35 mm-Petri dishes and incubated for 3 days at 37 °C in a humefied air

Excitation-Contraction Coupling in Skeletal, Cardiac, and Smooth Muscle
Edited by G.B. Frank *et al.*, Plenum Press, New York, 1992

atmosphere containing 5% of CO_2 in growth medium (HAM F12 with 10 % fetal calf serum and 10 % heat-inactivated horse serum). Fusion was induced after 3 days by exchanging the growth medium with a Dulbecco modified Eagle medium (+ 5% heat inactivated horse serum). The age of cells was always calculated taking the replacement of growth medium by the fusion permissive medium as time zero.

The whole-cell configuration of the patch-clamp technique (Hamill et al., 1981) was used to record membrane currents. The pipettes were connected to the headstage of the patch-clamp amplifier (RK300 Biologic) driven by a PC-AT compatible microcomputer through a Labmaster TM40 conversion board.

The contractile activity was simultaneously recorded with membrane currents by means of a photomultiplier tube (Hamamatsu IP28), as light transmission changes induced by cell shortening. The photomultiplier was mounted on an auxiliary light-path exit of an inverted microscope (IMT2 Olympus). The patch-clamped cell was thus helded between two fixed points: the membrane sticking at the bottom of the culture dish and the pipette tip. The pipette approach and the contractile activity could be followed too through a video camera.

Calcium transients were recorded using a pulsed laser cytofluorimeter (ACAS 570 Meridian) and Indo-1 as fluorescent probe. Cells were loaded for 30-60 min with Indo-1/AM before a 10 min period (for desesterification). The calcium signal was estimated as the ratio of the 405/485 nm emission of the 2 forms of excited Indo-1. It was generally measured at a fixed point in the cell. The calcium concentration was estimated using the following formulae (Grynkiewicz et al., 1985):

$$[Ca^{2+}]_{int} = \beta .\ K_d\ [(R-R_{min})/(R_{max}-R)]$$

where K_d = 250 nM is the apparent dissociation constant for the dye-Ca^{2+} complex; β is the ratio obtained at 485 nM with free and saturating level of Ca^{2+} respectively. R_{max} is the maximum fluorescence ratio (Ca^{2+}-ionomycin) and R_{min}, the minimal fluorescence ratio (Ca^{2+} free/EGTA-ionomycin).

For electron microscopy, cultures were fixed in 2 % glutaraldehyde (cacodylate buffer 0.1 M) at 4°C for 1 hour followed by 1 hour in cacodylate buffer and postfixed in OsO_4 (osmium tetroxyde). The specimen were dehydrated in alcohol and infiltrated in a mixture 1:1 alcohol-Epon and then 100 % Epon. Ultrafine sections of 50 nm were contrasted with heavy metal salts: 1 min in 1 % uranyl acetate in 50 % alcohol then 10 min in lead citrate. The observations were done with a JEOL 100C electron microscope.

All experiments were performed at room temperature (20-22 °C) on myoblasts, young myotubes (2-9 nuclei) during the 1st day postfusion and then between the 2nd and the 3rd day (more than 9 nuclei) and finally on spontaneously-formed or colchicine-induced (10-30 nM) myoballs (more suitable for electrophysiology than large myotubes). The culture medium was exchanged before each experiment by the basic saline solution containing (mM): 135 TEACl; 2.5 $CaCl_2$; 0.8 $MgCl_2$; 5.6 glucose; 10 HEPES; the pH was adjusted to 7.4. For patch-clamp experiments, pipettes contained 145 $CsCl_2$; 1 $MgCl_2$; 0.005 $CaCl_2$; 1 EGTA; 2 ATP; 10 HEPES; 5.6 glucose; pH 7.2. For cytofluorimetry 135 TEACl were replaced by 130 NaCl + 5.4 KCl.

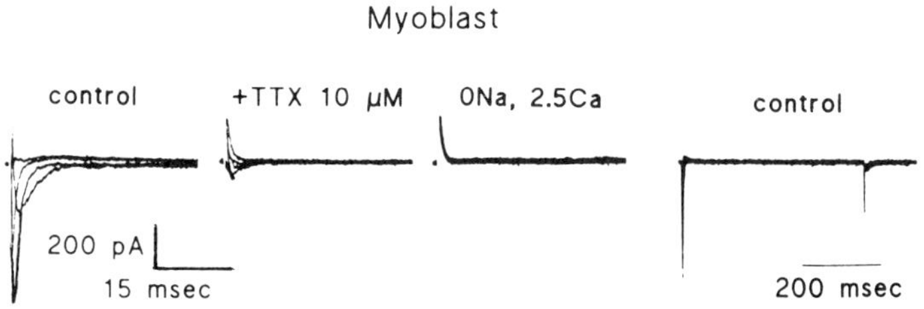

Figure 1. Whole-cell currents elicited on a myoblast by different depolarizations (from -90 mV to -50, -30, -10, +20 mV). From left to right in basic solution (for this particular experiment, the basic solution is the same as for cytofluorimetry), in the presence of TTX, in absence of external Na and finally in control at a slower time base.

RESULTS

Evolution of currents and contraction

As illustrated in figure 1, the only ionic current that was observed in myoblasts is a tetrodotoxin(TTX)-sensitive current which disappeared in absence of external sodium. In agreement with results of Frelin et al. (1984) the sensitivity to TTX of this sodium current is weak since 10 μM TTX did not fully block it. Calcium currents and/or contraction have never been observed before fusion.

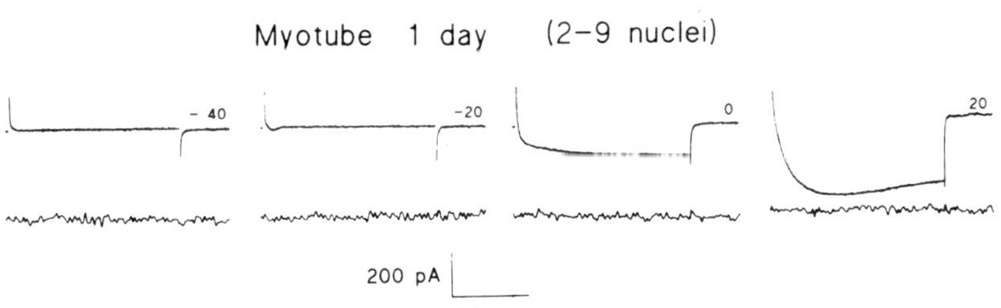

Figure 2. Calcium currents in a young myotube: from h.p. -90 mV whole-cell calcium currents induced by depolarizations to values indicated above current traces (upper traces). Note at - 20 mV the presence of a weak T-type current and the absence of contraction at all potentials (lower traces).

The high-threshold calcium current appears soon after fusion since it was observed in 75 % of myotubes during the 1st day postfusion (2 to 9 nuclei) as shown in figure 2, with a threshold near -30 mV and a peak at +20 mV. Sometimes (not more than 10 % of myotubes) a low-threshold T-type current was present as obvious at -20 mV but this type of current was never observed in cells lacking L-type current. Contraction was absent in all cases (table 1).

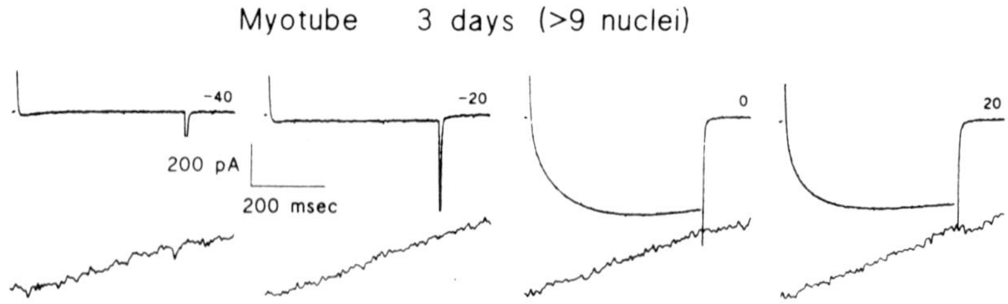

Figure 3. In a 3 days-old myotube, the calcium currents (upper traces) and the ascending phase of the contractile response (lower traces) elicited by depolarizations from -90 mV to the values indicated above current traces.

Contraction appeared later in myotubes with some tens of nuclei (less than 4 days post fusion) as shown in figure 3. However only 20 % of myotubes exhibited contraction, mainly during the 3rd day postfusion. Calcium currents were present in all cells but T-type only in 67 % (table 1).

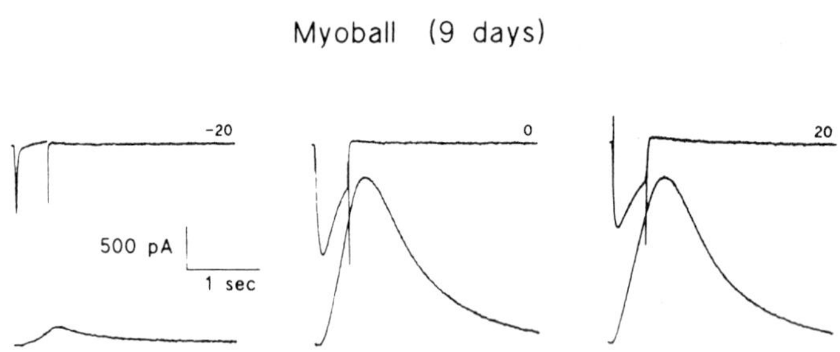

Figure 4. In a 9 days-old myoball, simultaneous recordings of calcium currents (upper traces) and contraction (lower traces) induced by depolarizations to -20 , 0, 20 mV.

In older cells such as the 9-days myoball of figure 4 a strong contraction was observed in about 85 % of cases. When present (more than 80 % of cells) the T-type current was important as obvious at -20 mV.

To summarize, the graph of figure 5 shows the occurence of T-type (open boxes), L-type (hatched boxes) currents and contraction (black boxes) as a function of age. $I_{Ca,L}$ was fully present after the 2nd day while the occurence of $I_{Ca,T}$ increased continuously to stabilize at 9 days at 86 % (in agreement with results from Cognard et al., 1986). Contraction appeared between day 2 and day 3 and stabilizes around 7-9 days at 85 %. It must be noticed that contraction was never recorded in cells lacking L-type calcium current and that there was no correlation between the presence of T current and that of contraction.

Role of L-type calcium current in contraction activation

As previously published (Rivet et al., 1989; Cognard et al., 1990) the contraction of cultured muscular cells depends on two mechanisms (figure 6):

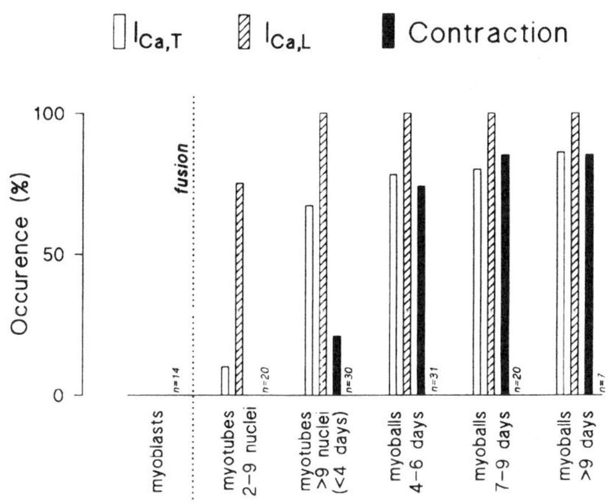

Figure 5. The percentage of observations of $I_{Ca,T}$, $I_{Ca,L}$ and contraction as a function of age.

Table 1. Occurence of calcium currents and contraction as a function of age

	no I_{Ca}	only $I_{Ca,T}$	only $I_{Ca,L}$		$I_{Ca,T}$ and $I_{Ca,L}$	
			contraction			
			without	*with*	*without*	*with*
Myotube 1 day	25 %	0 %	65 %	0 %	10 %	0 %
Myotube 2-3 days	0 %	0 %	33 %	0 %	46 %	21 %
Myoball 4-6 days	0 %	0 %	0 %	22 %	26 %	52 %
7-9 days	0 %	0 %	0 %	20 %	15 %	65 %
>9 days	0 %	0 %	0 %	14 %	15 %	71 %

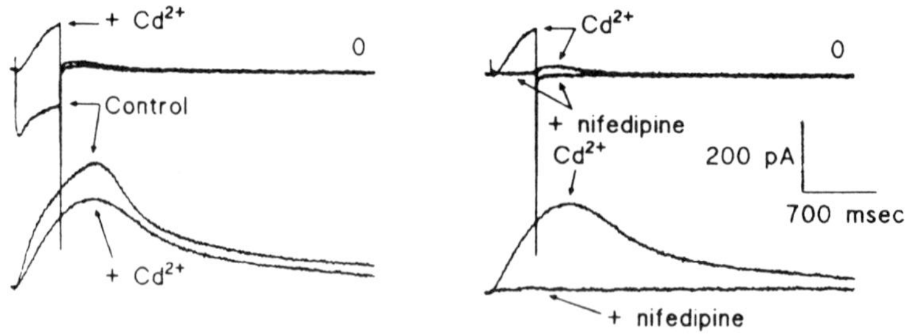

Figure 6. Two components of contraction in a 10 days-old myoball. On the left: superimposed recordings of currents (upper traces) and contraction (lower traces) induced by a depolarization to 0 mV, in the absence and in presence of 2 mM Cd^{2+}. On the right: in the presence of Cd^{2+}, the addition of 10^{-6} M nifedipine fully blocks the current-independent component of contraction. Note the simultaneous block of the contraction-linked outward current (Constantin et al., 1991) demasked by I_{Ca} disappearance.

- One is related to the L-type calcium current; it is blocked by 2 mM Cd^{2+} as shown in the left pannel of figure 6.
- The other is more purely voltage-dependent; it is abolished by nifedipine (here at 10^{-6} M), in the presence of cadmium (right pannel in figure 6). It seems attributable to the presence of a tubular voltage-sensor, controlling the opening of SR calcium channels.

The relative importance of each of these contraction components can be estimated through the action of Cd^{2+} as illustrated in figure 7 for 3 cells at different times after fusion.

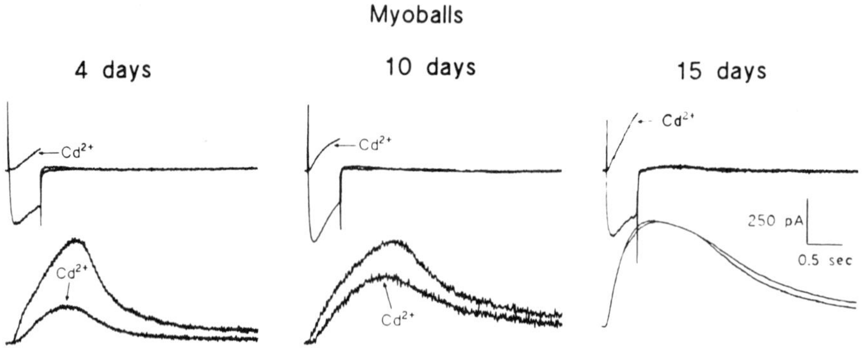

Figure 7. The evolution of the current-dependent component of contraction with age of cells. The effects of 2 mM Cd^{2+} on currents (upper traces) and contraction (lower traces) in 3 myoballs of different ages.

At an early stage (4 days) 2 mM Cd^{2+} drastically depress contraction (≈ 60 %). This relative effect decreases with age (30 % at 10 days) and nearly disappears in old myoballs (15 days).

This is illustrated in the graph of figure 8: before 7 days the current-dependent component represented around 50 % (43.7 ± 8.4 %) of the total contraction. This percentage fell to 20 % at 10-12 days and to 7 % at 13-16 days.

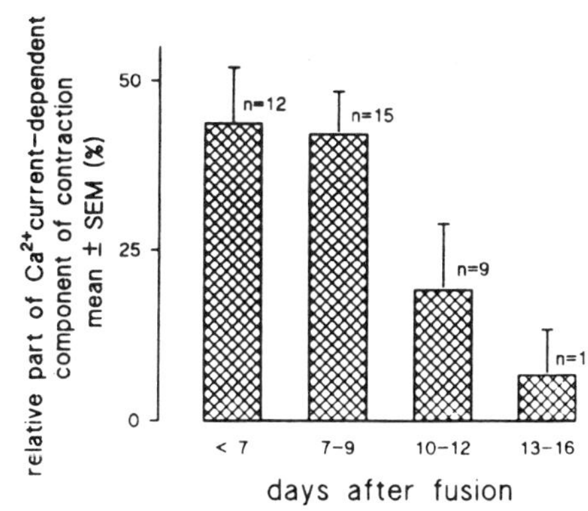

Figure 8. The evolution of the current-dependent component of contraction as a function of age of myoballs.

Thus the role of L-type calcium current in activation of contraction decreases during in vitro myogenesis while the voltage-dependent component increases (≈ 50 % before 7 days, more than 90 % after 12 days). It must be noticed that in mouse myotubes, Romey et al. (1989) had not observed any current dependence for contraction . This could be due to some differences in *in vitro* myogenesis, attested by the slowness of maturation of mouse myotubes compared to that of rat muscle cells. Alternatively some species differences could be involved.

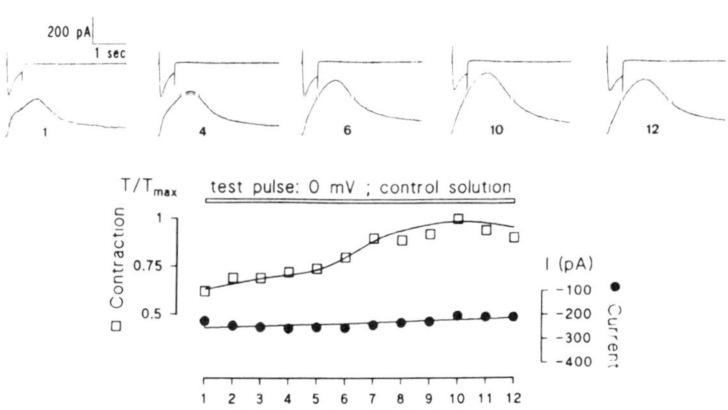

Figure 9. The effects of iterative pulsing on calcium current and contraction of a 7 days-old myoball. Upper panel: simultaneous recordings of calcium current (upper traces) and contraction (lower traces) upon repetitive pulsing at a frequency of 1 stimulation every 30 seconds. Lower panel: evolution of contraction (open squares) and $I_{Ca,L}$ (filled circles) amplitudes as a function of the number of successive pulses.

Role of L-type calcium current in SR loading

As previously reported (Cognard et al., 1988) an iterative pulsing to a potential inducing a large calcium current (from -90 mV to 0 mV in figure 9) led to a progressive increase in contraction amplitude while the current did not drastically change throughout pulsing. In the illustrated example, which corresponds to a 7 days-old myoball, the contraction amplitude grew to be maximum at the 10th pulse in such a way that the maximal response increased by 60 % compared to the initial one. This staircase phenomenon probably reflects loading of SR by calcium ions entering the cell through $I_{Ca,L}$.

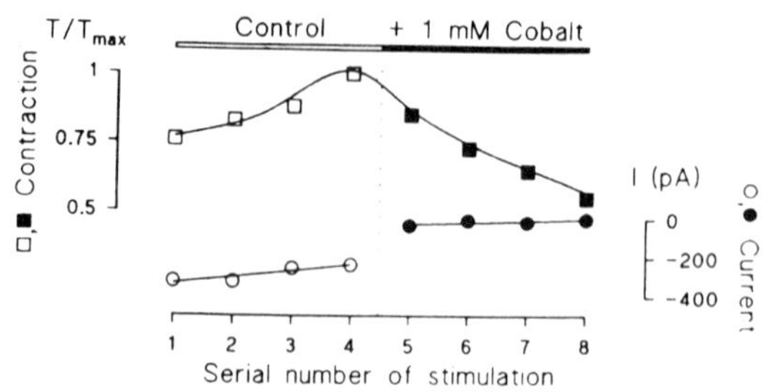

Figure 10. Effects on current and contraction amplitude of 1 mM Co^{2+} applied during repetitive pulsing (1/30 s) on a 9 days-old myoball.

It can be reversed by Cd^{2+} or Co^{2+} as shown in figure 10 in which 1 mM Co^{2+} was applied after a 35 % increase by repetitive pulsing. The addition of Co^{2+} led to decrease in contraction amplitude which was lower than the initial one at the 8th pulse that suggested a progressive decrease in the amount of Ca^{2+} available for contraction in SR.

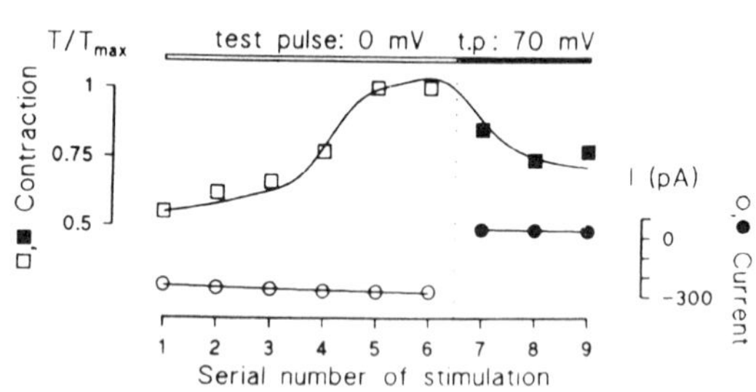

Figure 11. During a repetitive pulsing from -90 mV to 0 mV, reversion of the staircase phenomenon by a sudden increase of the depolarization amplitude to + 70 mV to suppress the calcium current (filled symbols). Age of myoball: 7 days.

Another way to reverse the positive staircase was to suddenly increase the depolarization amplitude to a value for which no inward current was elicited as shown in figure 11 at +70 mV. That induced an immediat negative staircase which was probably underestimated since the voltage-dependent component of contraction would have been enhanced by the larger depolarization. In effect, in our experimental conditions, the amplitude of the voltage-dependent component of contraction has been shown to be maximum at + 20 mV (Rivet et al., 1989).

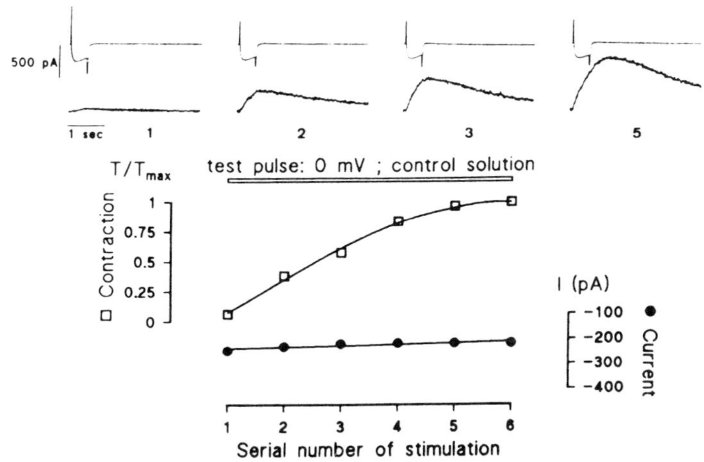

Figure 12. The effects of repetitive pulsing on calcium current and contraction in a 2 days-old myotube which did not (or weakly) contract at the beginning of experiment. Upper panel: simultaneous recordings of currents (upper traces) and contraction (lower traces) induced by depolarization to 0 mV. (From left to right, 1st, 2nd, 3rd and 5th stimulation). Lower panel: The evolution of contraction (open squares) and current (filled circles) amplitudes against the number of pulses.

The role of $I_{Ca,L}$ in loading SR with Ca^{2+} is better illustrated in figure 12 in a 2 days-old myotube which exhibited a very weak contraction but a large calcium current in response to 0 mV-depolarization (left record of upper panel). Repetitive pulsing alter $I_{Ca,L}$ but induced a progressive increase in contractile amplitude which stabilized at a level 14 times larger than the initial one. Such an effect was observed on several young myotubes which did not contract at 2 days. They developed a progressive contractile activity as the number of pulses increased, provided that calcium current was present. In effect this has never been observed in the presence of Cd^{2+}.

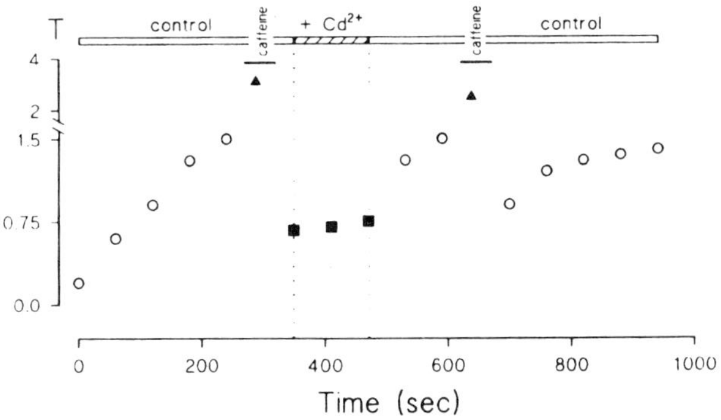

Figure 13. The effects of repetitive pulsing on the contraction induced by 0 mV-depolarization in a 2 days-old myotube which did not contract at the beginning of the experiment. As indicated above the graph, different external solutions were successively applied: control, 5 mM caffeine, 2 mM Cd^{2+}, control, 5 mM caffeine, control.

The absence of contraction in these cells was not due to some defect in SR Ca^{2+} loading capability as shown in figure 13: Following a repetitive pulsing to 0 mV to induce a "normal contraction", a 5 mM caffeine contracture was elicited to partly empty the SR before returning to repetitive pulsing in the presence of Cd^{2+} to block I_{Ca}. The subsequent contraction was decreased by 50 % but its amplitude was noticeably larger than the initial one (x5) and it remained stable for the 2-3 next pulses indicating a good functionality of the SR. Withdrawing Cd^{2+} allowed contraction to increase again upon pulsing. A second caffeine contracture followed by calcium current induction led to the development of a staircase but contraction stabilized at a lower amplitude than precedingly, probably as a consequence of some deleterious effect of successive applications of caffeine (Duncan and Smith, 1978).

Throughout myogenesis the relative amount of extra calcium storable in SR as a consequence of repetitive entry of Ca^{2+} via calcium channels, tended to diminish as attested by the progressive decrease in the staircase phenomenon (T_{max}-T/T): more than 500 % at 2 days postfusion, it fell to ≈ 100 % at 4-6 days and then to 60 % at 6-9 days. At later stages the staircase phenomenon became difficult to observe. This could agree with a progressive basic charge in Ca^{2+} in SR during myogenesis as the junctional SR differentiates since *in vitro* the SR differentiation seems slower than *in vivo* (Schiaffino and Margreth, 1969; Schiaffino et al., 1977).

Thus it appears that an important functional role for $I_{Ca,L}$ during myogenesis could be to load the sarcoplasmic reticulum with Ca^{2+}. In the early stages of *in vitro* maturation it might allow initial loading and later on it probably serves to maintain constant the intracellular amount of stored Ca^{2+} then compensating the Ca^{2+} lost during the efflux that follows the excitation (Nayler 1966).

Evolution of calcium signal

As reported above $I_{Ca,L}$ appeared soon after fusion since it was present in 75 % of myotubes in the 1st day postfusion whereas contraction appeared later (2-3 days postfusion). Was this contraction lag due to failure in intracellular calcium storage, to some defect in myofibrils or to uncoupling between excitation and internal calcium release system? To clarify this problem we have followed the intracellular calcium signal throughout myogenesis in response to different kinds of stimulation:

- application of 5 μM acetylcholine to stimulate nicotinic cholinergic receptors,
- hyperkaliemic solutions (100 mM K^+) to depolarize the surface and tubular membrane,
- caffeine (10 mM) to open SR calcium channels.

In mononucleated cells (myoblasts) the resting calcium concentration was 52.6 ± 4.9 nM (n =19). The only calcium signal that could be observed (45 % of cells) was in response to acetylcholine of which maximum amplitude ranged between 50 and 750 nM. It lasted about 30 seconds and peaked at 10 seconds. That agrees with results from Decker and Dani (1990) who concluded from single patch-clamp experiments that the calcium influx through nicotinic acetylcholine receptors is significant. However it seems surprising that calcium influx through cholinergic receptors could be sufficient by itself to increase the internal calcium concentration to such a level.

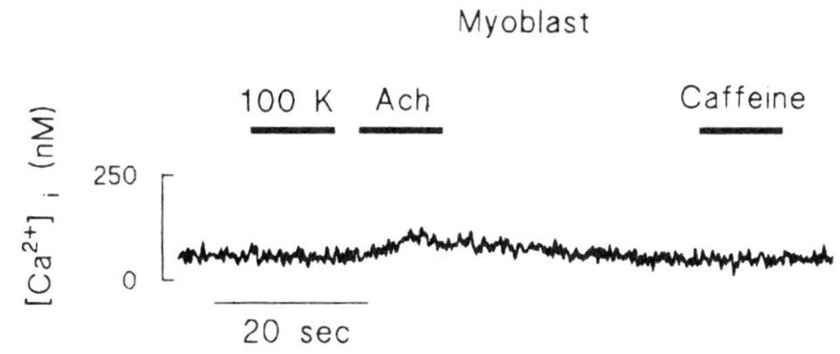

Figure 14. In a myoblast loaded with Indo-1, temporal evolution of the internal free calcium concentration in response to 5 μM acetylcholine. Note the absence of response to high K^+ and caffeine.

In the 1st day postfusion the resting calcium concentration was not different from that in myoblasts (51.0 ± 6.8 nM; n = 17). This resting calcium level agrees with estimations of Grouselle and Georgescauld (1990) obtained in rat myotubes from video-imaging with Fura-2, but it is smaller than that observed in human myotubes by Mongini et al., (1988) that remains to be explained.

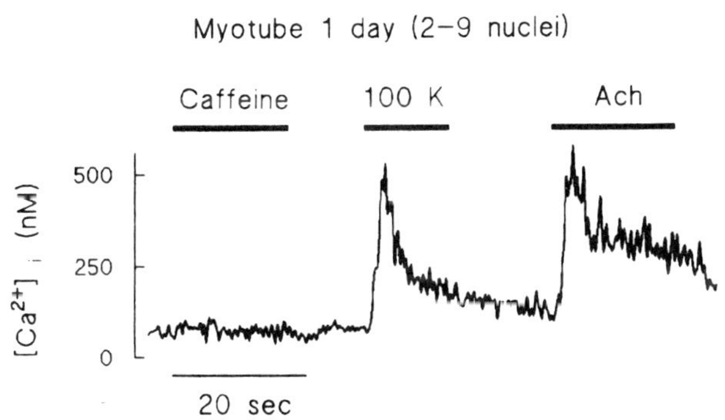

Figure 15. In a 1 day-old myotube high K^+ and acetylcholine but not caffeine induced a transient increase in cytoplasmic free calcium . Note the monophasic time course of the high K^+ response.

The acetylcholine response (figure 15) was present in 88 % of cells (mean value 450 nM ± 27 nM). The calcium signal that was the more present was the response to high potassium solution which was observable in almost all cells (611 nM ± 45 nM). At this stage the high K^+-induced calcium signal exhibited a monophasic time course with relatively rapid kinetics (peak at 2.5 s). No response to caffeine was ever observed.

At a later stage (ten to some tens of nuclei) the resting myoplasmic free calcium concentration was not very different (62.2 ± 10.6 nM, n = 15). All cells (figure 16) responded to acetylcholine (895 ± 32 nM) and to high potassium solutions (1154 ± 78 nM). 88 % of cells exhibited a caffeine-induced calcium signal which was significantly smaller than the Ach and K^+ responses (460 ± 31 nM).

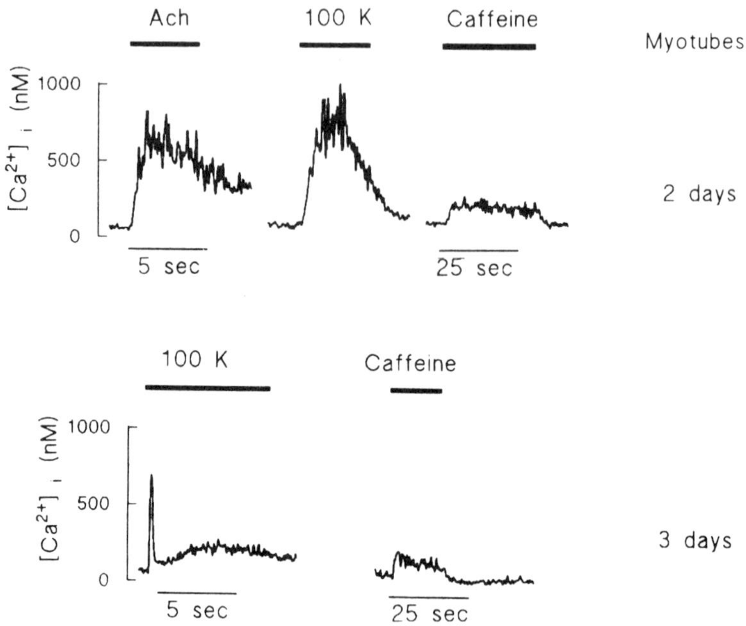

Figure 16. Calcium transients in response to acetylcholine, high K^+ and caffeine in a 2 days- (upper panel) and a 3 days-old (lower panel) myotubes. Note the appearance of a biphasic response to high K^+ at 3 days.

It must be noticed that at 3 days the response to high K^+ sometimes exhibited a biphasic time course as illustrated in the lower panel of figure 16: a fast transient response peaking at 200 ms and lasting 500 ms was followed by a second one of which the amplitude (200-300 nM) and the kinetics (peak at 4-5s, duration > 10 s) were more or less comparable to those of the caffeine response.

Later on, in myoballs, the resting free calcium concentration was surprisingly lower: 29.4 ± 3.8 nM (n = 21). This is difficult to explain but one possibility could be the reinforcement of activity of systems controlling calcium homeostasis due to the particular architecture of myoballs. Such a result remains to be confirmed in a larger number of experiments since the difference with other groups of cells did not appear to be significant.

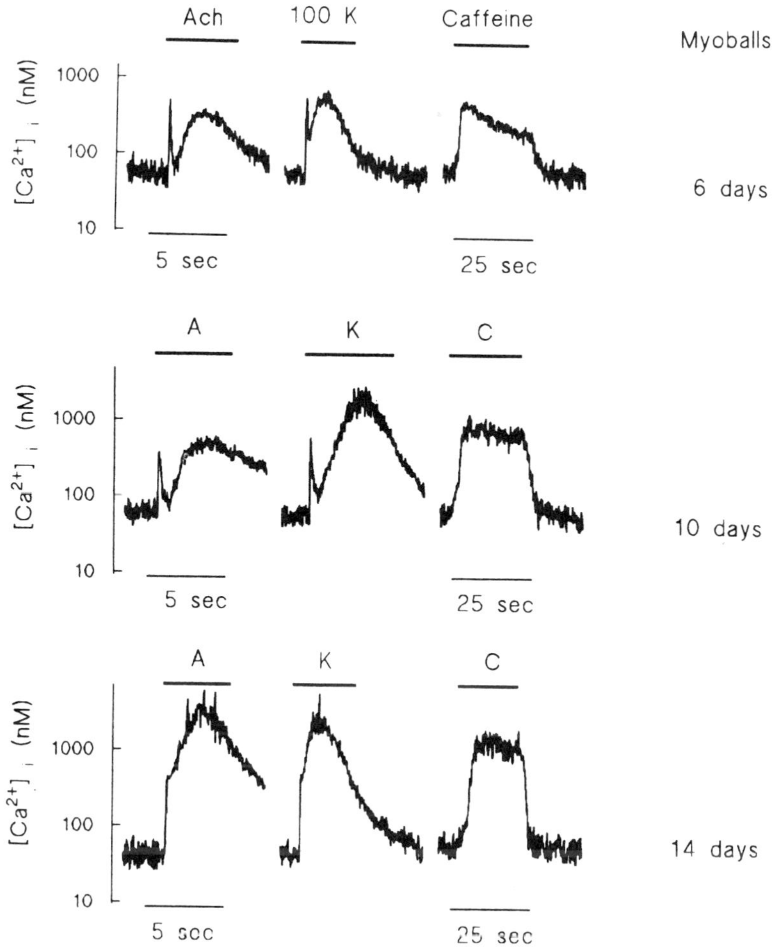

Figure 17. Calcium transients elicited by acetylcholine, 100 mM K^+ and caffeine in myoballs at different times postfusion. Note the progressive and simultaneous increase in the caffeine and the slow phase of high K^+ responses.

All cells responded to acetylcholine, high K^+ and caffeine figure 17). The transient response for acetylcholine and high K^+ did not increase contrary to that of caffeine (1120 ± 98 nM) compared to the preceding stage. In most cases the responses to high K^+ and also acetylcholine exhibited a biphasic time course. The amplitude of the second phase increased with age as that of the caffeine response did as shown at 10 days (middle panel) to finally overtake the first one in older myoballs (for example at 14 days in the lower panel).

Evolution of ultrastructure

In myoblasts, T-tubules were always absent, sarcoplasmic reticulum seemed to be present and sparse and inorganized myofibrils too.

In the first day following the fusion induction, membrane expansions were observed between cells which had not fully fused and myofibrils begun to arrange longitudinally. Sarcolemmal invaginations and caveolae were present in the neighbouring of the fusion area indicating T-tubule formation. There was no sign of triad or diad formation.

At the next stage (3 days) myofibrils were better organized but Z lines were rare. Diads and triads in formation were numerous; well-formed triads were sometimes encountered but not properly disposed relatively to myofibrils.

Table 2. Summary of events observed during in vitro myogenesis

AGE EVENT	Myoblasts	Myotubes 2-9 nuclei	Myotubes >10 nuclei	Myoballs > 4 days
CURRENTS				
I_{Na}	+++	+++	+++	+++
$I_{Ca,T}$	-	+	++	+++
$I_{Ca,L}$	-	++	+++	+++
CONTRACTION	-	-	+	+++
I_{Ca}-dependent	-	-	++	+
V-dependent	-	-	++	+++
Staircase	-	-	+++	+
CALCIUM SIGNAL				
Acetylcholine	+	++	+++	+++
High K^+_{fast}	-	+	++	+++
High K^+_{slow}	-	-	+	+++
Caffeine	-	-	+	+++
ULTRASTRUCTURE				
Myofibrils	+/-	+	++	+++
Z line	-	-	+/-	++
$SR_{long.}$	+	++	++	+++
$SR_{junctional}$	-	-	+	++
Caveolae	-	++	++	++
T-tubules	-	+	+++	+++
Triads/diads	-	-	+	++
SR Ca channels	-	-	+	++

In late myotubes and myoballs, sarcomere were definitively organized with Z lines which were aligned, even in myoballs. Triads, inverted triads and pentads were numerous and properly disposed on each part of the myofibrills.

Such an ultrastructural evolution during *in vitro* myogenesis is in agreement with previous observations by Schiaffino et al., (1977) on rat myotubes and Romey et al., (1989) in mouse myotubes.

CONCLUSIONS

As reported in table 2, before fusion (in myoblasts) among signals involvable in an excitation-contraction coupling cycle, only sodium currents and nicotinic cholinergic calcium signals are present. This calcium signal was probably related to extracellular calcium entry through calcium channels since no functional intracellular storage site was evidenced.

During the 1st day following fusion, the two calcium currents $I_{Ca,T}$ and mainly $I_{Ca,L}$ were present, concomittantly with numerous caveolae and forming T-tubules and also the appearance of a depolarization-induced calcium transient. This calcium transient was blocked by nifedipine, therefore it probably reflects calcium entry through L-type calcium channels and did not seem to involve intracellular storage sites which had not yet accumulate calcium (or couldnot release it) as attested by the absence of caffeine response. Moreover the organization of myofibrills and/or internal architecture seemed insufficient to allow contraction proceeding although the internal calcium transient in response to a high potassium stimulus might be sufficient to activate contraction proteins.

The next stage led to contraction appearance in relation to internal calcium loading of the junctional reticulum (caffeine response), the better (even still incomplete) myofibrillar organization, the beginning of sarcotubular arrangement and the development of a second phase in the depolarization-induced calcium signal. That was noticeable at this stage is the fact that the calcium-dependent contraction which now represented 50 % of the total response appeared simultaneously with the potential-dependent one. In the debate concerning the dual role of the dihydropyridine receptor as calcium-channel and/or voltage-sensor (see Agnew, 1987; Lüttgau et al., 1987; Rios and Brum, 1987; Rios and Pizarro, 1988) this is in favour of a single complex with voltage-sensing and calcium channel properties rather than two different complexes with similar properties. However in a previous paper the differential recovery of contraction and $I_{Ca,L}$ following nifedipine application (Cognard et al., 1990) led us to hypothesized the existence of two different molecules. Moreover if there is a single complex acting as calcium channel and voltage sensor the two functions must have different voltage-dependence since contraction can be observed at potentials below the threshold for $I_{Ca,L}$ (see figure 3 at -30 mV).

At a later stage, the voltage-dependent part of the contraction prevailed as the triad organization was achieving. The second calcium signal induced by a high potassium solution (and often by acetylcholine) increased as the caffeine signal did to the same amplitude, that seems to confirm the link between them. The staircase phenomenon was more difficult to observe indicating that the basic SR calcium loading was achieved.

Thus during myogenesis, the calcium influx through L-type calcium channels seems to have a dual role: at the early stages it serves to fill with calcium the intracellular sites whose organization is in process and then to assure a part of the contraction when the sarcotubular arrangement is incomplete. Later on it must play some role in refilling SR with calcium to compensate calcium wasting and also to regulate the calcium release at the SR calcium channel level especially during durable depolarizations (see results on adult muscle by Potreau and Raymond, 1980; Ildefonse et al., 1985) or trains of APs which are the natural response of the skeletal muscle . This is in agreement with the model of excitation-contraction coupling of Brandt et al. (1990). In addition or alternatively, calcium ions entering the cell through calcium channels might interfere with the phosphatidylinositol metabolism in relation with the glycolytic pathway as suggested by Heilmeyer et al. (1990).

ACKNOWLEDGMENTS

This work was financially supported by grants from C.N.R.S. (URA 290), I.N.S.E.R.M. (CRL 875013), University of Poitiers, Association Française contre les Myopathies, Fondation pour la Recherche Médicale, Fondation Langlois.

REFERENCES

Agnew, W.S. (1987) Dual roles for DHP receptors in excitation-contraction coupling. **Nature** 328: 297.

Brandt, N.R., Caswell, A.H., Wen, S-R. and Talvenheimo, J.A. (1990) Molecular interactions of the junctional foot protein and dihydropyridine receptor in skeletal muscle triads. **J. Membrane Biol.** 113: 237-251.

Cognard, C., Lazdunski, M. and Romey, G. (1986) Different types of Ca^{2+} channels in mammalian skeletal cells in culture. **Proc. Natl. Acad. Sci. USA** 83:517-521.

Cognard, D., Raymond, G. and Rivet, M. (1988) On the role of calcium current in the development of contraction of patch- clamped rat skeletal muscle cells in culture. **J. Physiol. Lond.** 406: 104P.

Cognard, C., Rivet, M. and Raymond, G. (1990) The blockade of excitation-contraction coupling by nifedipine in patch-clamped rat skeletal muscle cells in culture. **Pflügers Arch.** 416: 98-105.

Constantin, B., Cognard, C., Rivet, M. and Raymond, G. (1991) Characterization of a slow outward current activated by internal calcium release during contractile activity of cultured rat skeletal muscle cells. **J. Physiol. Lond.**, under press.

Decker, E.R. and Dani, J.A. (1990) Calcium permeability of the nicotinic acetylcholine receptor: the single-channel calcium influx is significant. **J. Neurosci.** 10: 3413-3420.

Duncan, C.J. and Smith, J.L. (1978) The action of caffeine in promoting ultrastructural damage in frog skeletal muscle fibres. Evidence for the involvment of the calcium-induced release of calcium from sarcoplasmic reticulum . **N-S. Arch. Pharmacol.** 305: 159-166.

Frelin, C., Vijverberg, H.P.M., Romey, G., Vigne, P. and Lazdunski, M. (1984) Different functional states of tetrodotoxin sensitive and tetrodotoxin resistant Na+ channel occur during the in-vitro development of rat skeletal muscle. **Pflügers Arch.** 402: 121-128.

Grouselle, M. and Georgescauld, D. (1990) Fura-2 imaging of intracellular free calcium dynamics in excitable cells. In "Water and ions in biological systems". Advances in Life Sciences. Birkhäuser Verlag Basel. pp 229-240 1990.

Grynkiewicz, G., Poenie, M. and Tsien, R.Y. (1985) A new generation of Ca2+ indicators with greatly improved fluorescence properties. **J. Biol. Chem.** 260: 3440-3450.

Hamill, O.P., Marty, A., Neher, E., Sakmann, B. and Sigworth, F.J. (1981) Improved patch-clamp techniques for high-resolution current recording from cells and cell-free membrane patches. **Pflügers Arch.** 391: 85-100.

Heilmeyer, L.M.G., Han, J.W., Thieleczek, R., Varsanyi, M. and Mayr, G.W. (1990) Relation of phosphatidylinositol metabolism to glycolytic pathway in skeletal muscle membrane. **Mol. Cell. Biochem.** 99: 111-116.

Ildefonse, M., Jacquemond, V., Rougier, O., Renaud, J.F., Fosset, M. and Lazdunski, M. (1985) Excitation-contraction coupling in skeletal muscle: evidence for a role of slow Ca2+ channels using Ca2+ channel activators and inhibitors in the dihydropyridine series. **Biochem. Biophys. Res. Commun.** 129: 904-909.

Jacobs, A.E.M., Benders, A.G.M., Oosterhof, A., Veerkamp, J.H., Wevers, R.A. and Joosten, E.G.M. (1991) Effect of various agents on the cytoplasmic calcium concentration in cultured human muscle cells. **Int. J. Biochem.** 23: 93-99.

Lüttgau, H.Ch., Gottschalk, G. and Berwe, D. (1987) The effect of calcium and Ca antagonists upon excitation-contraction coupling. **Can. J. Physiol. Pharmacol.** 65: 717-723.

Mongini, T., Ghigo, D., Doriguzzi, C., Bussolino, F., Pescardona, G., Pollo, B., Schiffer, D. and Bosia, A. (1988) Free cytoplasmic Ca++ at rest and after cholinergic stimulus is increased in cultured muscle cells from Duchenne muscular dystrophy patients. **Neurology** 38: 476-480.

Nathanson, M.A. (1989) Differenciation of musculoskeletal tissues. **Int. Rev. Cytol.** 116: 89-164.

Nayler, W.G. (1966) Influx and efflux of Ca++ in the physiology of muscle contraction. **Clin. Orthop.** 46: 157-182.

Potreau, D. and Raymond, G. (1980) Calcium-dependent electrical activity and contraction on voltage-clamped frog single muscle fibres. **J. Physiol. Lond.** 307: 9-22.

Rios, E. and Brum, G. (1987) Involvement of dihydropyridine receptors in excitation-contraction coupling in skeletal muscle. **Nature** 325: 717-720.

Rios, E. and Pizarro, G. (1988) Voltage sensors and calcium channels of excitation-contraction coupling. **NIPS** 3: 223-227.

Rivet, M., Cognard, C. and Raymond, G. (1989) The slow inward calcium current is responsible for a part of the contraction of patch-clamped rat myoballs. **Pflügers Arch.** 413: 316-318.

Romey, G., Garcia, L., Dimitriadou, V., Pincon-Raymond, M., Rieger, F. and Lazdunski, M. (1989) Ontogenesis and localization of Ca2+ channels in mammalian skeletal muscle in culture and role in excitation-contraction coupling. **Proc. Natl. Acad. Sci.** 86: 2933-2937.

Schiaffino, S. and Margreth, A. (1969) Coordinated development of sarcoplasmic reticulum and T-system during post-natal differentiation of rat skeletal muscle. **J. Cell. Biol.** 41: 855-875.

Schiaffino, S., Cantini, M. and Sartore, S. (1977) T-system formation in cultured rat skeletal tissue. **Tissue and Cell Res.** 9: 437-446.

Schneider, M.D. and Olson, E.N. (1988) Control of myogenic differentiation by cellular oncogenes. **Molec. Neurobiol.** 2: 1-79.

VOLUME CHANGES DURING CONTRACTION OF ISOLATED FROG MUSCLE FIBERS

Stuart R. Taylor, Ian R. Neering*, Laura A. Quesenberry, and V. Arlene Morris

Department of Pharmacology, Mayo Foundation, Rochester, MN, 55905, USA and *School of Physiology and Pharmacology, University of NSW Kensington, NSW 2033, Australia

SUMMARY

A microscope objective and electronic imaging system were used to determine how isolated frog skeletal muscle fibers adjust their volume during an isometric tetanus. Cross-sectional area and volume of the middle third of a fiber increased rapidly with the development of active tension, which indicates that contraction produced components of force perpendicular to the long axis. The extreme ends are known to shorten whether or not the middle of a fiber is isometric or stretched. Shortening of the ends may shift water towards the middle, which could account for the volume changes we observed. The cytoskeletal matrices of muscle evidently adjust rapidly during contraction to maintain a dynamic equilibrium between the axial and radial forces that stabilize the whole cell. The Z disks have been shown to expand during active, but not passive, tension development. Z disks might be the elastic elements of the muscle cytoskeleton primarily involved in rapid balancing of the radial components of active force.

INTRODUCTION

Striation spacing is shorter near the ends of a stretched fiber than it is in its middle (Huxley and Peachey, 1961). The ends contract more strongly and shorten more than the middle (Figure 1). Therefore, to closely relate myofilament overlap to contractile tension, it usually is necessary to control length in the middle of a fiber where the resting striation spacing is uniform (Gordon, Huxley, and Julian, 1963).

There are several important implications of axial non-uniformities and the re-distribution of axial segment lengths during contraction. These have been considered by many investigators insofar as they aid understanding how

Excitation-Contraction Coupling in Skeletal, Cardiac, and Smooth Muscle
Edited by G.B. Frank *et al.*, Plenum Press, New York, 1992

sarcomeres produce force in the plane parallel to the long axis of a muscle fiber (e.g., Huxley, 1980; Edman and Reggiani, 1984; Edman et al., 1985; Huxley, 1988; Pollack, 1990). Dynamic changes in components of force directed radially with respect to the contractile filaments have been more difficult to measure directly, and thus are less well understood.

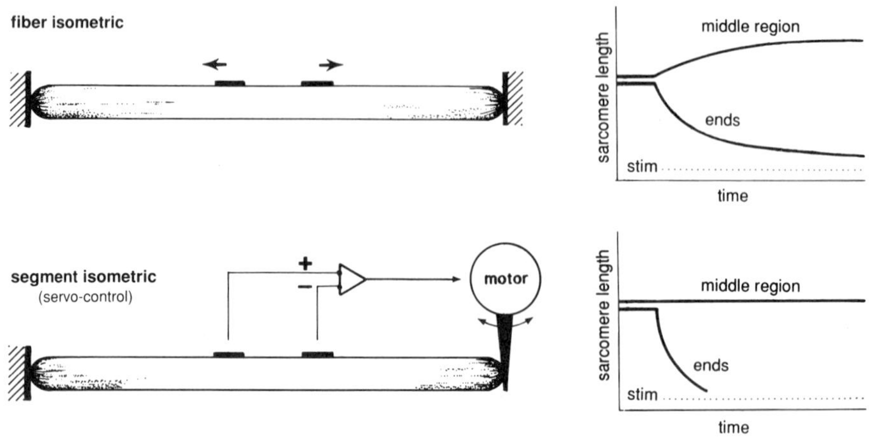

Figure 1. Axial movements of middle and end segments during tetani. The top diagram shows the shortening of the ends and lengthening of the middle that occurs during isometric contraction of a stretched fiber (Huxley and Peachey, 1961). Tendon to tendon length is held constant. The bottom diagram shows the shortening of the ends when one tendon is moved under servo-control to keep the length of a middle segment constant (Gordon, Huxley, and Julian, 1963). [Figure 12.14 from G. H. Pollack, 1990, with permission]

Live skeletal muscle cells are relatively thick, long objects for microscopy. Their cross-sectional shape varies so unpredictably that measurement of width alone leads to gross errors in calculating area and volume (Blinks, 1965). Hence we used an imaging system that makes no assumptions about cross-sectional shape or changes in cross-sectional area (Quesenberry et al., 1991) and measured these parameters while simultaneously measuring contractile force (Neering et al. 1991).

Some of the advantages, limitations, and sources of possible artifacts of our methods are discussed in an article where representative results were first shown (Neering et al., 1991). Here we focus on two additional points: (i) a comparison with a related study (Sato, 1954) that appeared before segment-to-segment variation was a well-known phenomenon; (ii) an expansion of some issues discussed only briefly or implicitly (Neering et al., 1991; Quesenberry et al, 1991), which support our conclusion that the results are not owing to possible artifacts of the methods.

METHODS

Image Acquisition

The opto-mechanical apparatus used to perform the experiments was loaned to us by J.R. Blinks (1965). Cross-sectional images along 1 mm segments of vertically-mounted frog fibers were made at rest, and for 10 seconds during the plateau of isometric tetani. Resting fibers and tetanized fibers were moved up or down under remote control through a slit of xenon light. When tetanic force developed by a fiber reached a plateau, a micro-stepper moved the fiber through the slit of light, while one image was taken for each 25 μm of movement. Movement was synchronized with the taking of images and simultaneous recording of force. The distance between the immersion objective and sensor was adjusted to keep a fiber constantly in the field of view if it shifted laterally as it was moved up or down, or moved during contraction. The spatial and temporal resolution of the imaging apparatus are described in Quesenberry et al. (1991).

Image Enhancement

Contöur maps of regions were formed manually and automatically, before and after computer enhancement (Neering et al., 1991). However, to facilitate discussion of potential artifacts of the methods, none of the images shown in this manuscript was enhanced or processed. When we refer to original data or unprocessed images in a figure we mean data that have not been subjected to background subtraction, histogram equalization, normalization, spatial filtering, or thresholding, either at fixed or variable threshold values. The images that we used for the most critical inspection had the minimum and maximum brightness values determined automatically. Then all values between the two extremes were mapped in a linear fashion.

Limitations of Human Vision

Dim stimuli must be presented for long periods of time or be relatively bright for the human visual system to detect them. The eye also has difficulty distinguishing between an edge that varies slowly in brightness, compared with an edge with a very abrupt transition in a black and white image. Edge variation may be smooth and gradual and adjacent regions of uniform brightness. But the human visual system overshoots around boundaries. The apparent slope of the transition appears steeper than it really is, and adjacent regions will appear extra dark or extra bright to the observer of a monochrome image when each region is actually uniform. This arises from the Mach band effect. This tendency of the eye to overshoot is also inherent in many digital filters. But the artifacts introduced by oscillations in a computer-enhanced image are detectable and measurable, especially when the gray scale values are assigned different colors for each value rather than displayed merely in black and white (Inoué, 1986; Gonzalez and Wintz, 1987).

In pseudo-color coded images, brightness values from black to white (0 to 4,096) are assigned wavelengths that range from short to long (380 to 700 nanometers). The relative sensitivity of human perception increases with brightness in a pseudo-color display. But even after the eye adapts to bright conditions (photopic vision), it is less than one-tenth as sensitive to the blue-violet part of the spectrum as it is to yellow. Complement of pseudo-color (e.g., Figure 3), maps the same brightness values with the wavelength range inverted. This makes the **dimmest** values in the image correspond to wavelengths that the human eye detects with the greatest sensitivity (Ballard and Brown, 1982; Inoué, 1986). Black and white images can be very unreliable for purposes of measurement by either human vision or a computer. Some processing was required to produce the values for area plotted in our previous publication (Neering et al., 1991). However, our basic findings are qualitatively evident without any processing (Figures 2 and 3) and, therefore, were not created by manipulation of the original data.

RESULTS

Figure 2 is a monochrome display of a fiber at rest and during contraction. It shows some of the limitations of using human vision alone to evaluate monochrome images. The top row shows three images along a resting fiber. These are the original data (4,096 possible gray scale values). Peak brightness in each image from left to right is 3,350, 3,347, and 3,317. The bottom row shows the same sections during the plateau of a tetanus, and the corresponding values for peak brightness are **lower**, 3,319, 3,309, and 3,282.

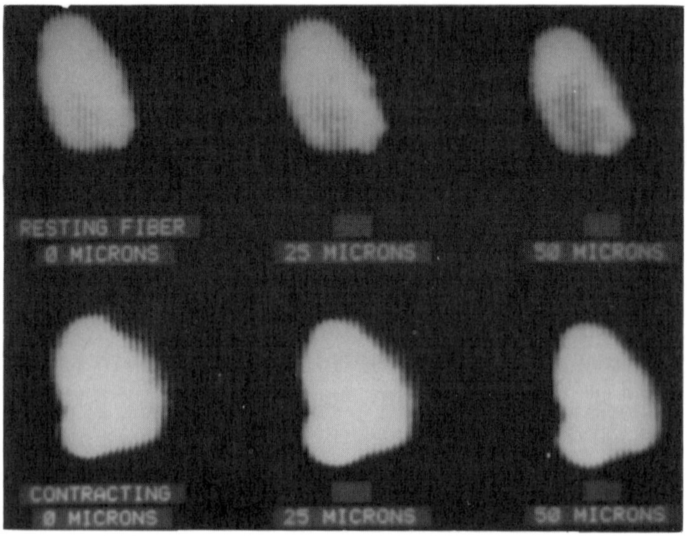

Figure 2. Monochrome images of the area near the middle of a muscle fiber. The top row shows the fiber at rest, at three consecutive planes each separated by 25 μm along the fiber's axis. The bottom row shows the same regions during the plateau of a tetanus with the tendons held stationary (Figure 6 from Quesenberry et al., 1991, with permission).

One can estimate contöur lines around regions in Figure 2 by eye, trace them manually, and compare areas. Alternatively, the total number of screen elements with gray scale values greater than zero can be counted automatically and assumed to represent the area. However, if one moves a measuring cursor through areas that appear black, or if one assigns colors to different gray scale values, it is immediately apparent that Figure 2 contains far more information and variation than is visible.

An automatic method of measuring un-processed images like those in Figure 2 would also count a substantial amount of low-level scattered light that varies markedly with no predictable pattern. A comparison of Figure 2 with Figure 3, allows one to see the change in shape, the increase in area, and the increase in **average** brightness of a contracting fiber, while also showing that simply counting all non-zero values in an unprocessed image introduces a large number of additional values that are not related to the cross-sectional area of the fiber. Separate regions in a color picture can be identified with far greater accuracy than regions in a black and white picture (Duda and Hart, 1973). We previously showed these same images in pseudo-color and compared them with images clipped at a fixed threshold (Neering et al., 1991). The comparison here between the same images in black and white with the inverse of pseudo-color and no thresholding, reaffirms our conclusion that neither scaling nor clipping by themselves can produce the apparent increase in cross-sectional area.

The human visual system now has more than just brightness information on which to make judgments (Figure 3). The observer now distinguishes areas on the basis of their saturation, that is, lack of whiteness in the color, and on the basis of hue, which is roughly proportional to the average wavelength of the color (Ballard and Brown, 1982; Inoué, 1986).

Some of the scatterers are particles that adhere to the fiber. Some scatterers are within the fiber itself. In these particular images, the brightest individual values are in the resting fiber. Hence the change in shape and increase in area during contraction can not be produced by simply scaling the top and bottom rows of images to the same peak value. The average brightness of the images often increased substantially in a contracting fiber and this change, produced by some unknown change in fiber properties, is qualitatively apparent even in Figure 2.

DISCUSSION

Comparison With a Related Study

Sato (1954) studied the changes in tetanic contractions of single frog fibers and concluded that contraction caused a volume **de**crease of unexpected magnitude. Sato chose to image a part of the fiber where the site of measurement did not change. Hence he always viewed the upper fixed end, though he noted this raised the question, do these results represent the volume of the whole fiber? Inspection of Table 1 in

Sato's paper, as well as images in his Figure 2, show clearly that the cross-sectional area increased substantially in every fiber he examined. He focused on the volume change calculated from measurements of the degree of shortening, and concluded that without a single exception the volume decreased during contraction. Sato did not have the benefit of the many subsequent studies that show the ends shorten substantially more than the middle. If he did, the points he raised in his own introduction might have led him to question his conclusion.

Sato noted that studies that do not involve direct imaging of an isolated single fiber can include water shifts between compartments that elude measurement. Our observations of the swollen middle of a fiber in tetanus, where the ends presumably shorten and increase in cross-sectional area but decrease in volume, might result from water shifts between the ends and middle of a fiber. A small net shortening of the middle occurs at optimum myofilament overlap. However, in contrast to this situation, there is an overall elongation of the middle and marked shortening of the ends when a fiber is stretched by about 20% beyond the length at which actin and myosin filament overlap is optimal (Edman and Regianni, 1984). Our experiments were done at the latter length, and water shifts presumably occurred in association with marked shortening of the ends (Neering et.al, 1991).

Water Shifts Among Compartments

There are at least two ways that different segments of a muscle cell can change volume rapidly and reversibly in contraction. One way involves water movement from an extracellular to an intracellular compartment. The other way involves movement of water among segments of a fiber.

The intracellular solute concentration might change. For example, a minute and a half of repetitive stimulation produces an increase in cross-sectional area that takes many minutes of rest to recover (Lännergren, 1989). Fluid may enter the cell and increase its volume until the effects of increased metabolic activity are reversed by the return of the number of osmotically active particles to the pre-stimulus condition. The liquid associated with these volume changes probably comes from the bathing medium. The metabolically generated influx of extracellular water might have contributed to the relative size of our area changes when some segments were viewed 10 seconds later in one tetanus than another. However, this can not account for the reproducible variation in segments from different fibers we examined, nor can it account for the reproducible variation in different segments of the same fiber (Figure 4).

Periods of intense mechanical activity, such as those induced by repeated tetanic stimulation, noticeably depress force in skeletal muscle. We allowed periods of rest between tetani that were at least 10 minutes long, sufficiently long to recover from any measurable effect on maximum tetanic tension. Nevertheless, several possible changes could have developed between the beginning and end of a 10 second imaging period on the plateau of a given tetanus (Figure 4). Among the possibilities suggested to account for fatigue are

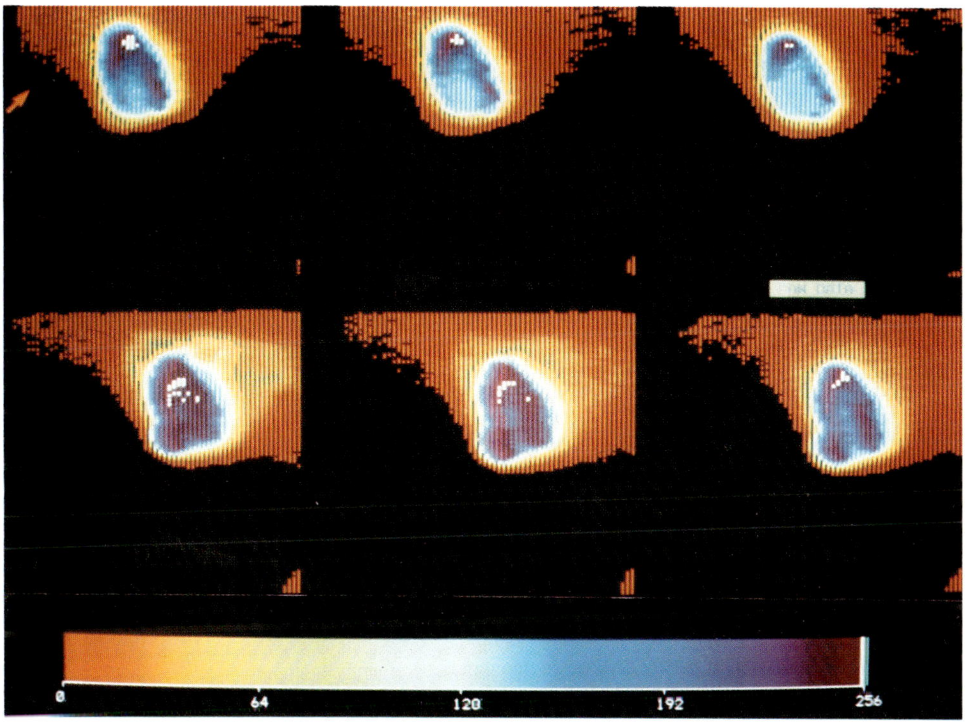

Figure 3. Inverse of pseudo-color display of the same images in Figure 2. The bar = 100 μm. The original data have not been altered in any way. A different hue was assigned to each gray scale value in the images and inverted so the dimmest values had wavelengths the human eye detects with the greatest sensitivity. This reveals substantial complexity in the patterns of scattered light in and around the fiber, which is not detectable in the same images displayed as they are in Figure 2.

the deleterious effects of accumulated metabolic products, and T-tubule conduction failure owing to changes in ionic composition in the T-tubule lumen (Howell and Oetliker, 1987; Westerblad and Lännergren, 1991). A small rise in extracellular [Ca^{2+}] and [K^{+}] or a decrease in [Na^{+}] in the transverse tubular system is accompanied by buckling of myofibrils in the center of shortening fibers. This change is consistent with the idea that the action potential may spread without decrement from the surface to the center at the beginning, but not the end, of a long tetanus (Howell et al., 1987). We assumed that the pattern of intracellular calcium release had reached a steady state during trains of low-frequency stimulation applied after long periods of rest (Blinks et al., 1978). We can not rule out the possibilities that changes related to the accumulation of ions in the T-system and decremental conduction of the action potential, or changes related to metabolic activity and an increase in osmotically active particles, played **some** role in determining the variation in the relative increase in cross-sectional area among segments. However, if differences in the kinetic properties from one region to another along the length of the fibers are involved, these differences are not owing to gradients of activation or regional differences in the composition of destabilizing solutes produced by fatigue (Edman et al., 1985).

The general pattern of change for a given segment was similar whether we began imaging that segment near the beginning or near the end of the 10 second period (Figure 4). The changes were unique for each fiber and each fiber segment. The pattern of optical changes during a scan in one direction, compared with a scan in the opposite direction, was virtually the same for a given segment in back-to-back contractions over a period of several hours. Hence we concluded that the volume increase of the middle of a fiber is not markedly time dependent, at least within the order of time of our measurements.

What Parts of a Fiber Might Balance Dynamic Radial Forces?

The diversity of volume changes (Figure 4) suggests that even if rapid re-distribution of water between regions or compartments is involved, some difference in elastic elements spanning or surrounding the fiber may account for the re-distribution of forces. The cytoskeleton, which evidently stabilizes the structure of sarcomeres at rest (Pierobon-Bormioli et al., 1989; Funatsu et al., 1990), may also stabilize a fiber as force changes dynamically during repeated, rapid cycles of contraction and relaxation.

The unique, unpredictable, axial translations that occur during contraction may reflect inherent differences in contractile strength and width along a fiber (Edman and Regianni, 1984; Edman et al., 1985). The possibility that the performance of individual segments in isolated living muscle cells may be related to differences in radial components of contractile force arose when we scanned fibers transversely down their length, rather than along their length only. Contraction may produce outwardly directed radial forces that cause the middle of a fiber's cross-sectional area to

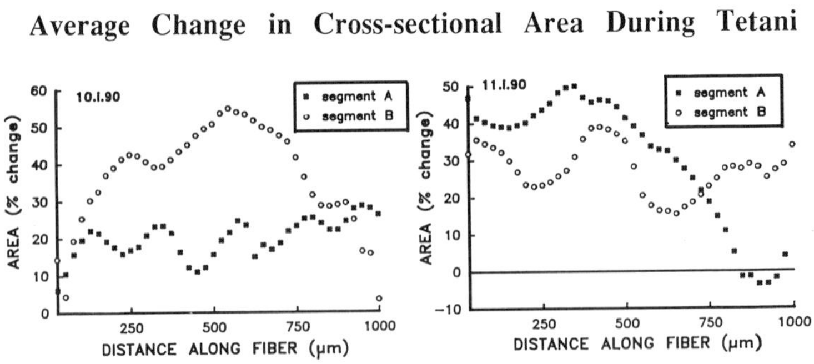

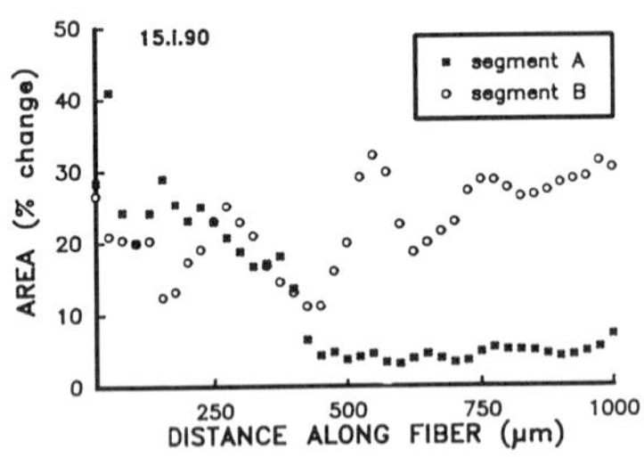

Fiber	Segments Either Side of a Fiber's Center	Tetanus (time)
10.i.90	segment A " A " A " A segment B " B	17:04 17:29 18:05 18:32 13:33 13:44
11.i.90	segment A " B " B	18:43 14:21 14:45
15.i.90	segment A " A segment B " B " B	15:42 16:38 13:29 13:41 13:54

Figure 4. Diversity of volume changes: patterns of change during contraction of two separate regions in three different fibers. The table shows the time of each tetanus, the number of tetani averaged to produce the data points, and the arbitrary designation of a segment as A or B to indicate they were different regions in the middle 2 mm of each fiber. Each fiber was imaged for a distance of 1 mm (= 10 seconds) along its axis in each tetanus. Images were taken in the opposite direction after a period of rest and this procedure was repeated several times. The fiber was then moved in the slit of light to image a different segment. The total length viewed was ± 1 mm on either side of the approximate middle.

increase without shortening (Neering et al., 1991). This is consistent with a conclusion drawn from optical diffraction studies of the Z-disk lattice that show muscle cells develop outwardly directed radial forces during maximal contraction, but not during passive stretch (Goldstein et al., 1988).

Figure 5 shows the actin filament lattice in the zone where actin overlaps myosin, compared with the actin lattice in the zone where actin approaches attachment to the Z disk in relaxed muscle (Hoyle, 1983). As the thin filaments extend from the overlap zone towards Z disks at the ends of each sarcomere, their lattice arrangement changes from a hexagonal pattern to one that is predominantly small squares. Connecting filaments that run from the end of myosin to the nearest Z disk evidently keep the thick filaments centered. They stabilize the resting sarcomere. But passive stretch of the connecting filaments does not change the Z disk in a manner that accounts for the changes during contraction. Expansion of the small-square actin lattice in contracting

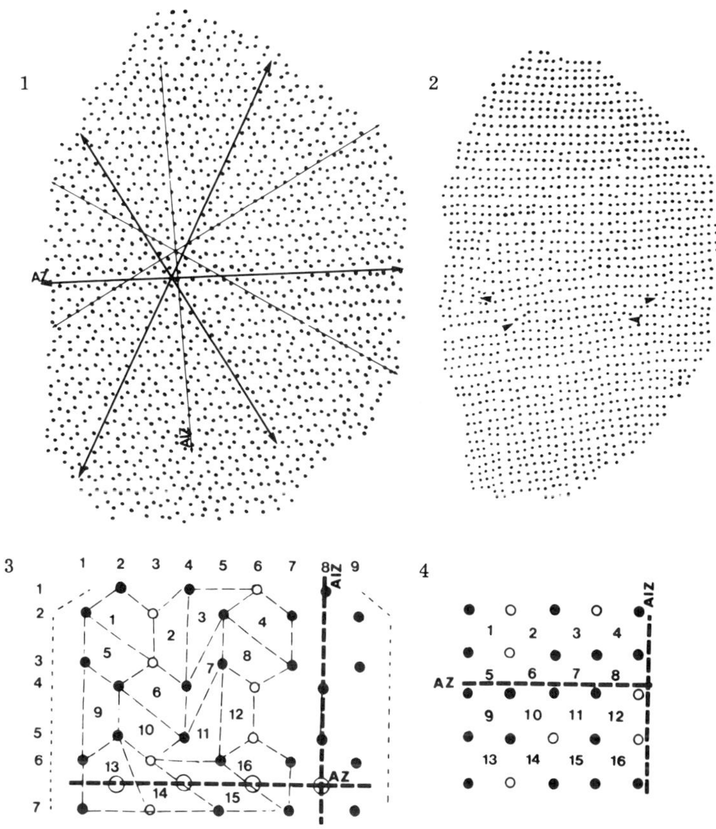

Figure 5. Comparison of the actin filament lattice in the basically hexagonal array of the myosin overlap region (upper left), with the square array just outside the Z disk at one end of a sarcomere (upper right). Diagrammatic interpretation of the transition between numbered sets of four actins in the overlap zone (lower left), and the same sets in the Z disk region (lower right). [Figure 3.27 from G. Hoyle, 1983, with permission]

muscle suggests that a tetanus **must** induce forces with perpendicular components that are not produced by passive stretch (Goldstein, et al., 1988). These observations can be explained if cross-bridge attachment and the power stroke in portions of the myosin molecule that push actin past myosin, causes oblique pulling on both sides of each Z disk by actins that are not co-linear.

The possible functional significance of the Z disk has long been appreciated, and the relationship between active tension development and slow changes in Z-disk width are well known (e.g., Eisenberg, 1983). Z-disk width is evidently one of the best discriminators of a functional relationship between a microscopic anatomical system in fixed muscle and the contractile system. The changes we have seen in individual contractions of living muscle that we attribute to a quick-acting mechanism of auto-regulation in the cytoskeletal matrices, may also have long-term functional implications. As has been suggested, perhaps when the mechanical stresses and strains borne by the Z disks are known, it will be possible to understand how these structures adapt to them and become specialized in a variety of circumstances such as during chronic stimulation, after tenotomy, in different fiber types, and during growth and exercise when myofibril splitting into discrete, contrasted with branching structures is related to fast versus slow contraction (Eisenberg, 1983; Goldspink, 1983).

REFERENCES

Ballard, D. H., and Brown, C. M., 1982, "Computer Vision," Prentice-Hall, Englewood Cliffs, NJ.

Blinks, J. R., 1965, Influence of osmotic strength on cross-section and volume of isolated single muscle fibres, J. Physiol. 177, 42.

Blinks, J. R., Rüdel, R. and Taylor, S. R., 1978, Calcium transients in isolated amphibian skeletal muscle fibres: Detection with aequorin, J. Physiol. 277:291.

Duda, R. O., and Hart, P. E., 1973, Region analysis; extensions, in: "Pattern Classification and Scene Analysis", John Wiley and Sons, New York, NY.

Edman, K. A. P., and Regianni, C., 1984, Redistribution of sarcomere length during isometric contraction of frog muscle fibres and its relation to tension creep. J. Physiol., 351:169.

Edman, K. A. P., Reggiani, C., and te Kronnie, G., 1985, Differences in maximum velocity of shortening along single muscle fibres of the frog, J. Physiol., 365:147.

Eisenberg, B. R., 1983, Quantitative ultrastructure of mammalian skeletal muscle, in: "Handbook of Physiology: Skeletal Muscle", American Physiological Society, Bethesda, MD.

Funatsu, T., Higuchi, H., and Ishiwata, S., 1990, Elastic filaments in skeletal muscle revealed by selective removal of thin filaments with plasma gelsolin, J. Cell Biol. 110:53.

Goldspink, G., 1983, Alterations in myofibril size and structure during growth, exercise, and changes in environmental temperature, in: "Handbook of Physiology: Skeletal Muscle", American Physiological Society, Bethesda, MD.

Goldstein, M. A., Michael, L. H., Schroeter, J. P., and Sass, R. L., 1988, Structural states in the Z band of skeletal muscle correlate with states of active and passive tension, J. Gen. Physiol. 92:113.

Gordon, A. M., Huxley, A. F., and Julian, F. J., 1963, Apparatus for mechanical investigations on isolated muscle fibres, J. Physiol. 167:42P.

Gonzalez, R. C., and Wintz, P., 1987, "Digital Image Processing", Addison-Wesley Co. Inc., Reading, MA.

Howell, J. N., and Oetliker, H., 1987, Effects of repetitive activity, ruthenium red, and elevated extracellular calcium on frog skeletal muscle: implications for t-tubule conduction, Can. J. Physiol. Pharmacol. 65:691.

Hoyle, G., 1983, Structural architecture of complete cross-striated sarcomeres, in: "Muscles and Their Neural Control", John Wiley & Sons, NY.

Huxley, A. F., 1980, Uncertainties about "independent force-generators", in: "Reflections on Muscle," Princeton University Press, Princeton, NJ.

Huxley, A. F., 1988, Prefatory chapter: Muscular contraction, Ann. Rev. Physiol. 50:1.

Huxley, A. F., and Peachey, L. D., 1961,The maximum length for contraction in vertebrate striated muscle, J. Physiol., 156:150.

Inoué, S., 1986, Physiological characteristics of the eye, in: "Video Microscopy," Plenum Press, New York, NY.

Lännergren, J., 1989, Volume changes of isolated *Xenopus* muscle fibres associated with repeated tetanic contractions, J. Physiol., 429:116P.

Neering, I. R., Quesenberry, L. A., Morris, V. A., and Taylor, S. R., 1991, Non-uniform volume changes during muscle contraction, Biophys. J., 59:926.

Pierobon-Bormioli, S., Betto, R., and Salviati, G., 1989, The organization of titin (connectin) and nebulin in the sarcomeres: an immunocytolocalization study, J. Musc. Res. Cell Motil., 10:446.

Pollack, G. H., 1990, Generators and sustainers: the tension-length relation, in: "Muscles and Molecules: Uncovering the Principles of Biological Motion," Ebner and Sons, Seattle, WA.

Quesenberry, L. A., Morris, V. A., Neering, I. R., and Taylor, S. R., 1991, Reduced defocus degradation in a system for high-speed 3D digital microscopy, Proc. Internatl. Soc. Photo-Optical Instr. Eng., 1428:177.

Sato, T. G., 1954, Volume change of a muscle fiber on tetanic contraction, Annot. Zool. JPN., 27:165.

Westerblad, H. and Lännergren, J., 1991, Slowing of relaxation during fatigue in single muscle fibres. J. Physiol., 434:323.

ANALYSIS OF FACTORS INVOLVED IN THE TISSUE SELECTIVITY OF CALCIUM ANTAGONISTS

Théophile Godfraind

Laboratoire de Pharmacologie,UCL 7350, Université Catholique de Louvain, Avenue E. Mounier 73, B - 1200 Brussels, Belgium

INTRODUCTION

In functional studies, the term tissue selectivity is used for an agent showing varying degrees of potency among tissues with a preferential action in a given one. Such a selectivity is well known in the field of β-adrenoceptors, where β_1 and β_2 agonists and antagonists, differ by the ratio of activity at the level of a given tissue. Usually, this is related to their respective affinities for β_1 and β_2 adrenoceptors. Selectivity is not specificity which means that a given agent interacts with only one type of receptor. Such an interaction is largely documented in the field of histaminergic receptors where H_1 antagonists do not interact with H_2 receptors and vice versa. A question that has been raised several times (Godfraind et al., 1986), is the existence of a possible tissue selectivity of calcium antagonists. In other words, does any selectivity exist in the pharmacology of calcium channels ?

When examining the effects of calcium antagonists in various isolated preparations, several pharmacological profiles may be delineated considering the main determinants: reported on Table 1.

It is obvious that several factors are involved, which needs an extensive analysis. This analysis is beyond the scope of this paper.

Excitation-Contraction Coupling in Skeletal, Cardiac, and Smooth Muscle
Edited by G.B. Frank *et al.*, Plenum Press, New York, 1992

Therefore, this presentation will address some typical examples of this diversity including : a) the contribution in vascular and cardiac preparations of various Ca pools to final response, b) the influence of membrane potential on the interaction of some Ca^{2+} antagonists with their receptors, c) the evidence that inhibition by Ca^{2+} antagonists of vascular contraction evoked by noradrenaline is related to their interaction with potential operated calcium channels, d) the evidence that the large difference in sensitivity between cardiac and vascular tissues, documented for a dihydropyridine such as nisoldipine, may be related to the kinetic parameters describing the interaction of Ca^{2+} antagonists with Ca^{2+} channels.

Table 1

1. Characteristics of the drug :
   Ca channel selectivity (L,T,N type), Site of action within the channel, Selectivity window, Voltage and/or frequency dependency, Ratio Ca antagonistic/Ca agonistic action, Physicochemical properties (pKa, partition coefficient), Susceptibility to tissue metabolism
2. Properties of the tissue :
   Relative contribution of various Ca pools to final response, Density and types of Ca channels, Level of resting potential, Gating control of Ca channels
   Pharmacokinetic determinants
3. Characteristics of the stimuli :
   Level, duration and frequency of depolarization, Nature of activated kinase(s)

1. Role and importance of calcium pools

a) Ca^{2+} antagonists and Ca channels

Calcium antagonists are agents which inhibit the contraction evoked by calcium in depolarized smooth muscle in a manner similar to the action of antagonists observed in receptors studies. Such an effect was initially described in my laboratory in the 1960's with agents that may be considered as the pharmacological prototypes of the drugs used now in therapy. At that time, we studied the action of polyvalent antagonists, that

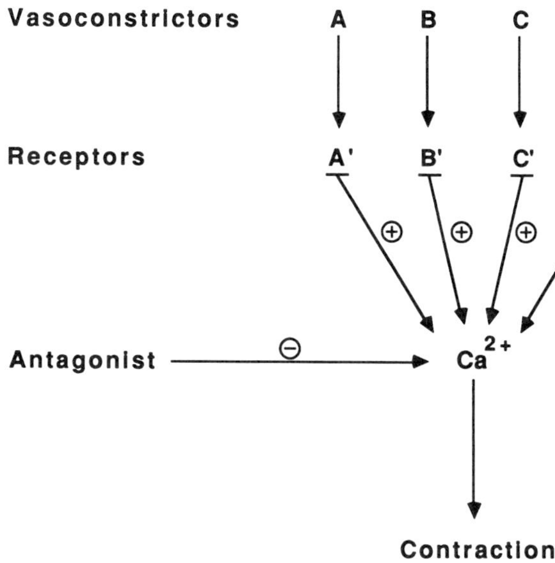

Fig. 1. Inhibition of Ca^{2+} entry: mechanism of drug action (Godfraind et al. 1966, 1968, 1969,...).

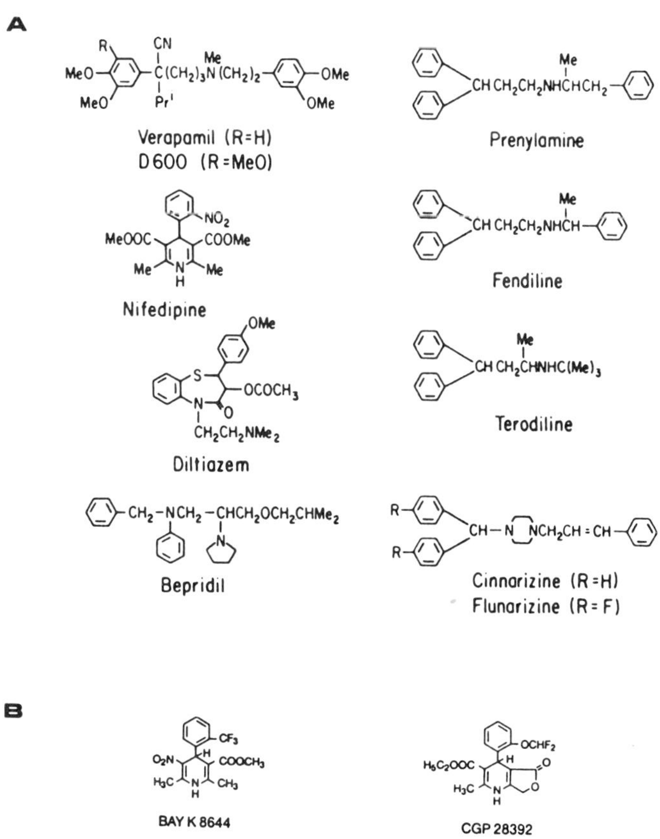

Fig. 2. Structure of some calcium entry blockers (A) and calcium agonists (B). *Me*, methyl; Pr^i, isopropyl.

are able to block in a same concentration range the vasoconstriction evoked by agents acting on different receptors. We proposed the hypothesis of a common calcium entry process activated by various vasoconstrictor agonists (Fig. 1). As illustrated by Fig. 2, those agents belong to different chemical classes which are apparently unrelated one from the other. Also among of dihydropyridines, two groups of agents with opposite action have been identified (for references, see Godfraind et al., 1986).

Opening of calcium channels is a crucial step of excitation-contraction coupling both in cardiac and smooth muscle tissues. In these tissues, depolarization opens two different kinds of voltage gated calcium channels : L and T channels which differ by their kinetic and voltage sensitivity of activation and inactivation (Tsien, 1983). L type channels are blocked by calcium antagonists belonging to the 1,4- dihydropyridines (nifedipine), to phenyalkylamines (verapamil), to benzodiazepines (diltiazem) and to diphenylpiperazines (flunarizine). T-type channels are relatively insensitive to those agents, flunarizine being the most active. These drugs do not act by physical plugging L-channels but rather they decrease the probability of active channels. Ca^{2+} agonists, such as BAY K 8644 do increase this probability (for ref. : Godfraind et al., 1986).

A third sub-type of voltage gated calcium channel identified in patch clamp studies is the N-channel which is found in neuronal tissues and seems to be involved in neurotransmitter release. It is blocked by Omega conotoxin. P-channels have been differentiated on the basis of sensitivity to funnel web spider toxin and are the least well characterized. As suggested by Mc Kenna et al. (1990), the existence of other channels and/or of subtle modifications in the structure of the already identified calcium channels which means the existence of isoforms of calcium channels is indicated by molecular biology studies. But at the present time, there is not enough information to discuss the significance of iso-channels in terms of functional experiments.

L-channels identified by their capacity to bind selective ligands are present in membranes isolated from various tissues. In vascular smooth muscle, it may be assumed that they represent the major calcium pathway opened by tissue depolarization. Stereoselective, high affinity binding sites for the main chemical classes of calcium antagonists have been demonstrated in membranes isolated from various excitable tissues including smooth muscle. If we consider the various chemical classes, it

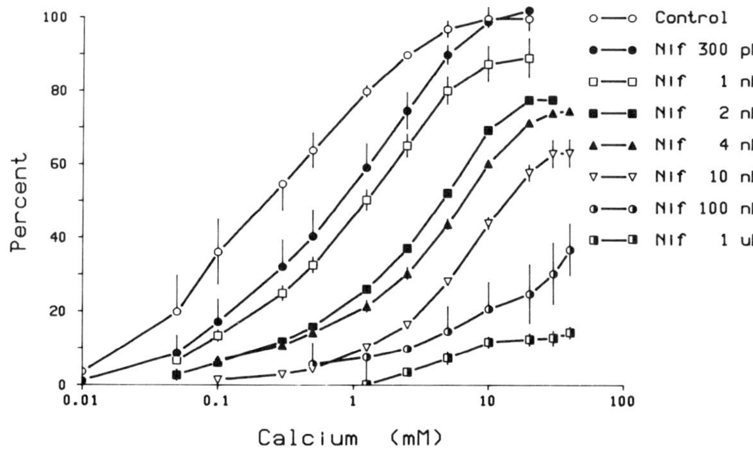

Fig. 3. Effect of nifedipine on contractions evoked by Ca^{2+} in K^+-depolarized rat aorta. Cumulative concentration-effect curves were obtained before and after addition of nifedipine (Nif) at the concentrations indicated. Responses are expressed as the percentage of maximal contraction evoked in the absence of nifedipine.

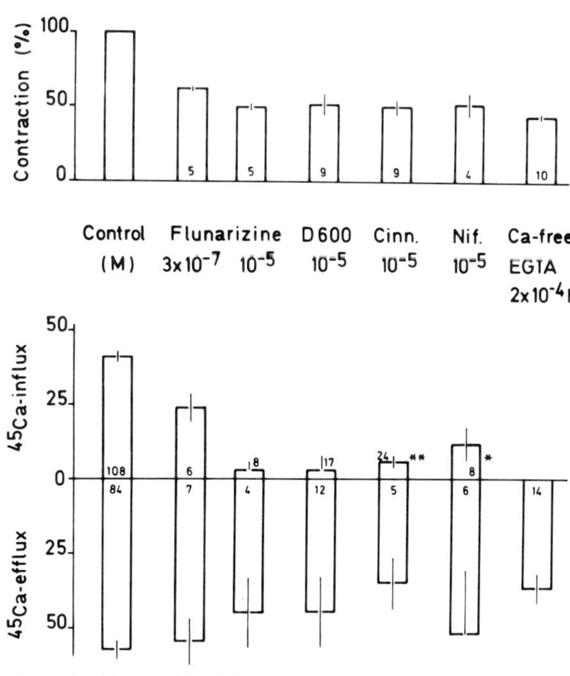

Fig. 4. Effect of calcium entry blockers and of Ca^{2+}-free medium on contraction (top) and ^{45}Ca influx and efflux (bottom) evoked by norepinephrine (10 µM) in rat isolated aorta. Tissues were preincubated with blockers for 90 min. or with EGTA for 5 min., before addition of norepinephrine. Contractions are expressed as the percentage of the contraction induced in the absence of drug (control). *Numbers in columns*, number of determinations; *vertical bars*, SE. *Cinn.*, cinnarizine; *Nif.*, nifedipine.

appears that their binding sites on the calcium channels are distinct but interact allosterically.(Godfraind at al., 1986)

b) <u>The action of calcium antagonists in isolated arteries</u>

Figure 3 illustrates a typical experiment performed with nifedipine in isolated rat aorta. The tissue was immersed in a depolarizing calcium-free solution and the contraction was elicited by increasing the concentration of calcium in the perfusion fluid before and after treatment with nifedipine. The dose/effect curves obtained by this cumulative increase of calcium in the muscle chamber are displaced to the right with their maximum being progressively depressed by increased concentrations of nifedipine. That inhibition of calcium-evoked contraction in depolarized vascular smooth muscle is related to inhibition of calcium entry, as illustrated by the superimposition of the dose/effect curves relating the action of nifedipine on ^{45}Ca entry and on contraction evoked by calcium in depolarized arteries (Godfraind, 1983).

As initially reported by Godfraind and Kaba, 1969, the contractile response of vascular smooth muscle to noradrenaline may be divided into a component that is sensitive and a component that is resistant to calcium antagonists, an observation confirmed several times (Dunn et al. 1991) Figure 4 illustrates that unidirectional calcium efflux evoked by noradrenaline is barely affected by the presence of those agents. In the absence of external calcium, the release of calcium from the tissue evoked by noradrenaline has the same magnitude as the one observed in the presence of calcium antagonists. By contrast, noradrenaline-evoked calcium entry is depressed dose-dependently by calcium entry-blockers but a small fraction of calcium entry is resistant to blockade by calcium antagonists.(Godfraind and Miller, 1983).

In rat mesentric arteries, the contraction resistant to calcium antagonists is much smaller than in rat aorta indicating heterogeneity of calcium control within the vascular bed. Dunn et al (1991) have shown that the capability to produce activation of adrenoceptor and calcium entry is not related to the subtype of a-adrenoceptors, since α_1- and α_2-adrenoceptors produce similar effects.

The contractile activity of neonate rat hearts is more depressed by calcium antagonists than that of adult rat hearts. This difference in sensitivity may be related to the observation that in neonate hearts

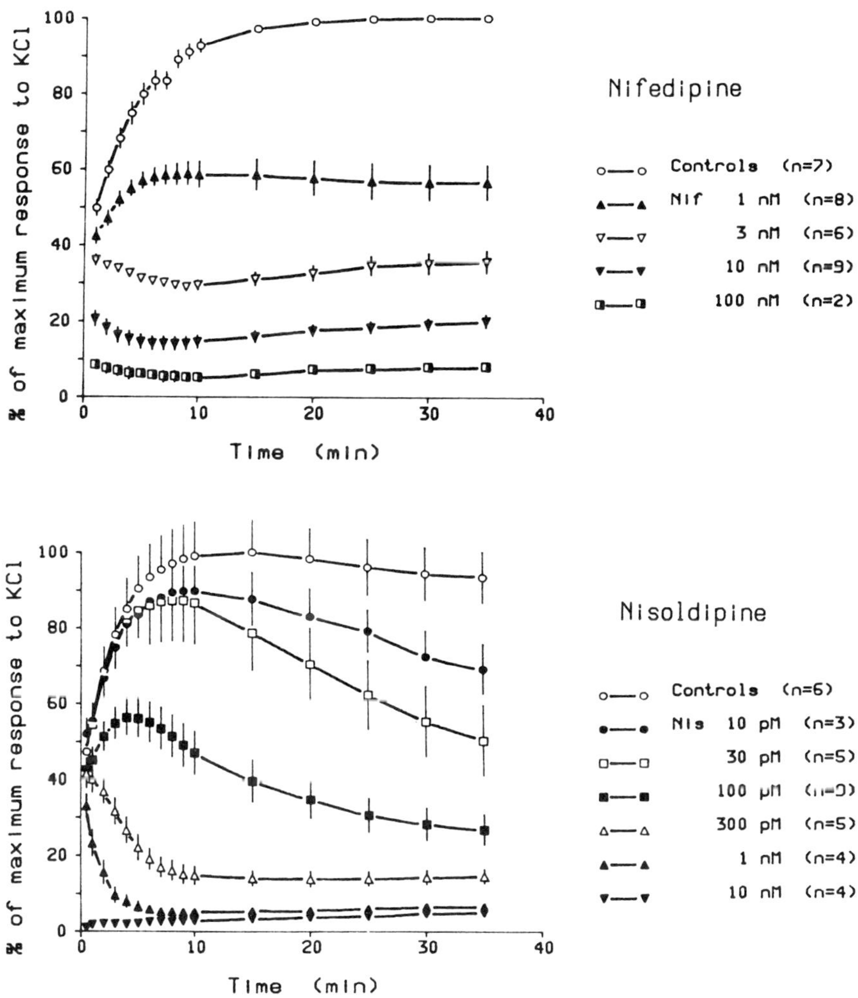

Fig. 5. Time-course of the contraction of rat aorta evoked by K-depolarization before (control) and after 90 min. preincubation with various concentrations of nifedipine and nisoldipine. Note the difference in time course pattern between the two dihydropyridines.

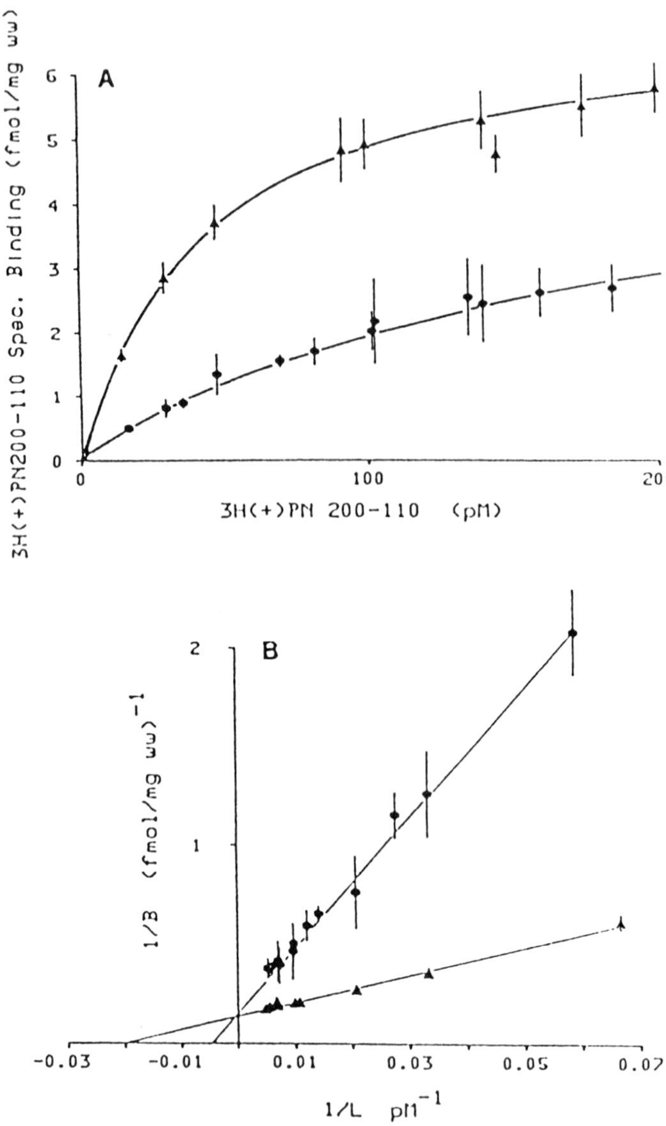

Fig. 6. Binding of (+)-[^{3}H]PN 200-110 to isolated arteries. A, saturation study of the specific binding of (+)-[^{3}H]PN 200-110 in polarized (●) and depolarized (▲) arteries. Each point represents measurements from 6 to 18 rings ± S.E.M. Curves are least squares fit of the function $B = B_{max}L/(L + K_D)$ with B = specific binding, L = concentration of free (+)-[^{3}H]PN 200-110 in the medium, K_D = apparent dissociation constant, obtained with curve fitting computer program for fitting nonlinear models (MLAB). Values of the binding parameters are given in Table 1. B, double-reciprocal plot of (+)-[^{3}H]PN 200-110 specific binding data in polarized (●) and in depolarized (▲) arteries. Binding parameters calculated by linear regression were for arteries bathed in physiological solution: K_D, 220 pM, B_{max}, 6.7 fmol/mg wet wt. (correlation coefficient, 0.997); for arteries bathed in K^+-depolarizing solution: K_D, 49 pM, B_{max}, 7.2 fmol/mg wet wt. (correlation coefficient, 0.996).

calcium channels are associated with the plasma membrane, whereas in adults, Ca channels are predominantly associated with sarcoplasmic reticulum from which originates activator calcium (Wibo et al, 1991).

2. THE INFLUENCE OF MEMBRANE POTENTIAL ON THE INHIBITION BY CALCIUM ANTAGONIST

Figure 5 illustrates the pattern of the contractions of isolated rat aorta stimulated by potassium depolarization before and after pretreatment by various concentrations of either nifedipine or nisoldipine. The pattern with nifedipine ressembles the pattern oberved with verapamil and diltiazem and the pattern with nisoldipine is similar to that one oberved with other dihydropyridines such as (+)PN 200-110, amlodipine and lacidipine. With nifedipine, the inhibition of the contraction is immediate and sustained as long as the depolarization is maintained. By contrast, with nisoldipine, the inhibition requires some time of depolarization to be observed and it increases with the duration of the depolarization, to reach a steady state.

We have studied how membrane potential influences the specific binding of (+)PN 200-110 to intact arteries (Morel and Godfraind, 1987). Dose-effect curves for inhibition measured 2 and 30 min after initiation of depolarization are characterized by IC_{50} values respectively equal to 270 ± 65 pM and 33 ± 3 pM. (-)PN 200-110 is less potent than (+)PN 200-110 and its inhibitory potency is also time-dependent, IC_{50} values measured at 2 and 30 min in K^+-solution being respectively equal to 16 ± 4 nM and 5 ± 3 nM. Thus, the effect of PN 200-110 is stereoselective, the (+) enantiomer being markedly more potent than the (-) enantiomer ($P < 0.01$). (+)[^{3}H]PN 200-110 binding was determined in mesenteric arteries bathed in normal and high K^+-solutions. Non-specific binding is not different in polarized and in depolarized arteries. On the contrary, as shown in Fig. 6, specific binding is markedly enhanced in depolarized arteries. Experimental estimates of specific binding are well fitted by one hyperbolic curve indicating a one-to-one binding of (+)PN 200-110 to a single class of sites. Computer least-squares fitting of the data reveals that membrane depolarization significantly lowers apparent K_D ($P < 0.01$) while maximum binding capacity is not significantly affected ($P > 0.1$). In the reciprocal plot of the same data it clearly appears that both curves intersect at the same point on the 1/B axis, confirming that B_{max} value is not different in arteries bathed in physiological and in K^+ depolarizing solutions. K_D value obtained in depolarized arteries (44 ± 2 pM) is similar to that reported for (+)PN 200-110 in membrane preparations from other tissues and is close to the IC_{50} measured after 30 min of depolarization (see Table 2). These data show that like its pharmacological effect, the binding of PN 200-110 to intact tissue is stereospecific and voltage-dependent. Data reported on Table 2 illustrate the absence of voltage dependency of nifedipine also observed with verapamil and with diltiazem (Morel and Godfraind, 1989).

Table 2. Rat mesenteric artery : Comparison of pharmacological and radioligand binding data

Drug	Intact artery				Membranes
	IC_{50} (nM)		K_i (nM)		K_i (nM)
	2 min	30 min	Physiol. solution	K-rich solution	
(+)PN 200-110	0.270	0.033	0.410	0.025	0.055
(-)PN 200-110	16.0	5.0	23.0	1.0	-
Nifedipine	4.0	1.9	3.1	1.2	3.0
Flunarizine	37.0	2.0	unpercieved	900.0	500.0

K_i values (binding studies) and IC_{50} values (KCl-evoked contractions, 2 and 30 min), were reported by Morel and Godfraind (1987, 1988)

The dissociation constant estimated in K-depolarized intact arteries and in membrane corrolate well with value obtained in membrane and with the IC_{50} value measured after 30 minutes depolarization. Furthermore, the kinetics of inhibition during depolarization, is similar to the kinetics of binding to membranes isolated from the rat aorta (Fig.7) (Wibo et al., 1988). This indicates that depolarization induces a conformation of calcium channels, with increased affinity for dihydropyridines, and that voltage dependant dihydropyridines interact with this form in depolarized tissues rather than with Ca channels in resting tissues. On the other hand, the absence of a marked time-dependence in the effect of nifedipine can be accounted for by the interaction of nifedipine with calcium channels in resting tissues.

This interaction of calcium antagonists with L-channels is consistant with a modulated receptor model, according to which the channel can exist in three inconvertible states: the resting state (*R*), in which the channel is closed but available for opening, predominates in polarized cells; the open or activated state (*O*) is promoted by depolarization pulses beyond a certain threshold; and the inactivated state (*I*), in which the channel is closed but unavailable for opening, is favored by prolonged depolarization, even to potentials below the threshold for activation. A 1,4-dihydropyridine such as nisoldipine bind preferentially to the inactivated state of L-channels, since depolarized holding potentials that increase the proportion of inactivated channels, but fail to open them, enhance their inhibitory potency. As resting and inactivated channels are in equilibrium, drug binding to inactivated channels decrease the proportion of channels that are available for activation. If K_L and K_H are dissociation constants for L (low-affinity) and H (high-affinity) state, respectively, and L is the fraction of channels in state R in the absence of drug, the model predicts that, at any given holding potential, the concentration dependence of drug binding, and thus of Ca^{2+} influx inhibition, follows a simple adsorption isotherm with an apparent dissociation constant that is given by :

$$K_{app} = 1/[(L/K_L)+(1-L)/K_H] \qquad (1)$$

(Godfraind et al., 1990, Morel and Godfraind, 1991).

Results reported in this section are in agreement with previous results describing voltage sensitive binding of dihydropyridines in cardiac cell (Kokobun et al, 1986, White et al, 1989), in mesenteric artery (Morel and Godfraind, 1987), coronary artery (Sumimoto et al., 1988) and cerebral microvessels (Morel and Godfraind, 1989).

Quantitative analysis of electrophysiological data supports this model and indicates that in canine ventricle myocytes K_L and K_H for nitrendipine are respectively equal to 730 and 0.25 nM (Bean, 1984). In smooth muscle cells from rat mesenteric artery (Bean et al., 1986) and dog saphenous vein (Yatani et al., 1987), K_L values were estimated to be 222 and 10 nM and K_H values 0.46 and 0.25 nM, respectively. Apparent affinity of voltage-dependent Ca-antagonists to receptors on Ca channels in intact tissues is therefore related to resting membrane potential. The degree of initial inhibition of a vasoconstrictor stimulus by a given Ca antagonist may therefore be different among various vessels, since resting membrane potential is varying along the vascular tree. For instance, cerebral microvessels show a resting potential much lower than peripheral vessels (Morel and Godfraind, 1989).

3. Evidence that the inhibition by calcium antagonists of the response to vasoactive agents is related to their interaction with voltage operated calcium channels

In many tissues, calcium antagonists have been reported to be less potent inhibitors of the contraction induced by agonist stimulation than by depolarization (Cauvin et al., 1983; Marriot, 1988, Godfraind et al. 1986). This observation lead to the hypothesis of the existence of two different types of calcium channels, one controlled by the occupation of specific receptors and the other controlled by membrane potential (Cauvin and van Breemen, 1985).

We have recently analyzed the effect of nisoldipine in rat aorta and have observed that although agonists and depolarization-induced responses have different sensitivities to calcium-entry blocking drugs, this does not necesseraly means that responses are mediated by different calcium channels. It is worth noting that while the potency of nisoldipine is higher against a 100 ml KCl induced-contraction than against a noradrenaline-induced contraction when determination of inhibition is performed after prolonged contraction, there is no significant difference between IC_{50} values of nisoldipine on ^{45}Ca entry or contraction-induced by 100 ml KCl solution or by noradrenaline, when the responses are measured after a short duration of stimulation (Morel and Godfraind, 1991).

The inhibitory potency of nisoldpine on the response evoked by a short application of noradrenaline could be increased by pre-incubation of the tissue with the antagonist in depolarizing solution just as the contractile response induced by a short exposure to KCl-depolarizing solution was.

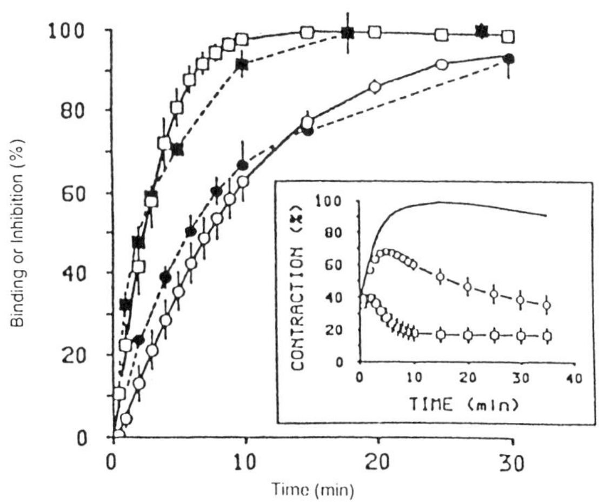

Fig. 7. Comparison between kinetics of $[^3H](+)$PN 200-110 association to its specific binding site [broken line, ●:0.1 nm; ■:0.38 nm] and (+)PN 200-110 inhibition of K^+-evoked contraction of rat aorta [solid line; o:0.1 nm; □:0.3 nm (+)PN 200-110]. **Inset:** Mechanical responses to KCl-depolarization of rat aortic rings. Control (upper curve), rings treated with 0.1 nm (O) or 0.3 nm (□)(+)PN 200-110.

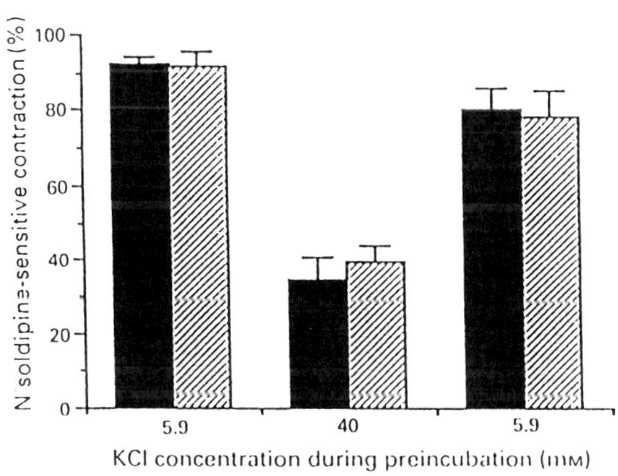

Fig. 8. Inhibition produced by nisoldipine on noradrenaline-induced contraction (hatched columns) and on KCl-induced contraction (solid columns) after preincubation in physiological solution or in 40 mM KCl depolarizing solution. After recording of control contractions, aortic rings were preincubated for 60 min. in the presence of nisoldipine (0.1 nM) in physiological solution. They were then incubated for 30 min. in zero Ca^{2+} physiological solution with nisoldipine (0.1 nM) before contraction was elicited by exposure to 10^{-5}M noradrenaline or to 100mM KCl-solution for 1 min., in the continuous presence of nisoldipine (run 1). Aortic rings were thereafter incubated for 30 min. in zero Ca^{2+}, 40mM KCl-depolarizing solution containing the same concentration of nisoldipine (0.1 nM) before a second run of contraction was recorded. At the end of this contraction, aortic rings were further incubated in zero Ca^{2+} physiological solution with nisoldipine (0.1nM) and a third run of contraction was recorded. Nisoldipine resistant contraction was determined in the presence of nisoldipine 100nM in the same conditions and subtracted from the total response to give the nisoldipine-sensitive contraction. All contractions were corrected for control responses recorded in the same conditions. Data are means from 4 determinations with s.c.mean shown by vertical bars.

In Fig. 8, we have compared the response to KCl and the dihydropyridine sensitive part of the contraction evoked by noradrenaline. Obviously, preincubation in KCl depolarizing solution increases in the same extend inhibition by nisoldipine for both noradrenaline and KCl. This suggests that noradrenaline-evoked contraction depends at least partly on the same nisoldipine sensitive calcium entry pathway as depolarization-induced contraction. This may be interpreted in the light of voltage-dependence of the binding of nisoldpine to binding sites on calcium channels characterized by an increase in affinity without change of maximum binding capacity.

We have observed that noradrenaline may modulate the voltage-dependent binding sites of dihydropyridine. It produces an increase in the affinity of isradipine and nisoldipine to the specific binding sites on calcium channels. We have also observed that under the same conditions, noradrenaline increases the affinity of nisoldpine. This effect of noradrenaline 10^{-6} M is comparable of that of 20 mM KCl and it may be suggested that it is the result of a small depolarization produced by noradrenaline (Morel and Godfraind, 1991). Mechanisms underlying the depolarization evoked by noradrenaline are still conflicting. Recently, however, evidence have been provided which suggest that the noradrenaline-evoked depolarization is mediated by an increase in chloride and cation membrane conductance (Amedee and Large, 1989; Pacaud et al 1989). It therefore appears that in rat aorta, noradrenaline activates voltage operated calcium channels that contain the specific voltage sensitive binding sites for calcium antagonistic dihydropyridines. In addition, there is a fraction of noradrenaline- stimulated calcium entry that is resistant to nisoldpine and to other calcium antagonists blockade, suggesting that another entry calcium pathway is also opened by the agonist (Morel and Godfraind, 1991)

4. EVIDENCE THAT THE LARGE DIFFERENCE IN SENSITIVITY BETWEEN CARDIAC AND VASCULAR TISSUE MAY BE RELATED TO THE KINETIC PARAMETERS DESCRIBING THE INTERACTION OF CALCIUM ANTAGONISTS WITH CALCIUM CHANNELS

At the tissular level, the selectivity of calcium antagonists is obvious. This has been demonstrated in animal studies comparing the action of diltiazem, nifedipine and verapamil, the representatives of the three main groups of calcium antagonists used for the management of

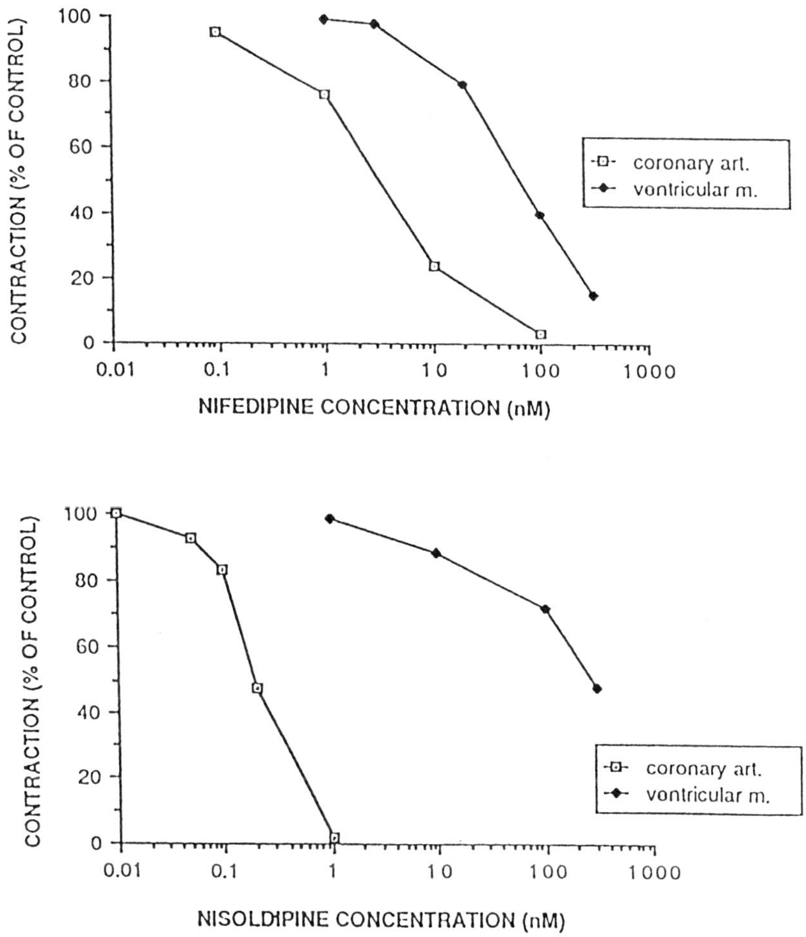

Fig. 9. Vascular selectivity of dihydropyridines: depol. human coronary art. vs. electr. paced human ventric. m. (from Godfraind et al., 1987, Pharmacol. Toxicol. 61:79).

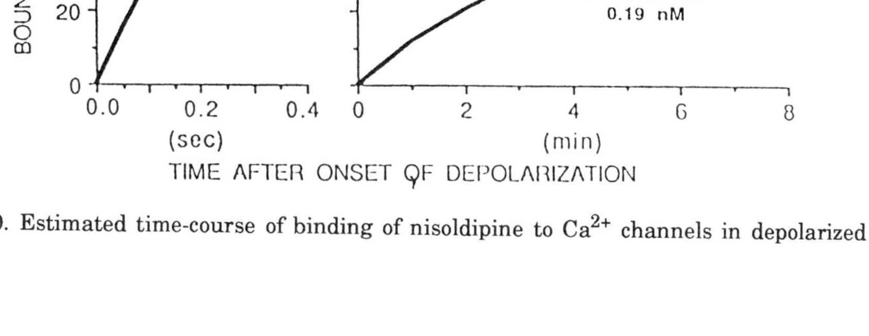

Fig. 10. Estimated time-course of binding of nisoldipine to Ca^{2+} channels in depolarized tissue.

cardiovascular disease. Bristow et al (1984) have shown that the ratio of sensitivity between respectively smooth muscle and cardiac muscle is equal to 3 with verapamil, to 72 with diltiazem and 243 with nifedipine. This support clinical reports showing that verapamil is more cardio-depressive than nifedipine.

Differences in the vascular selectivity of calcium antagonists are also obvious in the class of dihydropyridines, the most prolific class from the point of view of chemistry. For instance, as illustrated in Fig. 9, in vitro mechanical studies on human tissues (arterial and cardiac preparations) have shown that nisoldipine present a pronounced vascular selectivity which is 100-fold higher than that of its parent compound nifedipine. Both the difference between nisoldipine and nifedipine and the difference between the sensitivity of the artery contraction and the cardiac contraction to nisoldipine illustrated on Fig. 9 may be explained by their interaction with calcium channels. Indeed, although there are obvious molecular differences between α_1 subunit of Ca channels from vascular and cardiac tissues, there is no indication that the affinity of nisoldipine is much different for those two isoforms. However, the concentration of nisoldipine required to reduce by 50% the contraction of human myocardium in vitro is 1,500-fold higher than the concentration producing 50% inhibition of the contraction of isolated human coronary artery. Such a large difference may be explained if it is assumed that at active concentrations, drug binding to channels of polarized cells is negligible and that drug binding does occur on inactivated channels according to pseudo-first order association kinetics such as

$$B = B_E(1\text{-exp}\,[-k_1\, L_T\, B_{max}\, t\, B_E]) \quad (2)$$

where B_E is drug binding at equilibrium, k_1 the association rate constant and L_T the total drug concentration.The assumption that binding to resting channels is negligible is in agreement with an extimate of K_L value of 1.3 µM in calf Purkinje fibers (Bean, 1984; Wibo, 1989), to be compared with K_H value of 58 pM (Morel and Godfraind, 1991). Triggle and Janis (1984) have reported that k_1 of nisoldipine in rat heart membranes at 37°C was equal to 0.67 nM^{-1} min^{-1}. We have observed that the K_i value of nisoldipine in human heart membranes was similar to its K_i value in rat heart membrane. It is therefore likely that k_1 values are similar. Figure 10, illustrating the simulated time-course of binding of nisoldipine to Ca^{2+} channels in depolarized tissue, was drawn on the basis of equation (2). The time-scale of the left panel corresponds to systolic depolarization times in human myocardium (about 0.2 sec) and in the right panel to the time required to reach steady-state tonic contraction in arteries. At nisoldipine concentrations that are effective on coronary artery (0.2 nM), the time spent by calcium channels in their high-affinity conformation is much too short for significant drug binding to occur. Calculation shows that, when used at a concentration of 300 nM, the IC_{50} in electrically paced ventricular muscle, nisoldipine should occupy about 50 % of the channels by the end of the systolic contraction. The vascular selectivity reported for cinnarizine and flunarizine (Van Nueten and Janssen, 1973) may also be described by such a model since they also present a high affinity for inactivated channels (Godfraind et al., 1990).

The rather low vascular selectivity of phenylalkylamines is in agreement with this model since their affinity for Ca channels does not appear to be enhanced by depolarization, an observation also made with diltiazem (Morel and Godfraind, 1989). Also in the group of dihydropyridines, voltage-dependency may be absent, as for instance with nifedipine which shows a much lower vascular selectivity than nisoldipine (Morel and Godfraind, 1988)

In summary, the vascular selectivity of calcium antagonists that present a much higher affinity for inactivated than for resting Ca^{2+} channels, may be accounted for both by the proportion of inactivated Ca^{2+} channels and by drugs rate of association with inactivated Ca^{2+} channels.

CONCLUDING REMARKS

Experimental observations summarized above illustrate the diversity of action of calcium antagonists according to the tissue considered and to its mode of activation. An important parameter is the state of Ca channels which modulates the affinity of receptor sites for dihydropyridines. Therefore, characteristics of the tissues to be considered do not only deal with the localisation of activator calcium, but also with the level of the membrane potential. Such factors are involved in the tissue selectivity of calcium antagonists.

REFERENCES

Amédée, T., and Large, W. A., 1989, Microelectrode study on the ionic mechanisms which contribute to the noradrenaline-induced depolarization in isolated cells of the rabbit portal vein, Br. J. Pharmacol. , 97:1331.

Bean, B. P., 1984, Nitrendipine block of cardiac calcium channels : high-affinity binding to the inactivated state, Proc. Natl. Acad. Sci. , 81:6388.

Bean, B. P., Sturek, M., Puga, A. and Hermsmeyer, K., 1986, Calcium channels in muscle cells isolated from rat mesenteric arteries : modulation by dihydropyridine drugs, Circ. Res. , 59:229.

Bristow, M. R., Ginsburg, R., Laser, J. A., McAuley, B. J. and Minoba, W., 1984, Tissue response selectivity of calcium antagonists is not due to heterogeneity of [^{3}H]-nitrendipine binding sites, Br. J. Pharmacol., 82:309.

Cauvin, C., and Van Bremen, C., 1985, Different Ca^{2+} channels along the arterial tree, J. Cardiovasc. Pharmacol., 7:54.

Cauvin, C., and Van Bremen, C., 1985, ^{45}Ca fluxes in isolated rabbit mesenteric resistance vessels as compared to rabbit aorta, Prog. Appl. Microvasc., 8:47.

Chen, G., and Suzuki, H., 1989, Some electrical properties of the endothelium-dependent hyperpolarization recorded from rat arterial smooth muscle cells, J. Physiol. (Lond.), 410:91.

Dunn, W. R., Daly, C. J., McGrath, J. C. and Wilson, V. G., 1991, A comparison of the effects of angiotensin II and Bay K 8644 on responses to noradrenaline mediated via postjunctionnal α_1 and α_2-adrenoceptors in rabbit isolated blood vessels, Br. J. Pharmacol., 103:1475.

Féron, O., Wibo, M., and Godfraind, T., 1991, Interaction of Pinaverium bromide with the dihydropyridine binding site in rat ileum smooth muscle, Arch. Int. Pharm.

Godfraind, T., 1983, Actions of nifedipine on calcium fluxes and contraction in isolated rat arteries, J. Pharmacol. Exp. Ther., 224:443.

Godfraind, T., Egleme, C., Finet, M. and Jaumin, P., 1987, The actions of nifedipine and nisoldipine on the contractile activity of human coronary arteries and human cardiac tissue in vitro, Pharmacol. Toxicol., 61:79.

Godfraind, T. and Kaba, A., 1969, Inhibition by cinnarizine and chlorpromazine of the contraction induced by calcium and adrenaline in vascular smooth muscle, Br. J. Pharmacol., 35:354P.

Godfraind, T., Miller, R. C., and Wibo, M., 1986, Calcium antagonism and calcium entry blockers, Pharmacol. Rev., 38:321.

Kokubun, S., Prod'Hom, B., Becker, C., Porzig, H., and Reuter, H., 1986, Studies on Ca channels in intact cardiac cells voltage -dependent effects and cooperative interactions of dihydropyridine enantiomers, Mol. Pharmacol., 30:571.

Marriott, J. F., 1988, Comparison of the effects of the calcium entry blockers, verapamil, diltiazem and flunarizine against contractions of the rat isolated aorta and portal vein, Br. J. Pharmacol. , 95:145.

McKenna, E., Koch, W. J., Slish, D. F., and Schwartz, A., 1990, Toward an understanding of the dihydropyridine-sensitive calcium channel, Biochem. Pharmacol., 39:1145.

Morel, N., and Godfraind, T., 1987, Prolonged depolarization increases the pharmacological effect of dihydropyridines and their binding affinity for calcium channels of vascular smooth muscle, J. Pharmacol. Exp. Ther., 243:711.

Morel, N., and Godfraind, T., 1988, Selective modulation by membrane potential of the interaction of some calcium entry blockers with calcium channels in rat mesenteric artery, Br. J. PHarmacol., 95:252.

Morel, N., and Godfraind, T., 1989, Pharmacological properties of voltage-dependent calcium channles in functional microvessels isolated from rat brain, Naunyn-Schmiedeberg's Arch. Pharmacol., 340:442.

Morel, N., and Godfraind, T., 1991, Characterization in rat aorta of the binding sites responsible for blockade of noradrenaline-evoked calcium entry and contraction by nisoldipine, Br. J. Pharmacol., 102:467.

Pacaud, P., Loirand, G., Mironneau, C., and Mironneau, J., 1989, Noradrenaline activates a calcium-activated chloride conductance and increases the voltage-dependent calcium current in cultured single cells of rat portal vein, Br. J. Pharmacol., 97:139.

Sanguinetti, M. C., and Kass, R. S., 1984, Voltage-dependent block of calcium channel current in the calf cardiac Purkinje fiber by dihydropyridine calcium channel antagonists, Circ. Res., 55:336.

Sumimoto, K., Hirata, M., and Kuriyama, H., 1988, Characterization of [3H]nifedipine binding to intact vascular smooth muscle cells, Am. J. Physiol., 254:C45.

Tsien, R.W., 1983, Calcium channels in excitable cell membranes, Annu. Rev. Physiol., 45:341.

White, D. G., and Martin, W., 1989, Differential control and calcium-dependence of production of endothelium-derived relaxing factor and prostacyclin by pig aortic endothelial cells, Br. J. Clin. Pharmacol., 97:683.

Wibo, M., De Roth, L., and Godfraind, T., 1988, Pharmacological relevance of dihydropyridine binding sites in membrane from rat aorta : kinetic and equilibrium studies, Circ. Res., 62:91.

Wibo, M., 1989, Mode of action of calcium antagonists: voltage-dependence and kinetics of drug-receptor interaction, Pharmacol. Toxicol., 65:1.

Yatani, A., Codina, J. Imoto, Y., Reeves, J. P., Birnbaumer, L., and Brown, A. M., 1987, A G protein directly regulates mammalian cardiac calcium channels, Science, 238:1288.

THE FUNCTIONAL ROLE OF T-TUBULAR CALCIUM CHANNELS IN SKELETAL MUSCLE CONTRACTIONS

George B. Frank and Murat Öz

Department of Pharmacology, University of Alberta
9-70 Medical Sciences Building
Edmonton, Alberta, Canada, T6C 2H7

Nearly 35 years ago, when I began my studies on the role of calcium ions in excitation-contraction (e-c) coupling in skeletal muscle there seemed to be excellent evidence supporting the concept that the e-c coupling process was identical during twitches and depolarization contractures (Frank, 1961). For this reason in my initial studies I used high K^+-induced contractures in a Na^+-free solution as a model for the events occurring during e-c coupling in both types of muscle activity (Frank, 1958, 1960).

Fritz Buchtal, a Danish muscle physiologist, was the first to suggest the possibility to me that the e-c coupling mechanism might not be identical in the two response types. Although at the time I argued vigorously for the validity of the contracture model and for a single e-c coupling mechanism, he of course, was correct. It is now clear that the mechanisms are the same in that the contraction (or contracture) is initiated by an increase in the intracellular free Ca^{++} ion concentration; on the other hand they differ in that depolarization contractures require the entrance of extracellular Ca^{++} ions for e-c coupling but twitches do not (Frank, 1982a, b).

The Trigger Ca^{++} Hypothesis For E-C Coupling During the Twitch

Since I have reviewed in the first two E-C Symposia (Frank, 1982a, 1987) and elsewhere (Frank, 1980) the evidence which led to the development of 'trigger Ca^{++}' hypothesis by C.P. Bianchi and me and the supporting evidence subsequently obtained, at this meeting I will limit my comments to a brief description of this mechanism and to some newer developments in this area.

Very briefly then this hypothesis is that when the action potential which produces a twitch enters the t-tubules of skeletal muscle fibers, it causes the release of Ca^{++} ions bound to the intracellular surface of the t-tubules (the 'trigger Ca^{++}' ions) into the triadic junctional region. These Ca^{++} ions then diffuse across the triadic junction to reach the surface membrane of the terminal cisternae where by a 'Ca^{++}-induced Ca^{++}-release' mechanism, a much larger quantity of Ca^{++} ions is released into the myoplasm. The increased myoplasmic concentration of free Ca^{++} ions thus produced causes the contractile proteins to produce a mechanical response.

Most of this mechanism is now accepted. However, still in dispute is the exact means by which the action potential in the t-tubule produces the release of the Ca^{++} ions stored

Excitation-Contraction Coupling in Skeletal, Cardiac, and Smooth Muscle
Edited by G.B. Frank *et al.*, Plenum Press, New York, 1992

in the terminal cisternae. Also unclear is the mechanism by which the action potential causes the release (or activation) of the Ca^{++} ions bound to the t-tubules. Possibly the Na^{+} ions which enter the muscle fiber during the action potential play an essential role in this process (see the papers by B. Curtis and by V.P. Nesterov in this volume).

In the first Symposium of this series I reviewed the physiological evidence supporting an essential role for t-tubular bound Ca^{++} ions in e-c coupling during the twitch and for extracellular Ca^{++} ions during high K^{+} or depolarization induced contractures (Frank, 1982a). An important observation on this point (Frank, 1982a, b) was made in studies showing that when isolated skeletal muscle fiber preparations were placed in a Ca^{++}-free solution, high-K^{+} contractures were eliminated in the few seconds it took for the Ca^{++} ions to diffuse out of the lumen of the t-tubules whereas twitches were only eliminated after several minutes (usually 2-5 min; occasionally 10 or more min). It takes hours in a Ca^{++}-free solution to remove sufficient amounts of Ca^{++} from the sarcoplasmic reticular stores to interfere with contractile activity (Frank, 1962), therefore the block of the twitch was due to the removal of Ca^{++} ions bound to the t-tubular membranes.

In my report to the second Symposium of this series I reviewed the pharmacological evidence supporting an essential role for extracellular Ca^{++} ions during e-c coupling during depolarization contractures, and for t-tubular bound Ca^{++} (i.e. 'trigger Ca^{++}') for e-c coupling during the twitch (Frank, 1987). There were two types of results I would like to mention here.

In the first type of study it was shown that the organic channel blocking drugs verapamil and gallopamil (D-600) blocked high K^{+} contractures but not twitches at concentrations which blocked only skeletal muscle calcium channels (Frank, 1984). Since then we have reported that the dihydropyridine Ca^{++} channel antagonist, nitrendipine (Frank *et al.*, 1988; Frank, 1990) blocks high K^{+} contractures and that the dihydropyridine calcium channel agonists, Bay K8644 and CGP-28392, could increase the size of submaximal high K^{+} contractures (Frank, 1990). In none of these studies were twitches reduced in size at drug concentrations which blocked the high K^{+} contractures. These results are consistent with the results obtained in an early study of the effects of dihydropyridine agonists and antagonists on tonic depolarization contractures (Ildefonse *et al.*, 1985). We also have made similar observations using two other types of organic calcium channel blockers; namely diltiazem and flunarazine (unpublished results).

The second type of study referred to in my previous report (Frank, 1987) concerned the effects of TMB-8. TMB-8 is an organic compound, known to block many calcium dependent processes by stabilizing intracellular membrane-bound calcium stores. It was shown that under appropriate conditions TMB-8 could block twitches without reducing the size of either high K^{+} contractures or caffeine contractures. Thus TMB-8 under these conditions did not block the entrance of extracellular Ca^{++} ions needed for the high K^{+} contracture nor the release of Ca^{++} ions from the sarcoplasmic reticulum needed for caffeine contractures. It was concluded that t-tubular bound Ca^{++} ions (i.e. trigger Ca^{++} ions) have to be released in order to produce e-c coupling during the twitch.

There are two other, more recently proposed mechanisms for e-c coupling during the twitch that I would like to discuss here. One is the proposition that inositol 1,4,5-trisphosphate produced by the hydrolysis of membrane bound phosphatidylinositol 4,5-biphosphate rather than 'trigger Ca^{++}' is the chemical transmitter in e-c coupling in skeletal muscle (Berridge, 1984; Volpe *et al.*, 1986; Horiuti, 1988). The other is a more elaborate mechanical mechanism proposed by Rios and Pizarro (1988). This involves a mechanical connection between the t-tubular membrane and the terminal cisternae of the sarcoplasmic reticulum. This mechanical link is presumably via the "junctional feet" which are though to be ryanodine receptors. The mechanical link is supposed to involve voltage-dependent charge movements as proposed some years ago by Schneider and Chandler (1973).

It has been suggested that there are far more dihydropyridine binding sites than calcium channels in skeletal muscle (Schwartz *et al.*, 1985; Curtis and Catterall, 1986). Since they, like the calcium channels, are concentrated in the t-tubules, it is thought that they might play a role in excitation-contraction coupling in skeletal muscle. They have been called "voltage sensors" (Rios and Brum, 1987) and are suggested to be associated, anatomically and functionally, with the "feet" processes which cross the triadic junction and which have been identified as ryanodine receptors (Chadwick *et al.*, 1988). Together these two receptors are suggested to couple the electrical activity in the t-tubule to the release of Ca^{2+} ions from the terminal cisternae of the sarcoplasmic reticulum (Rios and Pizarro, 1988).

Since, I assume the evidence supporting these two alternative theories will be ably discussed by other speakers at this Symposium, I will limit my remarks to some results which do not support them.

As interesting as the IP3 hypothesis is and although it seems to play an important role in e-c coupling in some muscle types (e.g. see the paper by Rojas *et al.*, in this volume), it does not appear to play this role in vertebrate skeletal muscle. For example, although in skeletal muscle evidence has accumulated that isolated t-tubules can produce IP3 (Hidalgo *et al.*, 1986; Corrasco *et al.*, 1988), in rabbit skeletal muscle t-tubules the addition of exogenous phosphatidylinositol 4-phosphate was required for the production of IP3 (Corrasco *et al.*, 1988). The ability of IP3 to release Ca^{++} ions from the isolated sarcoplasmic reticulum of vertebrate skeletal muscle is a disputed point with some workers finding this effect (Volpe *et al.*, 1985) and others, after considerable effort, failing (Micos and Snow, 1987). More studies have concentrated on the ability of applied IP3 to produce mechanical responses in skinned muscle fibres with most reports finding such a response (Volpe *et al.*, 1985; Donaldson *et al.*, 1987; Vergara *et al.*, 1985) and others not (e.g. Lea *et al.*, 1986). However, even when obtained, the responses to applied IP3 always developed at a very much slower rate than the responses that were produced by caffeine or by Ca^{++} ions in skinned fibres and at least one group had to inject the IP3 directly into the myofibril lattice in order to obtain responses at reasonably low IP3 concentrations (Donaldson *et al.*, 1987). The slowness of the response to IP3 and its small size has been emphasized by Fabiato in his studies in skinned cardiac muscle fibres (Fabiato 1985a, b, 1986) leading him to the conclusion that IP3 is not the chemical messenger in cardiac muscle. This problem and other experimental results have led some workers to propose that IP3 plays a modulator rather than a chemical messenger role in e-c coupling in skeletal muscle (Thieleczek and Heilmeyer, 1986; Noske *et al.*, 1986). Finally in an elegant, and apparently definitive study, Walker *et al.*, (1987) concluded that IP3 could well be the chemical transmitter (or second messenger) in smooth muscles but in skeletal muscle the response to IP3 was far too slow (by at least 3 orders of magnitude) and the concentrations required were far too high for it to play a primary role in e-c coupling in skeletal muscle.

Interesting as the speculative hypothesis presented by Rios and Pizarro (1988), is, it should be regarded with considerable caution. A crucial point in this theory is their naming the dihydropyridine binding sites found in isolate t-tubule preparations 'voltage sensors', and presuming that they have such a function and play an essential role in e-c coupling. A major problem with this proposal is that "binding sites" are not necessarily functioning biological receptors. For example, no one would consider the stereospecific opioid receptors found on glass filters (Synder and Pert, 1975) to be functioning biological receptors. However, the problem becomes more subtle and more difficult when such binding sites are found on proteins isolated from tissues. In such a case one is tempted to think that such binding sites must have a physiological role. However, they might not have a function or their function, when it is discovered, might be very different than the role ascribed to them by speculation. (It was even suggested at this meeting by E.D. Lamb that this presumed surplus of dihydropyridine binding sites does not really exist.)

In addition, this speculative theory ignores considerable evidence supporting roles for extracellular and t-tubular bound Ca^{2+} ions in the e-c coupling processes in skeletal muscle, and as well this hypothesis is inconsistent with other experimental evidence. For example there is an irreducible, small but definite latent period between the action potential in the t-tubule and the start of the release of Ca^{2+} ions from the terminal cisternae (Frank, 1980, 1987; Vergara and Delay, 1986; Zhu *et al.*, 1986). This delay precludes a direct mechanical link between the action potential and the release of Ca^{2+} from the terminal cisternae. Another difficulty pointed out by Vergara and Delay (1986) is the discrepancy between the Q_{10} for the kinetic rate constants of the charge movements and the Q_{10} for the triadic delay between the start of the action potential in the t-tubule and the start of the release of Ca^{2+} ions from the terminal cisternae of the sarcoplasmic reticulum. The former is in the range of 1.65 to 2.0 and the latter is near to 2.7. The only way to get around these problems would be to speculate that the mechanical action initiates some hypothetical intracellular chemical event which after a delay, starts the release of Ca^{2+} ions from the terminal cisternae of the sarcoplasmic reticulum.

Finally, another problem is the observations that the organic calcium channel blocking drugs greatly depress or block the "charge movements" (Hui *et al.*, 1984; Lamb, 1986; Hui and Milton, 1987; Lamb and Walsh, 1987; Mikos and Snow, 1987; Rios and Brum, 1987; Walsh *et al.*, 1987). However, as we and others have found, these drugs do not reduce or at times potentiate twitches in amphibian (Frank, 1984; Bechem *et al.*, 1988) or mammalian skeletal muscle (Gallant and Goettl, 1985; Frank *et al.*, 1988) at concentrations which block the voltage sensitive, slow calcium channels; and presumably the 'voltage sensors' if they exist. Therefore it would seem that these "charge movements" are not required for e-c coupling during the twitch.

On the other hand, it should be pointed out that much of our evidence supporting the 'trigger Ca^{++}' hypothesis does not rule out either of these two alternative hypothesis just discussed. For example, it is known that membrane bound Ca^{++} would be needed for the production of IP3 from phosphotidylinositol (Abdel-Latif, 1986). It has also been proposed that calcium ions are required for the 'voltage sensors' to function in e-c coupling (Brum *et al.*, 1988).

In addition, it should be pointed out that both of these alternative theories incorporate recent findings on t-tubular structures and activities not included in our 'trigger Ca^{++}' hypothesis. Although this could be viewed as a strength of these two alternatives, it more properly should be viewed as a weakness. Thus according to the rule of logic called 'Ockham's razor', the explanation most acceptable is the simplest one that can adequately explain the mechanism for a process consistent with the known evidence about the process. Of the three hypothesis, only the 'trigger Ca^{++} meets this criterion. The other structures and events included in two alternative hypotheses might well play a modulatory role in e-c coupling but there is no evidence that these additional factors are required for the e-c coupling process to produce a twitch.

<u>Role Played by the T-tubular Slow Calcium Channels During Tetanic Responses</u>

Since the action potential lasts only 2 to 2.5 msec and the negative after potential averages about 15 to 20 mV at its start and declines exponentially with a time constant of 15 msec (Frank, 1957), there is not sufficient time nor sufficient depolarization to open the voltage sensitive, slow calcium channels which activate slowly reaching a maximum at about 100 to 200 msec (Beaty *et al.*, 1987). Thus the small amount of Ca^{++} ions which does enter with each action potential (Bianchi and Shanes, 1959) presumably enters via the voltage sensitive, fast Ca^{++} channels which are not blocked by the organic Ca^{++} channels agonists (Beaty *et al.*, 1987). However, even this small amount of extracellular Ca^{++} ion entry is not needed to produce a twitch (Armstrong *et al.*, 1972; Frank 1982 a, b, 1984, 1990).

These and similar findings have led some workers to speculate that slow inward calcium currents via the slow calcium channels play no role in e-c coupling (e.g. Gonzales-Serratos *et al.*, 1982). Since these slow calcium channels are concentrated in the t-tubules which in turn are essential for e-c coupling in skeletal muscle fibers, this would be a surprising situation. However, in our recent studies we have found that the slow inward calcium currents are required to maintain the tension during tetanic responses (Oz and Frank, 1991). Inasmuch as isolated twitches are rarely if ever seen in functioning skeletal muscle *in situ,* and bursts of action potentials lasting more than 0.5 sec are common (Kernell, 1965), we would suggest that these voltage-sensitive, slow Ca^{++} channels play an important role in the normal physiological functioning of vertebrate skeletal muscles.

Our studies were based on the demonstration some time ago that during rapid, repetitive stimulation of skeletal muscle there is an accumulation of K^+ ions in the t-tubules that produces a large, late negative afterpotential (LAP) (Freygang *et al.*, 1964; Gage and Eisenberg, 1969; Kirsch *et al.*, 1977). Thus, it seemed possible that the small depolarization which develops during the tetanic stimulus train and lasts for a sec or two after the stimulation has stopped, might open the slow calcium channels causing an entry of Ca^{++} ions. There is evidence that small elevated K^+ concentrations, which are subthreshold for producing contractures, result in a slow but continued influx of Ca^{++} ions (Bianchi and Shanes, 1959; Weiss and Bianchi, 1965) resulting in a low level of activation of the contractile apparatus characterized by an increase in resting 0_2 consumption and heat production (the Solandt effect; Solandt, 1936), an increased resistance to stretch and an increase in the mechanical response produced by slightly higher K^+ concentrations (Vos and Frank, 1972a, b). Furthermore, in many studies, it has been shown that there is a sustained increase in Ca^{++}-related Fura II signals (Suda and Kurihara, 1991), and in the twitch tension for several seconds after a tetanic contraction (Blinks *et al.*, 1978; Lopez, 1982). Although the dissociation of Ca^{++} from parvalbumin binding sites has been thought to be responsible for this sustained increase in intracellular Ca^{++} (which lasts tens of seconds), this mechanism has been seriously questioned by different groups (Ogawa and Tanokura, 1986; Fryer and Nearing, 1988). On the other hand, the entrance of extracellular Ca^{++} ions during tetanic responses and the LAP could explain these long lasting increases of intracellular free or unbound Ca^{++}. In a recent study, with the use of electron probe studies in frog muscle fibers, the increase of the total Ca^{++} level after a tetanic contraction was demonstrated, and the possibility that the entrance of extracellular Ca^{++} into the muscle fibers was the source of this increased Ca^{++} level was mentioned (Somlyo *et al.*, 1985). Furthermore, in our recent study, using t-tubule membrane vesicles, we also found that there is a large Ca^{++} flux produced by a membrane depolarization of the size of the LAP (Oz, *et al.*, see this volume).

In our studies we investigated the effects of extracellular calcium removal and of the effects of nitrendipine and verapamil on tetani (100 Hz for 2 sec) and on twitches recorded 15 sec to 1 min either before or usually after the tetanic response in frogs' toe muscles.

When a toe muscle is placed in a 0-Ca^{++} Ringer's solution some EDTA has to be added to the Ca^{++}-free solution in order to block the twitch response (Frank, 1982a). Because the EDTA does not enter the muscle fibers and is not required to block the twitch in isolated fiber preparations (Frank, 1982a,b), the EDTA is presumably required to chelate Ca^{++} ions that leak out of the muscle fibers into the extracellular fluid in the intact toe muscles during stimulation. For this reason a small amount of EDTA was added to our Ca^{++} free solutions in these studies.

With small EDTA concentrations and short exposure times, it was found that the twitches were increased in size due to after-discharging, whereas the latter part of the tetani were depressed (Fig. 1). In accord with these findings it was previously demonstrated that although in the absence of external Ca^{++}, the contractile responses to short depolarizing pulses (400 ms) were not modified, the area under prolonged contractures (approximately 10 s) were reduced considerably (Bolanos *et al.*, 1986).

In experiments with nitrendipine or verapamil added to Ringer's solution with Ca^{++} ions, the twitch amplitudes and the initial tetanic tensions were not noticeably reduced (Fig. 2) but the tensions during the latter part of the tetanii were clearly depressed (Fig. 3). The dose-response curves for the effect of these procedures on the tetanic areas (tension X time) at a 10 min drug exposure time are presented in Fig. 4.

From these results it was concluded that extracellular Ca^{++} ions need to enter the skeletal muscle fibers via the slow Ca^{++} channels in order for the tetanic tension to remain at its maximum for longer than about 300 to 500 msec (Oz and Frank, 1991).

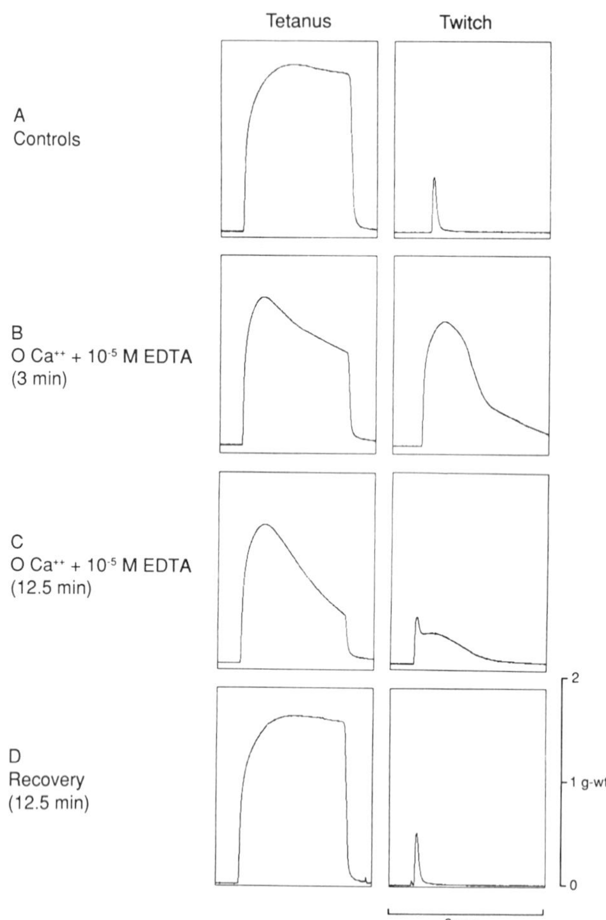

Fig. 1 Effects of removing extracellular Ca^{++} ions on twitches and tetani in an isolated frog's toe muscle. The 0-Ca^{++} solution contained a small amount of EDTA as indicated. Length of time the muscle was in 0-Ca^{++} solution is listed in B and C, and the time in Ringer's solution is listed in D. (From Oz and Frank, 1991).

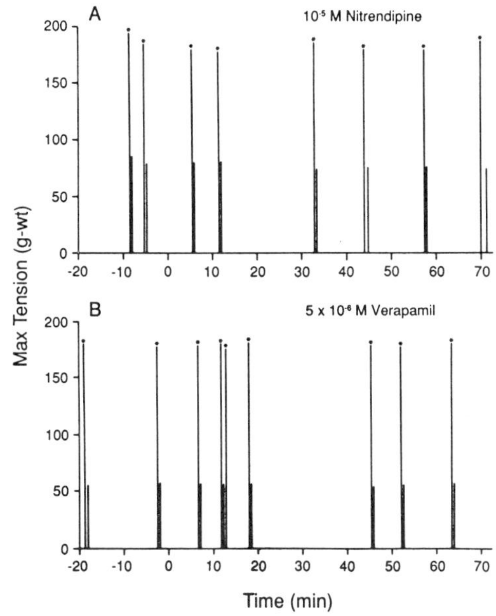

Fig. 2 Effects of nitrendipine (A) and verapamil (B) at the indicated concentration on the maximum tension development during twitches and tetanic responses (*). A and B separate experiments with different toe muscles. Muscles placed in drug containing solutions at time 0.

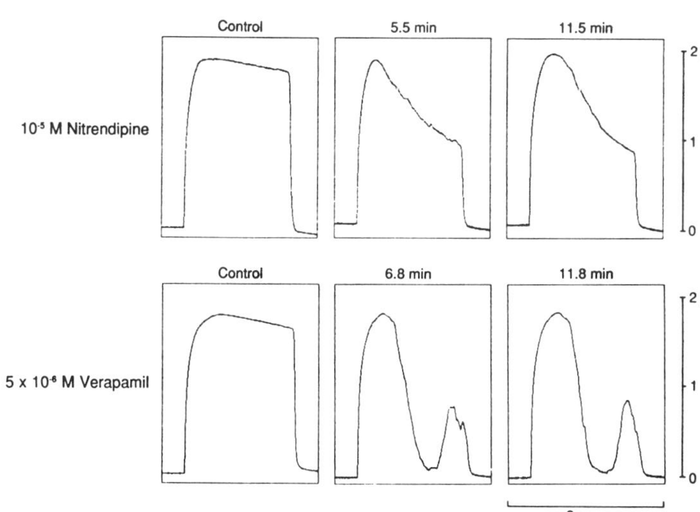

Fig. 3 Effects of nitrendipine and verapamil at the indicated concentrations on 2 sec tetani in frog toe muscle. These are some of the recordings made during the experiments shown in Fig. 2. Note the large decreases in tetanic responses produced by the drugs despite the lack of a decrease in the initial maximum tension.

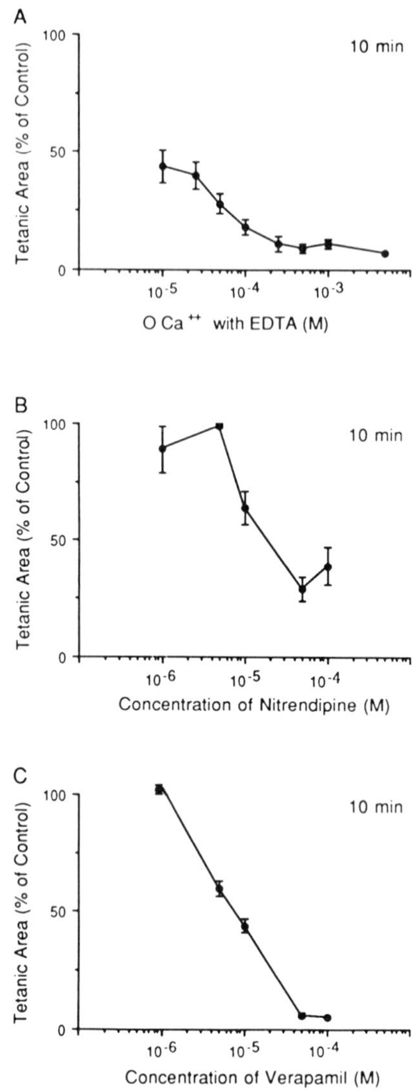

Fig. 4 Dose-response curves for the effects at 10 min. for various treatments on the tetanus areas. For nitrendipine (B) and verapamil (C) the number of experiments at each concentration were 3 to 6 and there were 3 to 8 for the 0 Ca^{++} experiments at different EDTA concentrations (A). The 10 min. treatment time was chosen because at this time the twitches were either unchanged or potentiated and the depressions of the tetanic areas were near or at the equilibrium effect.

In studies with intracellular microelectrodes (Oz and Frank, 1991) it was found that during a tetanic train the muscle fiber membrane becomes depolarized reaching a maximum of about 25 mV by the end of the 2 sec stimulus train. Treatment with nitrendipine concentrations (Fig. 5) which depressed the tetanic tension area or the lower verapamil concentrations which also depressed the tetanic tension area did not reduce the depolarizations produced by the stimulus trains. This is in accord with the suggestion made above that these drugs depressed the tetanus areas by blocking the voltage sensitive slow Ca^{++} channels and not the membrane electrical events.

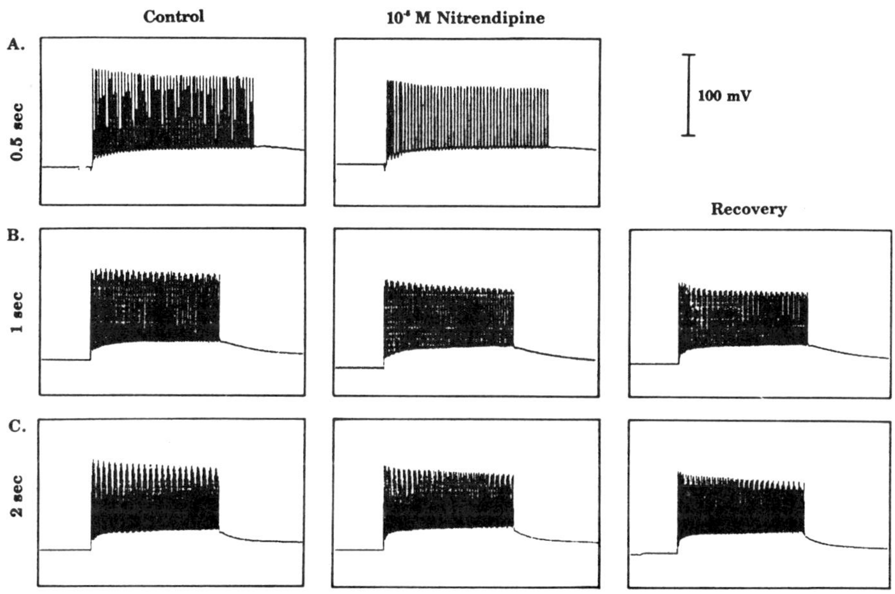

Fig. 5. Intracellularly recorded potentials for tetanic simulation trains in frog toe muscle fibers. Recordings from different single fibers are presented in A,B and C. The lengths of the stimulus trains at 100 Hz are listed at the left of the recordings. Muscles in the drug solution for 10 min or more before the recordings were made and back in Ringer's solution for at least 10 min before the recover recordings were made. (From Oz and Frank, 1991).

At higher verapamil concentrations (5 x 10^{-6} M to 10^{-4}M) we observed a use-dependent block of the sodium channels (Frank and Oz, 1991). This is similar to the well known use-dependent block of the Ca^{++} channels (Lee and Tsien, 1983; Frank, 1986). This was expressed by the loss of some action potentials during the stimulus train at 10^{-5}M to a complete block after about 300 msec with 10^{-4}M. Frequently a pattern such as shown in Fig. 6 was obtained. This pattern was an initial burst or two of action potentials followed by a block and then another burst of action potentials before the end of the stimulus train. This type of response pattern is undoubtedly responsible for the biphasic tetanic response patterns often recorded at the same verapamil concentrations (Fig. 6). A similar but less intense frequency-dependent, local anesthetic-like effect of verapamil at a 10^{-4} M concentration on slow mammalian muscle fibers was reported a few years ago (Kotsias and Muchnik, 1985).

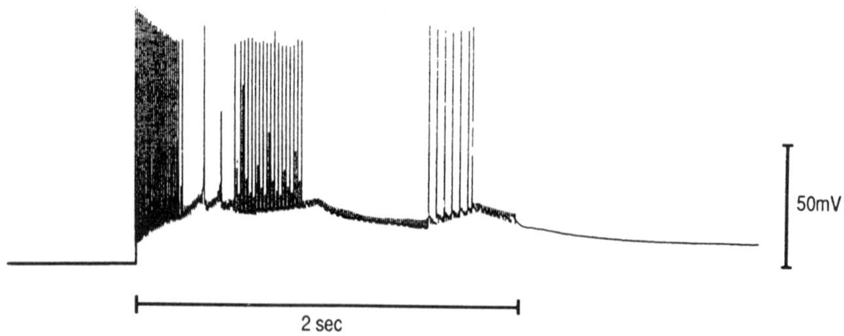

Fig. 6 Action potential block and bursting during a 2 sec stimulus train at 100 Hz recorded in a toe muscle fiber exposed to 10^{-5} M verapamil for 10 min. Small stimuius artifacts can be seen between these action potentials during the first two groups of action potentials. Between the bursts these artifacts have been reduced in size by whiting them out (From Frank and Oz, 1991).

CONCLUSIONS

In conclusion, it is clear that despite an overall similarity, there is a difference in the e-c coupling process during twitches and depolarization contractures in vertebrate skeletal muscle. During the twitch, Ca^{++} ions are released from the sarcoplasmic reticulum (S.R.) without requiring the entrance of extracellular Ca^{++} ions. In contrast, the entrance of extracellular Ca^{++} ions is required for e-c coupling during prolonged depolarization contractures.

It is clear also that during a twitch, calcium ions bound to the t-tubular membranes (the 'trigger Ca^{++}' ions) need to be released for e-c coupling to occur. Whether these released Ca^{++} ions directly cause the release of Ca^{++} from the S.R. as proposed by the 'trigger Ca^{++}' hypothesis of Bianchi and Frank or they are required for some other mechanism remains to be seen. The evidence, as opposed to speculation, still favors the 'trigger Ca^{++}' hypothesis. Also remaining to be determined is what role if any is played by the Na^{+} ions which enter during the action potential. Possibly they are required to release the 'trigger Ca^{++}' ions bound to the t-tubules.

Finally in our most recent studies we have found a physiological role for the voltage sensitive, slow Ca^{++} channels concentrated in the t-tubules of skeletal muscles. These open during a train of repetitive action potentials and permit a small influx of Ca^{++} ions. These Ca^{++} ions are required to maintain the full muscle tension during a tetanus. Since *in situ*, single action potentials rarely occur and trains of action potentials often lasting several sec. is the usual pattern seen, we would suggest that opening of these slow Ca^{++} channels is required for most muscle contractions under normal conditions in the intact animal.

REFERENCES

Abdel-Latif, A.A., 1986, Calcium-mobilizing receptors, polyphosphoinositides, and the generation of second messengers. Pharmacol. Rev., 38: 227-272.

Armstrong, C.M., Bezanilla, F.M. and Horowicz, P., 1972, Twitches in the presence of ethylene glycol bis(-aminoethylether)-*N,N'*-tetraacetic acid, Biochim. Biophys. Acta, 267: 605-608.

Beaty, G.N., Cota, G., Nicola Siri, L., Sanchez, J.A. and Stefani, E., 1987, Skeletal Muscle Ca^{++} channels, In Structure and Physiology of the Slow Inward Calcium Channel. 123-140, Alan R. Liss: New York.

Bechem, M., Hebisc, S. and Schramm, M., 1988, Ca^{2+} agonists: New, sensitive probes for $^{2+}$ chanels. TIPS, 9: 257-261.

Berridge, M.J., 1984, Inositol triphosphate and diacylglycerol as second messengers, Biochem. J., 220: 345-360.

Bianchi, C.P. and Shanes, A.M., 1959, Calcium influx in skeletal muscle at rest, during activity and during potassium contractures, J. Gen. Physiol., 42: 803-815.

Blinks, J.R., Rüdel, R., and Taylor, S.R., 1978, Calcium transients in isolated amphibian skeletal muscle fibres: Detection with aequorin. J. Physiol. (Lond.), 277: 291-323.

Bolanos, P., Caputo, C. and Velar, L., 1985, Effects of Calcium, Barium and Lanthanum on Depolarization-Contraction Coupling in Skeletal Muscle Fibres of Rana Pipiens. J. Physiol. (Lond.), 370: 39-60.

Brum, G., Rios, E. and Stefani, E., 1988, Effects of extracellular calcium on calcium movements of excitation-contraction coupling in frog skeletal muscle fibres. J. Physiol. (Lond.), 398: 441-473.

Carrasco, M.A., Magendzo, K., Jaimovich, E. and Hidalgo, C., 1988, Calcium modulation of phosphoinositicle kinases in transverse tubule vesicles from frog skeletal muscle. Arch. Biochem. Biophys., 262: 360-366.

Chadwick, C.C., Fleischer, S. and Inui, M., 1988, Identification and purification of a transverse tubule coupling protein which binds to the ryanodine receptor of terminal cisternae at the triad junction in skeletal muscle, J. Biol. Chem., 263: 10872-10877.

Curtis, B.M. and Catterall, W.A., 1986, Reconstitution of the voltage-sensitive calcium channel purified from skeletal muscle transverse tubules. Biochemistry, 11: 3077-3083.

Donaldson, S.K., Goldberg, N.D., Walseth, T.F. and Huetteman, D.A., 1987, Inositol trip phosphate stimulates calcium release from peeled skeletal muscle fibers. Biochim. Biophys. Acta, 927: 256-260.

Fabiato, A., 1985, Rapid ionic modifications during the aequorin-detected calcium transient in a skinned canine cardiac purkinje cell, J. Gen. Physiol., 85: 189-246.

Fabiato, A., 1985, Time and calcium dependence of activation and inactivation of calcium-induced release of calcium from the sarcoplasmic reticulum of a skinned canine cardiac purkinje cell. Simulated calcium current can both cause calcium loading in and trigger calcium release from the sarcoplasmic reticulum of skinned canine cardiac purkinje cell. J. Gen. Physiol., 85: 247-320.

Fabiato, A., 1986, Inositol (1,4,5) triphosphate-induced versus Ca^{2+}-induced release of Ca^{2+} from the cardiac sarcoplasmic reticulum. Proc. XXX Congress Int. Union of Physiol. Sciences, 16: 350.

Frank, G.B., 1957, Negative after-potential of frog's skeletal muscle. J. Neurophysiol., 20: 602-614.

Frank, G.B., 1958, Inward movement of calcium as a link between electrical and mechanical events in contraction. Nature, 182: 1800-1801.

Frank, G.B., 1960, Effects of changes in extracellular calcium concentration on the potassium-induced contracture of frog's skeletal muscle, J. Physiol., (London), 151: 518-538.

Frank, G.B., 1961, Role of extracellular calcium ions in excitation-contraction coupling in skeletal muscle. In Bio-physics of physiological and pharmacological actions. American Association for the Advancement of Science, Washington, D.C., 293-307.

Frank, G.B., 1962, Utilization of bound calcium in the action of caffeine and certain multivalent cations on skeletal muscle. J. Physiol. (London), 163: 254-268.

Frank, G.B., 1980, Commentary: The current view of the source of trigger calcium in excitation-contraction coupling invertebrate skeletal muscle. Biochem. Pharmacol., 29: 2399-2406.
Frank, G.B., 1982a, Roles of extracellular and 'trigger' calcium ions in excitation-contraction coupling in skeletal muscle. International Symposium on E-C Coupling, Banff, August, 1981. Can. J. Physiol. Pharmacol., 60: 427-439.
Frank, G.B., 1982b, The effects of reducing the extracellular calcium concentration on the twitch in isolated frog's skeletal muscle fibres. Jap. J. Physiol., 32: 589-608.
Frank, G.B., 1984, Blockade of Ca^{++} channels inhibits K^+ contractures but not twitches in skeletal muscle fibres. Can. J. Physiol. Pharmacol., 62: 374-378.
Frank, G.B., 1986, A pharmacological explanation of the use-dependency of the verapamil (and D-600) block of slow calcium channels. J. Pharmacol. Exp. Ther., 236: 505-511.
Frank, G.B., 1987, Pharmacological studies of excitation-contraction coupling in skeletal muscle. Can. J. Physiol. Pharmacol., 65: 711-716.
Frank, G.B., 1990, Dihydropyridine calcium channel antagonists block and agonists potentiate high potassium contractures but not twitches in frog skeletal muscle. Japanese J. Physiol., 40: 205-244.
Frank, G.B., Konya, L. and Sudha, T.S., 1988, Nitrendipine blocks high potassium contractures but not twitches in rat skeletal muscle. Can. J. Physiol. Pharmacol., 66: 1210-1213.
Frank, G.B. and Oz, M., 1991, Use-dependent block of sodium channels by verapamil in skeletal muscle during repetitive stimulation. Proceedings of the Western Pharmacology Society, 34th Annual Meeting.
Freygang, Jr., W.H., Goldstein, D.A. and Hellam, D.C., 1969, The after-potential that follows trains of impulses in frog muscle fibers. J. Gen. Physiol., 47: 929-952.
Fryer, M.W., Neering, I.R., 1986, Relationship between intracellular calcium concentration and relaxation of fastnd slow muscles. Neurosci. Lett., 64: 231-235.
Gage, P.W. and Eisenberg, R.S., 1969, Action potentials, after-potentials and excitation-contraction coupling in frog sartorius muscle fibres without transverse tubules. J. Gen. Physiol., 53: 298-310.
Gallant, E.M. and Goettl, V.M., 1985, Effects of calcium antagonists on mechanical responses of mammalian skeletal muscles. Eur. J. Pharmacol., 117: 259-265.
Gonzalez-Serratos, H., Valley-Aguilera, R., Lathrop, D.A., del Carmen Garcia, M., 1982, Slow inward calcium currents have no obvious role in muscle excitation-contraction coupling. Nature, 298: 292-294.
Hidalgo, C., Carrasco, M.A., Magendzo, K. and Jaimovich, E., 1986, Phosphorylation of phosphatidylinositol by transverse tubule vesicles and its possible role in excitation-contraction coupling. FEBS Lett., 202: 69-73.
Horiuti, K., 1988, Mechanism of contracture on cooling of caffeine-treated frog skeletal muscle fibres. J. Physiol. (Lond.), 398: 131-148.
Hui, C.S., Eisenberg, R.S. and Milton, R.L., 1984, Charge movement in skeletal muscle fibers paralysed by the calcium-entry blocker D600. Proc. Natl. Acad. Sci. U.S.A., 81: 2582-2585.
Hui, C.S. and Milton, R., 1987, Suppression of charge movement in frog skeletal muscle by D600. J. Muscle Res. Cell Motil., 3: 195-208.
Ildefonse, M., Fosset, M., Lazdunski, M., Renaud, J.F. and Rougier, O., 1985, Excitation contraction coupling in skeletal muscle: Evidence for a role of slow Ca^{2+} channel activators and inhibitors in the dihydropyridine series. Biochem. Biophys. Res. Commun., 129: 904-909.
Lamb, G.D., 1986, Components of charge movement in rabbit skeletal muscle: The effect of tetracaine and nifedipine. J. Physiol. (Lond.), 376: 85-100.
Lamb, G.D. and Walsh, T., 1987, Calcium currents, charge movement and dihydropyridine binding in fast- and slow-twitch muscles of rat and rabbit. J. Physiol. (Lond.), 393: 595-617.

Kernell, D., 1965, High-frequency repetitive firing of cat lumbosacral motorneurones stimulated by long-lasting injected currents. Acta Physiol. Scand., 65: 74-86.
Kirsch, G.E., Nichols, R.A. and Nakajima, S., 1977, Delayed rectification in the transverse tubules. J. Gen. Physiol., 70: 1-21.
Kotsias, B.A. and Muchnik, S., 1985, Frequency-dependent effect of verapamil on rat soleus muscle. Experientia, 41:1538-1540.
Lea, T.J., Griffiths, P.J., Tregear, R.T. and Ashley, C.C., 1986, An examination of the ability of inositol 1,4,5-triphosphate to induce calcium release and tension development in skinned skeletal muscle fibres of frog and crustacea. FEBS Lett., 207: 153-161.
Lee, K.S. and Tsien, R.W., 1983, Mechanism of calcium channel blockade by verapamil, D600, diltiazem and nitrendipine in single dialysed heart cells. Nature (Lond.), 302: 790-794.
Lopez, E., 1982 Post-tetanic potentiation and inhibition in single fibres isolated from frog semitendinosus muscle. Jpn. J. Physiol., 32: 103-119.
Mikos, G.J. and Snow, T.R., 1987, Failure of inositol 1,4,5-trisphosphate to elicit or potentiate Ca^{++} release from isolated skeletal muscle sarcoplasmic reticulum. Biochim. Biophys. Acta, 927: 256-260.
Nosek, T.M., Williams, M.F., Zeigler, S.T. and Godt, R.E., 1986, Inositol triphosphate enhances calcium release in skinned cardiac and skeletal muscle. Am. J. Physiol., 250: C807-811.
Ogawa, Y. and Tanokura, M., 1986b, Kinetic studies of calcium binding to parvalbumin from bullfrog skeletal muscle. J. Biochem. (Tokyo), 99: 81-89.
Oz, M. and Frank, G.B., 1991, Decrease in the size of tetanic responses produced by nitrendipine or by extracellular Ca^{++} ion removal without blocking twitches or action potentials in skeleal muscle. J. Pharmacol. Exp. Ther., 257, No. 2., 575-581.
Rios, E. and Brum, G., 1987, Involvement of dihydropyridine receptors in excitation-contraction coupling in skeletal muscle. Nature, 325: 717-720.
Rios, E. and Pizarro, G., 1988, Voltage sensors and calcium channels of excitation-contraction coupling. N.I.P.S., 3: 223-227.
Schneider, M.F. and Chandler, W.K., 1973, Voltage dependent charge movement in skeletal muscle: A possible step in excitation-contraction coupling. Nature, 242: 244.
Schwartz, L.M., McCleskey, W. and Almers, W., 1985, Dihydropyridine receptors in muscle are voltage dependent but most are not functional calcium channels. Nature, 314: 747-751.
Snyder, S.H. and Pert, C.B., 1975, Opiate receptor mechanisms. Neurosci. Res. Prog. Bull., 13: 26-34.
Solandt, D.Y., 1936, The effect of potassium on the excitability and resting metabolism of frog's muscle. J. Physiol. (Lond.), 86: 162-170.
Somlyo, A.V., McClellan, G., Gonzalez-Serratos, H., Somlyo, A.P., 1985, Electron probe X-ray microanalysis of post-tetanic Ca^{2+} and Mg^{2+} movements across the sarcoplasmic reticulum *in situ*. J. Biol. Chem., 260: 6801-6807.
Suda, N. and Kurihara, S., 1991, Intracellular calcium signals measured with Fura-2 and Aequorin in frog skeletal muscle fibers. Jpn. J. Physiol., 41: 277-295.
Thieleczek, R. and Heilmeyer, Jr., L.M.G., 1986, Inositol 1,4,5-triphosphate enhances Ca^{2+}-sensitivity of the contractile mechanism of chemically skinned rabbit skeletal muscle fibers. Biochem. Biophys. Res. Commun., 135: No. 2, 662-669.
Vergara, J. and Delay, M., 1986, A transmission delay and the effect of temperature at the triadic junction of skeletal muscle. Proc. R. Soc. Lond. (Biol.), 229: 97-110.
Vergara, J., Tsien, R.Y. and Delay, M., 1985, Inositol 1,4,5-triphosphate: a possible chemical link in excitation-contraction coupling in muscle. Proc. Natl. Acad. Sci. USA, 82: 6352-6356.
Volpe, P., Di Virgilio, F., Pozzan, T. and Salviati, G., 1986, Role of inositol 1,4,5-triphosphate in excitation-contraction coupling in skeletal muscle. FEBS Lett., 197: 1-4.

Volpe, P., Salviati, G., Di Virgilio, F. and Pozzan, T., 1985, Inositol 1,4,5-triphosphate induces calcium release from sarcoplasmic reticulum of skeletal muscle. Nature, 316: 347-349.

Vos, E.C. and Frank, G.B., 1972a, The threshold for potassium-induced contractures of frog skeletal muscle. Potentiation of potassium-induced contractures by pre-exposure to subthreshold potassium concentrations. Can. J. Physiol. Pharmacol., 50: 37-44.

Vos, E.C. and Frank, G.B., 1972b, Events occurring in the region of the threshold for potassium-induced contractures of frog skeletal muscle. Changes in elasticity and oxygen consumption. Can. J. Physiol. Pharmacol., 50: 179-187.

Walker, J.W., Somlyo, A.V., Goldman, Y.E., Somylo, A.P. and Trentham, D.R., 1987, Kinetics of smooth and skeletal muscle activation by laser pulse photolysis of caged inositol 1,4,5-triphosphate. Nature, 327: 249-252.

Walsh, K.P., Bryant, S.H. and Schwartz, A., 1987, Suppression of charge movement by calcium antagonists is not related to calcium channel block. Pflugers Arch., 409: 217-219.

Weiss, G.B. and Bianchi, C.P., 1965, The effect of potassium concentration on Ca45 uptake in frog sartorius muscle. J. Cell. Comp. Physiol., 65: 385-392.

Zhu, P.H., Miledi, R. and Parker, I., 1986, Minimal latency of calcium release in frog twitch muscle fibres. Proc. R. Soc. Lond. (Biol.), 229: 39-46.

EFFECTS OF CALCIUM RELEASE FROM THE SARCOPLASMIC RETICULUM

ON INTRAMEMBRANE CHARGE MOVEMENT IN SKELETAL MUSCLE

László Csernoch*, Gonzalo Pizarro†, Jesús García‡, Géza Szücs* Enrico Stefani‡ and Eduardo Ríos†

*Dept. of Physiol., Univ. Med. Sch., Debrecen, Hungary, †Dept. of Physiol., Rush Univ., Chicago, IL, USA and ‡Dept. of Physiol., Baylor Col. of Medicine, Houston, TX, USA

INTRODUCTION

Skeletal muscle contraction is initiated by the increase in the intracellular calcium concentration ([Ca^{2+}]) following the depolarization of the surface- and transverse-(T-) tubular membranes (see the review by Ebashi, 1991). Calcium ions are released into the myoplasm from an intracellular store, the sarcoplasmic reticulum (SR), through the ryanodine receptor calcium release channel (Imagawa et al., 1987; Inui et al., 1987). The voltage sensitive step that links channel opening to depolarization is the displacement of permanent charges (charge movement, Schneider and Chandler, 1973) found in the T-tubular membrane and localized in the DHP receptor (Ríos and Pizarro, 1991).

Charge movement components in skeletal muscle were separated according to kinetic criteria as early as the mid 1970-s (Almers et al., 1975; Adrian and Peres, 1977). The slow, so called hump component of intramembrane charge movement, attracted much attention ever since, due to the fact that its kinetic behavior is hard to explain in simple terms (Duane and Huang, 1982; Huang, 1983) and because its voltage dependence closely resembled that of the tension (Huang, 1982). Moreover, it becomes apparent approximately at those voltages where the calcium release from the SR is first noticeable (Vergara and Caputo, 1982; Csernoch et al., 1988) which makes it to be a possible candidate for the voltage sensitive step in excitation-contraction coupling (reviewed by Huang, 1989).

Several different criteria have been used for separating intramembranous charge movement into its components. The most widespread classification is based on the effects of local anaesthetics, terming the tetracaine sensitive charge Q_γ, the tetracaine resistant but lidocaine sensitive charge Q_β, and Q_α

the charge that remains when both anaesthetics are present (Huang, 1982). The tetracaine sensitive charge and the hump component of charge movement current are closely related (Huang, 1981, 1982; Hui, 1983a,b), to the point that the two terms were and still are used interchangeably.

This study presents a reexamination of the hump component with experimental evidences in favor of the hypothesis that the calcium released from the SR during muscle activation influences the T-tubular voltage sensor and thus produces the hump component of intramembranous charge movement. Furthermore, computer simulations are shown that explain how the model works and how well the experimental results are reproduced.

METHODS

The experiments were performed on cut skeletal muscle fibers of the frog mounted into single- (Kovács and Schneider, 1973) or double vaseline-gap chambers (Kovács et al., 1983) and voltage clamped to -100 mV. The details of the experimental apparatus and data acquisition can be found elsewhere (Kovács and Szücs, 1983; Brum and Ríos, 1987; García et al., 1989). The internal and external solutions used were essentially the same for the two set-ups (containing, in mM, 75 TEA_2SO_4, 8 $CaSO_4$ and 10 Cs_2SO_4 or 126 TEA-CH_3SO_3 and 2 $CaCH_3SO_3$, 5 mM Tris-maleate buffer and 0.31 μM TTX for the external, and 108 Cs-glutamate, 5.5 $MgCl_2$, 5 ATP, 6.5 glucose, 0.1 EGTA, 0.0082 $CaCl_2$ and 17.7 Tris-maleate for the internal solution; pH=7; for further details see Csernoch et al., 1991; Szücs et al., 1991), they intended to minimize ionic currents through the surface- and T-tubular membranes. Calcium transients were measured using the metallochromic indicator dye antipyrylazo-III (AP III) added to the internal solution in 0.8 or 1 mM concentration (Kovács et al., 1983; Brum et al., 1988b). The calcium release flux from the SR was calculated from the calcium transients according to Melzer et al. (1987) using the modifications introduced by Brum et al. (1988a). Charge movements were calculated from the current records by subtracting the linear capacity component, derived from 100 ms long pulses taken at hyperpolarized membrane potentials (Csernoch et al., 1991).

RESULTS and DISCUSSION

The main goal of the experiments was to test whether interventions which intended to primarily affect the release of calcium from the SR interfered with charge movements.

Fig. 1A demonstrates the effect of a short, 50 ms, conditioning pulse (see inset) on the charge movement current. This pulse protocol should not, in principle, affect the charge movement process, but causes a marked reduction

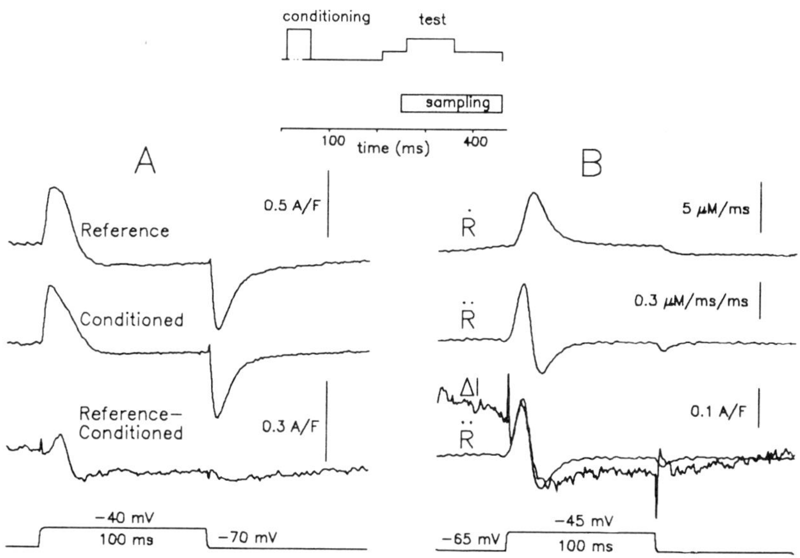

Fig. 1. A 50 ms conditioning pulse suppresses a delayed component of the charge movement current. (A) Charge movements recorded during the test pulse (see inset) with (Conditioned) and without (Reference) the conditioning pulse. The charge that was affected is shown as third trace (Reference-Conditioned) calculated from the traces above. (B) Kinetic comparison of the release flux (R), its first derivative (R) with the difference record (ΔI) obtained the same way as described in part A. R is repeated to show the striking similarity of its time course to the difference record. The lowest trace in both panel shows the recorded voltage during the sampling period, the prepulse level was selected to be just subthreshold for calcium release. Fibers 482 (A) and 492 (B), sarcomere length (sl)=4.0 μm, temperature (T)=13°C.

in calcium release during the test pulse due to calcium dependent inactivation of the release channel (Schneider and Simon, 1988; Simon et al., 1991). Unexpectedly, as demonstrated on the difference record calculated as reference test - conditioned test, the conditioning pulse suppressed a component that has a slow onset and a relatively late peak.

The difference record has some interesting features. It is obvious that the OFF component is very small and very slow, probably owing to a slow return of the charged particles to their resting position. There seems to be an undershoot on the ON part of the record which would correspond to an inward phase in charge movement. This undershoot was observed on the total membrane current in the presence of various channel blockers, in very different ionic media. Furthermore, it disappeared at higher voltages where

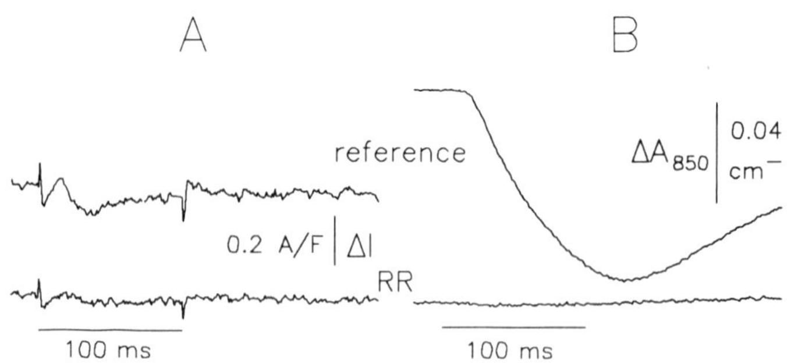

Fig. 2 Internally applied ruthenium red (RR) suppresses the same component as the conditioning pulse. Since RR gave a precipitate with AP III the fiber's intrinsic absorbance change was used as an indicator of calcium release from the SR. (A) Current suppressed by a conditioning pulse (determined as described in connection with Fig. 1) right after (reference) the application of the RR ([RR]=0.5 mM) containing internal solution and when the RR concentration reached 130 μM in the voltage clamp segment (RR). (B) Simultaneously recorded absorbance changes measured at 850 nm. Fiber 566, sl=4.0 μm, T=14°C, voltage during the test pulse (V_{test})=-55 mV.

the hump merged into the early part of the charge movement current and satisfied the charge conservation criteria (data not shown, see also Pizarro et al., 1991). We therefore believe that its a true feature of charge movement and it will be shown that the model readily accounts for it.

Fig. 1B demonstrates how the time-course of this isolated component compares to other myoplasmic events. The upper trace shows the calcium release flux from the SR, the second from the top is its derivative. The difference record is presented as the next trace. The lowest trace repeats the first derivative of the release record, shifted to have the same peak as the current record. The similarity between the derivative and the difference current will be discussed in connection with the model.

A component having similar characteristics can be isolated using chemical interventions that are supposed to be specific modifiers of SR calcium release. Ruthenium red (RR) was applied at a concentration of 0.5 mM at the open ends and its effects on the intrinsic optical change (ΔA_{850}) of the fiber and on the charge movement process were studied (Fig. 2). Together with the increase in ruthenium red concentration in the voltage clamped segment (130 μM at the time of the measurement shown in Fig. 2) the intrinsic signal (B) decreased indicating that the calcium release from the SR was blocked by the drug.

The charge movement records showed similar time-dependent changes. The difference records (A), calculated as reference - conditioned test (similar to that in Fig. 1), reveal that the slow component was almost completely eliminated as the ruthenium red concentration reached a value that was sufficient to suppress the release.

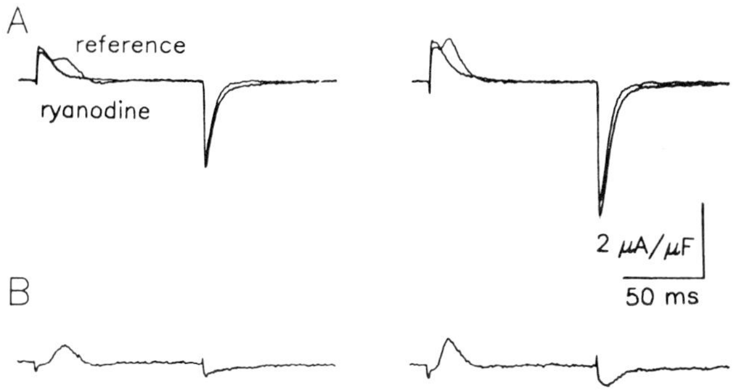

Fig. 3. 100 μM ryanodine blocks the late, hump, component of charge movement. (A) Charge movement currents recorded before (reference) and after (ryanodine) the application of the alkaloid. Test voltages, -40 (left) and -30 (right) mV, were selected so as the late component would be clearly visible on the current records. (B) Ryanodine sensitive currents, calculated as direct differences between the traces in panel A. Fiber 9003, sl=2.8 μm, T=17°C.

The plant alkaloid ryanodine was applied at a 100 μM concentration (Fig. 3), a large concentration where it inhibits the calcium release channel of the SR (Meissner, 1986). The charge movement currents were recorded at two voltages (-40 and -30 mV for the left and right column, respectively) where the late component was clearly visible before the application of the drug (Fig. 3A). At both voltages the hump was abolished after the addition of ryanodine. The difference records (Fig 3B) show that the component affected by the alkaloid has essentially the same characteristics as those described for the difference record in the previous figures.

Humps are readily observable in single-gap at the contraction threshold, as first demonstrated in cut fibers by Horowicz and Schneider (1981). Charge movement currents were recorded using a stepped pulse protocol (illustrated at the bottom of Fig. 4), to enhance the detection of the hump component, 5

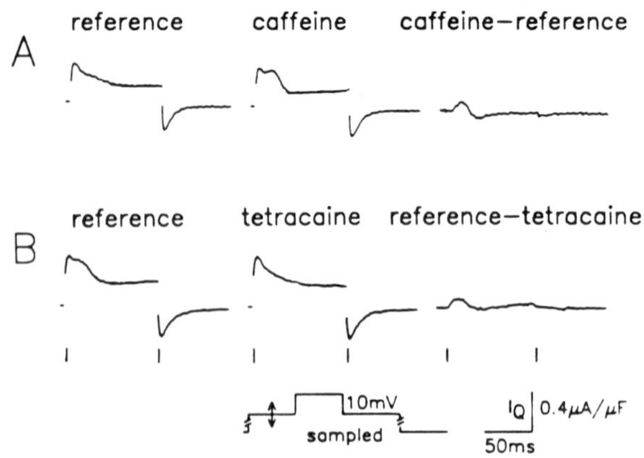

Fig. 4. Opposite effects of caffeine and tetracaine on the charge movements in single vaseline gap. (A) 0.75 mM caffeine shifted the contraction threshold by 12 mV to more negative membrane potentials and enhanced the hump detected at 5 mV below the reference threshold. (B) 25 μM tetracaine suppressed the hump recorded 5 mV above the threshold, measured in reference, together with shifting the threshold by 7 mV to more positive potentials. Though traces before (left) and after (middle) the application of the drugs were not corrected for baseline the difference records (right) show no steady level at the end of the ON proving that fiber conditions were hardly altered. Pulse protocol shown in inset. Fibers 81104A and 81130, sl=2.1 μm and 2.5 μm, T=6 and 5.5°C, V_{test}= -47 and -45 mV for caffeine and tetracaine, respectively.

mV below and above the contraction threshold in reference solution and in the presence of 0.75 mM caffeine (A) and 25 μM tetracaine (B), respectively.

The addition of 0.75 mM caffeine shifted the contraction threshold to more negative membrane potentials, from -42 mV to -54 mV in this fiber. The depolarization to 5 mV below the reference threshold thus became slightly suprathreshold in caffeine. The drug, which is a direct agonist of SR calcium release, shifted the hump component too, as shown in the middle row of the figure. There is a small undershoot below the baseline, as mentioned earlier, that seems to be associated with the hump since it is not present in the reference trace. The difference record shows a hump at the ON and little charge at the OFF.

To compare the component isolated with the above methods to the original definition of Q_γ the stepped pulse protocol was repeated in the presence of 25 μM tetracaine (Fig. 4B), a concentration that is 10x lower than that used in previous studies (Huang, 1982; Vergara and Caputo, 1983). Together with shifting the contraction threshold by 7 mV, the drug clearly suppressed the late component that was present on the reference trace, measured 5 mV above the reference contraction threshold (-50 mV). The difference record, calculated as reference-tetracaine, showed the characteristic hump at the ON with very little charge at the OFF. There is a striking similarity between the component isolated with 25 μM tetracaine and those isolated with caffeine and with the methods in the previous figures arguing that interventions specific for SR calcium release affect the tetracaine sensitive charge.

Ríos and Pizarro (1988) suggested and Jacquemond et al. (1991) provided evidence that the peak component of calcium release is caused by a calcium-induced release mechanism. Since tetracaine is known to inhibit calcium-induced calcium release in skinned fibers (Endo, 1985) we propose that tetracaine, at this low concentration, exerts its effect on the SR release channel and that the effect on the hump component of intramembrane charge movement is secondary to the inhibition of calcium release.

The results presented so far thus show that interventions that affect the release of calcium from the SR selectively suppress, or enhance in case of caffeine, the late component of charge movement. They also suggest a common site of action for these drugs, namely, the SR calcium release channel.

How can a simple model account for the results presented above? Assume that there are negatively charged groups close to the intracellular side of the molecule of the voltage sensor. The microscopic voltage that the charged particles will then experience is the sum of the macroscopic or command voltage and the voltage contributed by the fixed negative charges. Due to a depolarization of the T-tubular membrane the voltage sensor changes its conformation, this is detected as the Q_β part of charge movement. When calcium is, consequently, released into the junctional space, it binds to the fixed charges and decreases the negative voltage that they contribute. This appears as an extra depolarization and thus causes extra charges, Q_γ, to move (Pizarro et al., 1991).

Computer simulations were done to test the model outlined above. The release flux was derived from the myoplasmic calcium transient and a one dimensional diffusion model was used to calculate the changes in the junctional $[Ca^{2+}]$, that is, to simulate the diffusion of calcium from the junctional space into the myoplasm. The usual calcium removal model (Brum et al., 1988a) described the changes in the myoplasmic calcium. The junctional calcium transient was then used to calculate the calcium bound to the hypothetical site. The changes in site occupancy were converted into changes in voltage which was then added to the clamp voltage. The microscopic voltage

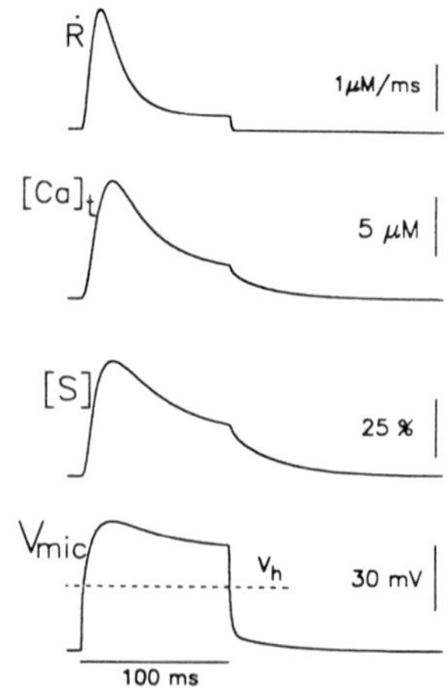

Fig. 5. Consecutive steps in the computer simulation of the calcium binding model of the hump component. First the release flux from the SR (R) is calculated from the measured calcium transient. A simple one dimensional diffusion model (Diff. coef.= 2.8 X 10^{-6} cm^2/s) is used to derive the calcium transient in the triadic junction ($[Ca]_t$). A calcium binding site is supposed to exist close to the intracellular face of the DHP receptor and its saturation ([S]) is calculated using first order kinetics (k_{on}=100 1/μM/s, k_{off}=2000 1/s). The negative charges of the binding site contribute (assumed value= -35 mV when all sites are free of calcium) to the microscopic voltage (V_{mic}) making it more negative than the holding potential (V_h). An increase in site occupancy appears as a decrease in this negative voltage and hence as a further depolarization, whereas a decrease in occupancy is experienced as a hyperpolarization by the voltage sensor. V_{test}= -60 mV.

thus obtained was used to calculate the rearrangement of the charged voltage sensors, assuming a single-barrier Eyring model.

Calculated time courses of different variables in the above cycle are represented in Fig. 5. The voltage-clamp pulse depolarized the membrane from -100 to -60 mV for 100 ms causing a marked calcium release from the SR (R) which inactivated during the pulse and returned to the base-line after repolarization. The time course of the junctional calcium transient ($[Ca]_t$) closely resembled that of the release, except that after repolarization the $[Ca^{2+}]$ in the junctional space was still elevated. Note that the microscopic voltage (V_{mic}) starts from a more negative value than the holding potential, due to the presence of the negative charges. It has a second, slower rising phase that is followed by a decline, and its return to the resting value following repolarization is slow. This can be understood from the time-course of the junctional calcium transient that has an early increase, due

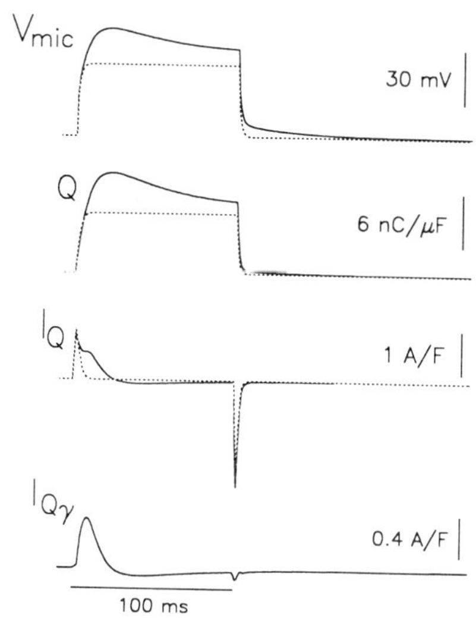

Fig. 6. The isolation of the hump component in the calcium binding model. The simulation was carried out using the release flux shown in Fig. 5 (solid lines) or setting the release arbitrarily to zero (dashed lines) to simulate the situation when the release of calcium from the SR is blocked. The charge in the activating position (Q) was calculated from the microscopic voltage (V_{mic}) using a one barrier Eyring rate model. The charge movement current (I_Q) was obtained by derivating the total charge. The hump component ($I_{Q\gamma}$) is the difference between the charge movement currents obtained in the presence (solid) and absence (dashed) of calcium release. Same parameters for the simulation as in Fig. 5.

to the release of calcium from the SR, and a decline, due to the inactivation of release and to the diffusion of calcium into the myoplasm. The occupancy ([S]) of the hypothetical sites driven by the junctional calcium transient therefore first increases then decreases. Consequently the microscopic voltage first depolarizes further and then repolarizes somewhat during the

ON. The slow return of the junctional calcium to its resting value causes a slow phase of repolarization of the microscopic voltage at the OFF.

The time-course of the charges in their new position (Fig. 6, Q) is determined by the changes in the microscopic voltage (V_{mic}), thus, it shows a secondary slower rising phase with a decline during the pulse and an additional slow part in the OFF. For comparison dotted lines on each trace show the case when the release flux was arbitrarily set to zero to simulate the case when calcium release was blocked. The first derivative of charge, the charge movement current (I_Q) has, consequently, a late component which, if isolated as in the fourth trace ($I_{Q\gamma}$), shows all the major characteristics presented earlier. It has a slow onset with a late peak, a clear undershoot and a very slow OFF. The undershoot is due to the slight repolarization of the microscopic voltage and the consequent inward movement of charges during the ON, while the slow OFF corresponds to the slow return of the microscopic voltage to its resting value.

Since the kinetics of the additional voltage follows that of the junctional calcium transient the kinetics of the extra charge does too. This in turn means that the extra current, the first derivative of charge, will resemble the first derivative of the junctional calcium transient or, which is almost the same, the first derivative of the release flux.

In conclusion, we've presented some experimental evidence that the alteration of the calcium release process affects the hump component of intramembrane charge movement. The proposed calcium binding model accounts for the effects of all drugs that influence calcium release and it describes the kinetic- and voltage- dependence of the hump component, too. The model also implies a positive feed-back in excitation-contraction coupling as has been pointed out elsewhere (Pizarro et al., 1990).

This work was supported by NIH grants to E. Stefani and E. Ríos, by an MDA grant to M.F. Schneider and an MDA fellowship to J. García.

REFERENCES

Adrian, R. H. and Peres, A., 1977, A 'gating' signal for the potassium channel? Nature (Lond.), 267:800.

Almers, W., Adrian, R. H. and Levinson, S. R., 1975, Some dielectric properties of muscle membrane and their possible importance for excitation-contraction coupling, Ann. NY Acad. Sci., 264:278.

Brum, G., and Ríos, E., 1987, Intramembrane charge movement in frog skeletal muscle fibres. Properties of charge-2, J. Physiol. (Lond.), 387:489.

Brum, G., Ríos, E., and Schneider, M. F., 1988a, A quantitative model of calcium removal from the myoplasmic solution (Appendix to Brum, G., Ríos,

E., and Stefani, E. (1988) Effects of extracellular calcium on calcium movements of excitation-contraction coupling in frog skeletal muscle fibers), J. Physiol. (Lond.), 398:441.

Brum, G., Ríos, E., and Stefani, E., 1988b, Effects of extracellular calcium on calcium movements of excitation-contraction coupling in frog skeletal muscle fibers, J. Physiol. (Lond.), 398:441.

Csernoch, L., Huang, C. L.-H., Szücs, G., and Kovács, L., 1988, Differential effects of tetracaine on charge movements and Ca^{2+} signals in frog skeletal muscle, J. Gen. Physiol., 92:601.

Csernoch, L., Pizarro, G., Uribe, I., Rodríguez, M., and Ríos, E., 1991, Interfering with calcium release suppresses I_γ, the "hump" component of intramembranous charge movement in skeletal muscle, J. Gen. Physiol., 97:845.

Duane, S. and Huang, C. L.-H., 1982, A quantitative description of the voltage dependent capacitance in frog skeletal muscle in terms of equilibrium statistical mechanics. Proc. R. Soc. Lond. B Biol., 215:75.

Ebashi, S., 1991, Excitation-contraction coupling and the mechanism of muscle contraction, Ann. Rev. Physiol., 53:1.

García, J., Amador, E., and Stefani, E., 1989, Relationship between myoplasmic Ca transients and Ca current in frog skeletal muscle, J. Gen. Physiol., 94:973.

Horowicz, P., and Schneider, M. F., 1981, Membrane charge moved at contraction thresholds in skeletal muscle fibres, J. Physiol. (Lond.), 314:595.

Huang, C. L.-H., 1981, Effects of local anaesthetics on the relationship between charge movements and contractile thresholds in frog skeletal muscle, J. Physiol. (Lond.), 320:381.

Huang, C. L.-H., 1983, Time domain spectroscopy of the membrane capacitance in frog skeletal muscle, J. Physiol. (Lond.), 341:1.

Huang, C. L.-H., 1989, Intramembrane charge movements in skeletal muscle, Physiol. Rev., 68:1197.

Hui, C. S., 1983a, Pharmacological studies of charge movement in frog skeletal muscle, J. Physiol. (Lond.), 337:509.

Hui, C. S., 1983b, Differential properties of two charge components in frog skeletal muscle, J. Physiol. (Lond.), 337:531.

Imagawa, T., Smith, J. S., Coronado, R., and Campbell, K. P., 1987, Purified ryanodine receptor from skeletal muscle sarcoplasmic reticulum is the Ca^{2+} permeable pore of the calcium release channel, J. Biol., Chem., 262:16636.

Inui, M., Saito, A., and Fleischer, S., 1987, Purification of the ryanodine receptor and identity with feet structures of junctional terminal cisternae of sarcoplasmic reticulum from fast skeletal muscle, J. Biol. Chem., 262:1740.

Jacquemond, V., Csernoch, L., Klein, M. G., and Schneider, M. F., 1991, Voltage-gated and calcium-gated calcium release during depolarization of skeletal muscle fibers, Biophys. J., 60:in press.

Kovács, L., Ríos, E., and Schneider, M. F., 1983, Measurement and modification of free calcium transients in frog skeletal muscle fibres by a metallochromic indicator dye, J. Physiol. (Lond.), 343:161.

Kovács, L., and Schneider, M. F., 1978, Contractile activation by voltage clamp depolarization of cut skeletal muscle fibres, J. Physiol. (Lond.), 277:483.

Kovács, L., and Szücs, G., 1983, Effect of caffeine on intramembrane charge movement and calcium transients in cut skeletal muscle fibres of the frog, J. Physiol. (Lond.), 341:559.

Meissner, G., 1986, Ryanodine activation and inhibition of the Ca^{2+} release channel of sarcoplasmic reticulum, J. Biol. Chem., 261:6300.

Melzer, W., Ríos, E., and Schneider, M. F., 1987, A general procedure for determining calcium release from the sarcoplasmic reticulum in skeletal muscle fibers, Biophys. J., 51:849.

Pizarro, G., Csernoch, L., Uribe, I., Rodríguez, M., and Ríos, E., 1991, The relationship between Q_γ and Ca release from the sarcoplasmic reticulum in skeletal muscle, J. Gen. Physiol., 97:913.

Pizarro, G., Rodriguez, M., Csernoch, L., and Ríos, E., 1990, Positive feedback in skeletal muscle E-C coupling, Biophys. J., 57:401a.

Ríos, E., and Pizarro, G., 1988, Voltage sensors and calcium channels of excitation-contraction coupling. NIPS, 3:223.

Ríos, E., and Pizarro, G., 1991, The voltage sensor of excitation-contraction coupling in skeletal muscle, Physiol. Rev., 71:849.

Schneider, M. F., and Chandler, W. K., 1973, Voltage dependent charge movement in skeletal muscle: a possible step in excitation-contraction coupling, Nature, 242:244.

Schneider, M. F., and Simon, B. J., 1988, Inactivation of calcium release from the sarcoplasmic reticulum in frog skeletal muscle, J. Physiol. (Lond.), 405:727.

Simon, B. J., Klein, M. G., and Schneider, M. F., 1991, Calcium dependence of inactivation of calcium release in frog skeletal muscle, J. Gen. Physiol., 97:437.

Szücs, G., Csernoch, L., Magyar, J., and Kovács, L., 1991, Contraction threshold and the "hump" component of charge movement in frog skeletal muscle, J. Gen. Physiol., 97:897.

Vergara, J., and Caputo, C., 1983, Effects of tetracaine on charge movements and calcium signals in frog skeletal muscle fibers, Proc. Natl. Acad. Sci. (USA), 80:1477.

Ca^{2+} ANTAGONISTS AS TOOLS IN THE ANALYSIS OF EXCITATION-CONTRACTION COUPLING IN SKELETAL MUSCLE FIBRES

H.Ch. Lüttgau, Th. Böhle and A. Schnier

Department of Cell Physiology
Ruhr-University of Bochum
D-4630 Bochum, F.R.G.

INTRODUCTION

Organic Ca^{2+} antagonists are specific substances which interfere with the gating of Ca^{2+} channels. They are of great interest as they are effective drugs in the therapy of different heart diseases. At present one distinguishes between three main classes of Ca^{2+} antagonists: dihydropyridines (DHP_s), phenylalkylamines (PAA_s) and benzothiazepines (BTZ_s). These classes appear to have different binding sites at the Ca^{2+} channel which show a reciprocal allosteric interaction among each other and with additional Ca^{2+} binding sites (Glossmann et al., 1985).

The voltage sensor in the T-tubular membrane of skeletal muscle fibres (Schneider and Chandler, 1973) which controls the release of Ca^{2+} from the sarcoplasmic reticulum (SR) via the Ca^{2+} release channel (ryanodine receptor) appears to be either a modified Ca^{2+} channel or directly identical with this Ca^{2+} channel (dihydropyridine (DHP) receptor) (Lamb and Walsh, 1987; Berwe et al., 1987; Rios and Brum, 1987; Tanabe et al., 1988).*

These associations were revealed with the help of Ca^{2+} antagonists which in this way became interesting tools in the analysis of excitation-contraction coupling. However, that Ca^{2+} antagonists modify the internal release of Ca^{2+} or contractile force can in many cases only be observed with special pulse programmes or modified external solutions. Frank et al. (1988) and Frank (1990), for example, showed that nitrendipine (DHP) remained without effect on twitches induced by action potentials in mammalian and frog muscles, even when applied at

*In the present article the term "voltage sensor" is used to describe the "remote control" of Ca^{2+} release from the SR. From the interconnections described here it follows that the sensor is more or less identical with the voltage sensing system of the T-tubular Ca^{2+} channel. (In addition there may exist L-type Ca^{2+} channels which are no voltage sensors.) The term dihydropyridine (DHP) receptor encloses voltage sensor and Ca^{2+} channel.

relatively high concentrations. A related experiment is shown in Fig. 1, in which a small muscle was activated by directly inducing action potentials at different frequencies. After the application of relatively high concentrations of the (-)isomer of D888 (PAA) or nifedipine (DHP), no alterations in force development could be detected. The experiment quite convincingly shows that these drugs (at concentrations which were much higher than therapeutical doses) have no effects on skeletal muscle fibres when they are stimulated for short periods by action potentials. Only when tetani or depolarizations are induced for a longer time or when the composition of the external solution is altered effects on force development or Ca^{2+} release become visible. These results reveal that Ca^{2+} antagonists interfere with the gating of the voltage sensor to control Ca^{2+} release from the sarcoplasmic reticulum.

The interest in Ca^{2+} antagonists as tools for studying of excitation-contraction coupling was initiated by the observation of Eisenberg et al. (1983), who showed that in the presence of the phenylalkylamine D600 skeletal muscle fibres could develop only one normal potassium contracture after which they became "paralysed". By emphasizing this effect the Ca^{2+} antagonists can be divided into two groups:
a. Those which upon depolarization transfer both, the Ca^{2+} channel and the voltage sensor, into a stable secondary in-

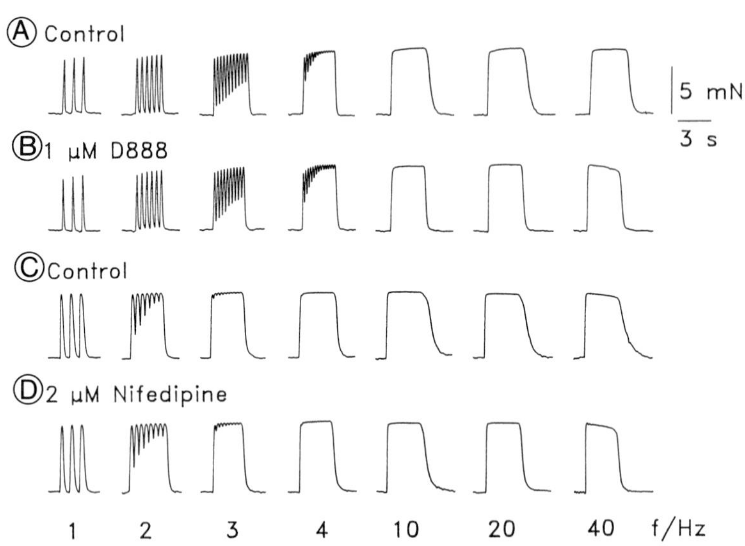

Fig. 1. Twitches and short tetani (3 s) at different frequencies induced under control conditions (A, C) and after the application of 1 µM of (-)-D888 (B) or 2 µM nifedipine (D) in a whole muscle (M. cutaneus pectoris of Rana temporaria). Isometric force was measured after direct stimulation (action potentials). Ringer's solution with 5 mM-K^+ at constant $K^+ \times Cl^-$ product (resting potential ca -84 mV); 10 °C. Similar results were obtained with 30 µM d-cis-diltiazem. (C. Ortland, unpublished)

activated state (paralysis), from which both systems recover rather slowly but completely after repolarization (PAAs, BTZs).

b. Those which accelerate primary inactivation but do not promote the transition into a stable inactivated state. The dihydropyridines belong to this group. They usually show a stronger blocking effect on the Ca^{2+} current through the T- tubular membrane than on Ca^{2+} release from the SR.

METHODS

The experiments shown in Figs. 1-5 were performed with skeletal muscle fibres of the frog (Rana esculenta and R. temporaria). Besides the test with a whole muscle presented in Fig. 1 two methods were applied:
1. The force measurements were performed with short toe muscle fibres (M. lumbricalis longissimus digiti IV). Isometric force and Ca^{2+} current of single fibres were estimated under voltage clamp control with two internal microelectrodes (for details see Gomolla et al., 1983). 2. Charge movements and Ca^{2+} transients were measured again under voltage clamp control in cut fibres (M. semitendinosus) in the double-vaseline-gap system (for details see Kovacs et al., 1983).

Benzothiazepines (PAAs) and phenylalkylamines (BTZs)

The characteristic effect of these Ca^{2+} antagonists can be best demonstrated in an experiment in which the fibre was repeatedly depolarized for a rather long duration (30 s) each time followed by a repolarizing phase of only 17 s (Fig. 2). Under normal conditions 17 s were sufficient for full restoration at a holding potential of -120 mV so that no decline in force took place. In the presence of diltiazem, a benzothiazepine derivative, however, force decreased during repeated stimulations until after 8 contractures it could no longer be induced. In the presence of the drug, a time of 17 s in the repolarized state became too short to ensure restoration. A subsequent long-lasting repolarization, however, resulted in a complete restoration of the ability to contract (not shown). These results can be described with the following state diagram of the voltage sensor:

$$R \leftrightarrow A \leftrightarrow I \leftrightarrow I_o$$

At negative membrane potentials the system is in the resting state R. Upon depolarization it turns transiently to the active (A) and then to the inactivated state (I). If depolarization continues the drug induces a transition into a stabilized inactivated or paralysed state (I_o), from which it recovers only slowly . The decline in Ca^{2+} current or force during repetitive stimulation has been described as "use dependent inactivation" (Kanaya et al., 1983; Böhle, 1991). However, in the present case it is not use dependence in its restricted original meaning. The decline depends on the relation of the time during which the fibre was depolarized or repolarized, respectively, rather than on the number of transitions between the resting and the inactivated states of the sensor (Böhle, 1991).

Erdmann and Lüttgau (1989) investigated the effects of

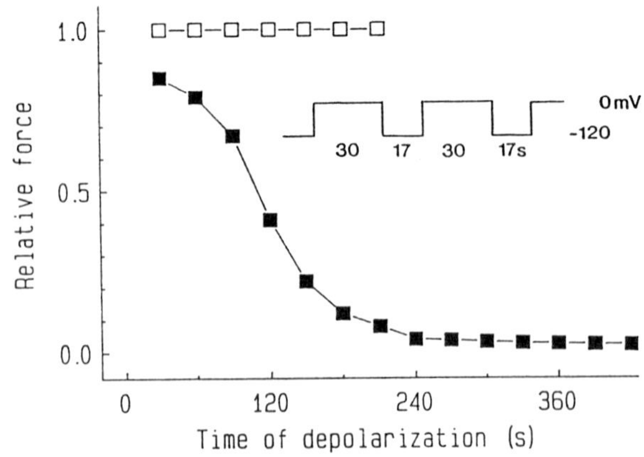

Fig. 2. The effect of repetitive stimulation with relatively long-lasting depolarization periods (the pulse programme is shown in the inset) on force development in a single muscle fibre (see METHODS, voltage-clamp in short toe muscle fibres). □, control, ■, 1 µM d-cis diltiazem; two different fibres; 10 °C. (From Th. Böhle, 1991)

the isomers of D888, a very potent phenylalkylamine, in greater detail. In Fig. 3A, 0.2 µM (-)-D888 was applied to the toe muscle preparation. A depolarization for 3 min transferred the voltage sensing system into the paralysed state I_o. Subsequently, restoration was initiated at -80 mV and its extent checked with short depolarizing pulses. While under normal conditions about 1 min was needed for restoration we observed an increase in restoration time by about 1-2 orders of magnitude in the presence of D888. In related experiments with D600, another phenylalkylamine derivative, Feldmeyer et al. (1990) showed that this drug retarded the restoration of both, intracellular Ca^{2+} transients and intramembrane charge movements (see Fig. 3B). The parallel disappearance and reappearance of both quantities strengthens the concept that these charge movements originate from the DHP receptor, which controls the release of Ca^{2+} from the SR. It is worth mentioning that, in the presence of derivatives of these two classes of Ca^{2+} antagonists, a full restoration of Ca^{2+} release and Ca^{2+} current through the slow Ca^{2+} channel in the T-system (DHP receptor) remains possible although long-lasting repolarization times are needed.

Erdmann and Lüttgau (1989) assumed that D888 can only bind to the inactivated sensor (I), thus transforming it into a stabilized secondary inactivated state (I_o). A selective binding of the drug to the inactivated state should reduce the probability of finding the sensor in the resting state R and should cause a shift of the steady state restoration curve to more negative potentials. Following the modulated receptor

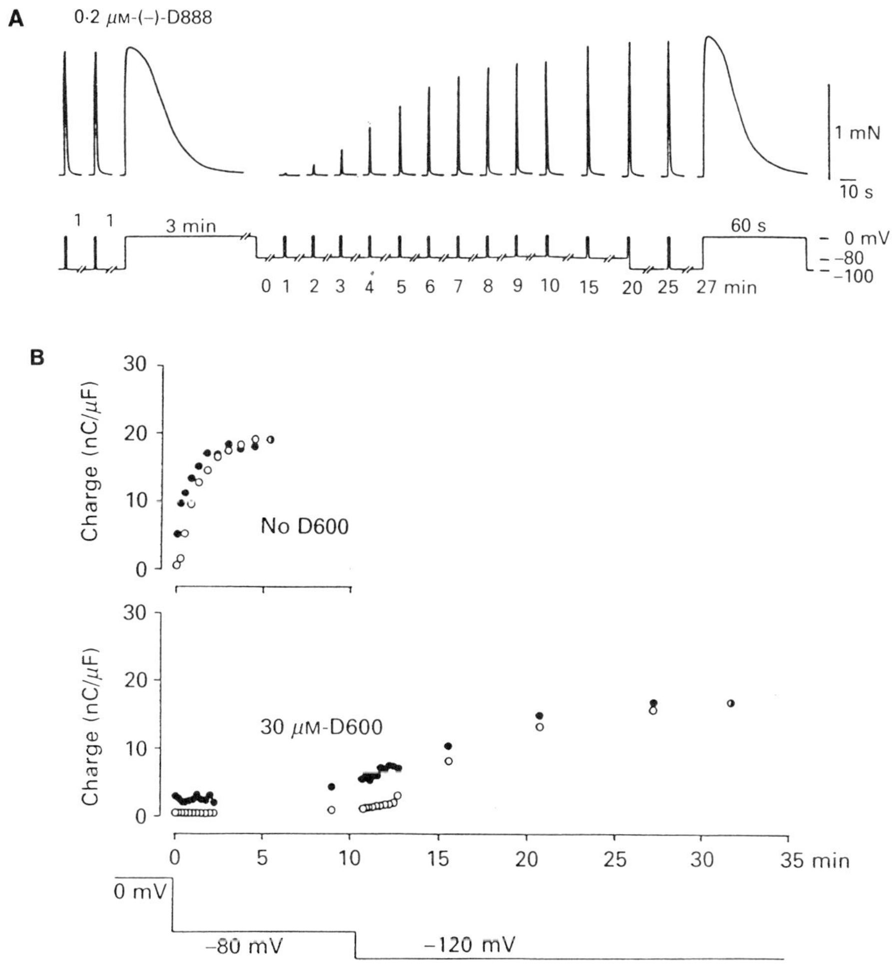

Fig. 3. A. Restoration of force in a single muscle fibre after the application of 0.2 μM (−)-D888 (see METHODS, voltage clamp in short toe muscle fibres); 10 °C. (From Erdmann and Lüttgau, 1989)
B. Time course of charge movement restoration and recovery of Ca^{2+} release. ●, charge movement, O, peak d Ca^{2+}/dt as a measure of the maximum rate of Ca^{2+} release. The values were arbitrarily scaled such that the last value coincided with the corresponding charge value. Above, control, below, in the presence of 30 μM D600 (see METHODS , voltage clamp in cut fibres). 8.5 °C. (Modified, from Feldmeyer et al., 1990)

hypothesis (Hille, 1977; Bean, 1984), the shift of the midpoint voltage of the restoration curve ($\Delta V_{0.5}$) in the presence of different concentrations of the drug S depends on the following equation

$$\Delta V_{0.5} = k \cdot \ln (1 + S/K_D), \qquad (1)$$

with k being a steepness factor and K_D the dissociation constant of the D888-receptor complex (for further details see Erdmann and Lüttgau, 1989). The steady state restoration curves for different concentrations of the (+) and (-) isomers were measured and a least-squares fit of equation 1 to the $\Delta V_{0.5}$ data obtained provided the following values for the free parameters in the fit: k = -6.47 mV; $K_{D(-)}$ = 1.71 nM and $K_{D(+)}$ = 12.9 nM. These results correspond to K_D values reached in binding studies by different authors (for literature see Erdmann and Lüttgau, 1989).

Force restoration experiments as those shown in Fig. 3 in the presence of diltiazem (1-30 µM d-cis and 30 µM l-cis diltiazem) were performed by Böhle (1991). He extended the restoration time of 15 min, used by Erdmann and Lüttgau (1989), to up to 40 min and obtained a potential-dependence of restoration with a midpoint voltage of -49 mV. Since this voltage-dependence corresponded to that which was also reached in the absence of the drug the author concluded that, upon depolarization, diltiazem merely accelerates a transition of the voltage sensor to the stable inactivated state ($I \rightarrow I_0$).

In former experiments (Lüttgau et al., 1986), it had indeed been shown that a paralysis-like stabilized inactivated state could also be induced in the absence of drugs during prolonged depolarizing conditioning pulses, in particular when external Ca^{2+} was replaced by Mg^{2+}. With long-lasting hyperpolarizations the effect could be reversed even in Ca^{2+}-free solutions.

When the potential dependence of inactivation (instead of restoration) is estimated with voltage steps from the resting potential to different conditioning potentials and subsequently to zero (Fig. 4) a midpoint voltage of ca -35 mV is usually measured if the conditioning pulse lasts for only one minute or less. With this pulse duration a transition from I to I_0 can probably be neglected. A further prolongation of the conditioning pulse shifts the midpoint voltage to more negative potentials (Lüttgau et al., 1986). In the steady state it should reach the restoration midpoint voltage of -49 mV obtained by Böhle (1991) in the absence and presence of diltiazem with R, I <u>and</u> I_0 in a stable steady state.

A secondary inactivated state has also been observed in ion channels. Thus, Ruff et al. (1988) measured similar shifts in midpoint voltage of inactivation during long-lasting depolarizations in Na^+ channels.

Although they bind to different sites at the DHP receptor (Glossmann et al., 1985) phenylalkylamines and benzothiazepines principally act in a similar way by facilitating a shift of both Ca^{2+} channel and voltage sensor to a secondary inactivated state. However, as described before different experimental approaches led to deviating interpretations of the mode of action of D888 and diltiazem. These discrepancies are of preeminent nature and need further comments. The modulated receptor hypothesis (Hille, 1977) was applied by Erdmann and Lüttgau (1989) to interpret the great difference between the low dissociation constants ((-)D888 = ~2 nM; d-cis diltiazem = 50 nM) and the concentrations of these drugs applied in the experiment that were by orders of magnitude higher. The hypothesis also explains why an increase of drug concentration far

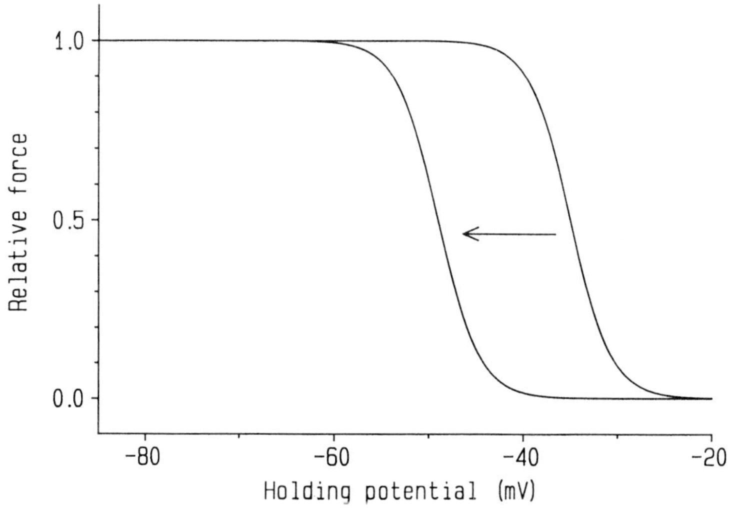

Fig. 4. Schematic drawing of the voltage dependence of inactivation after a short (< 1 min) and a longer-lasting (> 4 min) conditioning pre-pulse, representing steady states between R⟷I and R⟷I⟷I_o, respectively. (Modified, from Böhle, 1990)

beyond the dissociation constant further augmented the physiological effect. The finding by Böhle (1991), on the other hand, i.e. that during long-lasting restoration times the midpoint voltage $V_{0.5}$ approached a value which corresponds to that in the absence of the drugs, speaks against a modulated receptor concept with a specific binding to the inactivated state only. In addition, Erdmann and Lüttgau (1989) themselves admitted that they could not exclude the possibility that a steady state has not yet been reached after a restoration time of 15 min and that restoration might have proceeded at a slower rate. A possible way out of this dilemma - if one wishes to adhere to the modulated receptor hypothesis - would be to assume that the Ca^{2+} antagonists bind first to the inactivated state and, at a slower rate, also to other states, i.e. secondary resting states. This would ultimately shift the restoration curve to its potential dependence in the absence of the drug. Further experiments are needed to clearly distinguish between the two hypotheses or to find an alternative way of interpreting the apparently contradictory observations.

Besides the divergencies described before there is agreement as to the hypothesis that the two classes of Ca^{2+} antagonists facilitate the shift of the DHP receptor to a stable inactivated state I_o. This state can apparently be reached in the absence of the drugs, too, i.e. it is not really a "new" state. In this context it is of interest that Glossmann et al. (1985) found in binding studies that both the phenylalkylamine and the benzothiazepine binding sites are negative-allosterically coupled with a low affinity Ca^{2+} binding site on the Ca^{2+} channel. Since in the experiments by Lüttgau et al. (1986) the secondary inactivated state was reached faster in Ca^{2+}-free solutions, it appears reasonable to assume that the voltage-independent shift from I to I_0 is caused by the dissociation of Ca^{2+} from a binding site on the DHP receptor and that the two drugs allosterically facilitate this process (Böhle, 1991).

Dihydropyridines

Dihydropyridines are known for their high and specific affinity to slow Ca^{2+} channels. In binding studies with skeletal muscle fibres dissociation constants of ~1 nM were estimated. Cognard et al. (1986) measured Ca^{2+} currents with the whole cell mode of patch clamp in myoballs of skeletal muscle from newborn rats. They obtained a 50 % block of current with 13 or 0.15 µM of the dihydropyridine derivative PN 200-110 when inducing the current from a holding potential of -90 and -65 mV, respectively. These effects can be well described with the modulated receptor hypothesis by assuming a high affinity binding of this drug to the inactivated state only. Following equation 1, large shifts (-62 mV with 13 µM PN 200-110, a k-value of -6.5 mV and an assumed dissociation constant of 1 nM) in the potential dependence of inactivation can indeed be expected if this simple assumption is correct. Related results were obtained by Pizarro et al. (1988) in cut muscle fibres of the frog. They applied 1 µM nifedipine and observed no intracellular Ca^{2+} transients and a change in polarity of intramembrane charge movements, which indicated a transformation of the voltage sensor into the inactivated state. By hyperpolarizing the fibre from -110 to -130 mV they could partially restore Ca^{2+} transients and charge movements of normal polarity. Therefore nifedipine must have caused the expected large shift of the potential dependence of Ca^{2+} release (and force) to more negative potentials.

Different results were obtained with intact, i.e. non-cut muscle fibres. Neuhaus et al. (1990) showed that nifedipine caused a concentration-dependent block of the slow Ca^{2+} channel with a 50 % decrease at ~ 1 µM nifedipine and only a minor shift of the potential dependence of inactivation to more negative potentials. The time course of force development remained unaltered even at extremely high concentrations of the drug (5-10 µM). Since the Ca^{2+} current block could not be reversed by hyperpolarization, the effects of nifedipine might be explained with a low affinity and a state-independent binding to the DHP receptor. That nifedipine did not cause a shift of the inactivation curve does not necessarily exclude a high affinity binding of the drug. If the shift in the potential dependence of inactivation induced by dihydropyridines, as described above, rested on a specific binding to the inactivated state only, the absence of this shift in non-cut whole muscle fibres of the same animal species might then be explained by an additional high affinity binding to other states so that a voltage shift loses its foundation (Neuhaus et al., 1990). The question if the binding of dihydropyridines to different states of the receptor can be modified, e.g. by alterations in the composition of the internal solution, long-lasting depolarizations or other means, shows that the interpretation of therapeutical effects with results obtained under quite different conditions in isolated cells or in "unphysiological" solutions must be regarded with caution.

Of particular importance is the observation that dihydropyridines shorten contractures in frog muscles if external

Ca^{2+} is replaced by Mg^{2+} (Dulhunty and Gage, 1988; Neuhaus et al., 1990) or if Na^{+} remains the only metal ion in the external solution (Fig. 5). These results are similar to those obtained by Frank (1990) in mammalian muscles even in the presence of external Ca^{2+}. After the application of dihydropyridines he observed normal or even larger twitches with nifedipine. Contractures became shorter (nifedipine 10^{-4} M) or were suppressed almost completely (10^{-5} M nitrendipine). The stronger effects observed by Frank (1990) may partly be explained by the higher temperature (22 °C) in these experiments. In order to interpret the shortening of contractures, more must be said about the role of metal ions in excitation-contraction coupling, which is indeed the central topic of the present symposium.

In 1979, Lüttgau and Spiecker showed that the replacement of external Ca^{2+} by the same amount of Mg^{2+} causes a large shift of the voltage dependence of inactivation to more negative potentials, a phenomenon that is difficult to interpret

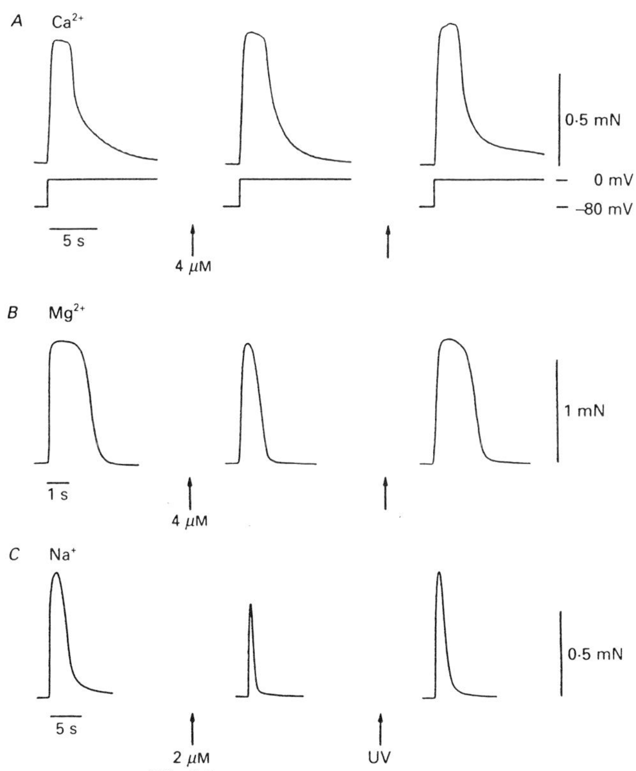

Fig. 5. Contractures of single muscle fibres induced by a depolarizing voltage clamp step (20 s in A and B, 10 s in C) from -80 to 0 mV with Ca^{2+}, Mg^{2+} or Na^{+} in the external solution. In each row a control is followed by a contracture in the presence of 4 (A, B) and 2 (C) μM nifedipine and after the photoconversion of the drug into an inactive form by U.V.-light (see METHODS, voltage clamp in short toe muscle fibres). Three different fibres; 15 °C. (From Neuhaus et al., 1990)

with the concept that divalent cations influence voltage-dependent parameters by screening surface charges at the external side of the membrane.

A hypothesis put forward by Pizarro et al. (1989) might explain this phenomenon. These authors measured intramembrane charge movements and Ca^{2+} transients in cut muscle fibres in the presence of different metal ions in the external solution and showed that they are needed for the normal function of the voltage sensor. They measured an efficiency sequence of the order $Ca^{2+} > Sr^{2+} > Mg^{2+} > Ba^{2+} > Na^{+} > K^{+} > Cs^{+}$. To explain their results they assumed that this sequence corresponds to the affinity of these metal ions to a specific external site at the voltage sensor and that a dissociation of the cation upon depolarization is the cause of inactivation. Schnier and Lüttgau (1991) confirmed this concept by measuring the voltage dependence of force inactivation at different concentrations of several metal ions (Ca^{2+}, Ba^{2+}, Na^{+}, Li^{+}) in the external solution. The observed shift in the midpoint voltage of inactivation $V_{0.5}$ to more negative potentials with decreasing concentrations of each of these metal ions could be well explained with the modulated receptor hypothesis by assuming a high and low affinity binding to a metal ion binaing site in the resting and activated states, respectively.

The effect of nifedipine on force development could be accounted for by an allosteric action on the Ca^{2+} (metal) ion binding site. When a metal ion with high affinity like Ca^{2+} is bound, nifedipine remains without any effect. However, when the site is occupied by metal ions with a lower affinity, nifedipine appears to accelerate the dissociation of these cations by allosteric means, thus causing a shortening of the contracture (Fig. 5).

CONCLUSION

Ca^{2+} antagonists bind to the so-called voltage sensor in the T-tubular membrane of skeletal muscle fibres which, in some way, controls the release of Ca^{2+} from the sarcoplasmic reticulum and which is likewise a Ca^{2+} channel. They interfere with the gating of this system by facilitating transitions to different inactivated states, most probably by promoting the dissociation of Ca^{2+} from specific binding sites at the DHP receptor. The main conclusions developed in the present survey are summarized in Fig. 6. It illustrates the different states of the voltage sensor in the T-tubular membrane with the assumed binding sites. A Ca^{2+} binding site (1) binds metal ions, preferably Ca^{2+} with high affinity in the resting and low affinity in the activated state. A dissociation of the metal ion shifts the system to the inactivated state, a process which is thought to be allosterically promoted by dihydropyridines (DHP_s). Phenylalkylamines (PAA_s) and benzothiazepines (BTZ_s) are bound to different sites and facilitate a further transition to a stable inactivated state. This is probably achieved by an allosteric promotion of Ca^{2+} dissociation from a second Ca^{2+} binding site (2). Contrary to the primary inactivated state, slow inactivation or paralysis can be neglected under normal conditions. However, it would be of some interest to find out whether in some muscle diseases a secondary inactivated state might develop (Böhle, 1990).

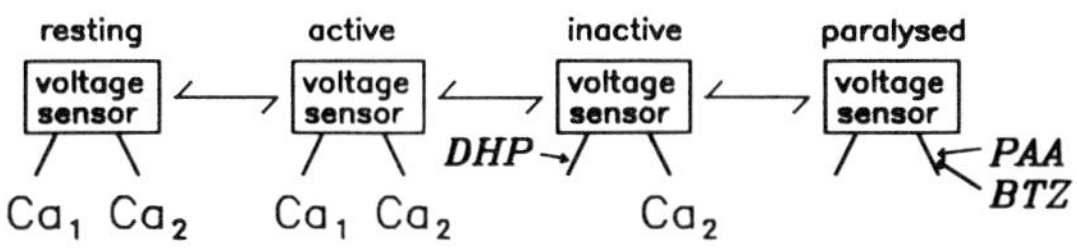

Fig. 6. State diagram of the voltage sensor, summarizing the possible actions of Ca^{2+} ions and Ca^{2+} antagonists on the transition from state R to state I_0 upon depolarization. Further details are explained in the text.

In 1987, Tanabe et al. worked out the primary structure of the α_1 subunit of the muscular DHP receptor by cloning and sequence analysis of DNA complementary to its messenger RNA. They suggested that two regions of the amino acid sequence, one situated at the outer and the other at the inner edge of the integral protein, resemble Ca^{2+} binding sites to some extent. Since the removal of metal cations from the external solution immediately blocks EC coupling, it is assumed that the Ca^{2+} binding site 1 is located at the external side. Following Tanabe et al. (1987), the Ca^{2+} binding site 2 should then be located at the internal side. Several findings favour an internal binding site for phenylalkylamines. In addition, the effect of Ca^{2+} removal on slow inactivation is not as dramatic as that on fast inactivation, which would likewise suggest that the second Ca^{2+} binding site is situated at the internal edge of the DHP receptor. Further experiments are needed to confirm the suggested role of Ca^{2+} in slow inactivation and its position on the DHP receptor.

The interactions between Ca^{2+} and Ca^{2+} antagonists derived from electrophysiological experiments as described here have many features in common with those derived from binding studies (e.g. Glossmann and Strießnig, 1988; Kanngießer and Pongs, 1989). The remaining differences may be explained by incomparable experimental conditions and uncertainties about the state of the DHP receptor in the binding studies. A detailed description is outside the scope of the present short survey.

In the presence of phenylalkylamines and benzothiazepines, force development and Ca^{2+} current disappear and reappear simultaneously with an identical time course on depolarization and repolarization. This may be regarded as an indication that the step from I to I_0 is identical for both systems. The situation is different with nifedipine, which interferes with fast inactivation (Neuhaus et al., 1990). In the presence of Ca^{2+} in the external solution, the Ca^{2+} current could be blocked completely while force development remained unchanged. This difference is of interest. However, it does not necessarily mean that there are two different gating controls. Several explanations are possible. Since e.g. Ca^{2+} release from the SR (controlled by the voltage sensor) is activated noticeably faster than the T-tubular Ca^{2+} current, the described phenomenon could be explained by assuming that nifedipine affects mainly the slower or more "sensitive" processes.

As Ca^{2+} ions are involved in the transition of the DHP receptor to both, the primary *and* the secondary inactivated state, it becomes difficult to clearly distinguish between these two steps. The shift towards the negative direction of the inactivation curve after replacing external Ca^{2+} by Mg^{2+} has formerly been described as an indirect effect (for further details see Lüttgau and Spiecker, 1979). It is now explained with the modulated receptor hypothesis by assuming a lower affinity of the receptor to Mg^{2+}, as compared to Ca^{2+} (Schnier and Lüttgau, 1991). This shift, however, should not be confused with the likewise negative shift of this curve during longer-lasting conditioning pulses (see Fig. 4), as explained by a new steady state in which a secondary inactivated state (I_0) becomes involved. The situation is even more complicated as the release channel (ryanodine receptor) needs myoplasmic Ca^{2+} for its activation, i.e. it possesses a Ca^{2+} binding site, too. Mg^{2+} inhibits Ca^{2+} release and it is assumed that activation of the voltage sensor may in some way lower the affinity of the release channel for Mg^{2+}, thereby removing the resting inhibition so that Ca^{2+} release can be achieved at lower myoplasmic Ca^{2+} concentrations (Lamb and Stephenson, 1991).

REFERENCES

Bean, B. P., 1984, Nitrendipine block of cardiac calcium channels: high-affinity binding to the inactivated state, *Proc. Natl. Acad. Sci. USA*, 81:6388.

Berwe, D., Gottschalk, G., and Lüttgau, H. Ch., 1987, Effects of the calcium antagonist gallopamil (D600) upon excitation-contraction coupling in toe muscle fibres of the frog, *J. Physiol.*, 385:693.

Böhle, Th. 1990, Quantitative Untersuchungen zur Wirkung von Diltiazem auf die elektro-mechanische Kopplung in Skelettmuskelfasern, Dissertation, Fakultät für Biologie der Ruhr-Universität Bochum.

Böhle, Th., 1991, The effect of the benzothiazepine diltiazem on force and Ca^{2+} current in isolated frog skeletal muscle fibres, *J. Physiol.*, in press.

Cognard, C., Romey, G., Galizzi, J.-P., Fosset, M. and Lazdunski, M., 1986, Dihydropyridine-sensitive Ca^{2+} channels in mammalian skeletal muscle cells in culture: electrophysiological properties and interactions with Ca^{2+} channel activator (BayK8644) and inhibitor (PN 200-110), *Proc. Natl. Acad. Sci. USA*, 83:1518.

Dulhunty, A. F., and Gage, P. W., 1988, Effects of extracellular calcium concentration and dihydropyridines on contraction in mammalian skeletal muscle, *J. Physiol.*, 399:63.

Eisenberg, R.S., McCarthy, R.T., and Milton, R.L., 1983, Paralysis of frog skeletal muscle fibres by the calcium antagonist D600, *J. Physiol.*, 341:495.

Erdmann, R., and Lüttgau, H. Ch., 1989, The effect of the phenylalkylamine D888 (Devapamil) on force and Ca^{2+} current in isolated frog skeletal muscle fibres. *J. Physiol.*, 413:521.

Feldmeyer, D., Melzer, W., and Pohl, B., 1990, Effects of Gallopamil (D600) on calcium release and intramembrane charge movements in frog skeletal muscle fibres, *J. Physiol.*, 421:343.

Frank, G. B., 1990, Dihydropyridine calcium channel antagonists block and agonists potentiate high potassium contractures but not twitches in frog skeletal muscle, *Jap. J. Physiol.*, 40:205.
Frank, G. B., Konya, L., and Subrahmanyam Sudha, T., 1988, Nitrendipine blocks high potassium contractures but not twitches in rat skeletal muscle, *Can. J. Physiol. Pharmacol.*, 66:1210.
Glossmann, H., Ferry, D. R., Goll, A., Striessnig, J., and Zernig, G., 1985, Calcium channels and calcium channel drugs: Recent biochemical and biophysical findings, *Arzneim.Forsch.*, 35(II), 12a.
Glossmann, H., and Striessnig, J., 1988, Calcium Channels, *Vitamines and Hormones*, 44:155.
Gomolla, M., Gottschalk, G., and Lüttgau, H. Ch., 1983, Perchlorate-induced alterations in electrical and mechanical parameters of frog skeletal muscle fibres. *J. Physiol.*, 343:197.
Hille, B., 1977, Local anaesthetics: hydrophilic and hydrophobic pathways for the drug-receptor reaction. *J. Gen. Physiol.*, 69:497.
Kanaya, S., Arlock, P., Katzung, B. G., and Hondeghem, L. M., 1983, Diltiazem and verapamil preferentially block inactivated cardiac calcium channels, *J. Molec. Cell. Cardiol.*, 15:145.
Kanngiesser, U., and Pongs, O., 1989, Binding Ca^{2+} to intracellular or to extracellular sites of dihydropyridine receptor of rabbit skeletal muscle discriminates between *in vitro* binding of Ca^{2+}-channel agonist and antagonist, Eur. J. Biochem., 181:467.
Kovacs, L., Rios, E., and Schneider, M.F., 1983, Measurements and modification of free calcium transients in frog skeletal muscle fibres by a metallochromic indicator dye, *J. Physiol.*, 343:161.
Lamb, G.D., and Stephenson, D.G., 1991, Effect of Mg^{2+} on the control of Ca^{2+} release in skeletal muscle fibres of the toad. J. Physiol., 434:507.
Lamb, G. D., and Walsh, T., 1987, Calcium currents, charge movement and dihydropyridine binding in fast- and slow-twitch muscles of rat and rabbit. *J. Physiol.*, 393:595.
Lüttgau, H. Ch., Gottschalk, G., and Berwe, D., 1986, The role of Ca^{2+} in inactivation and paralysis of excitation-contraction coupling in skeletal muscle, in: "Fortschritte der Zoologie, vol. 33, Membrane Control of Cellular Activity", H. Ch. Lüttgau ed., Gustav Fischer Verlag, Stuttgart.
Lüttgau, H. Ch., and Spiecker, W., 1979, The effects of calcium deprivation upon mechanical and electrophysiological parameters in skeletal muscle fibres of the frog, *J. Physiol.*, 296:411.
Neuhaus, R., Rosenthal, R., and Lüttgau, H.Ch., 1990, The effects of dihydropyridine derivatives on force and Ca^{2+} current in frog skeletal muscle fibres, *J. Physiol.*, 427:187.
Pizarro, G., Brum, G., Fill M. Fitts, R., Rodriguez, M., Uribe, I., and Rios, E., 1988, The voltage sensor of skeletal muscle excitation-contraction coupling: a comparison with Ca^{2+} channels, in: "The Calcium Channel: Structure, Function and Implications", M. Morad, , W. Nayler, S. Kazda, and M. Schramm, ed., Springer Verlag, Berlin, Heidelberg, New York.

Pizarro, G., Fitts, R., Uribe, I., and Rios, E., 1989, The voltage sensor of excitation-contraction coupling in skeletal muscle. Ion dependence and selectivity, *J. Gen. Physiol.*, 94:405.

Rios, E. and Brum, G., 1987, Involvement of dihydropyridine receptors in excitation-contraction coupling in skeletal muscle, *Nature*, 325:717.

Ruff, R. L., Simoncini, L., and Stühmer, W., 1988, Slow sodium channel inactivation in mammalian muscle: A possible role in regulating excitability, *Muscle & Nerve*, 11:502.

Schneider, M. F., and Chandler, W. K., 1973, Voltage dependent charge movement in skeletal muscle: a possible step in excitation-contraction coupling. *Nature*, 242:244.

Schnier, A., and Lüttgau, H.Ch., 1991, The effect of extracellular metal cations on excitation-contraction coupling in frog skeletal muscle fibres, *J. Physiol.*, in press.

Tanabe, T., Takeshima, H., Mikami A., Flockerzi, V., Takahashi, H., Kangawa, K., Kojima, M., Matsuo, H., Hirose, T., and Numa, S., 1987, Primary structure of the receptor for calcium channel blockers from skeletal muscle, *Nature*, 328:313.

Tanabe, T., Beam, K.G., Povel, J.A., and Numa, S., 1988, Restoration of excitation-contraction coupling and slow calcium current in dysgenic muscle by dihydropyridine receptor complementary DNA, *Nature*, 336:134.

REGULATION OF CALCIUM SLOW CHANNELS IN CARDIAC MUSCLE AND VASCULAR SMOOTH MUSCLE CELLS

Nicholas Sperelakis, Noritsugu Tohse, and Yusuke Ohya

Department of Physiology and Biophysics
University of Cincinnati, College of Medicine

I. INTRODUCTION

Considerable attention during the past few years has been given to phosphorylation of ion channels as a means whereby the activity of the ion channels can be regulated. This article will cover the evidence that cyclic nucleotides regulate the Ca^{2+} influx into the myocardial cells during each cardiac cycle and into vascular smooth muscle (VSM) cells. This regulation is presumable mediated by phosphorylation(s) of the Ca^{2+} slow channel protein and/or of associated regulatory protein(s). In myocardial cells, such phosphorylation (Fig. 1) presumably (a) increases the number of Ca^{2+} slow channels available for voltage activation during the action potential (AP); (b) increases the probability of their opening, and (c) increases their mean open time. A greater density of available Ca^{2+} channels increases Ca^{2+} influx and inward Ca^{2+} slow current (I_{si}) during the AP, and so increases the force of contraction of the heart. In some VSM cells, phosphorylation by cAMP-PK or cGMP-PK inhibits the Ca^{2+} slow channel activity and thereby produces vasodilation, whereas phosphorylation by PK-C stimulates the Ca^{2+} slow channel activity and produces vasoconstriction.

Besides the standard slow Ca^{2+} channel, a fast-type of Ca^{2+} channel has been found in cardiac muscle and VSM cells on the basis of kinetics (e.g., Bean, 1985; Nilius et al., 1985). These fast Ca^{2+} channels are much more rapidly inactivated than the slow Ca^{2+} channels, are active over a more negative voltage range, and are little affected by cAMP or Ca^{2+} antagonists (Table 1). Their function is as yet unclear.

In addition, a new type of Ca^{2+} channel was discovered in 18-day-old fetal rat ventricular cells (Tohse and Sperelakis, unpublished observations). This residual I_{Ca} remaining in the presence of a high concentration (3 μM) of nifedipine (nifedipine-resistant I_{Ca}) was not blocked by diltiazem, tetramethrine (T-type channel blocker), or ω-conotoxin (N-type channel blocker), and had a half-inactivation potential about 20 mV more negative than the nifedipine-sensitive (L-type channel) I_{Ca}.

II. CARDIAC MUSCLE

A. Regulation of the Slow Ca^{2+} Channels

1. Cyclic AMP dependence of the myocardial slow channels. Cyclic AMP (cAMP) modulates the functioning of the Ca^{2+} slow channels (Shigenobu & Sperelakis, 1972; Tsien et al., 1972; Sperelakis & Schneider, 1976; Schneider et al., 1976; Reuter & Scholz, 1977).

Excitation-Contraction Coupling in Skeletal, Cardiac, and Smooth Muscle
Edited by G.B. Frank *et al.*, Plenum Press, New York, 1992

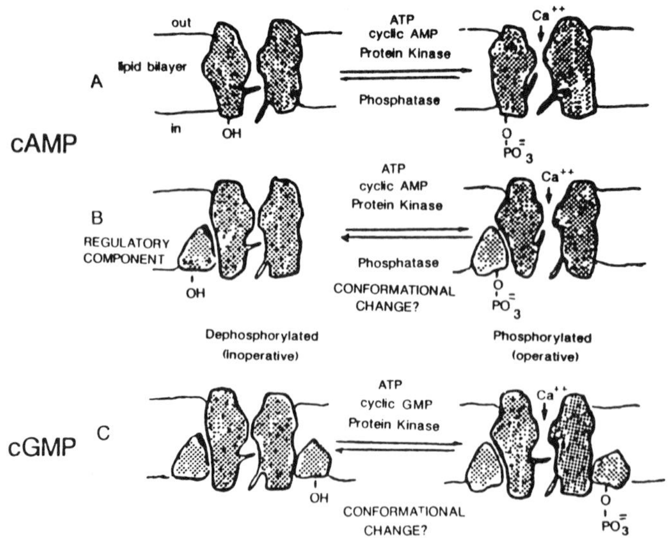

Figure 1

Schematic model for a Ca^{2+} slow channel in myocardial cell membrane in two hypothetical forms: dephosphorylated (or electrically silent) form (left diagrams) and phosphorylated form (right diagram). The two gates associated with the channel, an activation gate and an inactivation gate, are kinetically much slower than those of the fast Na^+ channel. The phosphorylation hypothesis states that a protein constituent of the slow channel itself (A) or a regulatory protein associated with the slow channel (B) must be phosphorylated in order for the channel to be in a state available for voltage activation. Phosphorylation occurs by a cAMP-dependent protein kinase (PK-A) in the presence of ATP. Presumably, a serine or threonine residue in the protein becomes phosphorylated. Phosphorylation may produce a conformational change that effectively allows the channel gates to operate, or increases the diameter of the water-filled pore so that Ca^{2+} can pass through. The slow channel (or an associated regulatory protein) may also be phosphorylated by a cGMP-PK (C), thus mediating the inhibitory effects of cGMP on the slow Ca^{2+} channel. (Modified from Sperelakis & Schneider, 1976).

Histamine and ß-adrenergic agonists, after binding to their specific receptors, lead to rapid stimulation of adenylate cyclase with resultant elevation of cAMP levels. The methylxanthines enter the myocardial cells and inhibit the phosphodiesterase, thus causing an elevation of cAMP. These positive inotropic agents also concomitantly induce Ca^{2+}-dependent slow APs, presumably by increasing I_{si}.

Additional evidence for the regulatory role of cAMP has been obtained. It was shown that the GTP analogue, GPP(NH)P, which directly and irreversibly activates adenylate cyclase, induced slow APs in chick heart cells (Josephson & Sperelakis, 1978). Forskolin, another highly potent activator of adenylate cyclase activity, exerts a strong positive inotropic effect in isolated guinea pig atrial muscle and induces and potentiates slow APs (Spah, 1984; Wahler & Sperelakis, 1986). cAMP iontophoretically microinjected into dog Purkinje fibers and

TABLE 1
Summary of Major Differences Between the Slow (L-Type) and Fast (T-Type) Ca^{2+} Channels

Properties	Ca^{2+} Channels	
	Slow (L-type)	Fast (T-type)
Duration of current	Long-lasting (sustained)	Transient
Inactivation kinetics	Slower	Faster
Activation kinetics	Slower	Faster
Threshold	High (ca. -30 mV)	Low (ca. -50 mV)
Half-inactivation potential	ca. -20 mV	ca. -50 mV
Single-channel conductance	High (18-26 pS)	Low (8-10 pS)
Regulated by cAMP and cGMP	Yes	No
Regulated by phosphorylation	Yes	No
Blocked by Ca^{2+} antagonist drugs	Yes	No (slight)
Opened by Ca^{2+} agonist drugs	Yes	No
Permeation by Me^{2+}	Ba > Ca	Ba ≃ Ca
Inactivation by $[Ca]_i$	Yes	Slight (?)
Recordings in isolated patches	Runs down	Rel. stable

guinea pig ventricular muscle cells induced slow APs in the injected cells (Vogel & Sperelakis, 1981). The effect of the injected cAMP occurred within seconds of the injection. Pressure injection of cAMP, GPP(NH)P, and cholera toxin also rapidly induced and potentiated slow APs (Li & Sperelakis, 1983) (Fig. 2). Cholera toxin has an effect on adenylate cyclase similar to that of GPP(NH)P, namely an irreversible activation of the enzyme. Liposome injection of cAMP into cultured chick heart cells also induced slow APs (Bkaily & Sperelakis, 1985), in agreement with the results obtained by iontophoresis and pressure injection.

The results from a number of other laboratories also support a role for cAMP in stimulating the slow inward Ca^{2+} current in myocardial cells (Table 2). For example, injection of cAMP enhanced I_{si} in isolated single cardiac cells (Irisawa & Kokobun, 1983). Similarly, a photochemical activation method for suddenly increasing the intracellular cAMP level enhanced I_{si} in bullfrog atrial cells (Nargeot et al., 1983). Noise analysis and patch clamp analysis (Cachelin et al., 1983; Trautwein & Hoffmann, 1983; Bean et al., 1984) suggest

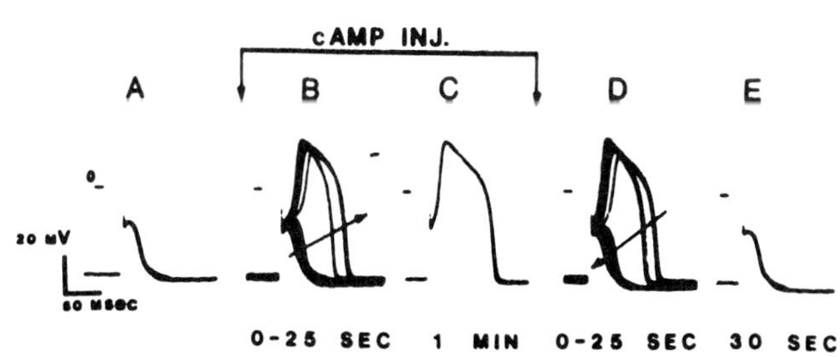

Figure 2

Introduction of Ca^{2+}-dependent slow action potentials (APs) in guinea pig papillary muscle by intracellular pressure injection of cyclic AMP. The muscle was depolarized in 22 mM $[K]_o$ to voltage inactivate fast Na^+ channels. A: Small graded response (stimulation rate 30/min). B: Superimposed records showing the gradual appearance of slow APs upon cyclic-AMP injection over a 25 s period. C: Presence of stable slow APs after injection for 1 min. D: Gradual depression of slow APs over a period of 25 s after stopping injection. E: Complete decay of slow APs 30 s after cessation of cyclic-AMP injection. All records were obtained from one impaled cell. (Data taken from Li & Sperelakis, 1983).

that cAMP increases the number of functional slow channels available in myocardial sarcolemma and/or the probability of opening of a given channel. The net results would be the same, i.e. an increase in the number of slow channels open at any instant of time. Reuter et al. (1982) demonstrated that isoproterenol increased the mean open time of single Ca^{2+} channels and decreased the intervals between bursts; the conductance of the single channel was not increased. Therefore, the increase in the slow inward current produced by isoproterenol could be produced by the observed increase in mean open time of each channel, as well as by an increase in the number of available channels.

2. Phosphorylation Hypothesis. Because of the relationship between cAMP and the number of available slow channels, and because of the dependence of the functioning of the slow channels on metabolic energy, it was postulated that a membrane protein must be phosphorylated in order for the slow channel to become available for voltage activation (Shigenobu & Sperelakis, 1972; Tsien et al., 1972; Sperelakis & Schneider, 1976; Brum et al., 1983). Elevation of cAMP by a positive inotropic agent activates a cAMP-dependent protein kinase (cA-PK), which phosphorylates a variety of proteins in the presence of ATP. One protein that is phosphorylated might be the slow channel protein itself or a contiguous regulatory type of protein (Fig. 1).

Phosphorylation could make the slow channel available for activation by a conformational change that either allowed the activation gate to be opened upon depolarization or effectively increased the diameter of the water-filled pore so that Ca^{2+} could pass through. In this model, the phosphorylated form of the slow channel is the active (operational) form, and the dephosphorylated form is the inactive (inoperative) form. The dephosphorylated channels are electrically silent. Thus, phosphorylation increases the probability of channel opening with depolarization. An equilibrium would exist between the phosphorylated and dephosphorylated forms of the slow channels for a given set of conditions. Thus, agents that act to elevate the cAMP level would increase the fraction of the slow channels that are in the phosphorylated form, and hence readily available for voltage activation.

A test of whether the regulatory effect of cAMP is exerted by means of the cA-PK and phosphorylation was made by intracellular injection of the catalytic subunit of the cA-PK. Such injections induced and enhanced the slow APs (Bkaily & Sperelakis, 1984) and potentiated I_{si} (Brum et al., 1983). Another direct test of the phosphorylation hypothesis was done by liposome injection of an inhibitor (protein) of the cA-PK into cultured chick heart cells (Bkaily & Sperelakis, 1984). The inhibitor inhibited the spontaneous slow APs. Similarly, this protein kinase inhibitor has been directly shown to inhibit I_{si} of cardiac cells (Kameyama et al., 1986). Thus, these results further support the phosphorylation hypothesis.

Based on the rapid decay of the response to injected cAMP (Fig. 2), the mean life span of a phosphorylated channel is likely to be only a few seconds at most, and it is possible that the channels are phosphorylated and dephosphorylated with every cardiac cycle (Li & Sperelakis, 1983). A phosphoprotein phosphatase would dephosphorylate the slow channel, and agents which affect or regulate the phosphatase would affect the life span of the phosphorylated channel. Channel stimulation can be produced either by increasing the rate of phosphorylation (by the cA-PK) or be decreasing the rate of dephosphorylation (inhibition of the phosphatase) (Vogel et al., 1977). For example, the Ca^{2+}-dependent phosphatase, calcineurin, inhibits slow APs in 3-day old chick hearts (Tripathi & Sperelakis, unpublished results). Phosphatases have been directly shown to decrease the Ca^{2+} current in neurons (Chad & Eckert, 1986). In guinea pig ventricular cells (Hescheler et al., 1987), the catalytic subunit of the protein phosphatases 1 and 2A inhibited the Ca^{2+} channel. Further, okadaic acid, a protein phosphatase inhibitor, enhanced the amplitude of the I_{si} pre-stimulated by ß-adrenergic agent, and prolonged the wash-out time of ß-adrenergic agonists (Hescheler et al., 1988). These results support the participation of phosphatase in regulation of the Ca^{2+} slow channels.

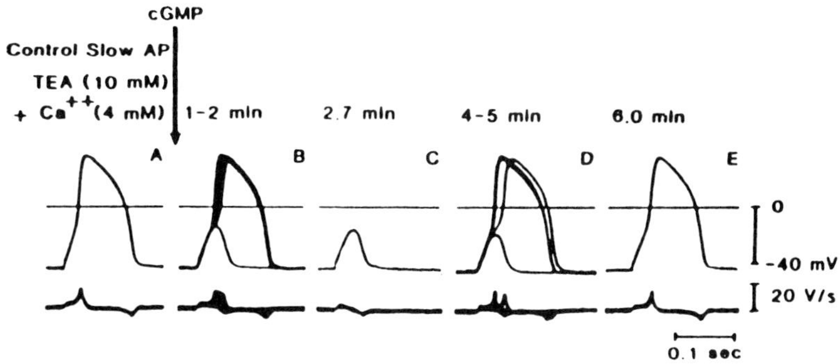

Figure 3

Transient abolition of Ca^{2+}-dependent slow APs by pressure injection of cGMP. A: Control slow AP induced by 10 mM TEA plus 4.0 mM $[Ca]_o$ (fast Na^+ channels inactivated by 25 mM K^+). B-C: 1-2 min following the onset of the cGMP pulse (10 s duration), the slow APs were depressed and then abolished. D-E: At 4-6 min, the slow APs recovered spontaneously to control levels. All records from the same cell. (Taken from Wahler & Sperelakis, 1985).

Consistent with the phosphorylation hypothesis, it has been found that the slow Ca^{2+} channel activity disappears within 90 sec in isolated membrane inside-out patches (Reuter, 1983), and was restored by applying catalytic subunit of cA-PK and ATP-Mg (Armstrong & Eckert, 1987). This is consistent with the washing away of regulatory components of the slow channels or of the enzymes necessary to phosphorylate the channel. Even in whole-cell voltage clamp, there is a progressive run-down of the slow inward current, which is slowed or partially reversed by conditions that enhance cA-PK phosphorylation (Chad & Eckert, 1986).

3. Cyclic GMP Inhibition of the Inward Slow Ca^{2+} Current. Superfusion of isolated guinea pig papillary muscles with 8-Br-cGMP abolished the Ca^{2+}-dependent slow APs and accompanying contractions within 7-20 min (Wahler & Sperelakis, 1985). A similar inhibition by cGMP was shown for the slow APs of guinea pig atria and canine Purkinje fibers (Kohlhardt & Haap, 1978; Mehegan et al., 1985). Intracellular pressure injection of cGMP into cells of guinea pig papillary muscle transiently depressed or abolished slow APs much more quickly (e.g. 1-2 min) (Wahler & Sperelakis, 1985) (Fig. 3). Injection of cGMP into cultured chick heart cells by the liposome method also abolished the slow APs (Bkaily & Sperelakis, 1985). It was also demonstrated that 8-Br-cGMP inhibits the basal I_{si} (unstimulated by cAMP) in voltage-clamped chick ventricular myocytes (Wahler et al., 1990) (Fig. 4).

We recently demonstrated cGMP inhibition of Ca^{2+} slow channel activity at the single-channel level (Tohse & Sperelakis, 1991). In 3-day-old embryonic chick heart cells, we found that the Ca^{2+} slow channels often exhibited long-lasting openings (e.g. for 300 ms) under normal conditions, especially at the more positive command potentials (Tohse & Sperelakis, 1990). That is, the Ca^{2+} slow channels naturally possessed mode 2 behaviour, in the absence of any added Ca^{2+} channel agonist such as the dihydropyridine, Bay-K-8644. Addition of Bay-K-8644 did not further prolong the open times, but appeared to recruit silent Ca^{2+} channels (Tohse et al., submitted). Long-lasting openings were much less frequently observed in 17-day-old embryonic cells (Tohse & Sperelakis, unpublished observations). Addition of 8-Br-cGMP to the bath completely inhibited Ca^{2+} slow channel activity (Fig. 5).

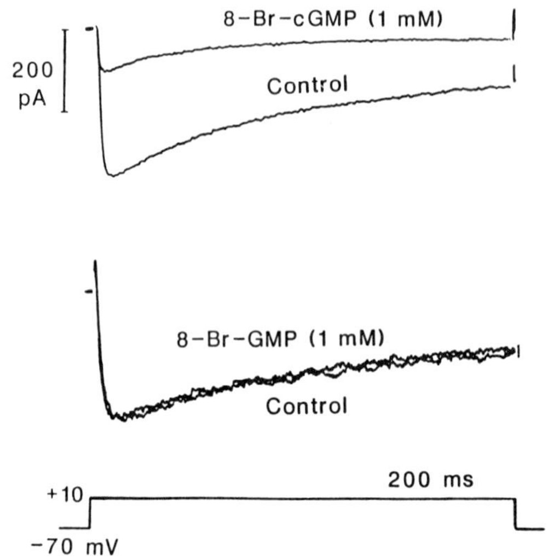

Figure 4

Effect of 8-Br-cGMP on the slow inward current in cultured embryonic chick ventricular myocytes. Upper Traces: Currents elicited by depolarizing pulses from -70 mV to +10 mV in the control bath solution and following 10 min superfusion with a solution containing 1 mM 8-Br-cGMP. Note the large inhibition of $I_{ca(s)}$. Middle Traces: Currents elicited by depolarizing pulses in the control bath solution and following 10 min superfusion with a solution containing 1 mM 8-Br-GMP, the non-cyclic analog of 8-Br-cGMP. the bath solution (20-22° C) included (in mM): 10 $BaCl_2$ and 135 TEA-Cl; the pipette solution included: 150 Cs-glutamate, 5 MgATP, 1 EGTA. (Modified from Wahler et al., 1991).

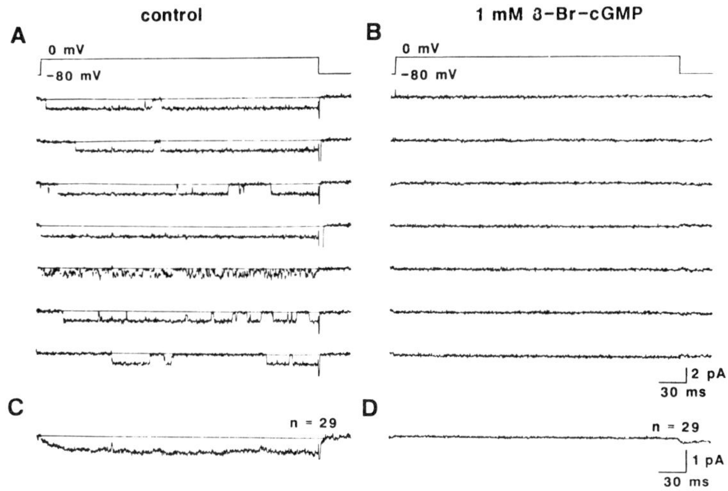

Figure 5

Current recordings from a cell-attached patch showing effect of 8-bromo-cGMP on the Ca^{2+} slow channel activity in a single myocardial cell isolated from a 3-day-old embryonic chick heart. Single-channel currents were evoked by depolarizing voltage pulses to 0 mV from a holding potential of -80 mV, at a repetition rate of 0.5 Hz and duration of 300 msec. A and B: Examples of original current recordings from the same patch, before (A) and after (B) superfusion with 1.0 mM 8-bromo-cGMP. C and D: Ensemble-averaged currents calculated from the current recordings (n - 29). The current tracings were low-pass filtered at 1 kHz and corrected for leakage and capacitive currents. Data taken from Tohse & Sperelakis, 1991.

Therefore, cGMP regulates the functioning of the myocardial Ca^{2+} slow channels in a manner that is antagonistic to that of cAMP (Fig. 6). It is possible that the Ca^{2+} slow channel protein has a second site that can be phosphorylated by cG-PK and which, when phosphorylated, inhibits the slow channel. Another possibility is that there is a second type of regulatory protein that is inhibitory when phosphorylated (Fig. 1).

Another mechanism proposed for cGMP inhibition is based on cGMP depression of the cAMP level. Intracellular application of cGMP inhibited I_{si} of frog ventricular myocytes, but only after the cAMP levels had been increased, i.e. there was no effect of cGMP on the basal I_{si} (unstimulated by cAMP) (Fischmeister & Hartzell, 1987). These authors concluded that cGMP inhibited I_{si} by stimulating a phosphodiesterase, resulting in increased degradation of cAMP. However, 8-Br-cAMP inhibition of slow APs in guinea pig papillary muscles occurs without a decrease in cAMP levels (Thakkar et al., 1988). Thus, it appears that in mammalian ventricular muscle, cGMP inhibits I_{si} directly through a cGMP-mediated phosphorylation (8-Br-cGMP is a potent activator of cG-PK) of some protein involved in the functioning of the slow Ca^{2+} channels (Fig. 1).

A single protein has been found to be specifically phosphorylated by cG-PK in guinea pig sarcolemmal preparations (Cuppoletti et al., 1988). In the presence of the purified kinase plus 10^{-5} M cGMP or 8-Br-cGMP, a protein of approximately 47 kD was phosphorylated. Thus, this substrate, the identity of which is unknown, may be a possible mediator of cGMP-mediated control of cardiac function.

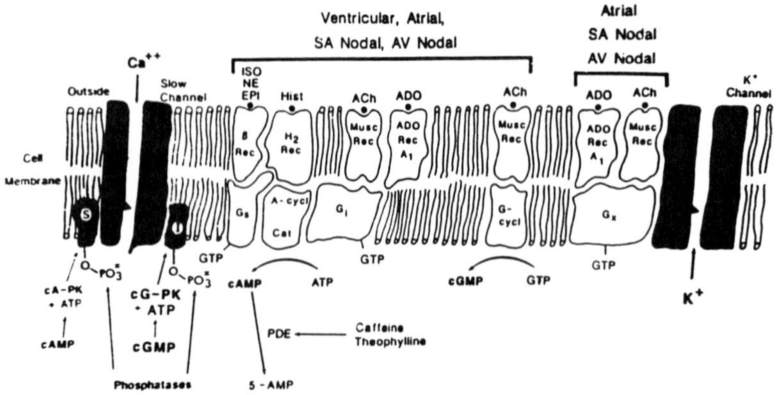

Figure 6

Diagrammatic summary of some of the regulation of the Ca^{2+} slow channels in the myocardial cell membrane and the mechanisms of action of some positive inotropic agents. The ß-adrenergic agonists and histamine H_2 agonists act on a regulatory component (GTP-binding protein) of the adenylate cyclase complex to stimulate cAMP production. The voltage-dependent myocardial slow Ca^{2+} channels are stimulated by cAMP, presumably because a protein constituent (or regulatory component) of the channel must be phosphorylated in order for it to be in a form that is available for voltage activation. cGMP-dependent phosphorylation also regulates the slow channel in a manner antagonistic to cAMP, namely producing inhibition. Thus, the muscarinic receptor activated by ACh can produce inhibition of Ca^{2+} influx by at least the three mechanisms depicted: (1) reversal of adenylate cyclase stimulation produced by B agonists or H_2 agonists; (2) stimulation of guanylate cyclase and production of cGMP; and (3) activation of a K^+ channel ($I_{K(ACh)}$) that produces an outward K^+ current which depresses excitability and terminates the AP earlier, and thereby voltage deactivates the Ca^{2+} slow channels earlier.

4. Acetylcholine. The parasympathetic neurotransmitter acetylcholine (ACh) exerts a negative inotropic effect on the ventricular myocardium that has been stimulated by ß-adrenergic agonists. Activation of the muscarinic receptor by ACh exerts an inhibitory effect on adenylate cyclase and cAMP levels, via the G_i (inhibitory) coupling protein, to reverse the stimulation of adenylate cyclase produced by means of the G_s coupling protein due to, for example, activation of the ß-adrenoceptor (Fig. 6). Thus, ACh may depress Ca^{2+} influx and contraction by reversing cAMP elevation produced by various agonists. For example, in voltage-clamp experiments on cultured chick ventricular cells, ACh depressed I_{si} that had been stimulated by isoproterenol (Josephson & Sperelakis, 1976). ACh also reverses the electrophysiological effects of direct adenylate cyclase stimulation by forskolin (Wahler & Sperelakis, 1986). Additionally, in ventricular cells, ACh may inhibit the slow Ca^{2+} channels due, in part, to elevation of cGMP levels (MacLeod & Diamond, 1986).

5. Calmodulin and Protein Kinase C. Inhibitors of calmodulin (trifluoperazine and calmidazolium) inhibit the slow APs of cultured chick heart cells (Johnson et al., 1983; Bkaily et al., 1984; Bkaily & Sperelakis, 1986). Subsequent injection of calmodulin reverses the inhibition produced by calmidzolium. It appears that maximal activation of the slow channels requires two separate phosphorylation steps (calmodulin-dependent and cAMP-dependent). These may be on the same protein or on two separate proteins.

High concentrations of the α-adrenergic agonist, phenylephrine, have been shown to cause a positive inotropic effect and to increase I_{si} in bovine cardiac muscles (Bruckner & Scholz, 1984). There is still controversy with respect to the effect of activation of the α-adrenoceptors of myocardial cells on the level of cAMP, some investigators reporting increases and others finding no changes. However, the α-adrenoceptor stimulates the phosphatidyl inositol cycle and generation of inositol trisphosphate (IP_3) and diacyl glycerol (DAG) (Brown et al., 1985). IP_3 has been implicated as a second messenger to act on the sarcoplasmic reticulum (SR) to release stored Ca^{2+}. DAG and Ca^{2+} activate PK-C, which phosphorylates a number of proteins. It s presently controversial whether PK-C is involved in regulation of the myocardial slow Ca^{2+} channels.

6. Action of Some Cardioactive Agents on Phosphorylation/Dephosphorylation. Some agents that affect the force of contraction of the heart may do so without changing the levels of the cyclic nucleotides. For example, fluoride ion (<1 mM) is a potent positive inotropic agent and potentiates the Ca^{2+}-dependent slow APs and Ca^{2+} influx (I_{si}), but yet does not elevate cAMP (Vogel et al., 1977). Fluoride is known to be a potent inhibitor of phosphatases. Therefore, fluoride may act by inhibiting the phosphatase which dephosphorylates the slow channel protein (or associated regulatory protein). This would prolong the life span of the phosphorylated channel, resulting in potentiation of I_{si} and contraction.

It is possible that some negative inotropic drugs may depress the rate of phosphorylation. It might be difficult to distinguish between a drug that depressed phosphorylation of the slow Ca^{2+} channel and one that physically blocked the channel.

B. Summary and Conclusions

The voltage-and time-dependent Ca^{2+} slow channels in the myocardial cell membrane are the major pathway by which Ca^{2+} ions enter the cell during excitation for initiation and regulation of the force of contraction of cardiac muscle. The slow channels have some special properties, including functional dependence on metabolic energy, selective blockade by acidosis, and regulation by the intracellular cyclic nucleotide levels. Because of these special properties of the slow channels, Ca^{2+} influx into the myocardial cell can be controlled by extrinsic factors (such as autonomic nerve stimulation or circulating hormones) and by intrinsic factors (such as cellular pH or ATP level).

TABLE 2

Comparison of Properties of Ca^{2+} Slow Channels in Myocardial Cells and Vascular Smooth Muscle Cells

	Myocardial Cells	VSM Cells
ATP	stim.	stim.
cAMP	stim.	inhib.†
cGMP	inhib.	inhib.†
acidosis	inhib.	?
Ca antagonist drugs	inhib.	inhib.
Ca agonist drugs	stim.	stim.

†In some VSM cells, e.g., rat aortic.

The slow Ca^{2+} channels of the heart are stimulated by cAMP. Elevation of cAMP produces a very rapid increase in the number of slow channels available for voltage activation during excitation. The probability of a slow channel opening and the mean open time of the channel are increased. Therefore, any agent that increases the cAMP level of the myocardial cell will tend to potentiate I_{si}, Ca^{2+} influx, and contraction.

The myocardial slow Ca^{2+} channels are also regulated by cGMP, in a manner that is opposite to that of cAMP. The effect of cGMP is presumably mediated by means of phosphorylation of a protein, as for example, a regulatory protein (inhibitory-type) associated with the slow channel. Preliminary data suggest that calmodulin also may play a role in regulation of the myocardial slow Ca^{2+} channels, possibly mediated by the Ca^{2+}-calmodulin-protein kinase and phosphorylation of some regulatory-type of protein.

Thus, it appears that the slow Ca^{2+} channel is a complex structure, including perhaps several associated regulatory proteins, which can be regulated by a number of extrinsic and intrinsic factors (Fig. 6).

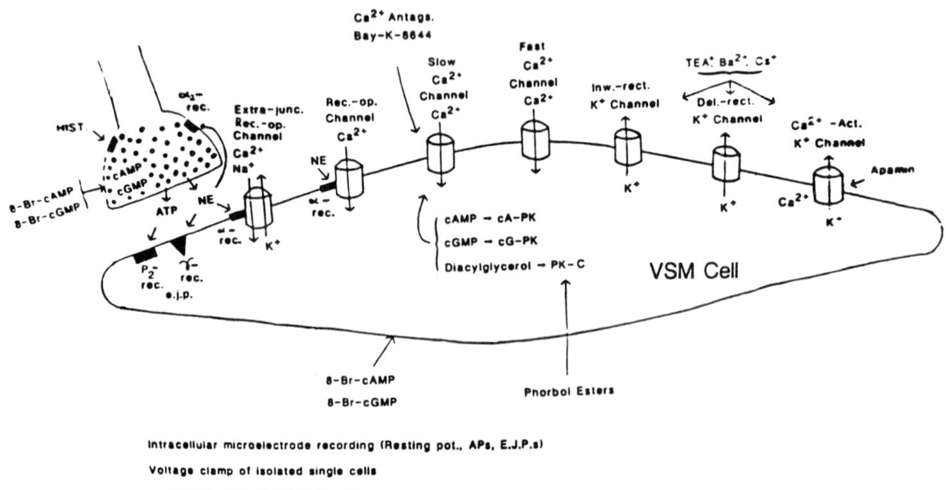

Figure 7

Diagrammatic representation of various ion channels in a vascular smooth muscle (VSM) cell and a number of agents that may either activate or inhibit these channels. Depicted are three different types of Ca^{2+} channels (fast, slow, and receptor-operated) and a nonselective ion channel (extrajunctional, receptor-operated), which allows Ca^{2+}, Na^{+}, and K^{+} to pass through. The voltage-dependent Ca^{2+} slow channels are blocked by Ca^{2+} antagonists and enhanced by Ca^{2+} agonists (Bay-K-8644). The three different K^{+} channels (inward rectifier, delayed rectifier, and Ca^{2+} activated) are blocked by TEA^{+}, Ba^{2+}, and Cs^{+}. Also shown is an adrenergic nerve terminal from which norepinephrine (NE) and ATP are released to activate a- or Q-receptors and purinergic (P_2) receptors, respectively, on the postsynaptic membrane. Release of neurotransmitters is modulated by cyclic nucleotides and phorbol esters.

VSM CELLS

A. Regulation of Ca^{2+} Slow Channels by Cyclic Nucleotides

Figure 7 depicts some of the types of ion channels in VSM cells, including the two types of Ca^{2+} channels: slow (L-type) and fast (T-type). The properties of these two types of Ca^{2+} channels are rather similar to those in myocardial cells described in the preceding section, with the exception of the regulatory mechanisms (Table 2).

Both cAMP and cGMP elevation have been implicated in the relaxation of smooth muscle tissue in response to some vasodilators, such as EDRF, ANF, and nitroprusside. We examined the effects of cyclic nucleotide analogs and related agents on the electrical properties of cultured rat aortic reaggregates (Ousterhout & Sperelakis, 1987). Agents that activate adenylate cyclase, such as isoproterenol and forskolin, depressed or abolished the Ca^{2+}-dependent APs and hyperpolarized the membrane. These effects are mimicked by membrane permeable analogs of cAMP (dibutyryl or 8-bromo-cAMP) (Fig. 8). 8-Bromo-cGMP also depressed or abolished the APs. Synthetic atrial natriuretic factor (ANF), which activates the membrane-bound guanylate cyclase, had inhibitory effects, whereas nitroprusside, which stimulates only cytosolic G-cyclase, had little effect. These results are summarized in Table 3. They suggest that cAMP and cGMP decrease the inward Ca^{2+} current and/or increase an outward K^+ current in the VSM cells.

In voltage-clamp experiments on cultured single cells from rabbit aorta, the dibutyryl analogs of cAMP and cGMP suppressed the inward Ba^{2+} current carried through the Ca^{2+} slow channels (Bkaily et al., 1988b). The 8-bromo analogs of cAMP and cGMP and nitroprusside also enhanced the delayed-rectifier outward K^+ current (Bkaily et al., 1988b). Thus, these effects of cyclic nucleotides on the Ca^{2+} and K^+ channels can explain their inhibitory effects on the APs.

We propose a phosphorylation model for regulation of Ca^{2+} slow channels in VSM cells, similar to that for myocardial cells, except in the case of VSM cells, cAMP and cGMP have the same effect, namely both inhibit the Ca^{2+} slow channels (see Fig. 9). A second regulatory protein, when phosphorylated by protein kinase C (PK-C), stimulates the Ca^{2+} slow channels, as depicted (Fig. 9).

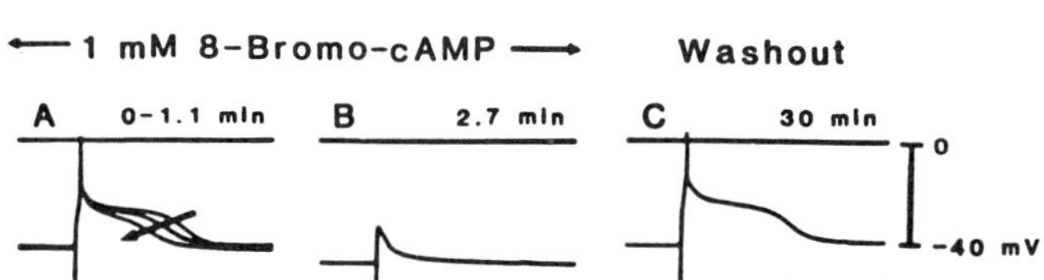

Figure 8

Depression and abolition of TEA-induced APs. APs in cultured rat aortic smooth muscle cells by 8-bromo-cAMP. A: Superimposed traces of a control AP (in 15 mM TEA) and depression of the AP within 1.1 min after addition of 1 mM 8-bromo-cAMP. B: Abolition of the AP after 2.7 min. C: Recovery of the AP upon washout for 30 min. (Records A-C were from the same cell.) The stimulation frequency was 0.04 Hz. Data taken from Ousterhout & Sperelakis, 1987.

TABLE 3

Summary of the Effects of Cyclic Nucleotides and Related Agents on the Action Potentials of Cultured Rat Aortic Cells

Compound	Effect on APs
8-Bromo cyclic AMP (10^{-3} M)	Abolished
Dibutyryl cyclic AMP (10^{-3} M)	Abolished
Isoproterenol (10^{-6} M)	Abolished
Forskolin (10^{-6} M)	Abolished
8-Bromo cGMP (10^{-4} - 10^{-3} M)	Abolished or depressed
Nitroprusside (10^{-6} - 10^{-5} M)	Depressed slightly
Atrial Natriuretic Factor (10^{-8} - 10^{-7} M)	Abolished

Taken from Ousterhout and Sperelakis, 1987.

Figures 10 and 11 summarize the roles played by cyclic nucleotides and PK-C in regulating the intracellular Ca^{2+} concentration ($[Ca]_i$) in VSM cells. Activation of A-cyclase or G-cyclase by appropriate membrane receptors (e.g. beta-adrenergic, prostacyclin, ANP) and G coupling proteins or directly (by agents like nitroprusside, nitric oxide free radical, or forskolin) produces elevation of cAMP and cGMP and activation of cA-PK and cG-PK.

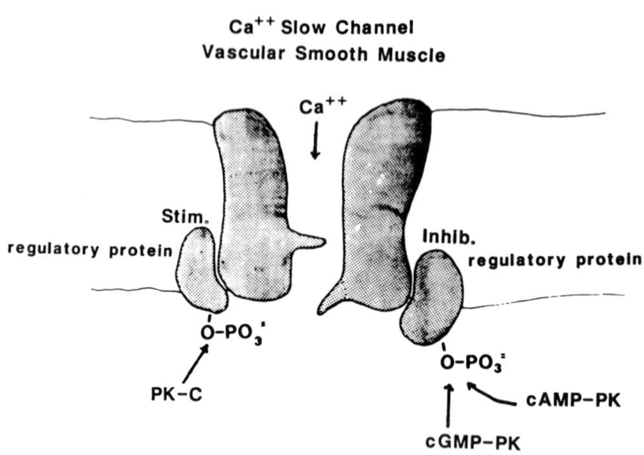

Figure 9

Cartoon model of a Ca^{2+} slow channel protein imbedded in the lipid bilayer of the cell membrane of a vascular smooth muscle cell. Depicted are two regulatory proteins, inhibitory and stimulatory, associated with the channel protein. As shown, both regulatory proteins have phosphorylation sites which affect the function of the regulatory proteins. Phosphorylation of the inhibitory protein by cAMP-protein kinase (PK-A) or by cGMP-PK (PK-G) inhibits the function of the ion channel, e.g. decreases probability of channel opening at depolarized potentials. Phosphorylation of the stimulatory protein by PK-C stimulates the Ca^{2+} slow channel, e.g. increases probability of opening. Thus, channel function is regulated by cyclic nucleotides and by diacylglycerol.

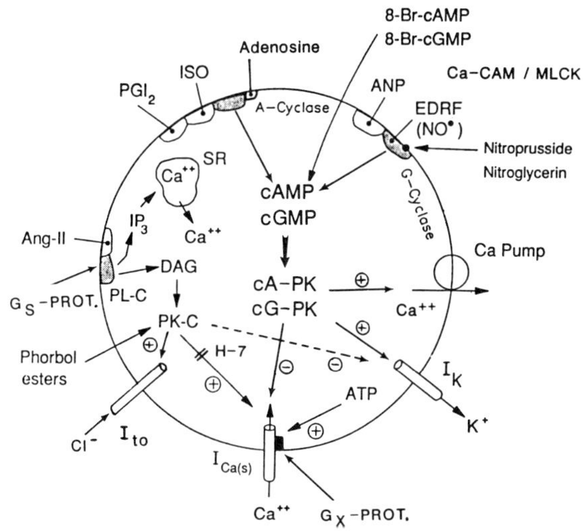

Figure 10

Diagram of a vascular smooth muscle cell to illustrate one mechanism whereby elevation of cyclic nucleotides can lead to vasodilation. Depicted are membrane receptors for stimulation of adenylate cyclase and guanylate cyclase, resulting in elevation of either cAMP or cGMP. As shown, cAMP activates PK-A and cGMP activates PK-G, which phosphorylate at least two types of ion channels and a Ca pump in the sarcolemma. Phosphorylation of the Ca^{2+} slow channel (or an associated inhibitory-type regulatory protein) produces inhibition of the channel, e.g. decreases probability of opening at depolarized potentials. Also depicted is the fact that phosphorylation by PK-C stimulates the Ca^{2+} slow channel. The delayed rectifier K^+ channel is stimulated by phosphorylation with cA-PK and cG-PK, as illustrated. Also depicted is stimulation of the sarcolemmal Ca^{2+} pump by cA-PK and cG-PK.

These kinases can phosphorylate the Ca^{2+} slow channel (to inhibit channel opening) and the K^+ channels (to stimulate channel opening). Both mechanisms would inhibit the Ca^{2+}-dependent APs, and thereby diminish Ca^{2+} influx and lower $[Ca]_i$. cG-PK and cA-PK also phosphorylate the sarcolemmal Ca^{2+}-ATPase and stimulate the Ca^{2+} pump in the surface membrane, thus lowering $[Ca]_i$. Lowering $[Ca]_i$ inhibits contraction and produces vasodilation.

Activation of PL-C by appropriate membrane receptors (e.g. Ang-II receptor) and G coupling proteins stimulates PI turnover with IP_3 and DAG production. DAG activates PK-C to phosphorylate the Ca^{2+} slow channels and stimulate Ca^{2+} influx and raise $[Ca]_i$. Phorbol esters act, like DAG, to directly activate PK-C. IP_3 acts on the SR to release Ca^{2+} (via the Ca-release channels) and elevate $[Ca]_i$. Raising $[Ca]_i$ potentiates contraction and produces vasoconstriction.

The cyclic nucleotides also phosphorylate the myosin light-chain kinase (MLCK), which diminishes its Ca^{2+} sensitivity. In addition, the cyclic nucleotides have been reported to inhibit phospholipase C (PL-C), and therefore production of IP_3 and diacylglycerol (DAG) is inhibited. These factors would contribute to the vasodilation.

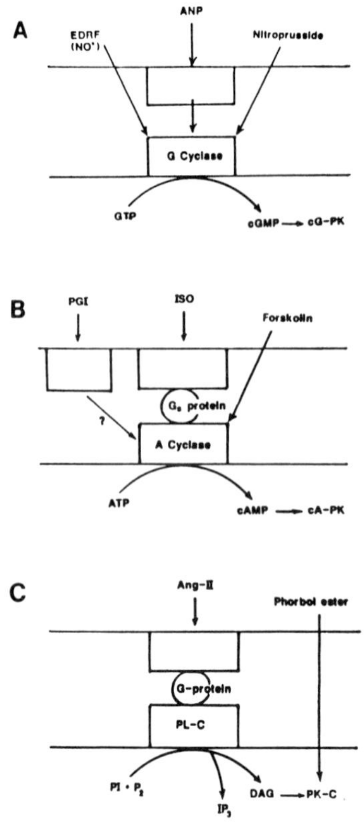

Figure 11

Diagram of expanded regions of the sarcolemma of a VSM cell to better illustrate the relationship between various membrane receptors and the cyclases and phospholipase C (PL-C). A: A receptor for atrial natriuretic peptide (ANP) is depicted, along with the guanylate (G) cyclase. Stimulation of G-cyclase by ANP, or directly by nitroprusside or EDRF (presumably nitric oxide free radical) elevates cGMP and activates cG-PK. B: Receptors for prostacyclin (PGI_2) and isoproterenol (beta-adrenergic agonist) are depicted, along with the adenylate (A) cyclase. the beta receptor is coupled to the A-cyclase by a GTP-binding protein (G_s). Activation of A-cyclase by agonists, or directly by forskolin, elevates cAMP level and thereby activates cA-PK. C: Depicted is the angiotensin-II receptor, coupled by a G-protein to PL-C. Stimulation of PL-C stimulates PI turnover and leads to production of IP_3 and DAG. DAG activates PK-C which phosphorylates several proteins including the Ca^{2+} slow channel. Phorbol esters directly stimulate PK-C (like DAG). Inositol trisphosphate (IP_3) causes Ca^{2+} release from the SR by activation of Ca release channels.

B. Regulation of Ca^{2+} Channels by ATP

Two distinct types of voltage-sensitive Ca^{2+} channels are found in VSM cells (Bean et al., 1986; Yatani et al., 1987; Pacaud et al., 1987; Friedman et al., 1986), as in myocardial cells. The two types of Ca^{2+} currents can be distinguished by differences in their kinetics, voltage ranges of activation and inactivation, and sensitivities to pharmacological agents (Table 1). The slow (sustained) Ca^{2+} channel is sensitive to dihydropyridine Ca^{2+} agonists and antagonists, prefers Ba^{2+} over Ca^{2+} as charge carrier, and activates at more positive membrane potentials (high threshold) than the fast (transient) Ca^{2+} channel. The fast Ca^{2+} channel activates at more negative membrane potentials (low threshold), has an equal preference for Ca^{2+} and Ba^{2+}, and is insensitive to the dihydropyridines.

We examined the properties of the two types of Ca^{2+} channels in freshly isolated cells from guinea pig small mesenteric arteries (resistance vessels), using the whole-cell voltage clamp technique (Ohya & Sperelakis, 1989a). Fast and slow Ca^{2+} channel currents were recorded, and basal characteristics of these channels were the same as previously reported (Fig. 12). Injection of ATP (0.3-5 mM), using an intracellular perfusion method, modified only the slow current, but did not affect the fast current (Fig. 13). When the production of ATP was inhibited by adding cyanide and 2-deoxy-D-glucose to the bath (in absence of glucose), the Ca^{2+} slow current was abolished within 10 min, whereas the fast Ca^{2+} current was only slightly affected (Table 2). These results indicate that only the slow Ca^{2+} channel is metabolically dependent. The ATP dependence of the activity of the Ca^{2+} slow channels was also shown at the single-channel level (Ohya & Sperelakis, 1989b) (Fig. 14).

AMP-PNP and cAMP could not substitute for ATP in sustaining the slow Ca^{2+} current in intestinal smooth muscle cells (Ohya et al., 1987). Since studies on guinea pig portal vein showed that AMP-PNP did not have similar effects as ATP, but ATP-γ-S did (Ohya & Sperelakis, 1989b), ATP may act via phosphorylation of ion channels or associated proteins. Alternatively, there may be an obligatory binding site for ATP on the inner surface of the channel which must be occupied for channel activity.

C. Modulation of Ca^{2+} Channels by Agonists

Agonists, such as norepinephrine (NE) or angiotensin II (A-II), may induce contraction of VSM via several mechanisms, including (a) stimulation of Ca^{2+} influx through receptor-operated channels (ROCs); (b) stimulation of Ca^{2+} entry through voltage-dependent Ca^{2+} channels, opened indirectly by the depolarization resulting from an increase in membrane conductance for other ions; and (c) release of intracellular Ca^{2+} from store sites (for review: Bolton, 1979; Johansson & Somlyo, 1980; Kuriyama et al., 1982; Sperelakis & Ohya, 1988). It is an attractive hypothesis that some agonists (NE, AII, etc.) may modify the voltage-dependent Ca^{2+} channels of VSM cells, as occurs in cardiac cells by beta-adrenergic agonists. The reported results have been contradictory, perhaps due to the use of different protocols and experimental conditions. For example, one group (Aaronson et al., 1986) reported that NE enhanced the Ca^{2+} channel current in rabbit ear artery, whereas another group (Droogmans et al., 1987) reported that NE inhibited the Ca^{2+} channel. Pacaud et al., (1987) reported that NE enhanced the fast-type of Ca^{2+} channel, but inhibited the slow-type in cultured rat portal vein cells. Others could not observe a significant change in mesenteric artery and dog saphenous vein (Bean et al., 1986; Yatani et al., 1987). Agonists, such as NE may affect the ion channel activity in VSM cells via a number of different mechanisms, some direct and some indirect. Further, some intracellular factor which regulates the response to agonists might be lost by perfusion of the cell during whole-cell voltage clamp.

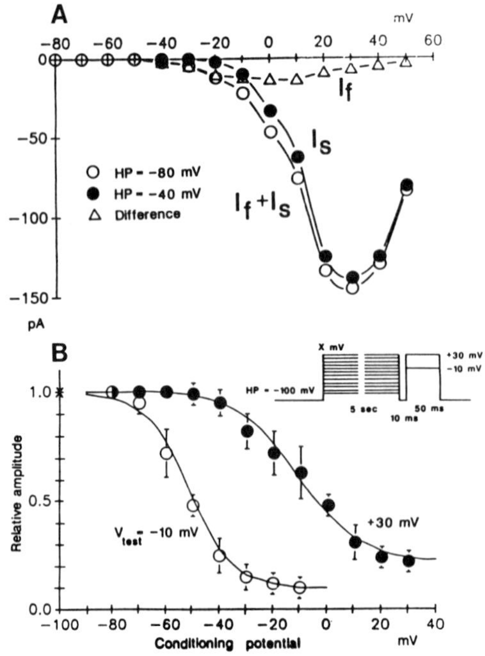

Figure 12

Two types of Ca^{2+} channel currents recorded from freshly-isolated single smooth muscle cells of guinea pig mesenteric arteries. A: I/V curves of the peak current amplitudes obtained from HPs of -80 mV (unfilled circles) and -40 mV (filled circles), and the difference curve between these two currents (unfilled triangles). B: Steady-state inactivation curves for the fast current (unfilled circles) and slow current (filled circles). Various levels of conditioning pulses (5 sec) were applied before application of the test pulse (to -10 mV for the fast current, and to +30 mV for the slow current); 10 msec intervals were used between the conditioning pulse and test pulse. The HP was kept at -100 mV. Data are shown as mean ± SD (n = 3-5). Continuous curves were drawn by fitting the data to a Boltzmann distribution. The pipette contained high Cs^+ and 5 mM ATP, and the bath contained isotonic Ba^{2+} solution. Data taken from Ohya & Sperelakis, 1989a.

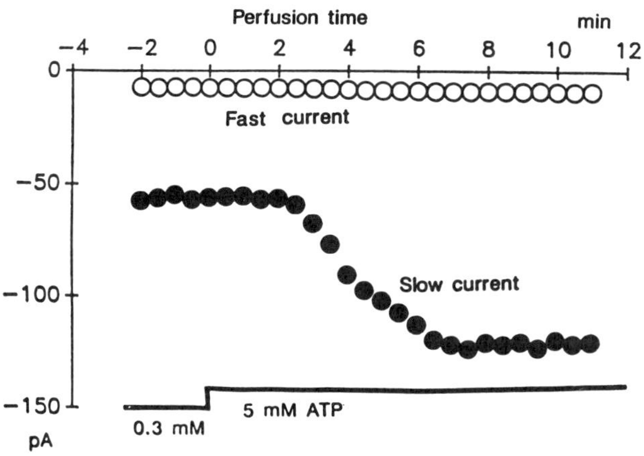

Figure 13

Effects of intracellularly-perfused ATP on the separated fast and slow Ca^{2+} channel currents. The concentration of ATP was increased from 0.3 mM to 5 mM using the intracellular perfusion technique. the time after beginning of the perfusion is indicated on the abscissa. The slow current was isolated by using a HP of -40 mV to voltage-inactive the fast channels; command potentials to + 30 mV were applied every 30 sec. the fast current was recorded using a HP of -80 mV and a command potential to -20 mV. The peak magnitudes of the fast (open circles) and slow (filled circles) Ca^{2+} channel currents were plotted as a function of time following perfusion with 5 mM ATP. The fast and slow currents were obtained from two different cell.s Data taken from Ohya & Sperelakis, 1989a.

We previously examined the effects of A-II on the electrical activity of cultured VSM cells from rat aorta (Zelcer & Sperelakis, 1981). A-II, applied by bolus perfusion, elicited a transient depolarization of up to 30 mV, which sometimes triggered an AP (Fig. 15, top). The A-II-induced depolarization disappeared in Na^{+}-free solution, suggesting that it was primarily due to an increase in Na^{+} conductance, with resulting Na^{+} influx. Continuous superfusion of cultured aortic reaggregates with A-II (10^{-6} M) also produced depolarization, which triggered a long-lasting AP (Johns & Sperelakis, 1990). The membrane resistance was decreased by A-II, consistent with an increase in conductance for an inward depolarizing current. Low concentrations of A-II (10^{-9} M) caused a marked prolongation of the Ca^{2+}-dependent APs without depolarizing the membrane (Fig. 15, bottom).

The whole-cell voltage clamp technique was used to examine the effects of A-II on Ca^{2+} and K^{+} channels in cultured VSM cells from rabbit aorta (Bkaily et al., 1988a). When the bath solution contained 50 mM Ba^{2+} (with TEA and 4-AP to block the outward current), depolarizing voltage steps from a holding potential of -80 mV elicited a small inward current. Addition of A-II caused a substantial increase in this current at all depolarizing steps, and an

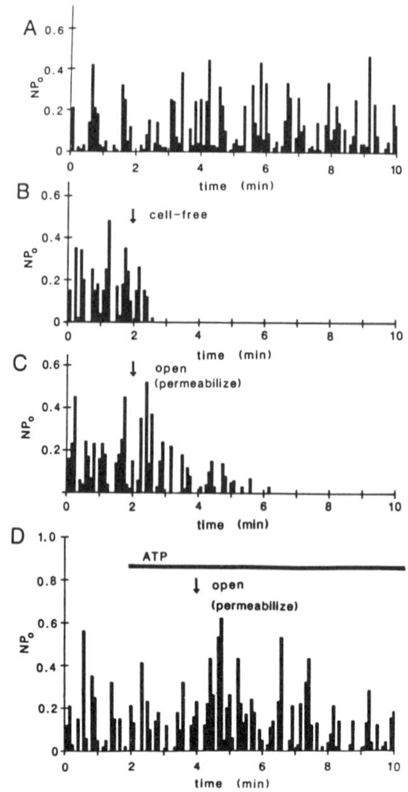

Figure 14

Histogram of Ca^{2+} slow channel activity to show changes over time in cell-attached patch (A) isolated inside-out patch (cell-free) (B) and cell-attached patch with cell opened distally without ATP (C) and with ATP in the bath (D). The values of NP_o (ordinates) were calculated for each sweep from 5 to 100 ms after onset of the depolarizing pulse from a holding potential of -70 mV to 0 mV (at 0.2 Hz), and were plotted against time (abscissa). In B, the channel activity was recorded in the cell-attached configuration initially, and, at the arrow, the patch was excised (isolated patch). In C, the cell-attached configuration was used, and, at the arrow, one end of the cell was mechanically disrupted by using another electrode to permeabilize the membrane. In D, 5 mM ATP was applied to the bath solution before the membrane was permeabilized (at the arrow). The pipette contained Ba^{2+} (100 mM) and the bath was K^+-rich. Data taken from Ohya & Sperelakis, 1989a.

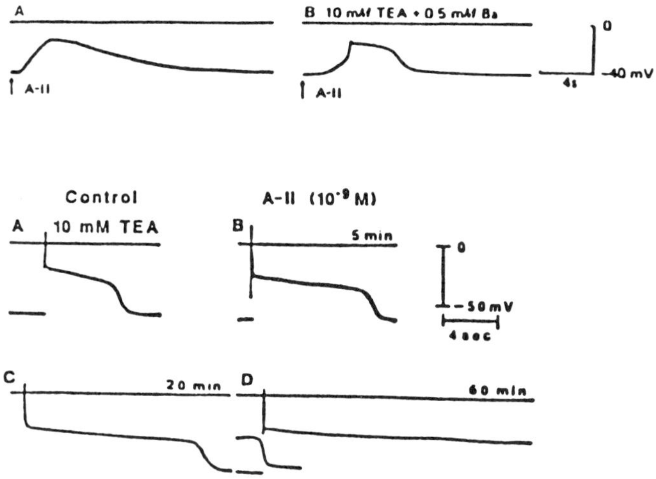

Figure 15

Effects of angiotensin (A-II) on the membrane potential and TEA-induced action potentials (APs) in cultured smooth muscle cells from rat aorta. Top: Bolus exposure (SHR preparations). A: A bolus of A-II (20 ml of 10^{-6} M) rapidly depolarized the cell. B: In the presence of 10 mM TEA and 0.5 mM Ba^2, an A-II bolus depolarized and induced an AP. (Taken from Zelcer & Sperelakis, 1981.) Bottom: Steady-state exposure. A: Control AP induced by electrical stimulation in the presence of 10 mM TEA. B-D: Bath addition of 10^{-9} M A-II (continuous exposure) progressively increased the AP duration at 5 min (B), 20 min (C), and 60 min (D). All records were from the same impalement. Data taken from Johns & Sperelakis, 1990.

A-II antagonist blocked this effect. A-II (1-5 x 10^{-8} M) suppressed the delayed-rectifier outward K^+ current. These effects of A-II on the Ca^{2+} and K^+ channel currents can explain the depolarization and AP prolongation found in cultured rat aortic cells. These observations suggest that agonists control vascular tone, at least in part, by modifying the Ca^{2+} channel and K^+ channel activities, and thereby changing $[Ca]_i$.

We also demonstrated that A-II greatly stimulated the Ca^{2+} slow channel current in isolated single cells from guinea pig portal vein (Ohya & Sperelakis, 1989b) (Fig. 16). This effect of A-II was mediated via a G-protein, based on inhibition of the A-II effect by GDP-ß-S and stimulation of the basal I_{Ca} by GTP-γ-S. However, this G-protein was not sensitive to pertussis toxin or to cholera toxin, hence is not G_i-like or G_s-like.

D. Possible Mechanisms for Agonist Modulation of Ca^{2+} Channel Function

1. Regulation by C-kinase. Many neurotransmitters and hormones, whose action involve Ca^{2+} mobilization, promote the breakdown of membrane phospholipids and the formation of two putative second messengers, inositol-1,4,5-trisphosphate (IP_3) and 1,2-diacylglycerol (DAG) (see review by Nishizuka, 1984). One of the major roles of IP_3 is to release Ca^{2+} from intracellular stores (see review: Berridge and Irvine, 1984), including the SR of VSM cells (Somlyo et al., 1985; Suematsu et al., 1985). It has been suggested (Rasmussen et al.,

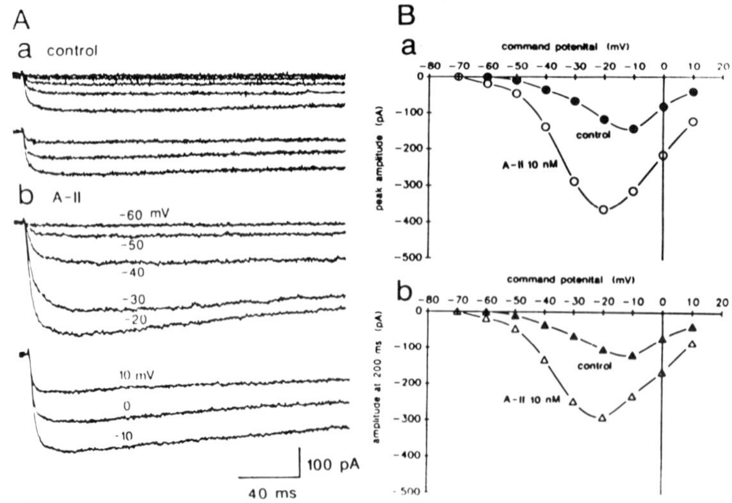

Figure 16

Effects of angiotensin-II (A-II) on the Ca^{2+} channel current recorded from guinea pig portal vein. A: Ba^{2+} current recorded before (a) and 3 min after application of 10 nM A-II. Step command pulses were applied over the voltage range of -60 mV to 10 mV from a holding potential (HP) of -90 mV. B: Peak current amplitude (a) and current amplitude measured at 200 ms (b) measured before (filled symbols) and after (unfilled symbols) 10 nM A-II application, were plotted against the command potentials. Data in A and B were collected from the same cell. Bath solution contained 2 mM Ba^{2+} as the charge carrier. Pipette solution contained high Cs with 5 mM ATP and 0.1 mM GTP. (A), (B), and (C) were taken from three different cells.

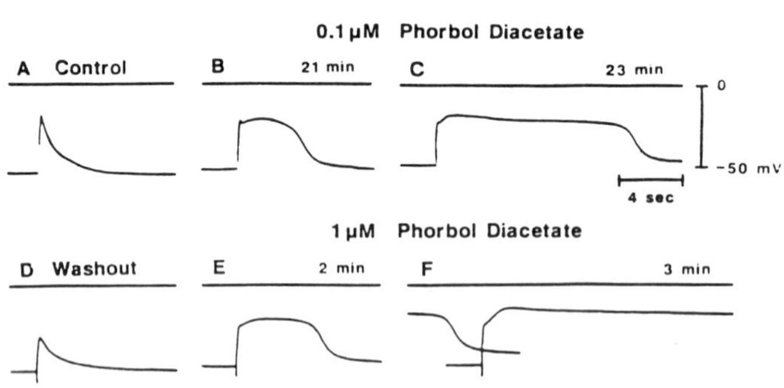

Figure 17

Induction of APs by phorbol diacetate in cultured rat aortic cells. A: Control record showing a lack of response to electrical stimulation in normal physiological salt solution. B: Induction of an AP after exposure to 10^{-7} M phorbol diacetate for 21 min. C: Further prolongation of the plateau component (23 min). D: Lack of response to stimulation after washout for 2.5 hr. E: Induction of the AP after exposure to 10^{-6} M phorbol diacetate for 2 min. F: Further prolongation of the plateau component (repolarization shown in subsequent sweep superimposed). Data taken from Ousterhout & Sperelakis, 1991.

1987) that activation of PK-C may be involved in agonist-induced Ca^{2+} influx, which is responsible for the tonic phase of smooth muscle contraction. The phorbol esters, which activate PK-C, produce a slowly-developing sustained contracture of VSM that is partially dependent on $[Ca]_o$ (Danthuluri & Deth, 1986; Gleason & Flaim, 1986; Chiu et al., 1987; Itoh & Lederis, 1987).

To examine a possible role of PK-C in the modulation of electrical activity in VSM cells, the effects of a phorbol ester, phorbol 12,13-diacetate (PDA), were determined on the resting potential and APs in cultured aortic reaggregates (Ousterhout & Sperelakis, 1991). Superfusion with PDA (0.1-5 μM) produced a gradual depolarization and allowed the appearance of an active (plateau-type) AP response to electrical stimulation (Fig. 17). This response was abolished by verapamil. When APs were elicited in the presence of 15 mM TEA, PDA also depolarized and prolonged the AP duration. However, after prolonged exposure (20-30 min) to higher concentrations of PDA, the APs became depressed, perhaps due to depolarization-induced inactivation of the Ca^{2+} channels.

Thus, the effects of phorbol esters on the ionic currents in VSM cells may be complex. The depolarization produced by the phorbol esters could be due to inhibition of K^+ channels, and the induction of APs could result from this plus the activation of a Ca^{2+} conductance, possibly mediated by phosphorylation. For example, in cultured aortic (A7r5 cell line) cells and neonatal rat cardiac myocytes, phorbol esters increased the slow (sustained) Ca^{2+} channel current (Fish et al., 1988; Dosemeci et al., 1988). Werz and MacDonald (1987) observed dual effects of phorbol esters on the Ca^{2+}-dependent APs in cultured neurons, and suggested that either Ca^{2+} or K^+ conductances could be depressed depending on the membrane potential. Our observations suggest that C-kinase may be involved in the regulation of ion channels in VSM cells.

2. Direct Regulation by G-Protein. Direct regulation of ionic channels by GTP-binding protein (G-protein) has been identified in several tissues (for review: Brown & Birnbaumer, 1988; Rosenthal et al., 1988). In some neurons, G-protein (G_o) is thought to mediate the modulation (inhibition) of Ca^{2+} channels produced by receptor activation (Hescheler et al., 1987). In cardiac cells, G-protein (G_s) was reported to directly enhance the activity of Ca^{2+} channels (Yatani et al., 1987).

We examined the effects of GTP-γ-S on Ca^{2+} channels in freshly-isolated single cells from guinea pig portal vein (Ohya & Sperelakis, 1988). Single Ca^{2+} channels (conductance of 18-22 pS) were recorded in cell-attached patch configuration with 100 mM Ba^{2+} and Bay-K-8644 in the pipette and high K^+ solution in the bath. After permeabilization of one end of the cell, GTP-γ-S (0.1 mM) applied to the bath (access into the cell interior) enhanced the channel activity. This observation suggests that a G-protein may be one of the factors regulating Ca^{2+} channels in VSM cells also. However, it is not known whether G-protein acts directly or indirectly on the Ca^{2+} channel.

E. Conclusions

Vascular tone is regulated by a variety of neurotransmitters, vasoactive hormones and autacoids, and vasoactive drugs. These actions are mediated, at least in part, by actions on the membrane ion channels, exerted either directly or indirectly. In this section, we described evidence that three different protein kinase systems (protein kinases A and G, and C-kinase) act on and modulate the Ca^{2+} slow channels in VSM cells, and that ATP is required for activity of these channels (Fig. 18).

Acknowledgements

These research projects were supported by N.I.H. Grants HL-31942 and HL-22619. We would like to thank Rhonda S. Hentz for typing this manuscript, and Anthony Sperelakis for making the figures.

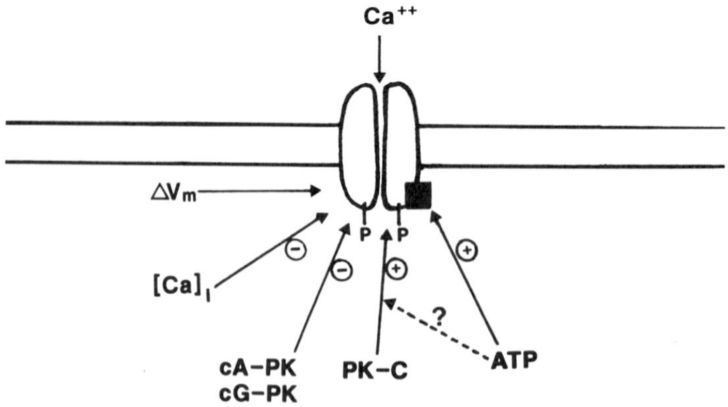

Figure 18

Diagram depicting some of the factors that regulate the Ca^{2+} slow channels in VSM cells. As shown, phosphorylation of a site on the channel by cA-PK or cG-PK inhibits channel activity, whereas phosphorylation of a second site by PK-C stimulates channel activity. Therefore, agents that elevate cAMP or cGMP inhibit the channels, and therefore decrease Ca^{2+} influx, and produce vasodilation. Agonists that increase DAG productions stimulate the channels, and therefore increase Ca^{2+} influx, and produce vasoconstriction. As shown, ATP is required for activity of the channels ($K_{0.5}$ of ca 0.3 mM). There may be a binding site for ATP that must be occupied, or ATP may be acting via phosphorylation. Ohya, Kitamura and Kuriyama (1987) had shown that internal Ca^{2+} inhibits channel activity.

References

Aaronson, P.I., Benham, C.D., Bolton, T.B., Hess, P., Lang, R.J., and Tsien, F.W., 1986, Two types of single-channel and whole-cell calcium or barium currents in single smooth muscle cells of rabbit ear artery and the effects of noradrenaline, J. Physiol., 377:36.

Armstrong, D. and Eckert, R., 1987, Voltage-activated calcium channels that must be phosphorylated to respond to membrane depolarization, Proc. Natl. Acad. Sci., USA, 84:2518-2522.

Bean, B., 1985, Two kinds of calcium channels in canine atrial cells, Differences in kinetics, selectivity, and pharmacology, J. Gen. Physiol., 86:1-30.

Bean, B.P., Nowycky, M.C., and Tsien, R.W., 1984, Beta-adrenergic modulation of calcium channels in frog ventricular heart cells, Nature, 307:371-375.

Bean, B.P., Sturek, M., Puga, A., and Hermsmeyer, K., 1986, Calcium channels in muscle cells isolated from rat mesenteric arteries: Modulation by dihydropyridine drugs, Circ. Res., 59:229-235.

Berridge, M.J. and Irvine, R.F., 1984, Inositol trisphosphate, a novel second messenger in cellular signal transduction, Nature, 312:315-321.

Bkailey, G. and Sperelakis, N., 1984, Injection of protein kinase inhibitor into cultured heart cells blocks calcium slow channels, Am. J. Physiol., 246:H630-H634.

Bkaily, G. and Sperelakis, N., 1985, Injection of cyclic GMP into her cells blocks the slow action potentials, Am. J. Physiol. (Heart Circ. Physiol.), 248:H745-H749.

Bkaily, G. and Sperelakis, N., 1986, Calmodulin is required for a full activation of the calcium slow channels in heart cells, J. Cyclic Nucleo. Prot. Phosph. Res., 11:25-34.

Bkaily, G., Peyrow, M., Sculptoreanu, A., Jacques, D., Chahine, M., Regoli, D., and Sperelakis, N., 1988a, Angiotensin II increased I_{si} and blocks I_K in single aortic cell of rabbit, Pflugers Arch., 412:448-450.

Bkaily, G., Sperelakis, N., and Eldefrawi, M., 1984, Effects of the calmodulin inhibitor, trifluoperazine, on membrane potentials and slow action potentials of cultured heart cells, Europ. J. Pharm., 105:23-31.

Bkaily, G., Yamamoto, T., Peyrow, M., Sculptoreanu, A., Jacques, D., and Sperelakis, N., 1988b, Macroscopic Ca^{2+} and K^{+} currents in single heart and aortic cells, Mol. Cell. Biochem., 80:59-72.

Bolton, T.B., 1979, Mechanisms of action of transmitters and other substances on smooth muscle, Physiol. Rev., 59:606-718.

Brown, A.M. and Birnbaumer, L., 1988, Direct G-protein gating of ion channels, Am. J. Physiol., 254:H401-H410.

Brown, J.H., Buxton, I.L., and Brunton, L.L., 1985, Alpha 1-adrenergic and muscarinic cholinergic stimulation of phosphoinositide hydrolysis in adult rat cardiomyocytes, Circ. Res., 57:532-537.

Bruckner, R. and Scholz, H., 1984, Effects of alpha-adrenoceptor stimulation with phenylephrine in the presence of propranolol on force of contraction, slow inward current and cyclic AMP content in the bovine heart, Br. J. Pharmacol., 82:223-232.

Brum, G., Flockerzi, V., Hofmann, F., Osterrieder, W., and Trautwein, W., 1983, Injection of catalytic subunit of cAMP-dependent protein kinase into isolated cardiac myocytes, Pflugers Arch., 398:147-154.

Cachelin, A.B., dePeyer, J.E., Kokubun, S., and Reuter, H., 1983, Ca^{2+} channel modulation by 8-bromo-cyclic AMP in culture heart cells, Nature, 304:462-464.

Chad, J.E. and Eckert, R.J., 1986, An ezymatic mechanism for calcium current inactivation in dialysed Helix neurones, J. Physiol., 378:31-51.

Chiu, A.T., Bozarth, J.M., Forsyth, M.S., and Timmermans, P.B.M.W.M., 1987, Ca^{2+} utilization of the constriction of rat aorta to stimulation of protein kinase C by phorbol dibutyrate, J. Pharmacol. Exp. Therap., 242:934.

Cuppoletti, J., Thakkar, J., Sperelakis, N., and Wahler, G., 1988, Cardiac sarcolemmal substrate of the cGMP-dependent protein kinase, Memb. Biochem., 7:135-142.

Danthuluri, N.R. and Deth, R.D., 1986, Acute desensitization to angiotensin II: Evidence for a requirement of agonist-induced diacylglycerol production during tonic contraction of rat aorta, Eur. J. Pharmacol., 125:1103.

Dosemeci, A., Dhalla, R.S., Cohen, N.M., Lederer, W.J., and Rogers, T.B., 1988, Phorbol ester increases calcium current and stimulates the effects of angiotensin II on cultured neonatal rat heart myocytes, Circ. Res., 62:347.

Droogmans, G., Declerck, I., and Casteel, R., 1987, Effect of adrenergic agonists on Ca^{2+}-channel currents in single vascular smooth muscle cells, Pflugers Arch., 409:7 12.

Fischmeister, R. and Hartzell, R.C., 1987, J. Physiol., 398:1475-1481.

Fish, R.D., Sperti, G., Colucci, W.S., and Clapham, D.E., 1988, Phorbol ester increases the dihydropyridine-sensitive calcium conductance in a vascular smooth muscle cell line, Circ. Res., 62:1049.

Friedman, M.E., Suarez-Kurtz, G., Karzorowski, G.J., Katz, G.M., and Reuben, J.P., 1986, Two calcium currents in a smooth muscle cell line, Am. J. Physiol., 1986:H699-H703.

Gleason, M.M. and Flaim, S.F., 1986, Phorbol ester contracts rabbit thoracic aorta by increasing intracellular calcium and by activating calcium influx, Biochem. Biophysi. Res. Comm., 138:1362.

Hescheler, J., Mieskes, G., Ruegg, J.C., Takai, A., and Trautwein, W., 1988, Effects of a protein phosphatase inhibitor, okadaic acid, on membrane currents of isolated guinea pig cardiac myocytes, Pflugers Arch., 412:248-252.

Hescheler, J., Rosenthal, W., Trautwein, W., and Schultz, G., 1987, The GTP-binding protein, G_o, regulates neuronal calcium channels, Nature (London), 325:445-447.

Irisawa, H. and Kokobun, S., 1983, Modulation by intracellular ATP and cyclic AMP of the slow inward current in isolated single ventricular cells of the guinea pig, J. Physiol., 338:321-327.

Itoh, H. and Lederis, K., 1987, Contraction of rat thoracic aorta strips induced by phorbol 12-myristate 13-acetate, Am. J. Physiol., 252:C244.

Johansson, B. and Somlyo, A.P., 1980, Electrophysiology and excitation-contraction coupling, in: Handbook of Physiology, Sect. 2: The Cardiovascular System, Vol. 2, Bethesda, Am. Physiol. Soc., 301-323.

Johns, D.W. and Sperelakis, N., 1990, Angiotensin-II stimulation of Ca^{2+}-dependent action potentials in cultured rat aortic smooth muscle cells, Eur. J. Pharmacol., 187:183-191.

Johnson, J.C., Wittenauer, L.A., and Nathan, R.D., 1983, Calmodulin, Ca^{2+}-antagonists and Ca^{2+}-transporters in nerve and muscle, J. Neural Trans. Suppl., 18:97-111.

Josephson, I. and Sperelakis, N., 1978, 5'-Guanylimidodiphosphate stimulation of slow Ca^{2+} current in myocardial cells, J. Mol. Cell. Cardiol., 10:1157-1166.

Josephson, I. and Sperelakis, N., 1976, Local anesthetic blockade of Ca^{2+}-mediated action grafted and organ cultured *in vitro*, Eur. J. Pharmacol., 40:201-208.

Kameyama, M., Hescheler, J., Hofmann, F., and Trautwein, W., 1986, Pflugers Arch., 407:123-128.

Kohlhardt, M. and Haap, K., 1978, 8-Bromo-guanosine-3',5'-monophosphate mimics the effect of acetylcholine on slow response action potential and contractile force in mammalian atrial myocardium, J. Molec. Cell. Cardiol., 10:573-578.

Kuriyama, H., Ito, Y., Suzuki, H., Kitamura, T., and Itoh, T., 1982, Factors modifying contraction-relaxation cycle in vascular smooth muscle, Am. J. Physiol., 243:H641-H662.

Li, T. and Sperelakis, N., 1983, Stimulation of slow action potentials in guinea pig papillary muscle cells by intracellular injection of cAMP, Gpp(NH)p, and cholera toxin, Circ. Res., 52:111-117.

MacLeod, K.M. and Diamond, J., 1986, Effects of the cyclic GMP lowering agent LY83583 on the interaction of carbachol with forskolin in rabbit isolated cardiac preparations, J. Pharmacol. Exp. Therap., 238:313-318.

Mehegan, J.P., Muir, W.W., Unverferth, D.V., Fertel, R.H., and McGuirk, S.M., 1985, Electrophysiological effects of cyclic GMP on canine cardiac Purkinje fibers, J. Cardiovasc. Pharmacol., 7:30-35.

Tohse, N. and Sperelakis, N., 1990, Long-lasting openings of single slow (L-type) Ca^{2+} channels in chick embryonic heart cells, Am. J. Physiol., 259:H639-H642.

Tohse, N. and Sperelakis, N., 1991, Cyclic GMP inhibits the activity of single calcium channels in embryonic chick heart cells, Circ. Res., in press.

Tohse, N. and Sperelakis, N., 1991, Developmental changes in long-opening behaviour of L-type Ca^{2+} (slow) channels in embryonic chick heart cells, Submitted.

Trautwein, W. and Hoffman, F., 1983, Activation of calcium current by injection of cAMP and catalytic subunit of cAMP-dependent protein kinase, Proc. Internat. Union Physiol. Sci., 15:75-83.

Tsien, R.W., Giles, W., and Greengard, P., 1972, Nature, 240:181-183.

Vogel, S. and Sperelakis, N., 1981, Induction of slow action potentials microiontophoresis of cyclic AMP into heart cells, J. Mol. Cell. Cardiol., 13:51-64.

Vogel, S., Sperelakis, N., Josephson, J., and Brooker, G., 1977, Fluoride stimulation of slow Ca^{2+} current in cardiac muscle, J. Mol. Cell. Cardiol., 9:461-475.

Wahler, G.M., Rusch, N.J., and Sperelakis, N., 1990, 8-bromo-cyclic GMP inhibits the calcium channel current in embryonic chick ventricular myocytes, Can. J. Physiol. Pharm., 68:531-534.

Wahler, G.M. and Sperelakis, N., 1985, Intracellular injection of cyclic GMP depresses cardiac slow action potentials, J. Cyclic Nucleo. Prot. Phos. Res., 10:83-95.

Wahler, G.M. and Sperelakis, N., 1986, Cholinergic attenuation of the electrophysiological effects of forskolin, J. Cyclic Nucleo. Prot. Phosph. Res., 11:1-10.

Werz, M.A. and MacDonald, R.L., 1987, Phorbol esters: Voltage-dependent effects on calcium-dependent action potentials of mouse general and peripheral neurons in cell culture, J. Neurosci., 7:1639-1647.

Yatani, A., Condina, J., Imoto, Y., Reeves, J.P., Birnbaumer, L., and Brown, A.M., 1987, A G-protein directly regulates mammalian cardiac calcium channels, Science, 238:1288-1292.

Yatani, A., Seidel, C.L., Allen, J., and Brown, A.M., 1987, Whole-cell and single-channel calcium currents of isolated smooth muscle cells from saphenous vein, Circ. Res., 60:523-533.

Zelcer, E. and Sperelakis, N., 1981a, Angiotensin induction of active responses in cultured reaggregates of rat aortic smooth muscle cells, Blood Vessels, 18:263-279.

SUBCELLULAR CALCIUM COMPARTMENTATION AND EXCHANGE IN THE CARDIAC CELL: RELATION TO CONTRACTILE CONTROL

Glenn A. Langer and Terrell L. Rich

Cardiovascular Research Laboratory
UCLA School of Medicine
Los Angeles, California

It is the ultimate goal of studies underway at the UCLA Cardiovascular Research Laboratory to identify, in the intact functional cell, the content and exchange characteristics of the calcium (Ca) in the various subcellular organelles. Until this is done a final, comprehensive model of excitation-contraction coupling processes as they perform and interrelate *in vivo* cannot be realized.

The classic approach to compartmentation has been to use isotopic labeling and washout to identify subcellular Ca distribution. Quantitative information with respect to accurate definition of Ca compartment size and flux has, however, been limited by the difficulty of selecting the appropriate kinetic exchange model, e.g. series, parallel or series-parallel arrangement of the compartments. An accurate model is required in order that realistic correction for isotopic reflux from one compartment to another be made.

An alternative to uncertain modeling is to devise an experimental system which eliminates isotopic reflux or reduces it to a minimum. This has been done by development of a rapid perfusion system for monitoring of ^{45}Ca washout from a preparation of single cardiac cells (1). The flow is at a level which approaches a non-limiting rate for the cells' Ca exchange. This minimizes or eliminates reflux and, in effect, allows direct estimation of compartment sizes and fluxes. Probes specific for subcellular organelles can be applied under these conditions and the origin of the kinetically defined Ca compartments derived.

Kinetically-Defined Compartments and their Origin. Figure 1 indicates the initial 120 seconds of a 400 second non-perfusion-limited ^{45}Ca washout from previously labeled rat ventricular cells. The curve is broken down, by computerized curve stripping, to the minimum number of compartments which account for all of the cells' exchangeable Ca.

Slow Compartment - $t½$ = 3.6 minutes, content = 1.6 mmoles/kg dry wt. cells. It was suspected, given that it represents 24%

Excitation-Contraction Coupling in Skeletal, Cardiac, and Smooth Muscle
Edited by G.B. Frank *et al.*, Plenum Press, New York, 1992

of the cell's exchangeable Ca and its exchange rate was the slowest, that the origin of this compartment was mitochondrial. This assumption was tested by specifically augmenting the cell's mitochondrial Ca content by labeling the cells in the presence of NaH_2PO_4. Phosphate acts as a proton donor (2) and stimulates the "Ca uniporter" of the mitochondria (3). After labeled Ca has been specifically directed to the mitochondria ^{45}Ca washout is started and phosphate removed at 3.5, 5, 7 and 10.5 minutes' washout time. Phosphate removal releases the increased Ca from the mitochondria. Plotting the ^{45}Ca released versus time identifies the exchange rate of the mitochondria at t½ = 3.3 minutes, virtually the same as the steady-state exchange rate of the slow compartment. The Ca content, representing 24% of the exchangeable Ca and 20% of the cell's total (including inexchangeable) Ca is in close agreement with the recent electron microprobe study of rabbit ventricular muscle which indicated that 20% of total cell Ca was localized to the mitochondria (4). The contribution of the mitochondria to cellular flux, at $[Ca]_o$ = 1.0 mM, is 5 μm/kg dry wt./sec, or less than 1% of total non-perfusion-limited steady state Ca flux of the cell.

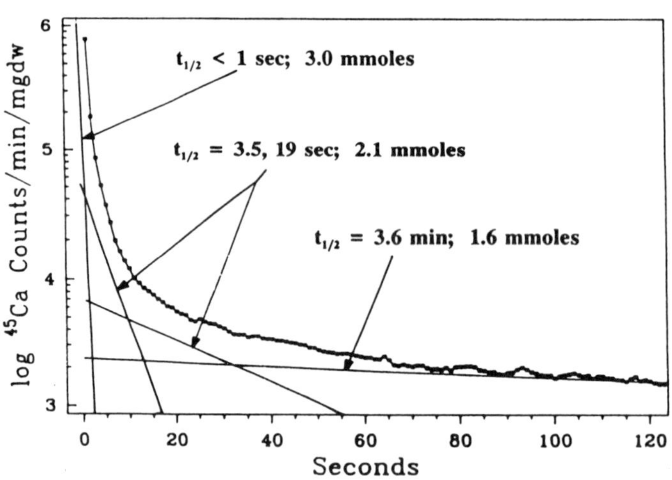

Fig. 1. The pattern of kinetically-derived cellular Ca compartmentation in intact rat ventricular cells as measured under non-perfusion-limited conditions. t½ and contents are indicated.

<u>Intermediate Compartment</u> - This is a biexponential compartment. The reason for placing both components in a single compartment will be evident below. Its t½'s = 3.5 and 19 seconds with a content of 2.1 mmoles which represents 31% of the cell's exchangeable Ca. Its origin was probed with caffeine and ryanodine. After ^{45}Ca labeling, washout was commenced and caffeine added (within 1 second) to the washout solution, in separate experiments, at ten points between 4 and 180 seconds'

washout. Caffeine produced ^{45}Ca release, the amount diminishing as it was added at later times. Pretreatment of the cell with 10^{-6}M ryanodine for 60 minutes before washout reduced the caffeine-releasable ^{45}Ca by at least 85% at each time point. The net ^{45}Ca activity released was plotted versus time and produced a biexponential curve with $t½$ = 2 and 19 seconds, virtually identical to the steady-state components of the intermediate compartment. The fact that caffeine and ryanodine are accepted as specific for sarcoplasmic reticulum (SR) Ca (5) and the ^{45}Ca released by caffeine exchanges at the same rate as the intermediate compartment indicate that this compartment originates in the SR.

The intermediate compartment contains 2.1 mmoles Ca/kg dry wt. cells. The maximum amount of Ca released by caffeine was 0.44 mmoles or 21% of the intermediate compartment. This indicates that caffeine does not release nearly all of SR Ca and is consistent with the recent study of Ondrias et al. (6) where it was found that caffeine released only 30% of Ca from an isolated SR vesicle preparation.

The contribution of the SR to steady-state cellular flux is 280 µmoles/kg dry wt./sec or 11% of total non-perfusion-limited flux.

Fast Compartment

(a) La displaceable Compartment. A component of the fast compartment was originally defined with the use of lanthanum (La), the trivalent cation. The addition of La to fully ^{45}Ca-labeled cells displaces 2.6 mmoles Ca/kg dry wt. The addition of La between the second and third second of washout fails to produce any ^{45}Ca displacement, which indicates that a compartment with $t½ < 1$ second contains the Ca accessible to La. With its content and rapid exchange its flux is > 1800 µmoles Ca/kg dry wt./sec. The specific localization of this La displaceable compartment, which accounts for 39% of the cell's exchangeable Ca, remains undefined. Given the compartment's rapid rate of exchange and its immediate accessibility to extracellularly applied La it is likely to reside within the sarcolemma. Recent studies from our laboratory on isolated cardiac sarcolemma indicate that virtually all Ca binding of the sarcolemma is to phospholipid (7). Ca would bind, primarily, to the anionic phosphatidylserine, -inositol and zwitterionic ethanolamine moieties of the membrane. These are in sufficient quantity to accommodate easily all of the La-displaceable component (8). There is, however, a problem. All of the anionic phospholipid and 75% of the phosphatidylethanolamine are localized at the inner leaflet of the sarcolemma and their Kd for Ca-binding is 1.1 mM, i.e. low affinity (7). Intracellular Ca concentration in the cell varies from 5×10^{-6} to 10^{-7} M. In order that Ca be bound to the inner leaflet in a quantity to account for the La-displaceable component Ca concentration at the inner membrane would have to be > 200 µmolar. This is not possible unless there is an unrecognized subsarcolemmal compartment of high Ca concentration. In short, the large rapidly exchangeable La-displaceable compartment has not been definitively localized but the sarcolemma remains the leading candidate.

(b) Na-Ca Exchange-Dependent Compartment - We suspected, on the basis of the Bridge et al. (9) and Crespo et al. (10) studies which indicated that Ca exchange via the Na-Ca exchanger had a t½ of a few hundred milliseconds, that a component of the fast compartment would be Na-Ca exchange dependent. This has proven to be so (11).

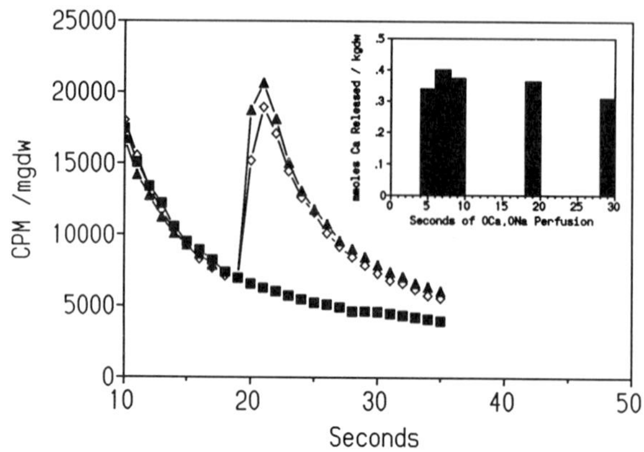

Fig. 2. The "Na-Ca dependent Ca compartment". Solid squares illustrate washout with 0 Na-0Ca in perfusate. Solid triangles illustrate washout with 0Na-0Ca for 19 seconds, with replacement of Na and Ca after the 19th second and for the remainder of the washout. Open diamonds represent washout with 0Na-0Ca for 19 seconds with replacement of Na-Ca after 19 seconds in cells exposed to 10^{-6} ryanodine prior to ^{45}Ca-labeling and throughout washout (11).

Figure 2 demonstrates the existence of a discrete Ca compartment which is absolutely dependent upon operation of the Na-Ca exchanger for its exchange. There are three ^{45}Ca washout curves depicted. Cells had been identically ^{45}Ca-labeled prior to washout. At start of the washout (0 time, not shown in figure) both Na and Ca are also removed from the washout solution, thus completely inhibiting operation of the Na-Ca exchanger. The washout curve indicated by solid squares represents the average of nine washouts in the absence of Na-Ca in the washout solution. The curve represented by solid triangles represents 0Na, 0Ca in the washout solution for 19 seconds with return to Na-Ca at the 20th second and continuing for the remainder of the washout. Note the immediate ^{45}Ca release, reaching a peak with 2 seconds followed by a gradual decay over the next 15 seconds. The total counts released over a 10 second-period after Na-Ca was returned represents 363 µmoles Ca/kg dry wt. cells. It is particularly noteworthy that net ^{45}Ca release was not significantly different when Na-Ca was returned after 5, 7, 9, 19 and 29 seconds of washout. The amount of Ca released at each time is illustrated in the inset (each bar represents the mean of 5-9 experiments). The set of experiments defines, for the first time, a cellular compartment of approximately 350 μmoles/kg dry wt. cells which has an absolute dependence for its exchange on the operation of the Na-Ca exchanger.

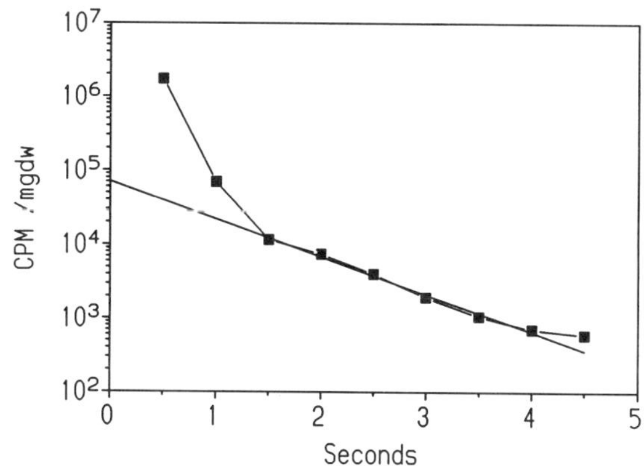

Fig. 3. Semilogarithmic plot of the difference in washout counts: Averaged washout with Na-Ca present minus averaged washouts without Na-Ca in the perfusate. The initial two points include counts from the extracellular fluid. Note mono-exponential representation of the Na-Ca dependent compartment with rate constant = 1.3/sec, t½ = 530 msec (11).

The turnover rate of the compartment was measured by plotting the time course of the difference in ^{45}Ca washout counts when Na-Ca is present and when it is not. The difference curve (Na-Ca present minus Na-Ca absent) is plotted semi-logarithmically each 0.5 sec for the initial 4.5 seconds of washout as illustrated in Figure 3. The plot is biphasic. The first two counts represent contribution of ^{45}Ca from the chamber but counts at 1.5 seconds and later represent counts derived predominantly from the cells. The slope of this line indicates a rate constant of 1.3/sec or t½ = 530 msec. At this exchange rate and a compartment content of 350 μmoles/kg dry wt. (6% of total exchangeable Ca), the flux is 460 μmoles/kg dry wt./sec or 18% of total non-perfusion-limited flux.

The subcellular origin of the Na-Ca exchange-dependent compartment has been investigated using a variety of probes. Treatment of cells with 10^{-6}M ryanodine does not significantly reduce the size of the compartment (see Figure 2, open diamonds). Exposure of the cells to 10 mM caffeine reduced the size of the compartment by 20%. This was a significant reduction but indicates that 80% of the compartment is not affected by this drug. In addition, exposure of the cells to La during 0Na-0Ca washout failed to produce a displacement from the compartment. This is consistent with an intracellular locus. The locus which comes first to mind is the junctional SR. Caffeine is believed to induce release of Ca from the SR and impair its reuptake (5, 12). As indicated above, caffeine reduced the amount of Ca in the Na-Ca dependent compartment by 20% or about 70 μmoles/kg dry wt. cells. However, the total caffeine releasable pool is 440 μmoles (1) or over 6 times the reduction in the Na-Ca dependent compartment produced by caffeine. This, along with the lack of effect of ryanodine, indicates that most of the Na-Ca dependent compartment is <u>not</u> in the SR or at least not in the component of the SR that is released by caffeine or affected by ryanodine.

As with the La-displaceable compartment, it is difficult to assign a subcellular locus for the Na-Ca exchange dependent compartment with any certainty. The requirement is a juxta-sarcolemmal region where Ca diffusion is restricted. It might be speculated that the "cleft" between the JSR and inner sarcolemmal leaflet, the region of the "feet" (13), might meet the requirement. But, as discussed under the La-displaceable compartment, binding of Ca to the low affinity inner leaflet phospholipid binding sites would, in order to account for the 350 μmoles Ca/kg dry wt. (90 μmoles/l cell water) of the Na-Ca dependent compartment, require that the "cleft" Ca concentration be maintained in the range of 25 μM. It is not apparent, at present, how such a concentration might be achieved.

<u>"Inexchangeable" Compartment</u> - Measurement of total cellular Ca (^{45}Ca labeled plus Ca which remains unlabeled after 60 minutes' ^{45}Ca exposure) indicates a content of 7.9 mmoles Ca/kg dry wt. Since exchangeable Ca = 6.7 mmoles, 1.2 mmoles is very slowly exchangeable or "inexchangeable". Its origin is not defined.

Table 1. Exchangeable Myocardial Ca^{++} (1mM$[Ca]_o$).

COMPARTMENT	SIZE (mmoles/Kgdw)	$t_{1/2}$	FLUX (μmoles/Kgdw/s)	ORIGIN
SLOW	1.6	3.6 min	5	MITOCHONDRIA
INTERMEDIATE	2.1	3.5, 19 sec	280	SARCOPL. RETIC.
Na^+-Ca^{++} DEP. (FAST)	0.3-0.4	500 msec	460	UNKNOWN
La^{+++} DISPL. (FAST)	2.6	< 1 sec	> 1800	? SARCOLEMMAL

Compartments and Contraction - Table 1 summarizes the distribution, localization and exchange characteristics of the cell's exchangeable Ca compartments described above. Figure 4 places these compartments within the cell in the context of our present conception of the excitation-contraction coupling (ECC) sequence in the mammalian heart.

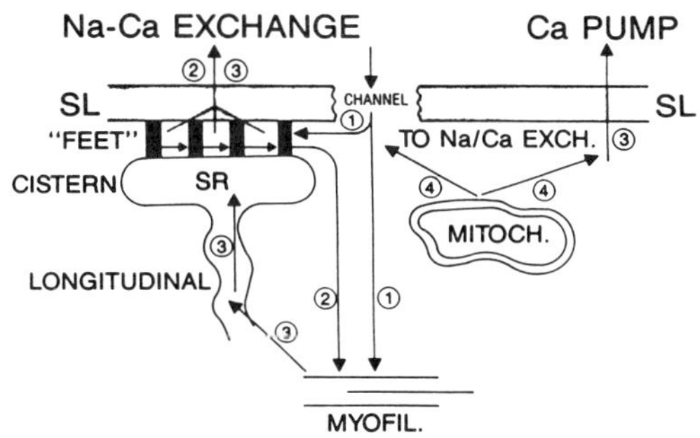

Fig. 4. Compartmentation and movement of Ca in the cardiac cell. See text for discussion. (Reproduced with permission from the Polish J. Physiol. and Pharmacol.).

Referring to Figure 4, upon SL depolarization "L"-type channels open to admit Ca to the "cleft" region where the "feet" are located and also directly to the myofilaments (process (1)). The Ca admitted to the "cleft" activates the opening of the channels in the "feet" (Ca-induced Ca release) and begins activation of the Na-Ca exchanger (process (2)). Thus, activation of the myofilaments is achieved via Ca-induced Ca release and via direct activation. In rat ventricle the former is dominant but in other species (e.g. rabbit) a more slowly developing component of force, most likely dependent upon direct activation, is also evident (14). The Ca which resides in the fast compartments (t½ of hundreds of milliseconds) participate in processes (1) and (2). The La displaceable compartment, likely bound to the sarcolemma, may contribute to direct activation upon depolarization though this is entirely speculative at present. The Na-Ca exchange-dependent compartment clearly feeds the exchanger to promote outward Ca flux across the sarcolemma. As contractile activation proceeds, Ca pumping into the longitudinal SR (with diffusion to the cistern) commences and increases to contribute to relaxation. In addition, the Na-Ca exchanger contributes increasingly to remove Ca from the cell (process 3). The intermediate compartment (t½ 3.5 and 19 sec) represents the turnover of the SR Ca with the extracellular space. The rate of SR turnover is that measured in quiescence. This may well increase with activation and release additional Ca into the cleft to be removed from the cell via the very fast and high capacity Na-Ca exchanger in order to maintain steady-state levels of intracellular Ca in the face of increased influx associated with activation. In the latter phase of diastole the SL Ca pump maintains the low level of resting $[Ca]_i$ (process 3). Its low capacity but high Ca affinity seems well-adapted to this role (16). The kinetic representation of this pump has not been achieved because there is, as yet, no known specific inhibitor of this pump. Finally, the slow mitochondrial exchange ($t½$ = 3.6 min) with the extracellular space is represented by process (4). The mitochondrial Ca exchange, though not contributing to the beat-to-beat regulation of contractile force adjusts, through Ca control of various mitochondrial dehydrogenases, the level of aerobic metabolism to energy demand (3). Mitochondria also serve as a large intracellular Ca buffer under conditions of Ca overload.

REFERENCES

1. G. A. Langer, T. L. Rich and F. B. Orner, Ca exchange under non-perfusion limited conditions in rat ventricular cells: Identification of subcellular compartments. Am. J. Physiol. 259:H592-H602 (1990)
2. A. L. Lehninger, Role of phosphate and other proton-donating anions in respiration-coupled transport of Ca^{2+} by mitochondria. Proc. Natl. Acad. Sci. USA 71:1520-1524 (1974)
3. M. Crompton, The role of Ca^{2+} in the function and dysfunction of heart mitochondria, in: "Calcium and the Heart", G.A. Langer, ed, Raven Press, New York (1990) pp. 167-198
4. L. G. Walsh and J. McD. Tormey, Cellular compartmentation in ischemic myocardium: indirect analysis by electron probe. Am. J. Physiol. 255:H929-H936 (1988)
5. E. Rousseau, and G. Meissner, Single cardiac sarcoplasmic reticulum Ca^{2+}-release channel: activation by caffeine. Am. J. Physiol. 256:H328-H333 (1989)
6. K. Ondrias, L. Borgatta, D. H. Kim and B.E. Ehrlich, Biphasic effects of doxorubicin on the calcium release channel from sarcoplasmic reticulum of cardiac muscle. Circ. Res. 67:1167-1179 (1990)
7. J. A. Post and G. A. Langer, Molecular origin of sarcolemmal calcium binding sites. FASEB J. 5:A1051 (1991)
8. J. A. Post, G. A. Langer, J. A. F. Op den Kamp and A. J. Verkleij, Phospholipid asymmetry in cardiac sarcolemma. Analysis of intact cells and "gas-dissected" membranes. Biochim. Biophys. Acta 943: 256-266 (1988)
9. J. H. B. Bridge, J. R. Smolley and K. W. Spitzer, The relationship between charge movements associated with I_{Ca} and I_{Na-Ca} in cardiac myocytes. Science 248:376-378 (1990)
10. L. M. Crespo, C. J. Grantham and M. B. Cannell, Kinetics, stoichiometry and role of the Na-Ca exchange mechanism in isolated cardiac myocytes. Nature 345:618-621 (1990)
11. T. L. Rich and G. A. Langer, Na-Ca exchange contribution to the "rapid" exchangeable Ca compartment of rat heart cells. FASEB J. 5: Part II, A1051 (1991)

12. A. Weber, The mechanism of action of caffeine on sarcoplasmic reticulum. J. Gen. Physiol. 52:760-772 (1968)
13. K. Nagasaki and S. Fleischer, Ryanodine sensitivity of the calcium release channel of sarcoplasmic reticulum. Cell Calcium 9:1-7 (1988)
14. B. Lewartowski, R. G. Hansford, G. A. Langer, E. G. Lakatta, Contraction and sarcoplasmic reticulum Ca^{2+} content in single myocytes of guinea pig heart: effect of ryanodine. Am. J. Physiol. 259:H1222-H1229 (1990)
15. E. Carafoli, Sarcolemmal calcium pump, in: "Calcium in the Heart", G. A. Langer, ed., Raven Press, New York (1990) pp. 109-126

CALCIUM RELEASE AND FORCE DEVELOPMENT

IN RAT MYOCARDIUM

H.E.D.J. ter Keurs, P.H. Backx, H. Banijamali, B. MacIntosh, and W.D. Gao

Department of Medicine and Medical Physiology, University of Calgary
3330 Hospital Drive N.W., Calgary, Alberta, T2N 4N1

INTRODUCTION

It is well known that cardiovascular control mechanisms which adjust cardiac output to the bloodflow required by the organs in the body continuously modulate force of the heartbeat. The balance between the vagal and sympathetic activities dictates the heart rate, and hence through the force-frequency and force-interval relationships the contractile state of the heart. In addition, adrenergic transmitters and hormones control both calcium transport and the sensitivity of the contractile filaments to calcium ions in the myocyte by their modulation of phosphorylation of calcium channels, calcium pumps and troponin-I. Adjustment of the venous tone by the cardiovascular control mechanisms dictates the work put out by the heart through the mechanism described by Starling, that the change of venous and hence end-diastolic pressure which results from activation of the control system leads to a change in stretch of the cardiac sarcomeres and thereby an increase or decrease of the response of the contractile filaments to calcium ions.

Control of cardiac force involves modulation both of calcium transport and of the response by the contractile apparatus; both are important aspects of excitation-contraction coupling. On the other hand, it is also important to realize how large the metabolic cost of the calcium transport process is, especially because it has been shown, now, by several groups employing different experimental techniques that the total turnover of Ca^{2+} ions amounts to several mmol per litre tissue water (cf the chapters by Drs. Langer and Wendt-Gallitelli). In this chapter, we will discuss studies designed to elucidate mechanisms involved in the process of excitation-contraction coupling. In particular, we will focus on the determinants of the $[Ca^{2+}]_i$ transient that follows the action potential in rat cardiac muscle related to the force response of the sarcomeres, and the factors that determine the force-interval relationship. We will also show that the energy required for Ca^{2+} turnover in isolated cardiac muscle amounts to approximately 30% of the total energetic expenditure per cardiac beat.

METHODS

Muscle Preparation and Dissection Procedure

Trabeculae, which were 2-3 mm in length, 100-150μm in width and 30-40μm in thickness, were carefully dissected from the right ventricle of 2-6 months old male and female Sprague-Dawley rats. The trabeculae were mounted between a force transducer and servo-motor controller (see below) in a perfusion bath (25x2x5 mm^3) located on the stage of a Nikon inverted microscope. Stimulation was accomplished by applying less than 10 Volts for 2-4ms via a pulse stimulator (DS2, Digitimer, UK) through two platinum electrodes running down both sides of the bath. The stimulator was triggered via a timing device (D4030, Digitimer, UK).

Excitation-Contraction Coupling in Skeletal, Cardiac, and Smooth Muscle
Edited by G.B. Frank *et al.*, Plenum Press, New York, 1992

The dissection and perfusion solutions were similar except for the concentration of potassium which was 20 mM in the dissection solution to arrest the heart and thus reduce possible damage to the muscle during dissection. The Krebs-Henseleit solution contained (in mM): Na^+ 141.9; K^+ 5.0; Cl^- 129.7; Mg^{2+} 1.2; SO_4^{2-} 1.2; $H_2PO_4^-$ 2.0; HCO_3^- 19; Ca^{2+} 0.7; and glucose 10. The solutions were equilibrated with 95% O_2 and 5% CO_2 to obtain a pH of 7.4 at 26 °C. Under these conditions, the muscles were stable for the duration of the experiments (as long as 6-7 hours).

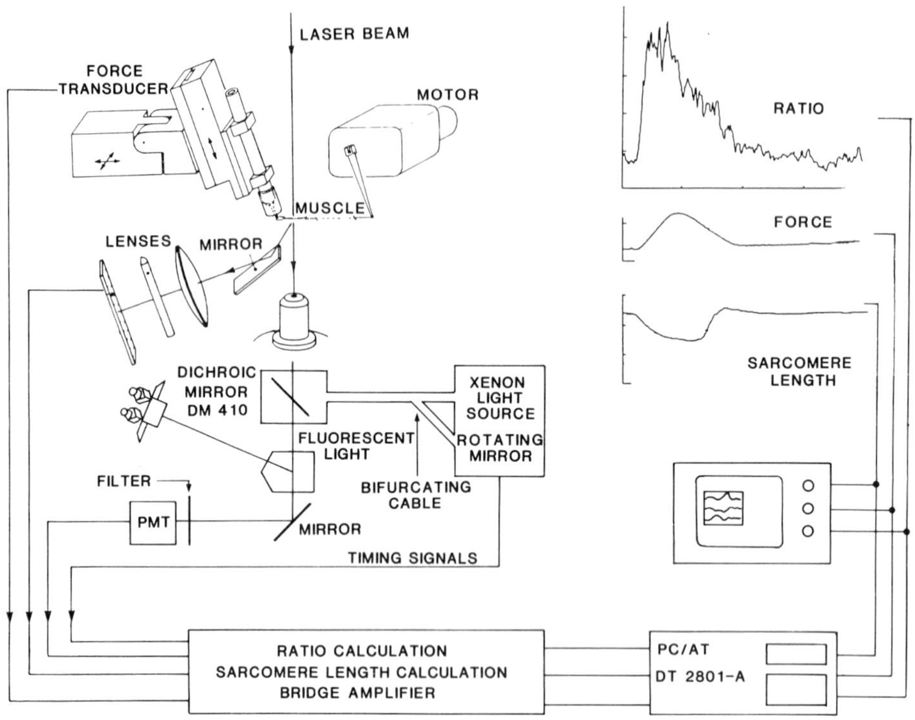

Figure 1. Illustration of the set-up used to measure the force, sarcomere length and fluorescence of rat cardiac trabeculae. Sarcomere length and fluorescence were not measured simultaneously due to physical constraints. The fluorescence was measured by manually changing the excitation filters (340 nm, 360 nm and 380 nm), or by sweeping the exciting light beam via filters over the faces of a bifurcated quartz fibre optic bundle by means of a rotating multifaceted surface mirror at a rate that could be varied between 150 and 1500 Hz. The filtered excitation light illuminated the preparation via an epifluorescence microscope using a dichroic mirror below a 10X UV objective with a high numerical aperture. The emitted light (510 nm) was collected by the objective and projected onto a low light level camera (not shown) and via the front camera port of the microscope onto a photomultiplier tube. The filtered output of the photomultiplier tube (1kHz, 3dB, 4 pole Bessel filter) was recorded, along with the force signal, on an electrostatic recorder. The signals were processed by means of a IBM-PC-computer after A/D conversion.

Force and sarcomere length measurements and muscle length control

Force was measured using a silicon strain gauge (AE801, AME, Sensonor, Horton, Norway) attached to a micromanipulator (see figure 1). A stainless steel wire was bent in a U-shape (1.5 mm wide) and the free ends were glued onto the transducer, such that U-loop was positioned in the horizontal plane. The closed end of the loop was bent 90 degrees upward and backward to form a basket [1].

Muscles were positioned in the basket in the following manner: i) the ventricular end was held and the valvular end was carefully directed through the basket and then gently pulled through the loop without stretching the muscle; ii) the remnant of the ventricular end was gently clamped between the top and bottom wires of the basket and the valvular end was attached to the stainless steel hook of a servomotor (Cambridge Technology, Cambridge, MA.). Muscles were mounted in such a way that examination under microscope showed neither twisting in the horizontal plane of the muscle nor any non-uniform shortening along the whole muscle. The muscles contracted while their length was kept constant.

Sarcomere length was measured using laser diffraction techniques. The muscle was illuminated via a 25 mW He-Ne laser (wave length 632.8 nm; Spectra-Physics) with a beam diameter at the preparation of about 300μm. The first order diffraction pattern on one side was projected onto a photodiode array (RC 105, Reticon, CA) which scanned the intensity pattern every 580μs; sarcomere length was computed electronically after correction for scattered light from the zero order. The method for calculating and calibrating sarcomere length dynamically has been discussed elsewhere [1].

Fluorescence measurements

The epifluorescence setup developed and constructed for these experiments, is shown in figure 1. The light output of a 150W Xenon lamp (Kratos) was passed through a 10 nm-bandpass filter with central wavelengths of 340 nm, 360 nm or 380 nm. The filtered ultra violet light was projected onto the muscle through a 10X Nikon Fluor objective via the epifluorescence port at the back of the inverted microscope (TMD-Diaphot, Nikon, Toronto, Canada) using a dichroic mirror (410DPLC, Nikon). The U-V excitation light illuminates an area about 1.5 mm in diameter in the focal plane which contains the trabecula. The use of a large field of view minimizes movement artifacts in thin uniform preparations when they are properly mounted (i.e. do not translate excessively) in a manner similar to that used with aequorin in papillary muscles. The fluorescent light was collected by the objective and transmitted through a bandpass filter at 510 nm (10 nm bandwidth) to a photomultiplier (3527, Hamamatsu, Japan) which was attached to the front camera port of the microscope by a light-tight housing. The photomultiplier output was filtered at 100Hz (3dB) and collected via an A/D converter (3801-A Data-Translation) and stored on computer for later analysis. The image of the muscle was observed using a low light level camera (DEP, Roden, The Netherlands).

Fura-2 Loading and Calibration

Fura-2 salt was loaded by dissolving 2 mM Fura-2 salt in a 10 mM Hepes buffered to a pH of 7.0. This solution was used to fill the tip of a micropipette. The remainder of the pipette was back filled with a solution containing 140 mM KCl and 10 mM Hepes buffered to a pH of 7.0. The pipettes used in this way had a resistance of 150-200 M Ohm (0.2μm tip diameter) when placed in a solution of modified Krebs Henseleit solution used to superfuse our trabeculae. After impalement, 4-10 nA current was passed for about 15 minutes; this resulted in a 5-10 mV hyperpolarization. The small hyperpolarization is probably a consequence of the large slope conductance of the inward rectifier at potentials negative to the reversal potential for K^+ and the large space constant for these trabeculae (> 5 mm) [2]. Fura-2 diffuses into the adjacent cells, which could be seen as a spread of fluorescence throughout the muscle. After injection, the preparation was stimulated at 1 Hz in order to accelerate the spread of Fura-2 from the site of injection. After one hour, preparations were uniformly loaded and the injection site was not detectable. At the level of loading employed in this study the force developed by these preparations decreased by about 40-60%. Much lower levels of loading are no doubt possible with this technique.

Calibration Procedure For Fura-2

In vitro calibrations were performed by placing 150 μL of a solution of known $[Ca^{2+}]$ into the perfusion bath and measuring the emitted fluorescence at 510 nm following excitation at wavelengths of 340 nm and 380 nm. The background fluorescence was determined by placing an EGTA solution without Fura-2 in the perfusion bath. The in vitro solutions contained (mM): KCl 120, MgCl 1.2, NaCl 5.8, Hepes 10, total EGTA 10, variable Ca^{++} added as CaEGTA, and KOH to adjust the pH to 7.00. Different $[Ca^{2+}]$ were achieved by mixing solutions containing 10 mM EGTA and 10 mM CaEGTA in various proportions. The relation between the ratio of intensities of fluorescence after subtraction of the background fluorescence and $[Ca^{2+}]$ is given by the following equation [3] (assuming 340/380):

$$[Ca^{2+}] = K' (R - R_{min}) / (R_{max} - R) \qquad (1)$$

where: $K' = (S^u/S^b)\ K_d$ = apparent dissociation constant

S^u/S^b = ratio of the fluorescence at 380 nm and at a $[Ca^{2+}]=0$ to that at a saturating $[Ca^{2+}]$.

K_d = dissociation constant of Fura-2 for calcium.

R_{max} = ratio of the fluorescence at 340 to that at 380 at a saturating $[Ca^{2+}]$

R_{min} = ratio of the fluorescence at 340 nm to 380 nm at a $[Ca^{2+}]$ of zero.

$[Ca^{2+}]$ calculated from equation (1) in vitro could be wrong in vivo if the properties of Fura-2 change in the cytosol. We, therefore, repeated calibration of Fura-2 in vivo at the end of three experiments by metabolically poisoning the muscle with a Hepes solution which was not equilibrated with 100% O_2 and which contained 1 mM $[Ca^{2+}]_o$ with 2 mM Iodoacetate and 4 mM CN^-. After rigor had developed (about 6 minutes) solutions containing 5μM of the calcium ionophore 4-Br-A23187 were applied to the preparation. The fluorescence in most preparations began to change immediately after the application of these compounds and reached a steady state after about 5-10 minutes. After applying solutions with different $[Ca^{2+}]$ a Hepes solution with zero calcium, 2 mM $MnCl_2$ and 5μM 4-B-A23187 was added to the preparation in order to determine the autofluorescence since this value could be different from that measured prior to loading of the preparation (see above).

RESULTS AND DISCUSSION

Cytosolic Ca^{2+} Transients and Force Development

Figure 2 shows force and the estimated cytosolic $[Ca^{2+}]$ as a function of time during a twitch at an external $[Ca^{2+}]_o$ of 1 mM in a preparation loaded by microinjection of Fura-2 salt. The results were obtained during contractions at long and short end-systolic sarcomere length by varying muscle length. Muscle length was kept constant during the twitch. At the different muscle lengths, the peak and time course of the calcium transients during the twitch were strikingly similar. The rise of $[Ca^{2+}]_i$ started 20 to 22 msec after the stimulus; peak $[Ca^{2+}]_i$ occurred 45-55 msec later; half relaxation of the calcium transient required 160 msec and 90 % relaxation of the transient required 450 to 600 msec depending on sarcomere length. The calcium transient at the longer sarcomere length relaxes more rapidly in the early phase of the relaxation than at the shorter sarcomere length. However, later during relaxation at long sarcomere length the calcium transient relaxes more slowly than at short sarcomere length. As a result the calcium transients at the two sarcomere lengths cross-over during relaxation period.

The relative time course of twitch force versus the calcium transient at the two muscle lengths was different which is a direct result of the prolongation of the twitch time with increasing muscle length [4]. 50% of peak twitch force during contractions in which muscle length was held constant coincided with the peak of the calcium transient at 70 to 80 msec; peak force always occurred later than the peak of $[Ca^{2+}]_i$ (120 msec at short sarcomere length and approximately 160 msec in the stretched muscle), while force relaxed to 10 % of peak twitch force more slowly than the calcium transient. The latter process varied strongly with sarcomere length (130 msec and 240 msec at short sarcomere length and long sarcomere length respectively). In the stretched muscle, the calcium transient has declined by 50% from its peak systolic value by the time the force has reached a maximum, and the relationship between the calcium transient and the time course of force is close to that of the calcium transient derived from aequorin luminescence [5]. The duration of the force transient during the twitch and the sarcomere shortening transient corresponded closely. The degree of shortening of the sarcomeres in the centre of the muscle was related to muscle length at which the contraction started and was much smaller in the stretched muscles than in the muscle at short length.

The delay between the stimulus and the beginning of the calcium transient, its initial rate of rise and the time to the peak of the transient (about 30msec) as inferred from the observations using Fura-2 were similar to that measured using aequorin [6]. In contrast, the decay phase measured with Fura-2 had a time constant of about 200msec at 22°C which is much longer than that derived from aequorin luminescence [7]. This could result from a number of different factors such as: slower calcium dissociation from Fura-2 than from aequorin; buffering of the cytosolic calcium by Fura-2 or; a higher sensitivity of Fura-2 than of aequorin to $[Ca^{2+}]$.

Buffering of calcium by Fura-2 must have taken place at the levels of loading reported here, as we observed a reduction of force as well as an increase of twitch duration (approximately 5 %). Further study at lower levels of Fura-2 loading are required in order to evaluate the contribution of the dissociation of Ca^{2+} from Fura-2 to the relaxation of the apparent calcium transient. The sensitivity of Fura-2 to calcium as reported here and in previous studies is indeed higher than that of aequorin and must have contributed to a longer lasting record of the decline of the calcium transient [3,5].

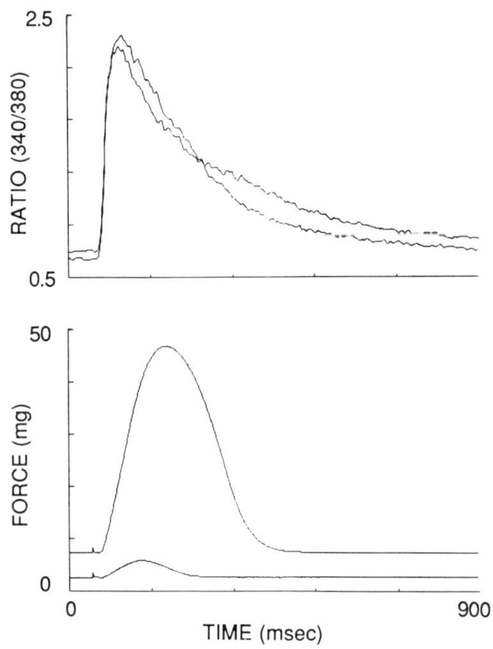

Figure 2. Panel A shows the ratio of the fluorescence at 340nm excitation to that at 380nm excitation during a twitch at two sarcomere lengths. The corresponding forces at the two sarcomere lengths are shown in panel B. where the larger force corresponds to that recorded at an end-systolic sarcomere length of 2.15 μm while the smaller force corresponds to a sarcomere length of 1.6 μm. The calcium transient at the longer sarcomere length relaxes more rapidly in the early phase of the relaxation than at the shorter sarcomere length. However, later in the relaxation process the calcium transient at the longer sarcomere length relaxes more slowly than at the shorter sarcomere length. As a result the calcium transients at the two sarcomere lengths cross-over during relaxation period.

Previously, it has been reasoned that the prolonged relaxation of the calcium transient as measured in single cells could be due to shortening deactivation [8,9,10,11]. However, this hypothesis can neither explain our results nor those obtained in whole hearts using Indo-1 [12]. These results provide at least a partial explanation for the discrepancy between calcium transients measured with aequorin in mechanically loaded multicellular preparations and those measured with the fluorescent indicators in single cells. At short sarcomere length and with a large degree of shortening in the central region of the trabecula, the peak calcium transient decayed much more slowly relative to the force trace than at greater muscle length close to L_{max}, see figure 2, which is commonly the length of the papillary muscles used in the aequorin studies. Notice that at greater length the calcium transient relative to the force trace is similar to that measured in aequorin. The unloaded shortening that takes place in single cells may correspond to conditions at short muscle lengths where we see a large amount of internal shortening and where the calcium transient returns to the baseline with a time course similar to the force trace.

The interpretation of the $[Ca^{2+}]_i$ transients and their relation to force development requires caution, since it is known that full activation of the contractile system requires approximately 70 μM of Ca^{2+}, with simultaneous binding of another 50 μM of Ca^{2+} to calmodulin [6]. Hence, even activation of the muscle at only 25% of maximum, such as in this study, is accompanied by Ca^{2+} turnover of near 30 μM. It is clear that only a small fraction of this Ca^{2+} is 'visible' with the aid of the fluorescent probe. With these considerations in mind we feel that still a number of pertinent conclusions can be drawn.

The calcium transients did vary systematically as a function of the muscle length, and hence sarcomere length. Figure 2 shows clearly that the rising phase is unaffected. It is therefore, unlikely that a major change in the rate and or amount of calcium release occurs as a result of stretch of the muscle, as has previously been suggested by Allen's group [7]. Furthermore, at steady-state the peak systolic $[Ca^{2+}]$ was slightly decreased at greater lengths. Conspicuous, albeit fairly small changes took place during decay of the calcium transient. Consistently the early phase of the relaxation process was faster while the later phase of relaxation was prolonged at longer sarcomere length relative to shorter sarcomere lengths. These changes occurred concomitantly with the development of a delay of relaxation of sarcomeres to their original diastolic length. Occasionally this delay took the character of a late secondary contraction. These observations would be expected if the early part of the decay of the calcium transient is governed by the rate of calcium binding to the contractile filaments and if binding in a stretched muscle is accelerated. This effect is consistent with the hypothesis that force-length relation (as shown in Figure 3) is determined principally by the length-dependent sensitivity of the contractile

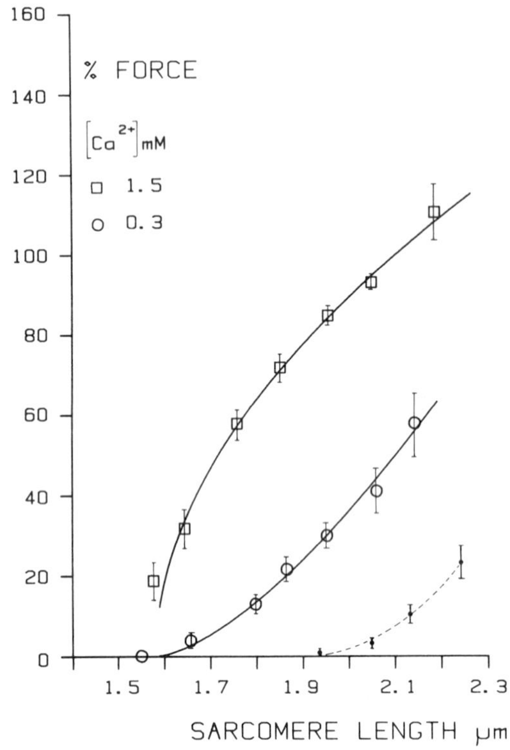

Figure 3. Illustrates the force sarcomere length relationship observed at a low and at a high $[Ca^{2+}]_o$ in intact rat cardiac trabeculae. The relationships coincide with the relationships predicted by the curves in Figure 4.

system, (see figure 4), which resides in calcium affinity of troponin for Ca^{2+} as a function of stretch [13,14,15,16]. This hypothesis, however, also implies that in the stretched myocardium much more Ca^{2+} is bound. Hence, one could explain the observation that the peak amplitude of the $[Ca^{2+}]_i$ transient is unchanged only if one assumes that also more Ca^{2+} has been released. The later decay of the calcium transient which occurs predominantly during relaxation of the twitch would be explained by dissociation of calcium from troponin-C. Similar to the reasoning explaining the acceleration of the early decay of the calcium transient by stretch, one would expect the later phase to be slowed in stretched cardiac muscle because of the enhanced binding of the contractile system to calcium ions.

The data presented in this study are consistent with the hypothesis that the Frank-Starling mechanism is determined principally by length-dependent sensitivity of the contractile system to Ca^{2+} ions and that this length-dependence resides predominantly in the affinity of troponin for Ca^{2+}. It cannot be excluded that the amount of Ca^{2+} released by' the sarcoplasmic reticulum varies substantially with sarcomere length.

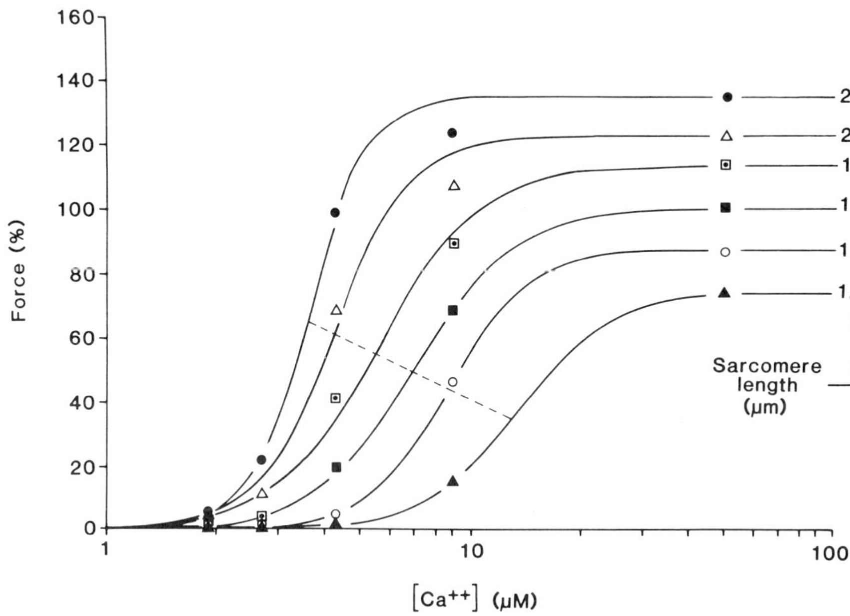

Figure 4. Relationship between active force and $[Ca^{2+}]$ in skinned muscles at various SL (given in μm next to the appropriate curves). Data obtained by replotting the curves of Fig. 1B. The solid lines are best-fits to the modified Hill equation:

$$\text{Force} = \text{Maximum force} \times [Ca^{2+}]^n/(K^n + [Ca^{2+}]^n)$$

The dashed line joins the points corresponding to half) maximal activation for each curve. It can be seen that increasing the SL raises both maximum force production and the Ca^{2+} -sensitivity of the myofibrils (Kentish et al 14).

THE TWITCH FORCE-INTERVAL RELATIONSHIP

Figure 5 shows the time course of twitch force (Ft), when a stimulus was inserted after the last contraction of a steady series (0.2 Hz) at a varied stimulus interval. The results are consistent with the Ft -interval relationship, which has been previously documented in detail [17]. At short intervals Ft was small, but recovered rapidly. Ft, then, increased more slowly to a maximum at approximately 100 sec. This phenomenon has been denoted rest potentiation. A decline in Ft, denoted as rest depression, at intervals > 100 sec was evident. With these long intervals (> 100 sec) we found neither signs of spontaneous sarcomere activity by microscopic observation nor evidence of spontaneous activity in the form of fluctuations of the intensity distribution of the laser diffraction pattern of the muscles; hence, the decline of Ft could not have been caused by spontaneous Ca^{2+} release prior to stimulation. Ryanodine diminished Ft at all intervals in a dose-dependent manner. The rapid recovery was less complete, but the negative inotropic effects of ryanodine were more pronounced at longer intervals as evident from the diminished rest potentiation and accelerated rest depression.

Contractures at near 0 °C

To examine the Ca^{2+} content of the SR available for contraction, and to test whether ryanodine alters the net Ca^{2+} content, we examined force of contractures (Fc) induced by rapid cooling of the muscle in the absence and presence of ryanodine. With rapid introduction of cold Krebs-Henseleit solution into the bath, the muscle developed a contracture as soon as the temperature reached near 0 °C. The Fc-interval relationship of three muscles is summarized in Figure 6. The magnitude of Fc (prior to ryanodine) following intervals < 1 sec was strikingly constant. This contrasts with the increase of the amplitude of the twitch from a negligible force to steady state twitch force during early recovery (Figure 5). Rest potentiation was also manifest in the contractures, i.e. Fc was 20 to 30 % larger at intervals of 100 sec than Fc at 5 sec interval. This increase of Fc was only slightly smaller than the increase of Ft in the control solution (32 to 48 %; cf Figure 5).

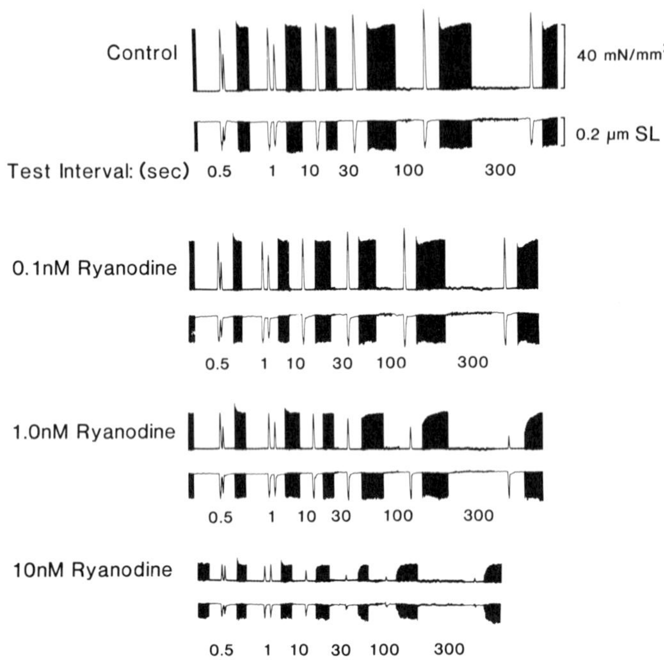

Figure 5. Record of twitch force (Ft) and sarcomere length (SL) at various intervals following the last twitch in steady state in one rat trabeculae. Note that the negative inotropic effect of ryanodine on Ft is increasing with increasing interval after the last steady twitch. $[Ca^{2+}]_o$ was 0.7 mM, frequency was 0.2 Hz, and temperature was 26 °C. The recording speed was increased to show the test twitches. The calibrations apply to all panels. (from 48)

Fc decreased in the muscles exposed to ryanodine, but the time to maximal force of the contracture induced by cooling was affected little. Ryanodine depressed Fc more at an interval of 100 sec than at 5 sec in a dose-dependent manner (cf. Figure 6). Ryanodine (1.0 nM) caused acceleration of rest depression of Fc and concomitantly shortening of the time to maximal rest potentiation. The magnitude of the effect of ryanodine on Fc was comparable to that on the Ft-interval relationship; at 10 nM, ryanodine reduced both Fc and Ft at a post stimulus interval of 5 sec by ~50% , eliminated rest potentiation and reduced both Fc and Ft at an interval of 100 sec by approximately 80%.

The results of this study confirmed the existence of: 1) Early recovery during the Ft-interval relationship; 2) Rest potentiation; and 3) Rest depression, as has been shown in rat cardiac muscle in detail [17]. Rest potentiation and rest depression of force development are also prominent in the force interval relationship of contractures induced by rapid cooling. A novel finding is that shortly after a twitch (<1sec) the amplitude of a rapid cooling contracture is independent of time following the twitch, whereas the amplitude of an electrically triggered twitch would increase from zero to steady state level as a function of time during the same interval. The effect of ryanodine consisted of a concentration dependent decrease of force of both twitches and rapid cooling contractures; the decline of force due to ryanodine increased progressively with increasing interstimulus intervals.

The effects of varying the interval between the last steady state twitch on the amplitude of a contracture induced by rapid cooling are consistent with previous studies and the proposed explanation of rest potentiation and rest depression [18,19,20]. Moreover, the Fc - interval relationship sheds further light on the process involved in the early rapid recovery of force development.

Recent studies have shown that rapid cooling of cardiac muscle to 0 °C rapidly 'freezes' the SR Ca^{2+} channels in an open state [21], which allows a large [22] and possibly complete [23,24] release of Ca^{2+} ions from the SR. Rapid cooling is neither accompanied by an action potential [18,25], nor by a depolarization of sufficient magnitude to cause gated Ca^{2+} entry into the cell [18]. The force of the contracture in response to cooling, therefore, is probably only determined by the amount of Ca^{2+} released by the SR during this procedure and by the sensitivity of the contractile filaments, which is lowered in the cold [24]. The amount of Ca^{2+} released by the SR is probably determined by the Ca^{2+} content of the SR as has been shown by the close correlation of the amplitude of the rapid cooling contracture and the SR-Ca^{2+} content, as measured by atomic absorption spectrophotometry [18,25]. Experiments by Dani et al [26] on isolated myocytes showed that SR-related Ca^{2+} sequestration amounted to ~ 300 μmol/kg wet weight. Bridge has shown that nearly all of this Ca^{2+} is lost from the tissue during a rapid cooling contracture, i.e. ~260μmol/kg wet weight [18]. In the same study a strong correlation between the magnitude of the rapid cooling contracture and the Ca^{2+} content of the tissue was documented [18]. Although part of the Ca^{2+} released may have some other origin, the similarity between SR Ca^{2+} sequestration in isolated myocytes and the amount of Ca^{2+} which is lost with cooling suggests that cooling causes a near complete Ca^{2+} release from the SR. We have, therefore, used measurement of the amplitude of the contracture (Fc) induced by rapid cooling to obtain an, albeit not necessarily linear, estimate of the Ca^{2+} content of the SR, as has been done in several previous studies [27-30].

Early recovery of twitch force

The following three candidates may underlie the early rapid recovery of Ft: 1) a delay in transport of Ca^{2+} ions from a hypothetical uptake compartment (probably the longitudinal component of the SR) to a release compartment (probably the terminal cisternae) with a transport time constant of ~ 700 msec in rat at 25 °C [see 17]; 2) recovery of Ca^{2+} current during action potential (Isi); 3) recovery of SR Ca^{2+} channels after the last release [31].

The properties of the rapid cooling contracture as well as the recovery of rapid cooling contractures following the twitch do not suggest the presence of a process that transports Ca^{2+} from an uptake compartment to a release compartment in the SR, i.e. transport of Ca^{2+} longitudinally through the SR. Figure 6 shows clearly that the time to peak force of the contractures induced by rapid cooling is approximately 350 msec. If opening of SR Ca^{2+} release channels [21] is the only effect of rapid cooling on the SR, it follows that release of all Ca^{2+} present in the SR [23] through these Ca^{2+} channels must have taken place in less than 350 msec. This is clearly longer than the time constant of recovery of the twitch, and inconsistent with a slow transport process between longitudinal SR and the terminal cisternae. It has been shown that the time course of Ca^{2+} transients during cooling contractures is even faster [22], and in fact it is likely that the limiting factor for the rate of rise of force during a rapid cooling contracture is determined by the response of the contractile filaments to Ca^{2+} ions [23]. The speed of the process of activation during rapid cooling is clearly inconsistent with the existence of a long delay (at 26 °C) in the transport between the hypothetical uptake and release compartments within the SR, especially because one would expect that the rate of transport would slow down due to the cooling before the contracture starts. It is possible, but less likely, that rapid cooling opens Ca^{2+} pathways everywhere along the SR including the ryanodine receptors [21] in the terminal cisternae [32,33]. If this were the case, the relationship between Fc and increasing intervals would not reveal a transport delay between the uptake and release sites in the SR. This possibility requires us, however, to postulate ad hoc a more complicated effect of rapid cooling. The hypothesis that a separate uptake compartment would exist in the SR, insensitive to rapid cooling, is ruled out because of the evidence suggesting that the rapid cooling contracture assays all of the Ca^{2+} in the SR [18,26]. For the same reason one would expect that other

Ca^{2+} sources (mitochondrial or extracellular), that potentially contribute to the recovery of the twitch, should be revealed by the rapid cooling contracture.

Studies of calcium-induced calcium release have shown that, provided that the SR is filled with Ca^{2+}, Isi causes a graded and variable release of Ca^{2+} depending on rate of rise and magnitude of Isi [31,34]. If the SR is Ca^{2+} depleted, the Ca^{2+} current will first load the SR and hence can not induce a substantial contraction by itself [35]. The importance of this loading has been shown in experiments where altering the duration of one action potential changed only the magnitude of subsequent contraction [36]. Furthermore, measurement of Ca^{2+} in the extracellular clefts of cardiac muscle during post rest contractions, have shown no link between Ca^{2+} entry and the magnitude of the corresponding contraction, especially at long rest intervals [37]. Since the results suggest that, at short intervals the SR Ca^{2+} load is large, a large Isi should induce a significant release of SR Ca^{2+}. It clearly does not. Hence the small magnitude of force at short intervals may reflect slow recovery of Isi or slow recovery of the SR Ca^{2+} channels [31]. Studies on recovery of Isi [38,39], however, suggest that this process is two to three times faster than recovery of Ft measured in our study. The time course of recovery of plateau duration of rat action potential [40] is consistent with rapid recovery of Isi and has also been shown to be substantially faster than the recovery of Ft [40].

Therefore, by elimination of the other candidates, we favour the hypothesis that slow recovery of SR Ca^{2+} channels determines the rate of recovery of Ft.

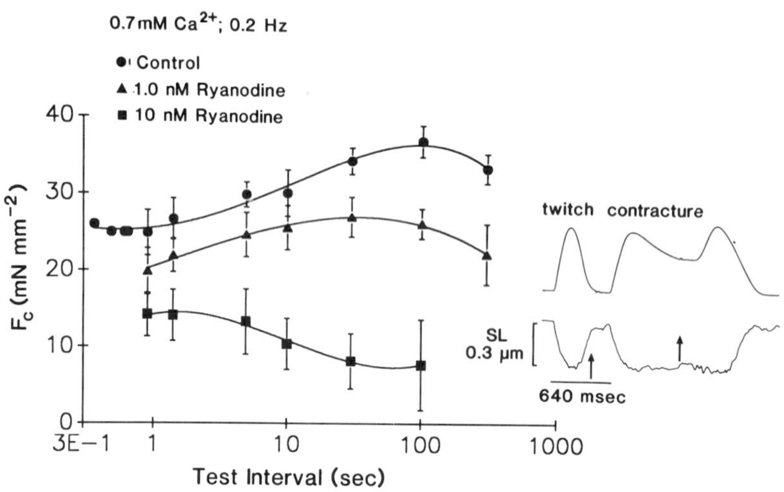

Figure 6. The average Fc-interval relationship. Rest potentiation and rest depression are manifest in the control solution. Ryanodine accelerated rest depression in a similar manner to the reduction of rest depression of the twitch in Figure 5. Note however, that unlike the twitch force (Figure 5), Fc is virtually constant at intervals < 1 sec, as is also illustrated in the **Inset** which shows a recording of a cold contracture induced after an interval of 640 msec. First arrow indicates the introduction of cold solution and the second one shows the rewarming. Fc: force of contracture. (from 48)

Rest potentiation and rest depression

Since there is an increase in the amplitude of Fc at intervals between 1 sec and 100 sec, it appears that the SR accumulates Ca^{2+} during these intervals. This supports the earlier postulates [17,41] that rest potentiation occurring in rat trabeculae is due to a net Ca^{2+} influx into the cell during rest, this Ca^{2+} is then taken up by the SR. Rest potentiation has been shown in several species including man [19,42] but is large in the rat, probably because of the relatively high Na^+ level in the rat myocyte [20], which allows for a net influx of Ca^{2+} ions during the first two minutes after a twitch.

The decrease in Fc at intervals > 100 sec (rest depression) can be interpreted as a reduction in the SR Ca^{2+} content at these intervals. Studies by Kitazawa have indeed demonstrated Ca^{2+} leak from the SR of skinned cardiac muscle with time [43]. Furthermore, measurements by means of atomic absorption spectrophotometry have shown a reduced Ca^{2+} content in rabbit papillary muscles following long rest intervals [18]. This also agrees with studies by means of Ca^{2+} sensitive microelectrodes, positioned in the extracellular clefts, that have revealed an increase of $[Ca^{2+}]_o$ during rest [44,45]. Taken together these studies suggest that Ca^{2+} leaks out of the SR during rest and leaves the cell by Ca^{2+} extrusion mechanisms.

We suggest that the loss of Ca^{2+} results from an increase in the background Ca^{2+} leak from the SR with ryanodine; evidently Ca^{2+} that has leaked out of the SR is expected to be extruded from the cell by the normal extrusion mechanisms [30]. Such a postulated Ca^{2+} leak is in agreement with previous studies, indicating that the appearance of Ca^{2+} in the extracellular space during the interstimulus interval, is faster in the presence of ryanodine than in control conditions [44,45].

It is of course attractive to assume that ryanodine at nanomolar concentrations exerts only one effect, i.e. to increase the probability that the Ca^{2+} channels in the SR occur in an open state with a low conductance. 'Partial' opening of SR-Ca^{2+} channels by ryanodine would lead to a progressive leak of Ca^{2+} from the SR as shown in Figure 6. However, ryanodine would be expected to exert little or no effect on the kinetics of Ca^{2+} release or recovery from Ca^{2+} release, as reflected by time to peak twitch force or on kinetics of early recovery of twitch force development, as is indeed shown by our data (cf. Figures 2,3,4, and results section). The latter would be consistent with the proposed mechanism for early recovery of twitch force which, as we have argued in the above, probably consists of recovery of SR-Ca^{2+} channels after the preceding twitch.

ENERGY REQUIRED FOR EXCITATION CONTRACTION COUPLING

Energy expenditure by the heart is partitioned between three processes: i) basal metabolism; ii) ion transport during excitation contraction coupling and ; iii) ATP hydrolysis by the contractile filaments. Basal metabolism is required for maintainance of the composition of the intracellular organelles, and for protein synthesis and is assumed to be independent of the heart rate. Recently, Suga and Sagawa have proposed that the energetic output of the heart consists of the sum of the externally performed work and a potential energy term which can be calculated from the End Systolic Pressure Volume relationship (ESPVR):

$$P_{es} = E_{max} (V_{es} - V_d)$$

in which P_{es} is the end systolic pressure; E_{max} is the apparent maximal stiffness of the ventricle; V_{es} is the end systolic volume; and V_d is the minimal volume that can be attained by the active ventricle. External work put out by the ventricle equals the area enclosed by the pressure volume trajectory during ejection; the potential energy term equals the area under the ESPVR between V_{es} and V_d. The sum of the two areas is generally denoted as the pressure volume area (PVA).

Suga has demonstrated, convincingly, a linear relationship between oxygen uptake (MVO_2) by the heart during each beat and PVA [46]. The relationship between MVO_2 and PVA has a succinct intercept with the MVO_2 axis. An important observation that followed from his work is that positive inotropic interventions shifted the relationship between MVO_2 and PVA up without a change of its slope. This is consistent with the concept that the slope of the relationship between MVO_2 and PVA reflects the amount of energy required for the mechanical process, while the intercept is composed of the energy required for basal metabolism together with the energy required for excitation contraction coupling. Resultantly the relationship between MVO_2 and PVA may be described as:

$$MVO_2 = HR.(k_w.PVA + k_{ecc}.ECC) + k_{bm}.BM.$$

where MVO_2 is the oxygen consumption of the heart per minute; HR is the heart rate in beats/min; k_w is the coefficient that relates MVO_2 to the total of generated mechanical energy (k_w is the inverse of the mechanical efficiency);PVA has the usual meaning; k_{ecc} is the coefficient or function that relates MVO_2 to excitation contraction coupling; ECC possibly reflects the number of calcium ions transported by the cell organelles per heartbeat. BM is basal metabolism; k_{bm} relates MVO_2 to basal metabolic processes.

The studies of Suga and collaborators suggest that the above coefficients are distributed in the following way: k_w amounts to approximately 30 % of the energy consumption of the heart, and the distribution of energy consumed by actual work vs potential energy depends on the loading conditions to which the heart is subjected. The sum of the energy required for excitation-contraction coupling and basal metabolism is comprises the remainder of the energy expenditure of the heart. 30-40% of the latter term is spent on excitation-contraction coupling. Recently, we have obtained estimates for the ATP consumption in isolated cardiac muscle for these processes during hypoxia and arrived at similar partitioning of the energetic expenditure in cardiac muscle [47].

Our results suggested that during hypoxia excitation-contraction coupling contributes as much to ATP consumption as twitch force. In addition we observed that ATP consumption by excitation-contraction coupling is in part due to increased Ca^{++} fluxes across the membranes as a result of endogenous catecholamine release.

Acknowledgement: This work was supported by MRC, AHFMR and the Heart and Stroke Foundations of Alberta and Canada.

REFERENCES

1. de Tombe PP, ter Keurs HEDJ. Force and velocity of sarcomere shortening in trebeculae from rat heart. Effects of temperature. Circ Res 66:1239-1254 (1990).
2. Daniels MCG, Fedida D, Lamont C, ter Keurs HEDJ. Role of the sarcolemma in triggered propagated contractions in rat cardiac trabeculae. Circ Res 68:1408-1421 (1991).
3. Grynkiewicz G, Poenie M, Tsien RY. A new generation of Ca^{2+} indicators with greatly improved fluorescent properties. J Biol Chem 260:3440 (1985).
4. Van Heuningen R, Rijnsburger WH, ter Keurs HEDJ. Sarcomere length control in striated muscle. Am J Physiol 242:H411-H420 (1982).
5. Allen DG, Blinks JR. Calcium transients in aequorin -injected frog cardiac muscle. Nature 273:509-513 (1978).
6. Wier WG, Yue DT. Intracellular calcium transients underlying the short term force-interval relationship in ferret myocardium. J Physiol 376:507-530 (1986).
7. Allen DG, Kurihara S. Calcium transients in mammalian ventricular muscle. Eur Heart J 1:5-15 (1980).
8. Spurgeon HA, Baartz G, Capogrossi MC, Raffaeli S, Stern M, Lakatta EG. The relationship between the cytosolic Ca^{2+} transient and cell length during a twitch in single adult cardiac myocytes. in: Biology of Isolated Adult Cardiac Myocytes. Editors: WA Clark, RS Decker, Borg TK. Elsevier Press, New York. pp350-353 (1988).
9. Krueger JW. Measurement and interpretation of contraction in isolated cardiac cells. In: Biology of Isolated Adult Cardiac Myocytes. Editors: WA Clark, RS Decker, Borg TK. Elsevier Press, New York. pp 172-186 (1988).
10. Housmans PR. The relation between contraction dynamics and intracellular calcium transient in mammalian cardiac muscle. In: Starling's Law of the Heart Revisited. Editors: HEDJ ter Keurs and MIM Noble. Kluwer Press, Dordrecht. pp 60-66 (1988).
11. duBell WH, Philips CM, Houser SR. A technique for measuring cytosolic free Ca^{2+} with Indo-1 in feline myocytes. in: Biology of Isolated Adult Cardiac Myocytes. Editors: WA Clark, RS Decker and TK Borg. Elsevier Publishing, New York. pp187-201 (1988).
12. Lee H, Mohabir R, Smith N, Franz MR, Clusin WT. Effect of ischemia on calcium dependent fluorescence transients in rabbit hearts containing Indo-1. Circ Res 1988;78:1047-1059.
13. Hibberd MG, Jewell BR. Calcium and length-dependent force production in rat ventricular muscle. J Physiol 329:527-540 (1982).
14. Kentish JC, ter Keurs HEDJ, Ricciardi,L, Bucx JJJ, Noble MIM. Comparison between the sarcomere length-force relations of intact and skinned cardiac trabeculae from rat right ventricle. Circ Res 58:755-786 (1986).
15. Hofmann PA, Fuchs F. Bound calcium and force development in skinned cardiac muscle bundles: effect of sarcomere length. J Mol Cell Cardiol 20:667-677 (1988).

16. Babu A, Sonnenblick E., Gulati J. Molecular basis for the influence of muscle length on myocardial performance. Science 240:74-76 (1988).
17. Schouten, V.J.A., Van Deen, J.K., de Tombe, P.P., Verveen, A.A. Force-interval relationship in heart muscle of mammals: A calcium compartment model. Biophys J 51:13-26 (1987).
18. Bridge, J.H.B. Relationships between the sarcoplasmic reticulum and transsarcolemmal Ca transport revealed by rapidly cooling rabbit ventricular muscle. J Gen Physiol 88:437-473 (1986).
19. Allen, D.G., Jewell, B.R., Wood, E.H. Studies on the contractility of mammalian myocardium at low rates of stimulation. J Physiol 254:1-17 (1976).
20. Shattock, M.J., Bers, D.M. Rat vs. rabbit ventricle: Ca influx and intracellular Na assessed by ion-selective microelectrodes. Am J Physiol 256:C813-C822 (1989).
21. Sitsapesan, R., Montgomery, R.A.P., MacLeod, K.T., Williams, A.J. Increased open probability of sheep cardiac sarcoplasmic reticulum calcium release channels induced by low temperatures. Proceedings of the Physiological Society. March 1990. 28p.
22. Bers, D.M., Bridge, J.H.B., Spitzer, K.W. Intracellular calcium transients during rapid cooling contractures in guinea-pig ventricular myocytes. J Physiol Lond 417:537-553 (1989).
23. Bers, D.M. SR calcium loading in cardiac muscle preparations based on rapid cooling contractures. Am J Physiol 256:C109-C120 (1989).
24. Harrison, S.M., Bers, D.M. Temperature dependence of myofilament calcium sensitivity of rat, guinea-pig, and frog ventricular muscle. Am J Physiol 258:C274-C281 (1990).
25. Kurihara, S., Sakai, T. Effects of rapid cooling on mechanical and electrical responses in ventricular muscle of guinea-pig. J Physiol Lond 361:361-378 (1985).
26. Dani, A.M., Cittadini, A., Inesi, G. Calcium transient and contractile activity in dissociated mammalian heart cells. Am J Physiol 237:C147-C155 (1979).
27. Bers, D.M., Bridge, J.H.B. Relaxation of rabbit ventricular muscle by Na/Ca exchange and SR calcium pump- Ryanodine and voltage-sensitivity. Circ Res 65:334-342 (1989).
28. Bers, D.M., Bridge, J.H.B., MacLeod, K.T. The mechanism of ryanodine action in rabbit ventricular muscle evaluated with Ca-selective microelectrodes and rapid cooling contractures. Can J Physiol and Pharmacol 65:610-618 (1987).
29. Bers, D.M., Bridge, J.H.B. Effect of acetylstrophanthidin on twitches, microscopic tension fluctuations and cooling contractures in rabbit ventricle. J Physiol Lond 404:53-69 (1988).
30. Hryshko, L.V. Stiffel, V., Bers, D.M. Rapid cooling contractures as an index of SR calcium content in rabbit ventricular myocytes. Am J Physiol 257:H1369-H1377 (1989).
31. Fabiato, A. Stimulated calcium current can both cause calcium loading and trigger calcium release from the sarcoplasmic reticulum of skinned cardiac Purkinje cells. J Gen Physiol 85:291-320 (1985).
32. Inui, M., Saito, A., Fleischer, S. Isolation of the ryanodine receptor from cardiac sarcoplasmic reticulum identified with the feet structures. J Biol Chem 262:15637-15642 (1987).
33. Inui, M., Saito, A., Fleischer, S. Purification of the ryanodine receptor and identification with feet structures of junctional terminal cisternae of sarcoplasmic reticulum from fast skeletal muscle. J Biol Chem 262:1740-1747 (1987).
34. Nabauer, M. Callewaert, G., Cleemann, L., Morad, M. Regulation of calcium release is gated by calcium current , not gating charge, in cardiac myocytes. Science 244:800-803 (1989).
35. Morad, M., Cleeman, L. The role of Ca^{2+} channel in development of tension in heart muscle. J Mol Cell Cardiol 19:527-553 (1987).
36. Antoni, H., Jacob, R., Kaufmann, R. Mechanische Reaktionen des Frosch- und Saeugetiermyokards bei Veraenderung der Aktionpotential-Dauer durch konstante Gleichstromimpulse. Pfluegers Arch 306:33-57 (1969).
37. Bers, D.M. Early transient depletion of extracellular calcium during individual cardiac muscle contractions. Am J Physiol 244:H462-H468 (1983).
38. Josephson, J., Sanchez-Chapula, J., Brown, A.M. A comparison of calcium currents in rat and guinea pig single ventricular cells. Circ Res 54:144-156 (1984).
39. Tseng, G.N. Calcium current restitution in mammalian ventricular myocytes is modulated by intracellular calcium. Circ Res 63:468-482 (1988).
40. Schouten, V.J.A. The negative correlation between action potential duration and force of contraction during restitution in rat myocardium. J Mol Cell Card 18:1033-1045 (1986).
41. Ragnarsdottir, K., Wohlfart, B., Johansson, M. Mechanical restitution in rat papillary muscle. Acta Physiol Scand 115:183-191 (1982).
42. Quagebeur, J.M. Schouten, V.J.A., ter Keurs, H.E.D.J. Mechanical restitution in human myocardium. J Physiol 377:120p (1986).

43. Kitazawa, T. Effect of extracellular calcium on contractile activation in guinea-pig ventricular muscle. J Physiol Lond 355:635-659 (1984).
44. Hilgemann, D.W. Extracellular calcium transients and action potential configuration changes related to post-stimulatory potentiation in rabbit atrium. J Gen Physiol 87:675-706 (1986).
45. Bers, D.M., MacLeod, K.T. Cumulative extracellular calcium depletions in cardiac muscle. Circ Res 58:769-782 (1986).
46. Sagawa K, Maughan L, Suga H, Sunagawa K. Energetics of the Heart, In: Cardiac Contraction and the Pressure-Volume Relationship. Oxford University Press, pp 171-231 (1988).
47. Gao WD, ter Keurs HEDJ. Determinants of ATP depletion during metabolic inhibition of rat cardiac muscle. Canadian Cardiovascular Society Meeting (abstract) 1991.
48. Banijamali HS, Gao WD, MacIntosh BR, ter Keurs HEDJ. The force-interval relationships of twitches and cold contractures in rat cardiac trabeculae: the effect of ryanodine. Circ Res (in press) 1991.

POTENTIATION OF CONTRACTION AS RELATED TO CHANGES IN FREE AND TOTAL INTRACELLULAR CALCIUM

Maria-Fiora Wendt-Gallitelli+ and Gerrit Isenberg*

+Institute of Physiology II, Gmelinstr. 5, University of Tübingen
*Department of Physiology, Robert-Koch-Str. 39
University of Cologne, Germany

SUMMARY

In voltage-clamped guinea-pig ventricular myocytes, we studied the potentiation of contraction in dependence on the concentration of intracellular calcium; ionized calcium $[Ca^{2+}]_c$ was measured by Indo-1 microfluospectroscopy and total calcium (ΣCa) by electronprobe microanalysis (EPMA). After a 15 min rest period, $[Ca^{2+}]_c$ was approx. 90 nM and ΣCa was below the detection limit (80 µM) in myoplasm (ΣCa_{myo}), junctional sarcoplasmic reticulum (ΣCa_{SR}) and mitochondria (ΣCa_{Mito}). Post rest, repetitive clamp steps (1 Hz) potentiated extent and rate of shortening by 300%. In the literature, post-rest potentiation is attributed to the replenishment of SR with releasable calcium; by EPMA the postulated increase in ΣCa_{SR} was measured directly. Post-rest, the peaks of systolic $[Ca^{2+}]_c$ transients increased, however only by 40%. In addition, a moderate increase of end-diastolic $[Ca^{2+}]_c$ was measured. In an other series of experiments, contraction was potentiated by 800% increase by means of paired voltage clamp pulses (1 Hz, 36 °C, 2 mM $[Ca^{2+}]_o$). In the potentiated state, end-diastolic $[Ca^{2+}]_c$ was 180 nM and ΣCa_{myo} was 0.65 mM. During systole, $[Ca^{2+}]_c$ peaked within 20 ms to 950 nM. ΣCa_{myo} rose within 20 ms to 1.4 mM and fell within 40 ms to 1.1 and within 90 ms to 0.8 mM. In contrast, the time course of contraction was slow and peaked at a time (130 ms) when the $[Ca^{2+}]_c$ and ΣCa_{myo} transients were finished. We suggest that Ca^{2+} bound to troponin C (TnC) controls only the onset but not the time course of myofilament interaction. From $[Ca^{2+}]_c$ and ΣCa_{myo} we estimated a Ca^{2+} buffering capacitance of 1.5 mmol ΣCa_{myo} per pCa change, only a fraction of which can be attributed to Ca^{2+} binding sites on TnC. A model explaining the results requires the assumption of 0.6 mM additional slow, high affinity Ca^{2+} sites and 2 mM fast, low affinity Ca^{2+} sites. We discuss that end-diastolic Ca^{2+} binding to these sites contributes to the potentiation of contraction. Junctional SR. At the end of diastole ΣCa_{SR} was 2.4 mM which is 4 times larger than ΣCa_{myo}. This difference disappeared 20 ms after depolarization (ΣCa_{SR} 1.1 mM), within another 20 ms it largely recovered (ΣCa_{SR} 2.0 mM). These properties suggest that the junctional SR is a compartment suitable not only for Ca^{2+} release but also for rapid Ca^{2+} re-uptake. Mitochondria. Paired-pulse potentiation increased end-diastolic ΣCa_{Mito} significantly (0.4 mM). During diastolic ΣCa_{Mito} followed the changes in ΣCa_{myo} with some delay (peak ΣCa_{Mito} of 1.2 mM after 40 ms). According to biochemical literature, we interpret the changes in ΣCa_{Mito} to stimulate the activity of dehydrogenases and to adapt the rate of ATP synthesis to the demands of muscle mechanics.

Excitation-Contraction Coupling in Skeletal, Cardiac, and Smooth Muscle
Edited by G.B. Frank *et al.*, Plenum Press, New York, 1992

INTRODUCTION

Contraction of cardiac muscle is activated when Ca^{2+} binds to troponin C (TnC). The concentration of Ca-TnC, usually, is extrapolated from the concentration of Ca^{2+} ionized in the cytoplasm ($[Ca^{2+}]_c$) as it is measured by aequorin-luminescence or by fluorescent dyes. Those extrapolations suffer from two problems: (1) The concentration of sites competing with TnC for Ca^{2+} binding is essentially un-known[30]. (2) Bound calcium can be calculated from $[Ca^{2+}]_c$ only in equilibrium and equilibrium is not reached during the fast contraction-relaxation cycle of the heart[21]. Bound calcium can be measured more directly by electron-probe microanalysis (EPMA, synonymous x-ray microprobe analysis, recent reviews[26,29]). As a matter of fact, EPMA measures the sum of bound and ionized calcium which is called "total Ca" (ΣCa). Since the concentration of bound calcium is several orders of magnitude larger than $[Ca^{2+}]_c$, extrapolation of bound from total calcium imposes only minor problems.

EPMA operates at the high lateral resolution of the electron microscope and can measure ΣCa in the myofibrillar space (ΣCa_{myo}), in the mitochondria (ΣCa_{Mito}) and inside the junctional sarcoplasmic reticulum (ΣCa_{SR}). With this lateral resolution we can test the question whether the jSR has indeed the properties that were postulated for the functional Ca^{2+} release compartment which is thought[5,8] to provide most of the activator Ca^{2+}. ΣCa_{myo} will be measured in the A-band region where the myofilaments overlap, i.e. in the very same place where calcium binds to TnC. We will compare ΣCa_{myo} with $[Ca^{2+}]_c$, measured by the fluorescence of Indo-1, the indicate being loaded from the patch electrode into the myoplasm. EPMA should attribute cellular functions to subcellular structures by means of their ΣCa. For such a correlation, ΣCa has to be measured in a defined functional state, for example after a long rest period, at the end of diastole, or at different times of the systole. A specimen, prepared with conventional fixation and sectioning, wuold not suite such a purpose since the calcium gradients would have disappeared by diffusion. To preserve the original calcium distribution all maneuvers have to be performed with cryo-techniques below -100 °C. For EPMA, the cell is rapidly frozen to -195 °C. The moment of shock-freezing ends the experiment, a new cell is required to obtain the next time point, hence, EPMA seems to be inadequate for analyzing time-dependent concentration-changes. Due to cell-to-cell variability, for each time point data from several cells must be averaged. The experimental effort seems to be justified by the result that ΣCa_{myo} increases to millimolar concentrations within a few milliseconds after depolarization and, more surprisingly, that large amounts of this ΣCa_{myo} are removed into the SR at a time when the cell is still contracting. Within the last 6 years, we developed a method that can shock-freeze ventricular myocytes under the conditions of the voltage-clamp[28]. During voltage-clamp, one can measure those trans-sarcolemmal Ca^{2+} fluxes as the Ca^{2+} channel current I_{Ca} or the current I_{ex} due to electrogenic Na/Ca-exchange. The comparison of the amount of transported Ca^{2+} with the increments in ΣCa_{myo} or ΣCa_{SR}, hopefully, leads to an understanding how this Ca^{2+} is bound or sequestered. In addition, the voltage-clamp method can be used to grade the Ca^{2+} influx and thereby contraction. For the interrelated changes in $[Ca^{2+}]_c$, ΣCa and contraction two examples will be presented here. (1) The post-rest adaptation induced by repetitive stimulation after long rest period. (2) Potentiation by paired voltage-clamp pulses.

METHODS

Cells and solution. Ca^{2+}-tolerant ventricular myocytes were isolated by means of the pronase technique[4] from the hearts of adult guinea-pigs (400 g). For the experiment the cells were continuously superfused with a 36°C warm physiological salt solution containing 150 mM NaCl, 5.4 mM KCl, 2 mM $CaCl_2$, 1.2 mM $MgCl_2$, 5 mM glucose, 5 mM HEPES/NaOH (pH 7.4). The cells voltage-clamped with a single patch electrode. The electrodes were filled with 140 mM KCl, 5 mM Na_2ATP, 6 mM $MgCl_2$, 0.05 mM EGTA, 5 mM HEPES/KOH (pH 7.3) and had resistances of approx. 2 MΩ. For $[Ca^{2+}]_c$ measurements, the 0.05 mM EGTA was substituted by 0.05 mM Na_5-Indo-1[10]. Contractile force of the single myocyte was measured in parallel experiments as described recently[22]. In the experiments leading to shock-freezing, contractions were monitored as un-loaded shortening of the cell[12].

Shock-freezing under voltage clamp[28]. An isolated myocyte was sucked up with the patch-pipette and transferred from the bottom of the chamber into a special silver holder[28]. Then, the patch was ruptured and the electrophysiological measurements were started. The moment of shock-freezing was timed by the computer that generated the voltage-clamp pulses. The events for shock-freezing started by horizontally sliding away the chamber and microscope, thereby positioning the coolant underneath the holder with the cell and were finished by pushing the container with the under-cooled propane (-195°C) upwards. Integrity of the sarcolemmal membrane was monitored by continuously recording the membrane current until the very moment of shock-freezing. The final contact with the coolant induced an artifact in the current trace (see below) that was evaluated for timing. For the "time course" of ΣCa, myocytes were frozen during systole 10, 15, 20, 25, 30, 35, 40, 45, 60, 90, 120, 125, 160, and 170 ms after start of the pulse to 5 mV. For diastole at -80 mV, cells were shock-frozen 560 ms after end of the second pulse (i.e. 40 ms before the next pulse pair should have been applied). In addition, cells were frozen 1.8 seconds and 3 minutes after the stimulation with paired pulses was ended in order to see how ΣCa, elevated by paired-pulsing, returned to the values evaluated for resting cardiac myocytes previously[27,28].

The cells were cut in 120 nm thin cryosections which were freeze-dried afterwards. Although the sections had not been stained chemically, sarcomeres, nucleus, mitochondria, z-line, t-tubuli and junctional SR were recognizable due to their differences in mass. The outer 4-6 µm were free of ice crystal damage, then, increasing cryo-artifacts tended to generate clefts and to clump myofilaments and membranes together. The absence of ice crystals in the superficial cell region suggests that the rate of freezing has been >10,000 °C/s[29]. EPMA was performed in a Philips electron microscope CM12/STEM equipped with a LINK-energy dispersive pentafet detector and with a full quantitative analysis-system.

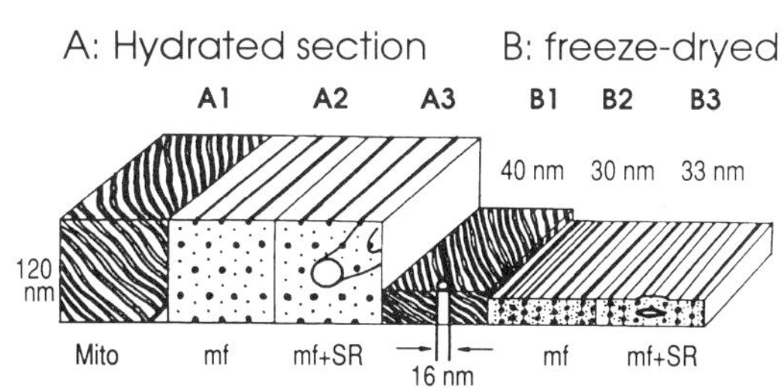

Fig. 1. Schematic showing the volume-loss of a cryosection due to freeze-drying. A: the 120 nm thick hydrated section goes through mitochondrium (A1), myoplasm with myofibrils (A2) and myoplasm containing longitudinal SR (A3). B:reduction of volume by loss of waster is thought to reduce exclusively the height of the section. The mitochondrial water content is assumed to 68%, here, the section shrinks to 39 nm (B1). In the myoplasm, the water content is 75%, and the section shrinks to 30 nm (B2).The freeze-dried section has a different total mass in mitochondria and myoplasm which provides contrast in the electron microscope. When a tubulus of free SR is included in the myoplasm, lipid bilayers with low water content may shrink less than the myoplasm (B3: 33 nm), however, the effect on total mass and on contrast is small.

Water loss and definition of concentrations. Since the cryosections have been freeze-dried, concentrations are measured in millimoles of an element per kg dry weight (mmol/kg d.w.) of the analyzed compartment. These values can be converted to units of molarity (mM) in the analyzed compartment: for example, there are good measurements of the water content of mitochondria[25] which is 66%, thus, 1 mmol/kg dry weight (d.w.) ΣCa_{Mito} can be converted into 0.33 mM (mmol/l mitochondria). Unknown hydration values, as in the SR, can be estimated on the basis of the known mitochondrial water: the x-ray continuum, measured the SR or in the A-band, was compared with the continuum in adjacent mitochondria using the same dose of electrons (i.e. the same count-time and same beam intensity). The statistical error of the single measurement of low calcium concentrations was kept low by analysis-times as long as 1000 seconds which provides a detection limit of 0.3 mmol/kg d.wt or 80 μM ΣCa. The spatial resolution of subcellular compartments is limited by the diameter of the electron beam obtainable in the microscope and by the thickness of the cryosection. In modern microscopes the diameter of the electron beam is not more limiting, for most measurements it was set to 16 nm. However, the 120 nm thickness of the cryosections may cause an underestimation of ΣCa when the compartment an thin as the 40 nm wide SR tubules; the analyzing beam can include some component of the surrounding myoplasm thereby artificially diminishing the true ΣCa_{SR} if ΣCa_{SR} were higher than ΣCa_{myo}.

RESULTS and DISCUSSION

Post-rest adaptation: filling of calcium stores

Changes in the unloaded shortening. A guinea-pig ventricular myocyte, being unstimulated for long rest periods (≥5 min), responds the first action potential or depolarizing voltage-clamp pulse with a contraction that shortens at a low and almost constant rate to a small and late peak. Subsequent repetitive stimulation potentiates the rate and extent of shortening by more than 300% (Fig. 2). Usually, this "positive staircase" has been explained with the following hypothesis[8]: During the rest period, the intracellular Ca^{2+} stores (the SR) deprive

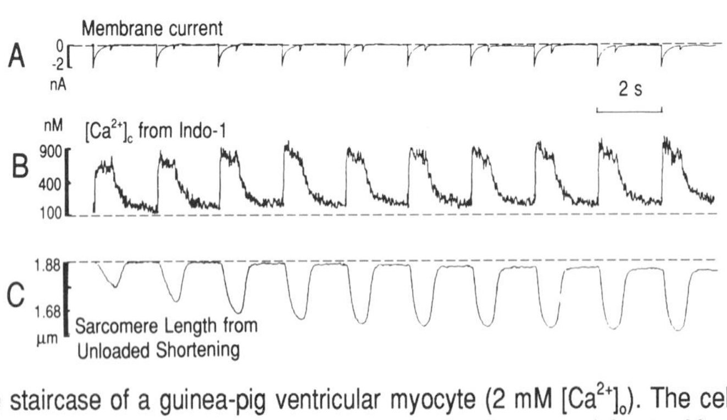

Fig. 2. Positive staircase of a guinea-pig ventricular myocyte (2 mM $[Ca^{2+}]_o$). The cell had remained unstimulated for 5 min before 1-Hz pulsing was resumed (holding potential -45 mV, pulse potential 0 mV, pulse duration 250 ms). A: Net membrane current which is mostly Ca^{2+} inward current. Capacitive currents have been removed for clarity. B: $[Ca^{2+}]_c$ as indicated by the Indo-1 fluorescence. C: Extent of shortening (in %) is defined as (diastolic - systolic)/ diastolic cell length. Note: Whereas peak I_{Ca} remained nearly constant, peak systolic $[Ca^{2+}]_c$ increases from 650 to 820 nM (e.g. by 26%) and extent of shortening from 5 to 13%. Note also the slight increase in diastolic $[Ca^{2+}]_c$ (95 to 145 nM) and diastolic shortening (1.7%).

of releasable Ca^{2+}. Resumption of stimulation and Ca^{2+} influx replenishes the stores with Ca^{2+} which, in turn, can release activator Ca^{2+} in increasing amounts. A surplus of activator Ca^{2+} via facilitated Ca^{2+} influx does not contribute, I_{Ca} does not increase when contractility is potentiated. Fig. 2 compares the positive staircase of contraction with the staircase in systolic peak $[Ca^{2+}]_c$ transients. Surprisingly, the 300% potentiation of contraction goes along with an only 40% enhancement of systolic $[Ca^{2+}]_c$. The discrepancy between the extent of mechanical and $[Ca^{2+}]_c$ staircase let us suggest that replenishment of Ca^{2+} stores is not the only inotropic mechanism. Later on, we will discuss a possible contribution of the "diastolic staircase"[15] which is shown in Fig. 2 as a moderate increase in diastolic $[Ca^{2+}]_c$ (from 95 to 180 nM) and reduction of diastolic sarcomere length (from 1.85 to 1.82 µm).

I_{Ca} increments cellular ΣCa. I_{Ca} transports Ca^{2+} ions through sarcolemmal L-type channels from the extra- into the intracellular space. The increase in cellular calcium content due to I_{Ca} can be estimated[12]. For such a calculation, we transform I_{Ca} into a Ca^{2+} flux (moles per second) by dividing the number of nA by the Faraday constant (96,500 A·s/mol) and the equivalence charge (2). We integrate this Ca^{2+} flux over time and divide it by the volume accessible for Ca^{2+} ions (10 pl or 48% of the cell volume estimated by light microscopy). The result suggests that I_{Ca} would have incremented the cellular ΣCa by approx. 100 µM (estimate for an 180 ms long depolarization to 0 mV in 2 mM $[Ca^{2+}]_o$). The increment in ΣCa due to I_{Ca} is reduced by Ca^{2+} efflux. Long rest periods, however, deprive the guinea-pig ventricular myocytes of calcium (see Tab. 1) and Ca^{2+} efflux becomes negligible small. Under those conditions I_{Ca} may be the most significant transmembrane Ca^{2+} flux, and the above approximation can be applied, not only for a single I_{Ca} but also for a train of several I_{Ca}. For example of Table 1, a series of 5 I_{Ca} (1 s long) was calculated to increment ΣCa by 830 µM.

Post-rest calcium load. When EPMA was applied to myocytes that had been unstimulated for 15 min, ΣCa in the myofibrillar space (ΣCa_{myo}) and in those compartments as SR and mitochondria was below the detection limit (0.1 mM). Since the SR had no calcium to release, this data proves the hypothesis that the first post-rest contraction received its activator Ca^{2+} by influx from the extracellular space[27]. The contraction's long latency (≥ 50 ms), low rate and extent of shortening are compatible with such an interpretation[12,27]. After the series of five voltage-clamp pulses (1 s duration) had been applied, shock-freezing was timed 800 ms after start of the fifth pulse, i.e. a time when the sarcomeres had re-

Table 1. Calcium distribution in guinea-pig ventricular myocytes. The cells were unstimulated for a 15 min rest period. Then they were stimulated with four 1 s voltage-clamp pulses from -45 mV to zero mV (0.5 Hz) and shock-frozen 813 ms after start of the fourth pulse. Calcium was measured over the myofilaments of the A-band, the sarcoplasmic reticulum (SR) facing T-tubules (jSR) and the peripheral sarcolemma (pSR) and in the mitochondria.

Rest	Myofil.	jSR	pSR	Mito
Mean (mmol/kg d.w.)	-2	1	-0.8	-2
± S.E.	2	1	2	1
n	13	12	10	10
4 Depolarizations	Myofil.	jSR	pSR	Mito
Mean (mmol/kg d.w.)	0.35	1.28	2.65	1.25
± S.E.	0.3	0.3	0.7	0.1
n	13	12	6	12
converted to µM	90	320	660	410

lengthened. In the myofibrillar space ΣCa_{myo} was 90 µM which is just above the detection limit. This measured ΣCa_{myo} is about 9 times lower than the increment of 830 µM calculated from I_{Ca} (see above); the difference suggests that most of the inflowing Ca^{2+} had been sequestered into intracellular compartments or extruded. By EPMA, the calcium filling of the intracellular compartments was measured. The calcium concentration ΣCa_{SR} was 320 µM in the junctional SR facing the T-tubules and 660 µM in the peripheral SR facing the inner side of the surface sarcolemma. Significant Ca-loading was also measured in the mitochondria (ΣCa_{mito} =410 µM). In all three compartments the ΣCa was significantly higher than at rest (at the 95% level, see Table 1). Thus, we concluded that the major part of the calcium transported by I_{Ca} has filled the intracellular compartments indeed. Such a "filling function" of I_{Ca} has been postulated for a long time, however, EPMA provides the first direct proof of it.

Time course of calcium during a potentiated contraction

Potentiation of contraction by paired voltage clamp pulses. In vivo, the heart of a guinea-pig is beating at approx. 3 Hz. For technical reasons, the computer-controlled experiments had to be done at a basic frequency of 1 Hz. In order to induce a degree of contractile potentiation comparable to the one during 3 Hz beating, a paired pulse protocol was applied. During diastole, the membrane was clamped to -80 mV, a potential which is close to the resting membrane potential of the non-clamped cell. A 20 ms pre-pulse to -45 mV activated

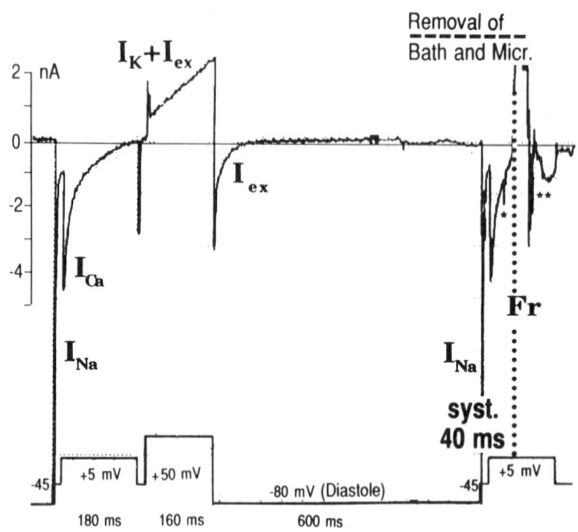

Fig. 3. Net membrane currents during shock-freezing. Lower panel: voltage clamp protocol. During diastole, -80 mV holding potential. Paired voltage-clamp pulses applied at 1 Hz. Pulse 1 starts with 20 ms step to -45 mV (activation and inactivation of I_{Na}, peak out of scale). Next, an 180 ms pulse to +5 mV activates and inactivates I_{Ca}. After 20 ms repolarization to -45 mV, pulse 2 steps for 160 ms to +50 mV and induces outward currents due to I_K and Ca^{2+} influx through Na/Ca-exchanger (positive I_{ex}). Step to diastolic -80 mV induces Ca^{2+} efflux through Na,Ca-exchanger (negative I_{ex}).For shock-freezing, the experimental chamber and the microscope are removed without disturbing the membrane current. When the preparation is removed from the chamber, I_{Na} and I_{Ca} are recored as before. Contact with the coolant results in a large outward current (dotted line). The abrupt stop of the movement induces vibrations which cause large a large current deflection. At the end, the current is zero. The recordings were taken from a cell superfused with 36 °C warm bath solution containing 2 mM $CaCl_2$.

and inactivated the fast sodium current. (The amplitude of I_{Na} is largely out off scale in Fig. 3). Then, a 180 ms pulse to +5 mV followed. This "first" pulse of the pair controlled the calcium current I_{Ca}. In Fig. 3, I_{Ca} peaked to -4.4 nA within 4 ms, then it decayed (inactivated) to the zero current line. According to present knowledge[5,7,8,29] this I_{Ca} triggers the Ca^{2+} release from SR, it fills the SR with Ca^{2+} (see above) and it contributes to some extent to the activator Ca^{2+}. We intended to enhance the calcium-loading of internal stores by adding a second pulse to the voltage-clamp protocol. The second pulse depolarized for 160 ms to +50 mV (see Fig. 3) which is thought[5] to activate Ca^{2+} influx through the Na/Ca-exchanger. The contraction induced by the second pulse was not a fast twitch but it rose at a low rate to a peak that occurred at the end of pulse (Fig. 4). Repolarization to diastolic -80 mV induced a large tail negative current which is thought[6] to represent Ca^{2+}-efflux via electrogenic Na/Ca-exchange (negative I_{ex} in Fig. 3). The change from single to paired pulses increased the amplitude of contractile force during the first pulse by 800% or more. When contraction was measured as unloaded shortening, the extent of potentiated shortening was 10.3% on average, which corresponds to an average systolic sarcomere length of 1.55 µm. In addition, potentiation slightly reduced the diastolic sarcomere length from 1.85 to 1.82 µm. When paired pulsing was ended, the sarcomere length recovered to the resting 1.85 µm with a half decay time of 1.6 ± 0.4 s.

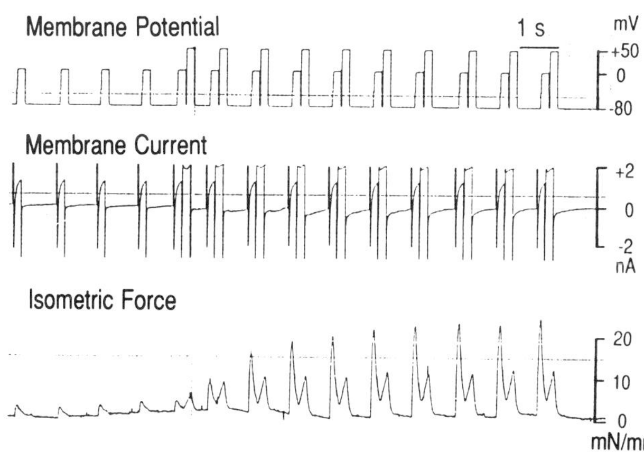

Fig. 4. Paired voltage clamp pulses potentiate contraction. A: voltage clamp protocol showing the change from single to paired pulses. (For details of the durations and voltages see Fig. 5.) B: membrane current. C: isometric force. Force traces were filtered at 30 Hz. Note that the second pulse to +50 mV potentiates the twitch during the first pulse to +5 mV by a factor of 8. (From Wendt-Gallitelli and Isenberg, J Physiol (Lond) 435, 349-372, with permission)

Effect of paired pulsing on the $[Ca^{2+}]_c$ transients. As a working hypothesis one can postulate that the second pulse induced additional Ca^{2+} influx through the Na/Ca-exchanger that filled the intracellular stores to a larger extent; the better filled stores would release more activator Ca^{2+} and thereby activate more force. Fig. 5 shows the effects of paired pulsing on the $[Ca^{2+}]_c$ transients. The cell was unstimulated for a 4 min rest period, and the resting $[Ca^{2+}]_c$ was 90 nM. The first depolarization increased $[Ca^{2+}]_c$ within 140 ms to 680 nM, during the second pulse the $[Ca^{2+}]_c$ signal did not visibly increase. During the 600 ms

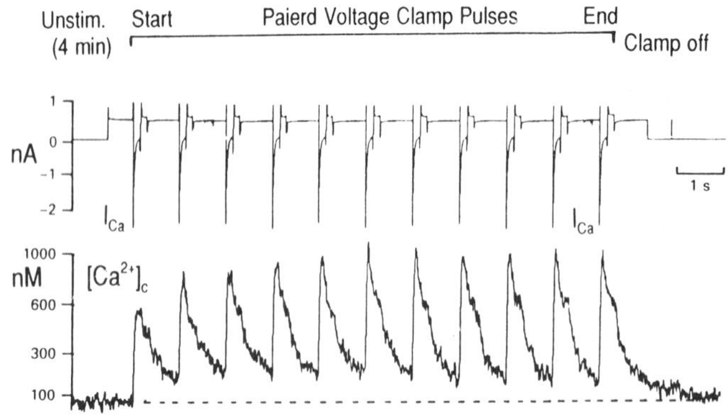

Fig. 5. Membrane currents (top) and changes in $[Ca^{2+}]_c$ (bottom) induced by 1 Hz paired pulsing. $[Ca^{2+}]_c$ of the unstimulated cell (rest period of 4 min) is about 95 nM. Paired pulsing increases diastolic $[Ca^{2+}]_c$ from 95 to 200 nM and the systolic peak of $[Ca^{2+}]_c$ from 600 to 1050 nM. On-line pen recording, $[Ca^{2+}]_c$ filtered at 50 Hz. An analogous circuit performed the subtraction of autofluorescence and the fluorescence ratio (410 nm / 480 nm) on-line. (From Wendt-Gallitelli and Isenberg, J Physiol. (Lond.) 435: 349-372, with permission)

repolarization to -80 mV $[Ca^{2+}]_c$ decayed to end-diastolic $\Sigma Ca_{myo\ of}$ 170 nM. A series of 6 paired pulses potentiated peak systolic $[Ca^{2+}]_c$, i.e. in comparison with the post-rest response the peak became larger (950 nM) and occurred earlier. When potentiation was steady, depolarization to +5 mV induced a rise of $[Ca^{2+}]_c$ that peaked within 25 ± 5 ms to 890 ± 220 nM (mean ± S.E., n=5). At on-going depolarization, $[Ca^{2+}]_c$ fell to 50% within about 140 ms. During the second pulse to +50 mV, the decay of $[Ca^{2+}]_c$ became retarded, and the followingrepolarization -80 mV let $[Ca^{2+}]_c$ fall at an enhanced rate. However, the diastole was too short for the return of $[Ca^{2+}]_c$ to the resting 90 nM, only when pulsing was ended, $[Ca^{2+}]_c$ promptly fell to the resting level with a half-decay time of 1.6 ± 0.2 s. The elevated diastolic $[Ca^{2+}]_c$ ("diastolic staircase"[15]) seems to correlate with the slightly reduced sarcomere length (see above). The increase in diastolic $[Ca^{2+}]_c$ should be distinguished from so-called "Ca^{2+} overload"; in Ca^{2+} overloaded cells, $[Ca^{2+}]_c$ fell transiently below the 180 nM and rose to 280 nM, eventually triggering a $[Ca^{2+}]_c$ transient and spontaneous activity. This phenomenon was observed in about 10% of the myocytes, these myocytes were not used for EPMA.

Potentiation increases total diastolic myofibrillar calcium. After potentiation, EPMA measured in the myofibrillar space at the end of diastole a total ΣCa_{myo} of 2.6 ± 0.4 (mean ± SEM) mmol/kg dry wt (n=62). This value can be translated into a molarity of 0.65 mM (0.65 mmol/liter myofibrillar space). The elevation of diastolic ΣCa_{myo} was reversible. When the period of diastole was prolonged from 0.6 s to 1.8 s, ΣCa_{myo} fell from 0.65 to 0.48 mM and after a 3 min rest period ΣCa_{myo} was 0.25 mM. The reversibility upon end of stimulation also shows that the potentiation-induced increase in ΣCa_{myo} is not an artefact but a phenomenon linked to the contractile potentiation. The increase in diastolic ΣCa_{myo}, induced by paired-pulse potentiation correlates not only with the increase in diastolic $[Ca^{2+}]_c$ and the shortening in diastolic sarcomere length but also with the potentiation of contractile force (see discussion below).

Diastolic ΣCa_{myo} is about 8 times higher than the indirect estimates[7,23] that calculated total ΣCa from $[Ca^{2+}]_c$ on the basis of the published Ca^{2+} ligands. Hence, one may wonder about possible artifacts. Contamination by ΣCa_{SR} can be excluded with the criterion of morphology and the low content of phosphorus[13,30]. Further, ΣCa_{myo} of 2.6 mmol/kg d.w. (0.65 mM) is not peculiar to single cell preparations; similar ΣCa_{myo} have been published for multicellular ventricular trabeculae of guinea-pigs[27]. In multicellular trabeculae, elevated diastolic ΣCa_{myo}, induced by potentiation, did not go along with elevated diastolic tension[27], in correspondence with the un-modified diastolic force in Fig. 4.

Time course of myofibrillar calcium during systole. The ΣCa measurements from 53 voltage-clamped myocytes, shock-frozen at different times during systole, were summarized to a time course of myofibrillar ΣCa_{myo}. Fig. 6 compares this time course of ΣCa_{myo} with the time course of $[Ca^{2+}]_c$ and the time course of isotonic cell shortening. About 20 ms after depolarization to +5 mV, both ΣCa_{myo} and $[Ca^{2+}]_c$ peaked. 40 ms after start of depolarization, the systolic ΣCa_{myo} transient had fallen to 66% of the peak value, and after about 90 ms to 20%. Generally, the decay of $[Ca^{2+}]_c$ was slightly slower than the decay of ΣCa_{myo} which could be an artefact from the slow rate (k_{off}) by which Ca^{2+} is released from Indo-1[2]. Both ΣCa_{myo} and $[Ca^{2+}]_c$ peaked at a time when shortening just started. The rate of shortening was maximal 42 ± 15 ms after start of depolarization, which is significantly later than the time to peak of ΣCa_{myo} or the time to peak $[Ca^{2+}]_c$. The contraction (unloaded shortening) peaked at 128 ± 25 ms (peak force at 110 ± 12 ms[22]), that is at a time where ΣCa_{myo} and $[Ca^{2+}]_c$ had largely returned to the diastolic values. Thus, the time course of ΣCa_{myo} precedes the time course of contraction. We interpret this result to suggest that Ca^{2+} bound to TnC controls only the onset of a series of reactions that finally lead to the actin-myosin interaction[1] but not directly the time course of this interaction.

Systolic ΣCa_{myo} values. 15-25 ms after start of depolarization ΣCa_{myo} increased to 5.5 ±0.2 mmol/kg d.w. which can be transformed into 1.4 mM. Nested analysis of variance shows that this systolic calcium concentration is significantly higher than during diastole. 30-45 ms after start of depolarization ΣCa_{myo} was 4.6 mmol/kg d.w. or 1.15 mM which is still significantly higher then in diastole. 60-120 ms after start of depolarization, when contraction had not yet reached its peak, ΣCa_{myo} had fallen to 3.1 mmol/kg d.w. or 0.8 mM. This value is no longer significantly different from the diastolic 0.65 mM ΣCa_{myo}. Since the ΣCa_{myo} transient lasts for only a short period, EPMA applied 90 ms after start of contraction may miss it[18].

From a comparison of ΣCa_{myo} with $[Ca^{2+}]_c$ we can estimate the Ca^{2+}-buffering capacitance in the myofibrillar space. In terms of molarity, diastolic ΣCa_{myo} of 0.65 mM compared with a $[Ca^{2+}]_c$ of 180 nM. That is, only 0.03% of ΣCa_{myo} was ionized and 99.97% was bound. During systole, ΣCa_{myo} increased to 1400 µM and $[Ca^{2+}]_c$ peaked to 890 nM. To calculate the Ca^{2+}-buffering capacitance, the systolic increment ΣCa_{myo} (750 µM) was divided by the correspondent increment in $[Ca^{2+}]_c$ (0.71 µM). This ratio was divided by the 0.69 pCa units by which $[Ca^{2+}]_i$ had changed (from 180 nM or pCa 6.74 to 890 nM or pCa 6.05). The result is a Ca^{2+}buffering capacity of 1.5 mmol ΣCa_{myo} per pCa unit for $[Ca^{2+}]_c$ between 0.2 and 0.9 µM. Similar millimolar values of the Ca^{2+} -buffering capacity have been reported for smooth muscle cells[3].

In comparison to published data about intracellular binding sites[7,21,24] the above Ca^{2+} buffering capacity is unexpectedly high. In the myofibrillar space of myocardial preparations

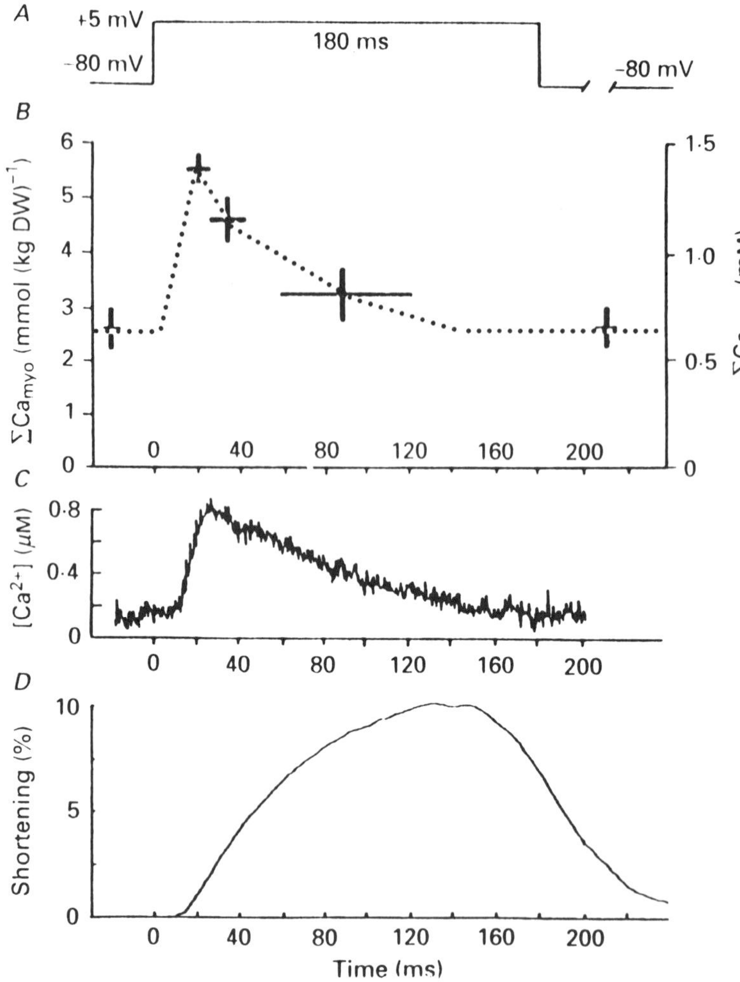

Fig. 6. Time course of total myofibrillar calcium (B) in comparison with $[Ca^{2+}]_c$ (C) and isotonic shortening (D). A shows the voltage clamp protocol (pre-pulse to -45 mV not shown for clarity). In B, vertical bars indicate S.E.M. of the mean, horizontal bars the period of time where measurements were grouped together. C and D are averages from ten sweeps. For $[Ca^{2+}]_c$, data are transformed into a linear scale by computer. (From Wendt-Gallitelli and Isenberg, J. Physiol. (Lond.) 435, 349-372: with permission)

Ca^{2+} may bind to calmodulin, troponin C and myosin. According to the published affinities and rate constants[21], the Ca^{2+},Mg^{2+}-sites of TnC would have properties suitable to account for the potentiation-induced increase in diastolic ΣCa_{myo} and $[Ca^{2+}]_c$. The concentration of TnC, however, provides only approx. 200 μM Ca^{2+},Mg^{2+}-sites whereas an increment of 650 μM ΣCa_{myo} was measured. Hence we postulate the myofibrillar space to contain, in addition to the Ca^{2+},Mg^{2+}-sites of TnC, sites with similar properties at a concentration of about 600 μM. With those additional sites we can adequately model the potentiation-induced increase in $[Ca^{2+}]_c$ and ΣCa_{myo} and its decay upon rest. The model predicts for the diastole of potentiated cells that 28% of the calcium specific sites on TnC were occupied by calcium, which could account for the observed modest reduction of averaged sarcomere length.

Table. 2. Comparison of ionized and total calcium concentrations in the myofibrillar space as they are measured in different functional states.

Ionized $[Ca^{2+}]_c$ (Indo-1)		Total Calcium (EPMA)
90 ± 30 nM	unstimulated myocyte	80 μM
180 ± 50 nM	potentiated, end-diastole	650 ± 100 μM
890 ±220 nM	potentiated, peak-systole	1400 ± 100 μM

Contribution of end-diastolic calcium to potentiation of contraction. The high affinity Ca^{2+} sites compete with the low affinity activator sites on TnC (Ca^{2+} specific sites) for binding of Ca^{2+}. When the cell is depleted of calcium and ΣCa_{myo} is low, most Ca^{2+} (from Ca^{2+} influx) will bind to the high affinity sites, correspondingly the probability of Ca^{2+} binding to the activator sites of TnC is low. Our data suggest a low "off-rate" of Ca^{2+} from the high affinity sites. Hence, repetitive stimulation by single or by paired pulses, increasing ΣCa_{myo} and $[Ca^{2+}]_c$, will saturate these sites to some degree. That is, after potentiation the probability of Ca^{2+} for binding to TnC and for activating contraction is enlarged. Hence, a similar apparent increment in ΣCa_{myo} is expected to generate more force when ΣCa_{myo} is high (during potentiation) than when ΣCa_{myo} is low (post rest). We suggest that occupancy of Ca^{2+} binding sites during diastole contributes to inotropy, in addition to the increased amount of released activator Ca^{2+}. The additional mechanism of "diastolic occupied Ca^{2+} binding sites" could explain that a 50% increase in the $[Ca^{2+}]_c$ transient is sufficient for an 800% increase in contractile force.

During systole, ΣCa_{myo} peaked to millimolar concentrations (1.4 mM). These numbers are not extraordinary high, for example ca. 2 mM was measured in the myofibrillar space of tetanized skeletal muscles[26]. The ratio between 1.4 mM ΣCa_{myo} and 0.9 μM $[Ca^{2+}]_i$ is >1000, or more than 10 times higher than the ratio calculated on the basis of the known Ca^{2+} ligands[7]. Regarding the fast time course of the systolic ΣCa_{myo} and $[Ca^{2+}]_c$ transients, the Ca^{2+}-specific site of TnC would have adequate kinetic properties. However, due to the low TnC concentration, these sites can bind 100 μM calcium at the best. Thus, we have to postulate the existence of a "fast" Ca^{2+} buffer in the myofibrillar space (about 2 mM[30]). Until now, the molecular identity of these sites is un-known.

Calcium in the sarcoplasmic reticulum. The rapid time course of the systolic $[Ca^{2+}]_c$ transient is well established by from earlier measurements with aequorin or fura-2[5]. But, the measurements of $[Ca^{2+}]_c$ alone could not answer whether the rapid decay of $[Ca^{2+}]_c$ was the result of a shift of Ca^{2+} from the indicator to TnC or due to re-uptake of Ca^{2+} into the SR. We think that the present EPMA measurements can answer this question unequivocally. Firstly, ΣCa_{myo} decreased at a similar time as $[Ca^{2+}]_c$ did, long before relaxation started. This result can exclude the idea that the decay is due to a shift from the indicator to the Ca^{2+} binding sites in the myofibrillar space (low affinity sites of TnC). Secondly, the total calcium concentration in the junctional SR (ΣCa_{SR}) at the end of diastole was about 4-fold higher (ΣCa_{SR} 8.6 mmol/kg d.w. of 2.4 mM) than the surrounding ΣCa_{myo}. Similar differences in the ΣCa of junctional SR and myofibrillar space have been published in multicellular trabeculae[18,27]. The differences between ΣCa_{SR} and ΣCa_{myo} disappeared 15-25 ms after start of depolarization, as if Ca^{2+} release from junctional SR were ended not only by inactivation of the Ca^{2+} release channels[8] but by the dissipation of the Ca^{2+} driving force. During later systole, ΣCa_{SR} increased rapidly, in reference to the end-diastolic ΣCa_{SR}, 65% filling were measured 40 ms after start of depolarization and 80% after 90 ms. The results suggest that the SR can remove a large fraction of ΣCa_{myo} within the short time of 30-45 ms, or, that Ca^{2+}

re-uptake into the SR is responsible for the falling phase of the systolic Ca^{2+} and ΣCa_{myo} transient indeed.

To increment ΣCa_{myo} during systole by 750 µM within 20 ms, large amounts of Ca^{2+} should be provided at a high rate. Influx of extracellular calcium has been estimated to reach 100 µm at the best (see above). Since Ca^{2+} influx is too low by one order of magnitude, the increment in ΣCa_{myo} was thought to be released from intracellular stores[5,7,8,27]. The analyzed time course of ΣCa_{myo} and ΣCa_{SR} favor the junctional SR as the morphological counterpart of the postulated release compartment. Quantitatively, however, the measured decrease in ΣCa_{SR} cannot account for the increment in ΣCa_{myo}. We follow the assumption[8] that not only junctional but also free SR contributes to the release of Ca^{2+}. In a first estimate, we multiply the fall of ΣCa_{SR} (1300 mM) with the SR volume fraction (3.5 volume%) and divide the product by the volume fraction of the myofibrillar space (48 vol%). The result suggests that release can increment ΣCa_{myo} by 94 µM which is much lower than the measured increment ΣCa_{myo} of 750 µM. This difference could suggest that the SR is not the only source for Ca^{2+} release. We think, however, that important information for such a conclusion is still missing. For example, ΣCa_{SR} will be contaminated by the lower ΣCa_{myo} if EPMA is applied to a section is thicker than the 40 nm diameter of the analyzed SR-compartment (see methods). Further, the volume fraction of SR is published for rat ventricle[19], in guinea-pig ventricular myocytes it might be significantly larger. We hope future studies by EPMA and morphometry may tackle those problems.

Mitochondria. For potentiated cells at diastole, ΣCa_{Mito} was 1.3 ± 0.2 mmol/kg d.w. or 0.4 mM. This diastolic ΣCa_{Mito} was significantly above the 80 µM in the mitochondria of un-stimulated myocytes (Table 1). When paired-pulsing was ended and the cell remained unstimulated, ΣCa_{Mito} decayed within 1.8 s to 0.4 ± 0.2 mmol/kg d.w. or 0.15 mM, and within 3 min to 0.15 mM. Thus, mitochondrial calcium seems to follow the slow, potentiation-mediated increase in diastolic myofibrillar calcium. During the initial 20 ms of systole, when ΣCa_{myo} peaked, ΣCa_{Mito} remained essentially constant (insignificant change from 0.4 to 0.5 mM). However, 30-45 ms after start of depolarization, ΣCa_{Mito} increased to 1.3 mM which is close to the 1.15 mM ΣCa_{myo}. During the late systole (60-120 ms), ΣCa_{Mito} fell to a value (0.5 mM) no longer significantly different from diastole. These results show the myofibrillar calcium has rapid access to the mitochondria. That is, ΣCa_{Mito} does participate in the rapid changes of intracellular calcium as they occur during the cardiac cycle. Since the increase in ΣCa_{Mito} followed the increase in ΣCa_{myo} with some delay one can conclude that ΣCa_{Mito} cannot be the cause for the systolic rise in ΣCa_{myo} but is rather its consequence.

The systolic increase in ΣCa_{Mito} was a short-lasting phenomenon, it would have been missed in preparations shock-frozen at the peak of contraction[18,27]. The rapid uptake and egress of mitochondrial Ca^{2+} has bees postulated from tracer experiments[9,16]. The systolic increase in ΣCa_{Mito} measured here suggests that the Ca^{2+} influx is about 5-fold and the egress 10- to 50-fold higher than Ca^{2+} flux studies in isolated mitochondria have indicated[11]. One could speculate that the Ca^{2+} fluxes or mitochondria "in situ" are stimulated by unknown "cytosolic factors" like spermin[11] which are lost when the mitochondria are isolated. Future studies have to clarify whether the present calcium shifts extrapolate to the myocytes in vivo. We like to speculate that the measured changes of ΣCa_{Mito} are large enough for the calcium regulation of those mitochondrial enzymes as pyruvate dehydrogenase[17]. Along this hypothesis, mitochondrial calcium is considered as a messenger that can adapt the rate of ATP-synthesis to the metabolic demands of the contractile cell.

REFERENCES

1. Ashley, C., Mulligan, I.P., and T.J. Lea, Ca^{2+} and activation mechanisms in skeletal muscle, Quart. Rev. Biophys. 24: 1-73 (1991).
2. Baylor, S.M. and S. Hollingworth, Fura-2 calcium transients in frog skeletal muscle, J. Physiol. (Lond.) 403: 151-192 (1988).
3. Becker, P.L., Walsh, J.V., Singer, J.J., and F.S. Fay, Calcium buffering capacity, calcium currents and $[Ca^{2+}]_i$ changes in voltage-clamped, fura-2 loaded single smooth muscle cells, Biophys. J. 53: 595a (1988).
4. Bendukidze, Z., Isenberg, G., and U. Klöckner, Ca-tolerant guinea-pig ventricular myocytes as isolated in the presence of 250 µM free calcium, Basic Res. Cardiol. 80: 13-18 (1985).
5. Beuckelmann, D.J. and W.G. Wier, Mechanism of release of calcium from sarcoplasmic reticulum of guinea-pig cardiac cells, J. Physiol. (Lond.) 405: 233-255 (1988).
6. Earm, Y.E., Ho, W.K., and I.S. So, Inward current generated by Na-Ca exchange during the action potential in single atrial cells of the rabbit, Proc. R. Soc. Lond. B 240: 61-81 (1990).
7. Fabiato, A., Calcium-induced release of calcium from the cardiac sarcoplasmic reticulum, Am. J. Physiol. 245: C1-C14 (1983).
8. Fabiato, A. Calcium-induced release of calcium from the Sarcoplasmic Reticulum, J. Gen. Physiol. 85: 189-320 (1985).
9. Fry, C.H., Powell, T., Twist, V.W., and J.P.T. Ward, Net calcium exchange in adult rat ventricular myocytes: an assessment of mitochondrial calcium accumulation capacity, Proc. R. Soc. Lond. B 223: 223-238 (1984).
10. Grynkievicz, G., Poenie, M., and R.Y. Tsien, A new generation of Ca^{2+} indicators with greatly improves fluorescent properties, J. Biol. Chem. 260: 3440-3450 (1985).
11. Gunter, T.E. and D.R. Pfeiffer, Mechanisms by which mitochondria transport calcium, Am. J. Physiol. 258: C755-C786 (1990).
12. Isenberg, G., Ca entry and contraction as studied in isolated bovine ventricular myocytes, Z. Naturforsch. 37c: 502-512 (1982).
13. Jorgensen, A.O., Broderick, R., Somlyo, A.P. and A.V. Somlyo, Two structurally distinct calcium storage sites in rat cardiac sarcoplasmic reticulum: an electron microprobe analysis study, Circ. Res. 63: 1060-1069 (1988).
14. Kitazawa, T., Shuman, H., and A.V. Somlyo, X-ray microprobe analysis: problems and solutions. Ultramicroscopy 11: 251-262 (1983).
15. Lee, H.C. and W.T. Clusin, Cytosolic calcium staircase in cultured myocardial cells. Circ. Res. 61: 934-939 (1987).
16. Lewartowski, B. and B. Pytkowski, On the subcellular localization of calcium fraction correlating with contractile force of guinea-pig ventricular myocardium, Biomed. Biochim. Acta 46: S345-350 (1987).
17. McCormack, J.G., Halestrap, A.P. and R.M. Denton, Role of calcium ions in regulation of mammalian intramitochondrial metabolism, Physiol. Rev. 70: 391-425 (1990).
18. Moravec, C.S. and M. Bond, Calcium is released from junctional sarcoplasmic reticulum during cardiac muscle contraction, Am. J. Physiol. 260: H989-H997 (1991).
19. Page, E., Quantitative ultrastructural analysis in cardiac membrane physiology, Am. J. Physiol. 235: C147-C158 (1978).
20. Polimeni, P.I., Extracellular space and ionic distribution in rat ventricle, Am. J. Physiol. 227: 676-683 (1974).
21. Robertson, S.P., Johnson, J.D. and J.D. Potter, The time course of Ca^{2+} exchange with calmodulin, troponin, parvalbumin, and myosin in response to transient increase in Ca^{2+}, Biophys. J. 34: 559-569 (1981).

22. Shepherd, N., Vornanen, M., G. Isenberg, Force measurements from voltage-clamped guinea-pig ventricular myocytes, Am. J. Physiol. 258: H452-H459 (1990).
23. Sipido, K. and W.G. Wier, Flux of Ca^{2+} across the sarcoplasmic reticulum of guinea-pig cardiac cells during excitation-contraction coupling, J. Physiol. (Lond.) 435: 605-630 (1991).
24. Solaro, R.J., Wise, R.M., Shiner, J.S. and F.N. Briggs, Calcium requirements for cardiac myofibrillar activation, Circ. Res. 34: 525-530 (1974).
25. Somlyo, A.V., Bond, M. and A.V. Somlyo, Calcium content of mitochondria and endoplasmic reticulum in liver frozen rapidly in vivo, Nature 314: 622-625 (1985).
26. Somlyo, A.V., Shuman, H. and A.P. Somlyo, Electron probe x-ray microanalysis of Ca^{2+},Mg^{2+}, and other ions in rapidly frozen cells, Methods in Enzymology 172: 203-229 (1989).
27. Wendt-Gallitelli, M.F., Ca pools involved in the regulation of cardiac contraction under positive inotropy. X-ray microanalysis of rapidly frozen ventricular muscles of the guinea-pig. Basic Res. Cardiol. 81: S1, 25-32 (1986).
28. Wendt-Gallitelli, M.F.and G. Isenberg, X-ray microanalysis of single cardiac myocytes frozen under voltage-clamp conditions. Am. J. Physiol. 256: H574-H583 (1989).
29. Wendt-Gallitelli, M.F. and G. Isenberg, X-ray microprobe analysis of voltage-clamped single heart ventricular myocytes. Methods in Neurosciences 4: 103-127 (1991).
30. Wendt-Gallitelli, M.F. and G. Isenberg, Total and free myoplasmic calcium during a contraction cycle: x-ray microanalysis in guinea-pig ventricular myocytes. J. Physiol (Lond.) 435: 349-372 (1991).

IMAGING OF CALCIUM TRANSIENTS DURING EXCITATION-CONTRACTION COUPLING IN SKELETAL MUSCLE FIBERS

Julio Vergara and Marino DiFranco[1]

Department of Physiology
University of California Los Angeles
Los Angeles, CA

INTRODUCTION

Action potential stimulation of skeletal muscle fibers activates the release of Ca^{2+} from the sarcoplasmic reticulum (SR), thus leading to a fast and transient change in myoplasmic calcium concentration ($[Ca^{2+}]_i$). Until very recently these calcium transients have been only recorded by means of space averaging photodetectors, such as photomultipliers and photodiodes, in combination with the use of several intracellular calcium indicators[1-8]. The high frequency response exhibited by this kind of photodevices is ideal to track the fast calcium transients, characteristic of the muscle cells; however, their physical construction make them inherently unable to provide information about the topological distribution of the release process. Given the structural complexity of the muscle cells, it is possible that the limited knowledge gained so far has omitted important features of this physiological process. In an attempt to surmount these limitations we have adapted video microscopy techniques to perform rapid Ca^{2+}-imaging in this preparation[9]. This has been a difficult task because the muscle fiber is a thick specimen (100-150 μm in diameter) able to undergo very fast changes in shape and volume as a consequence of its electrical activation.

Calcium-imaging demands the use of fluorescent calcium indicator dyes. From those available today, we prefer rhod-2 and fluo-3 over fura-2, perhaps the most popular calcium indicator, for several reasons: a) it has a low K_d[10] which limits the accuracy of Ca^{2+} detection at high $[Ca^{2+}]_i$; b) due to its short emission wavelength, it has traditionally been used in conjunction with SIT cameras which are not suited to follow rapid events; c) the deconvolution of fluorescence images into calcium images requires the acquisition of paired sets of images which also limits the frequency response of the imaging process. The long emission wavelength dyes rhod-2 and fluo-3 have excitation and emission spectra compatible with standard microscopy and photodetection by CCD cameras. In addition, they have K_d's in the micromolar range which are more compatible with Ca^{2+} detection of robust Ca^{2+} transients in skeletal muscle fibers. These dyes, nevertheless, do not

[1]Laboratorio de Fisiologia Animal, Escuela de Biologia, Universidad Central de Venezuela, Caracas, Venezuela

undergo spectral shifts upon Ca^{2+}-binding, thus demanding the use of non-ratiometric methods to deconvolve the fluorescence data[11,12].

We report here on recent results, obtained with new developments on Ca^{2+}-imaging techniques based on the use of rhod-2, about the Ca^{2+}-release process in skeletal muscle fibers.

METHODS

The electrophysiological methods utilized in the experiments described here were similar to those described elsewhere[4,13,14]. Briefly, short segments of single skeletal muscle fibers from the semitendinosus muscle of the bullfrog *Rana catesbeiana* were mounted in a triple vaseline gap chamber[13,15] situated on the stage of a compound microscope equipped with a standard epifluorescence attachment. The muscle fibers were extended across three vaseline partitions that separate four electrically isolated pools. The ends of the fibers were pinned down with glass pins to the bottom of the end pools lined with an elastomeric resin. The end pools were filled with an "internal solution" containing, in mM: 110 K-Aspartate; 20 K-MOPS; 3 K_2-ATP; 5 Na_2-PCr; 2 $MgCl_2$; and creatine kinase, 0.1 mg/ml; EGTA, 10-50 μM; 10 to 150 μM rhod-2; pH 7.0; pCa 7.0. Small incisions made in the fibers at these pools allowed for diffusion of solutes into the myoplasm. The fibers were externally perfused with Ringer's solution at controlled temperatures between 4 and 25 °C. The diffusion of rhod-2 was quantitated by periodically monitoring the fluorescence of the muscle fibers with a photodiode or a CCD camera; it usually reached equilibrium between 30 to 45 mins after the incisions were made at the end pools.

The image of the muscle fibers were directly formed, by a 20x fluorescence objective (N.A. 0.75), on the light sensitive area of a photodiode or the CCD array of a video camera (Sanyo VDC3800) placed at the focal plane of the trinocular head port of the microscope. These devices were routinely interchanged during the experiments. The output of the photodiode was baseline subtracted, amplified and acquired simultaneously with the transmembrane voltage by a data acquisition unit. The baseline fluorescence prior to stimulation of the muscle fibers was recorded separately and used to calculate $\Delta F/F$ values as shown in Figure 1.

The imaging techniques were described elsewhere[9]. Image data was acquired at standard video rate (60 Hz) and stored on-line on a video recording device. For analysis, video frames were delaced and processed to yield fluorescence and Ca^{2+} images which are displayed as halftone pictures by a video printer. The methodology used to deconvolve fluorescence data into $[Ca^{2+}]$ will be described below.

In-vitro calibrations were performed in microcuvettes of 0.1 mm path length containing Ca^{2+} standard solutions (of compositions described in references 16 and 17) placed on the stage of the fluorescence microscope. Solutions of intermediate $[Ca^{2+}]$ were controlled with Ca^{2+} selective electrodes[17,18].

RESULTS AND DISCUSSION

Photodiode detection of rhod-2 fluorescence transients

Figure 1 shows an action potential (trace **a**) and a fluorescence transient (trace **b**) recorded from a single skeletal muscle fiber stained intracellularly with rhod-2. The change in fluorescence of the dye is expressed in $\Delta F/F$ units. The fiber was illuminated episcopically with monochromatic light at a wavelength of 540 nm and the emitted light was filtered with a high pass barrier filter with a cut-on wavelength of 570 nm. It can be observed that, following a brief quiescent period of 1.5 ms ("propagational" plus "triadic delay", see reference 6), the action potential is followed by a transient increase of the rhod-2 fluorescence. The fluorescence signal rises rapidly, reaching a peak 8 ms after the onset of the action potential, and then decays slowly to pre-stimulus values, several

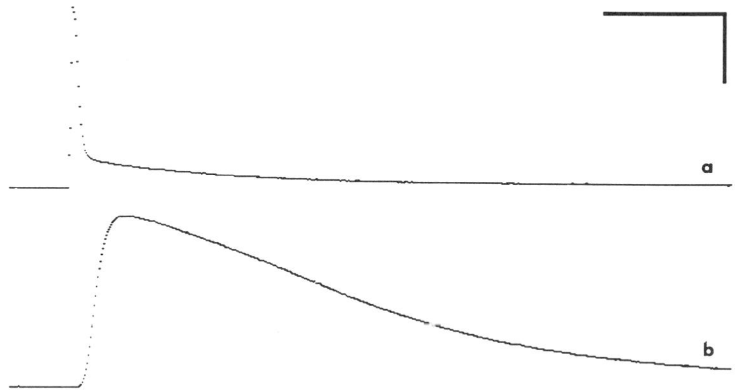

Figure 1. Temporal relations between action potential (trace **a**) and rhod-2 epifluorescence signal (trace **b**). The ends of the muscle fiber were cut in an internal solution containing 150 μM rhod-2 and 200 μM EGTA. The fiber was highly stretched to a sarcomere spacing of 3.2 μm. The resting membrane potential was held at -100 mV. The vertical calibration bar represents a fractional change in fluorescence ($\Delta F/F$) of 1.6 and a voltage of 58 mV. The temperature was set to 15 °C.

hundreds of ms later. It should be noted that the amplitude of the signal represents a significant increase in fluorescence of about 5-fold the resting fluorescence and about 200-fold the dark noise of the photodiode. We have previously shown that these virtually noiseless rhod-2 epifluorescence signals are not significantly contaminated by movement artifacts in stretched muscle fibers (sarcomere spacings between 2.6 to 3.6 μm) and that the signals are only slightly distorted (less than 10%) in fully contracting fibers[9,19]. Throughout this manuscript we will present data supporting the notion that the fluorescence changes detected from muscle fibers stained with rhod-2 arise primarily from the binding of Ca^{2+} ions to dye molecules in the myoplasm and not from volume changes associated with the mechanical response.

<u>Fluorescence imaging in skeletal muscle fibers</u>

<u>Visualization of the spot of fluorescence</u>. Figure 2, panel **a**, shows a bright field image, acquired with the video camera, of a quiescent muscle fiber stained with 8.5 μM rhod-2 and diascopically illuminated with 600 nm monochromatic light. The contour and the striation pattern of the muscle fiber can be clearly seen in this picture. By lowering the intensity of the diascopic illumination and sliding in the epifluorescence cube, it is possible to obtain a mixed image in which the epifluorescence of the dye in the muscle fiber is comparable with the transmitted diascopic light through the fiber (Figure 2, panel **b**). Panel **c** shows exclusively the epifluorescence image. These pictures illustrate how the fluorescence spot was centered on the muscle fiber in the experimental chamber.

<u>Image detection of rhod-2 fluorescence transients</u>. Figure 3 shows a sequence of eight consecutive fluorescence images acquired from a fiber stained intracellularly with rhod-2 in response to action potential stimulation. The images were captured at the

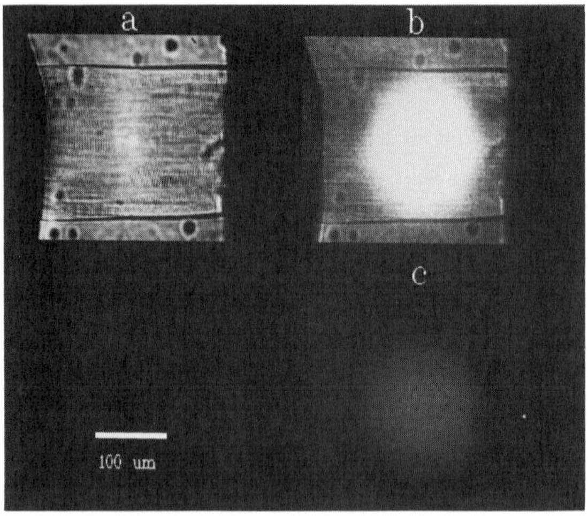

Figure 2. Panel **a**: Diascopic bright field image of a segment of a single skeletal muscle fiber stained intracellularly with 8.5 μM rhod-2 and illuminated at a wavelength of 600 nm. The fiber was stretched to a mean sarcomere spacing of 3.6 μm. The calibration bar represents 100 μm. Panel **b**: Image of the same segment of fiber but simultaneously illuminated diascopically with light of lower intensity than in panel **a**, and episcopically at wavelengths selected with the fluorescence cube. The diameter of the epifluorescence spot was adjusted to be similar to fiber diameter. Panel **c**: The same as **b** and **c**, but displaying only the epifluorescence from which the background was subtracted.

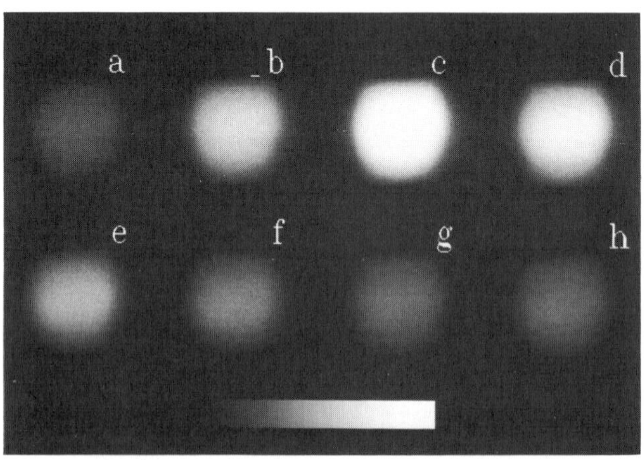

Figure 3. Sequence of consecutive fluorescence images delaced from video frames acquired during a Ca^{2+} transient. The internal solution at the end pools contained 7.5 μM rhod-2 and 100 μM EGTA. The sarcomeric spacing was 3.6 μm. The gray-scale calibration bar spans fluorescence values from 0 (black) to 128 (white). The fiber diameter was 93 μm.

standard video rate (60 Hz). The resting fluorescence image of the fiber prior to stimulation is shown in panel **a** and the onset of a 0.5 ms stimulus pulse is indicated by the white bar displayed in panel **b**, captured 17 ms after panel **a**. The sequence of images illustrate an homogeneous increase in fluorescence that reaches a maximum 33.4 ms after stimulation (panel **c**). Panels **d-h** show the progressive slow decrease in rhod-2 fluorescence which does not completely revert when the last image was acquired (122 ms). The sequence of images shown in Figure 3 emphasize the synchronic nature of the release process at every point of the muscle fiber. They do not show major heterogeneities suggesting local domains with different kinetics of release during the time course of the fluorescence transient.

At room temperature, the standard video rate allows for the acquisition of only one image during the raising phase of the fluorescence transient. We have previously shown that at lower temperatures, calcium transients in skeletal muscle fibers can be tracked by a standard CCD camera. Under these conditions, the time course of a fluorescence transient reconstructed using average intensity values from image sequences, like those shown in Figure 3, is almost superimposable with analog fluorescence transients acquired with a photodiode[9].

Fluorescence and mechanical changes during fiber activation. The dynamic correlation between the changes in the $[Ca^{2+}]_i$ and the mechanical activation of a muscle fiber can be readily studied using imaging techniques. Figure 4 shows sequences of images acquired during two consecutive stimulations of a muscle fiber that contracted vigorously, using episcopic illumination (upper row) and diascopic illumination (lower row), respectively. Column **A** shows images of the fiber at rest, while images in columns **B** to **D** were acquired 17, 33 and 133 ms, respectively, after electrical stimulation. As in Figure 3, the short horizontal line above the fluorescence image in column **b** indicates the onset of

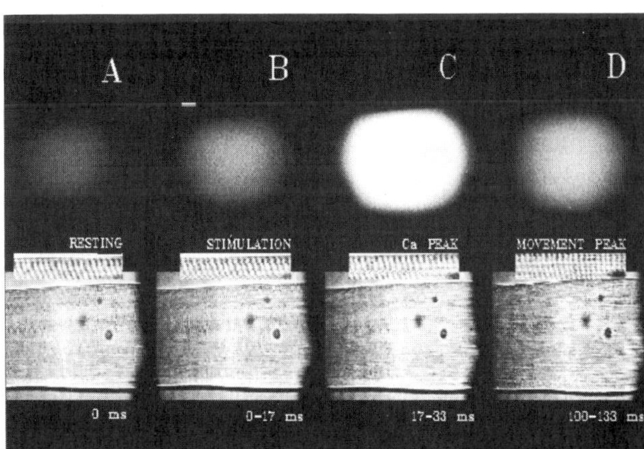

Figure 4. Fluorescence (upper row) and bright field (lower row) images at different stages of E-C coupling. The fiber diameter was 91 μm and the resting sarcomere spacing was 3.6 μm. See text for details.

the stimulus. The insets above each diascopic image correspond to an enlarged view (4x magnified) of the striation pattern of the same area of the muscle fiber. It can be readily appreciated from this sequence that while the fluorescence of the Ca^{2+} indicator raises very rapidly, reaching a peak in column **C**, the contractile response is maximal 100 ms later (column **D**). By this time, the rhod-2 fluorescence has already decayed significantly.

It should be pointed out, however, that during the early phase of the fluorescence increase (upper row, columns **B** and **C**) the mechanical activity is readily observable as a slight distortion of the sarcomere pattern of the muscle fiber (lower row, columns **B** and **C**). Movement, nevertheless, progressed slowly throughout a period of about 100 ms during which the myoplasmic [Ca^{2+}] gradually declined. The difference in kinetics between fluorescence and mechanical changes strongly suggest their cause-effect relationship and that the former cannot arise from volume changes in the fiber due to activation of the contractile machinery. This was verified in fibers stained with the dye rhodamine B (not sensitive to [Ca^{2+}]) which showed almost undetectable changes in fluorescence in response to vigorous contractions associated with gross changes in volume. The mismatch between the Ca^{2+}-related fluorescence increase and the mechanical response becomes very apparent at moderate and high frequencies of tetanic stimulation in which it is possible to detect large fluctuations in rhod-2 fluorescence intensity while the fiber is mechanically stationary during the tetanus (DiFranco and Vergara, unpublished observations).

Quantitative measurements of myoplasmic Ca^{2+} concentrations

The changes in rhod-2 fluorescence reported above qualitatively reflect rapid changes in $[Ca^{2+}]_i$ during the electrical activation of the muscle fibers. To further infer about the underlying physiological processes, it is important to quantitatively deconvolve these signals and obtain the actual $[Ca^{2+}]_i$. To this end, it is necessary to carefully determine the Ca^{2+}-binding characteristics of the dye *in vitro* and to validate its behavior *in situ*.

Rhod-2 *in-vitro* calibration. Figure 5 shows the normalized values of rhod-2 fluorescence (F_N) detected by the CCD camera and the photodiode as a function of the

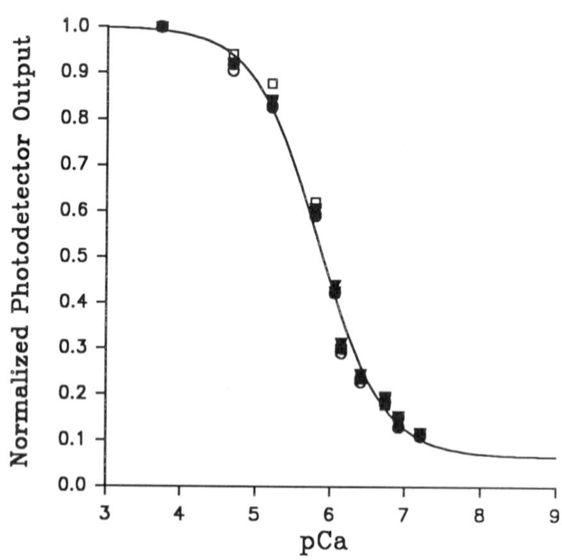

Figure 5. Rhod-2 Ca^{2+}-saturation curve. Fluorescence light was collected by the microscope objective from an homogeneously illuminated area of microcuvettes filled with solutions of known pCa containing 50 μM rhod-2. The filled circles were obtained with the photodiode; the other four symbols represent average pixel values obtained from areas of the images acquired with the CCD camera. The continuous curve was calculated from equation [1], using K_d= 1.56 μM and A= 15.

pCa of calibrating solutions in microcuvettes. The experimental points were fitted to a saturation function (continuous line) given by the formula:

$$F_N = 1/A + (A - 1)/\{A \cdot (1 + K_d/[Ca^{2+}])\} \quad [1]$$

in which $F_N =: F/F_{max}$, A is the fluorescence ratio, excess *vs.* zero Ca^{2+}, F_{max} is the maximal fluorescence of the saturated dye, and K_d is the dissociation constant. The fitted values for the parameters A and K_d were 15 and 1.56 μM, respectively. There are two important considerations that should be noted from Figure 5: i) the data points obtained with both photodetectors were identical, stressing the potential equivalence of imaging and space-averaged data for quantitative deconvolutions, and ii) the parameters K_d and A are intrinsic properties of the dye and sufficiently characterize its saturation equilibrium behavior.

Deconvolution formula. Quantitative estimations of the myoplasmic $[Ca^{2+}]_i$ in muscle fibers is based on the assumption that the equilibrium properties of the Ca^{2+}-sensitive dye *in-vitro* are preserved in the myoplasm. With this caveat, $[Ca^{2+}]_i$ can be calculated from rhod-2 fluorescence data (F) and *in-vivo* determinations of either F_{max} or F_{min} (fluorescence at zero $[Ca^{2+}]$). For technical reasons that will become clear later, we based our deconvolution in the application of the following formula derived from equation [1]:

$$[Ca^{2+}] = (K_d / A) \cdot (A \cdot F - F_{max}) / (F_{max} - F) \quad [2]$$

It can be seen that equation [2] requires *in-vivo* determinations of F_{max} to provide a unique value of $[Ca^{2+}]_i$ for each value of F.

In-vivo determination of F_{max}. To obtain values of F_{max} for rhod-2 dissolved in the myoplasm of skeletal muscle fibers, it is required to expose the dye to saturating values of $[Ca^{2+}]_i$ in the order of hundreds of μM while preserving the structural shape of the fiber. This is particularly difficult in cells which have the ability to contract in response to

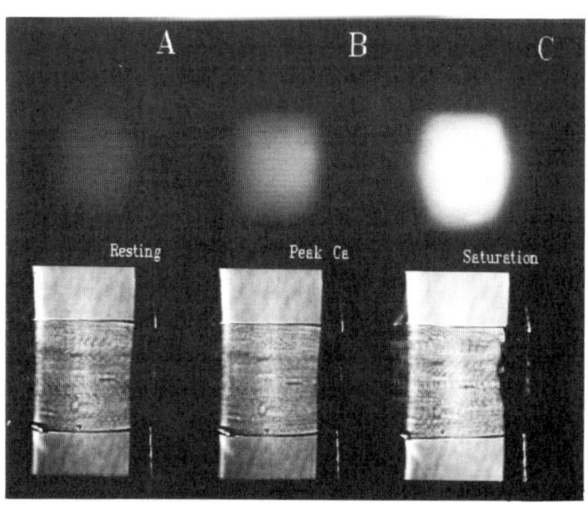

Figure 6. Imaging of rhod-2 *in-vivo* saturation. Upper row, fluorescence images; lower row, diascopic images. Columns **A** and **B** show images of a single muscle fiber in normal Ringer's at rest and at the peak of a Ca^{2+} transient, respectively. The images shown in column **C** were obtained after fixation and skinning during exposure to saturating $[Ca^{2+}]_i$. See text for details.

Ca^{2+}, thus undergoing dramatic volume changes and even disruption when exposed to high $[Ca^{2+}]_i$. In order to solve these problems it was necessary to effectively immobilize de contractile machinery of the muscle fibers; this was achieved by perfusing them for about 3-5 min with Ringer's solution containing 10 mM glutaraldehyde. Under these conditions, the excitability and Ca^{2+} release processes were preserved, but the contractile activation was fully blocked by the fixative. After the immobilization was attained, the sarcolemma was permeabilized by external perfusion with 1 mg/ml saponin dissolved in isotonic $CaCl_2$. This led to a rapid increase in fluorescence, until it reached F_{max}, which was followed by a slow decline to zero due to the washout of the dye from the fiber. Figure 6, column **C**, shows fluorescence (upper row) and bright field (lower row) images of a muscle fiber treated as described above. The preservation of the gross structure of the permeabilized muscle fiber at the peak Ca^{2+}-saturation condition can be appreciated comparing the images shown in columns **A**, **B**, and **C** of Figure 6. The gray levels of the three images shown in the upper row of Figure 6 report the differences in rhod-2 fluorescence intensity prevailing in the fiber at myoplasmic Ca^{2+} concentrations during resting, peak transient, and saturation conditions.

Deconvolution of fluorescence transients. The crude fluorescence transient shown in Figure 1 was deconvolved, using the values of K_d and A given in Figure 5 and equation [1] in a point-by-point algorithm, to yield the $[Ca^{2+}]_i$ transient shown in Figure 7, trace **b**. The value of F_{max}, also necessary for the deconvolution, was determined at the end of the experiment. This transient would reflect accurately the time course of the change in myoplasmic Ca^{2+} concentration, provided that the dye instantly equilibrated at each sampling interval. This assumption is likely not to be correct during the fast rising phase of the transient, but it probably holds true after the peak of the transient. The horizontal line below the dotted trace in Figure 7 represents the zero Ca^{2+} level and the vertical scale is linear in $[Ca^{2+}]_i$. It should be noted that the resting free $[Ca^{2+}]_i$, given by the pre-stimulus level of trace **b** was calculated to be 70 nM in this experiment, and the transient peaked at a value of 3.3 μM. When trace **b** of Figure 7 is compared with that of Figure 1, it is observed that the time course of the former is significantly narrower than the latter, but that both the time to peak and quiescent lag are preserved. The $[Ca^{2+}]_i$ at the end of trace **b** (110 ms) is about 200 nM.

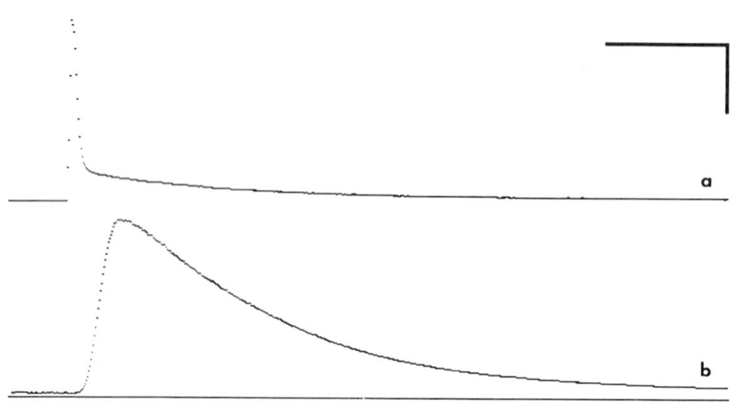

Figure 7. Action potential (trace **a**) and deconvolved $[Ca^{2+}]_i$ transient (trace **b**) of the same fiber of Figure 1. The vertical calibration bar represents 0.8 μM $[Ca^{2+}]_i$ and a voltage of 58 mV.

Deconvolution of fluorescence images. The same protocol used in Figure 7 to deconvolve the analog Ca^{2+} transient was applied to a sequence of images similar to that shown in Figure 3. In this case, the F and F_{max} data points correspond to the individual pixel values of the CCD array forming the image of the fiber at two different conditions. Figure 8 shows Ca^{2+}-images processed from fluorescence image data obtained from a

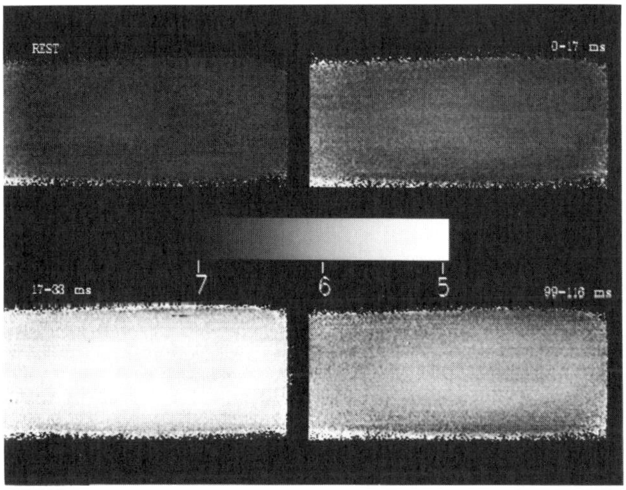

Figure 8. Ca^{2+}-images deconvolved from 17 ms fluorescence images of a muscle fiber at rest and at three intervals after its electrical stimulation. The gray-scale calibration bar shows the correspondence of pixels values to pCa. Diameter: 72 μm; sarcomere spacing: 3.6 μm. Temperature: 8 °C.

fiber at rest, and 17, 33, and 116 ms after electrical stimulation. The pixel elements of the images shown in Figure 8 were generated by a pixel-by-pixel deconvolution according to equation [2] in each fluorescence image and the saturating image of the fiber (not shown, but similar to Figure 5 **c**, upper row). The resting Ca^{2+}-image shows an average pCa value around 7; 33 ms later there is widespread increase in $[Ca^{2+}]_i$ to about pCa 5.8 which partially reverts back to pCa 6.3 within 100 ms.

ACKNOWLEDGEMENTS

This work was supported by NIH grant AR-25201 and by the Muscular Dystrophy Association. M. DiF. was partially supported by fellowships from CONICIT and Fundacion Polar, Venezuela. The results presented in this paper are part of M. DiF doctoral thesis project.

REFERENCES

1. Miledi, R., I. Parker, and G. Schalow. 1977. Measurement of calcium transients in frog muscle by the use of arsenazoIII. Proc. R. Soc. Lond. B. 198: 201-210.
2. Blinks, J.R.; Rudel, R.; and Taylor, S.R. 1978. Calcium transients in isolated amphibian muscle fibers: detection with aequorin. Journal of Physiology. 277: 291-323.

3. Baylor, S.M.; W.K. Chandler; and M.W. Marshall. 1982. Use of metallochromic dyes to measure changes in myoplasmic calcium during activity in frog skeletal muscle fibres. Journal of Physiology. 331: 39-177.
4. Palade, P. and J. Vergara. 1982. ArsenazoIII and antipyrylazoIII calcium transients in single skeletal muscle fibers. Journal of General Physiology. 79:679-707.
5. Kovacz, L., E. Rios, and M.F. Schneider. 1983. Measurements and modification of free calcium transients in frog skeletal muscle fibers by a metallochromic indicator dye. Journal of Physiology. 343: 161-196.
6. Vergara, J. and Delay, M. 1986. A transmission delay and the effect of temperature at the triadic junction of skeletal muscle. Proc. R. Soc. Lond. B. 229: 97-110.
7. Baylor, S.M.; and Hollingworth, S. 1988. Fura-2 calcium transients in frog skeletal muscle fibres. Journal of Physiology. 403: 151-192.
8. Hirota, A.; Chandler W.K.; Southwick, P.L.; and Waggoner, A.S. 1989. Journal of General Physiology. 94: 597-631.
9. Vergara, J., DiFranco, M., Compagnon, D., and Suarez-Isla, B.A. 1991. Imaging of calcium transients in skeletal muscle fibers. Biophysical Journal. 59: 12-24.
10. Grynkiewicz, G.; M. Poenie and R. Y. Tsien. 1985. A new generation of Ca^{2+} indicators with greatly improved fluorescence properties. Journal of Biological Chemistry. 260: 3440-3450.
11. Kao, J. P. Y.; A. T. Harootunian and R. Y. Tsien. 1989. Photochemically generated cytosolic calcium pulses and their detection by fluo-3. Journal of Biological Chemistry. 264: 8179-8184.
12. Minta, A.; Kao, J.P.Y.; and Tsien, R.Y. 1989. Fluorescent indicators for cytosolic calcium based on rhodamine and fluorescein chromophores. Journal of Biological Chemistry. 264: 8171-8178.
13. Vergara, J.; F. Bezanilla and B. M. Salzberg. 1978. Nile blue fluorescence signals from cut single muscle fibers under voltage or current clamp conditions. Journal of General Physiology. 72: 775-800.
14. Heiny, J. and J. Vergara. 1982. Optical signals from the surface and T-system membranes in skeletal muscle fibers. Journal of General Physiology. 80: 203-230.
15. Hille, B. and D. T. Campbell. 1976. An improved vaseline-gap voltage-clamp for skeletal muscle fibers. Journal of General Physiology. 67: 265-293.
16. Alvarez-Leefmans, F. J., Rink, T. J. and Tsien, R. Y. 1981. Free calcium ions in neurons of Helix Aspersa measured with ion-selective microelectrodes. Journal of Physiol. 315. 531-548.
17. Lopez, J. R., Alamo, L., Caputo, C., Dipolo, R., and Vergara, J. 1983. Determination of ionic calcium in frog skeletal muscle fibers. Biophysical Journal. 43: 1-4.
18. Simon, W., Amman, D., Oehme, M., and Morf, W.E. 1978. Calcium selective electrodes. Anna. N.Y. Acad. Sci. 307: 52-70.

ROLE OF CALCIUM CHANNELS OF THE SARCOLEMMA AND THE SARCOPLASMIC RETICULUM IN SKELETAL MUSCLE FUNCTIONS

C. Paul Bianchi

Dept. of Pharmacology, Jefferson Medical
College, Thomas Jefferson University
Philadelphia, PA 19107

Calcium channels of the sarcolemma and the sarcoplasmic reticulum play a key integrative role in determining and fine tuning the ionic free calcium concentration of the sarcoplasm and regulation of muscle function. Changes in the free ionic calcium concentration can lead to (1) force development and motion, (2) heat production during fever, and (3) breakdown of muscle protein reserves.

Calcium channels of the L-type present in the transverse tubule (TT) play an essential role in force development during stimulation of muscle at frequencies greater than 1 Hz. L-Calcium channels are also responsible for regulating aerobic heat production during membrane depolorizations between -90 mV and -50 mV. Calcium channels present in the sarcoplasmic reticulum, are (1) the feet operated calcium channels present in the junctional membrane of the terminal cisternae (TC) responsible for the single twitch, (2) calcium operated calcium channels, and (3) inositol triphosphate operated channels. These channels are involved in maintaining sarcoplasmic free calcium at concentrations above $1x10^{-7}$M but below $5x10^{-7}$M for muscle function other than movement. Present in the foot of the terminal cisternae (TC) are junctional ryanodine receptors that operate a calcium channel which can also be opened by voltage change across the T.T. membrane when linked to a dihydropyridine receptor modified L Channel (Tanabe et al. 1990, Hoffman et al. 1987, Rios and Pizzaro, 1988, Adams et al. 1990) and plays a primary role in the single twitch. Ryanodine binds to a receptor moiety present in the sarcolemmal junctional feet between the T.T. and the T.C. and to the Ca release of calcium channels present in the sarcoplasmic reticulum of cardiac ventricular cells (Fabiato 1985, Lakatta et al. 1985). Both of these modified channels are necessary for the coupling of excitation to contraction. Excitation-contraction coupling in skeletal muscle cannot be considered as a single process but involves more than one coupling process as we change function from a single twitch to a tetanus. The integrative role of the calcium channels of the transverse tubule (T.T.) and the terminal cisternae, both junctional and nonjunctional, is best demonstrated by the changes in calcium flux that occur during stimulation of muscle as a function of frequency. The calcium fluxes that occur during a single twitch are not the same as the calcium fluxes that occur as a function of time and frequency during a positive staircase and fatigue at 1 Hz and during a tetanus.

The complexity of the operation of the L channels of the T.T. (both junctional and non-junctional) and the ryanodine operated channels is revealed in the following studies. Initial studies performed by Bianchi and Shanes in 1959 revealed that stimulation of muscle at 1 Hz for 300 twitches increased calcium influx into what was termed the slow component

Excitation-Contraction Coupling in Skeletal, Cardiac, and Smooth Muscle
Edited by G.B. Frank *et al.*, Plenum Press, New York, 1992

of ^{45}Ca washout. The ^{45}Ca taken up by this component was revealed by washing out for 90 minutes the ^{45}Ca taken up by the muscle during stimulation and constructing a washout curve. Extrapolation of the slow component back to zero time provided the ^{45}Ca content of the muscle present following cessation of stimulation. The increase in the size of the intercept of the slow component in stimulated muscle was attributed to increased calcium influx during stimulation. Calcium influx per twitch calculated in this manner amounted to .02 pmol/cm^2 twitch when based on surface membrane area of 3000 cm^2/gm wet wgt muscle instead of 300 cm^2/g wet wgt as originally used in our calculations. The technique did not allow measurement of calcium influx into the fast or intermediate components of ^{45}Ca washout. Bianchi and Narayan in 1982 modified the technique to allow measurement of ^{45}Ca and ^{40}Ca changes in both the fast and slow components. The modified technique consisted of stimulating the muscles at 1 Hz during the last 5 minutes of a 10 minute equilibration with ^{45}Ca in Ringer solution and then washing out the ^{45}Ca and ^{40}Ca in Ringer solution in which 1 mM strontium replaced calcium. The new procedure allowed measurement of ^{45}Ca and ^{40}Ca in both the extraction Ringer and the ^{45}Ca and ^{40}Ca remaining in the muscle. The ^{45}Ca extracted in Sr Ringer was increased from 0.16 mmol/g to 0.24 mmol/g and surprisingly, the ^{40}Ca increased from 0.26 mmol/g to 0.55 mmol/g i.e. stimulation caused unlabeled calcium in the terminal cisternae to become mobilized from a bound form to a free form capable of being extracted rapidly in Sr Ringer. At this time Narayan and I attributed the increase in extractable calcium to a translocation of calcium from the T.C. to the T.T. We have since modified this position and now attribute the increase in extractable calcium due to stimulation to release of calcium from a bound form in the terminal cisternae from calsequestrin to a free form (Ikemoto, 1991) which can be readily extracted when strontium replaces calcium in the Ringer solution. An alternative explanation would be translocation of calcium from the T.C. to the paravalbumin between the fibrils. Since extraction of calcium from the intracellular compartment is rapid we prefer the terminal cisternae as the loci of rapidly extractable calcium.

Since the increase in extractable calcium was the net result of 300 twitches, we then modified the procedure to examine calcium changes after 30, 60, 120 or 300 twitches (Bianchi and Narayan, 1984). Table 1 summarizes our results and show changes in content of calsequestrin bound calcium and free calcium in the terminal cisternae and calcium in the transverse tubule.

Stimulation for 30 twitches at 1 Hz increases calcium content of both the calsequestrin bound fraction and the free fraction of the T.C. The increase in both these fractions can only be due to net calcium influx during the first 30 twitches. The net influx over the first 30

Table 1 Calcium content of Terminal Cisternae, Calsequestrin bound and free calcium and net calcium flux during stimulation at 1 Hz*

Condition	Terminal Bound μmol/g wet wgt.	Cisternae Free μmol/g wet wgt.	Net flux nmols/g twitch	Extracellular including T Tubule μmol/g wet wgt.
Rest	1.52	0.10	0	0.16
30 twitches	1.78	0.42	+16.7	0.16
60 twitches	1.62	0.43	0	0.16
120 twitches	1.31	0.65	-1.5	0.16
300 twitches	1.35	0.39	-1.4	0.16

*Each value is the mean of 12 frog sartorius muscles and is significant at P 0.05 or less.

Table 2. Net calcium flux during stimulation of muscle at 1 Hz

Twitches	pmols/cm2 twitch
	Influx
30	5.6*
60	0
	Efflux
120	0.5
300	0.5

*The value of 5.6 pmol/cm2 twitch is 280x greater than the value of 0.02 pmol/cm2 twitch corrected for 3000 cm2 membrane area/g wet wgt obtained by Bianchi and Shanes 1959.

twitches amounts to 5.6 pmol/cm^2 twitch which is 280 times greater than the 0.02 pmol/cm^2 twitch previously determined by Bianchi and Shanes in 1959. Calcium content of these two fractions does not change between 30 and 60 twitches indicating that net influx rapidly falls to zero and becomes reversed to a net efflux of 0.5 pmol/cm^2 twitch after 120 twitches (Table 2). The free calcium in the terminal cisternae is maximal after 120 twitches due to a translocation from the calsequestrin bound fraction to the free calcium fraction of the T.C. Associated with the translocation from bound to free is a marked increase in contractile force. The amount of calcium efflux from the terminal cisternae is thus not only determined by the opening of calcium channels but also by the state of calcium activity in the terminal cisternae (Ikemoto et al, 1989, 1991).

The changes in calcium distribution between the calsequestrin bound fraction and the free may be due to the magnesium and potassium taken up as calcium leaves. Magnesium is retained in the T.C. even after restoration of calcium to the TC 400-500 msec following a tetanus. The magnesium content remains high 2.1 seconds after relaxation from a tetanic stimulation (Somlyo, A.V. et al. 1985).

A schematic representation of the net changes in calcium between the extracellular volume including the T.T. and the T.C. is shown in figure 1.

According to the diagram the feet operated channel consisting of a skeletal type dihydropyridine receptor protein of 400 K (Hoffman et al) and a 450 K ryanodine receptor-sarcoplasmic reticulum calcium channel protein (Imazawa et al, 1987) is closed to both the T.T. and the T.C. at rest. The dihydropyridine receptor of the skeletal type as distinct from cardiac type is missing in muscular dystrophy. Its restoration to the transverse tubular elements of dysgenic mice restores excitation-contraction coupling to the dysgenic muscle (Tanabe, et al. 1990). The interaction between these two proteins present in the feet are necessary for excitation-contraction coupling in skeletal muscle (Tanabe et al., 1990, Adams et al., 1990).

At rest the voltage operated calcium channel present in the feet of the junctional membrane is envisioned as being in a closed configuration to both calcium influx from the T.T. and to calcium efflux from the T.C. During stimulation for 30 twitches at 1 Hz the channel is pictured as being open initially to calcium efflux from the T.C. to the sarcoplasm via an opening intermediate between the two junctional membranes, followed by slow opening to the T.T. allowing calcium influx after the first few twitches. Net calcium influx occurs and raises the calcium content of the calsequestrin bound fraction of the T.C. from 1.52 to 1.78 μmol/g wet wgt. and the free calcium to 0.42 μmol/g wet wgt. When strontium replaces calcium in the T.T. the channel is envisioned as being open at both the T.T. membrane and the T.C. membrane, but closed to the sarcoplasm, allowing the efflux of calcium from the T.C. and the appearance of 0.58 μmol of calcium in the Sr Ringer extraction solution. In order

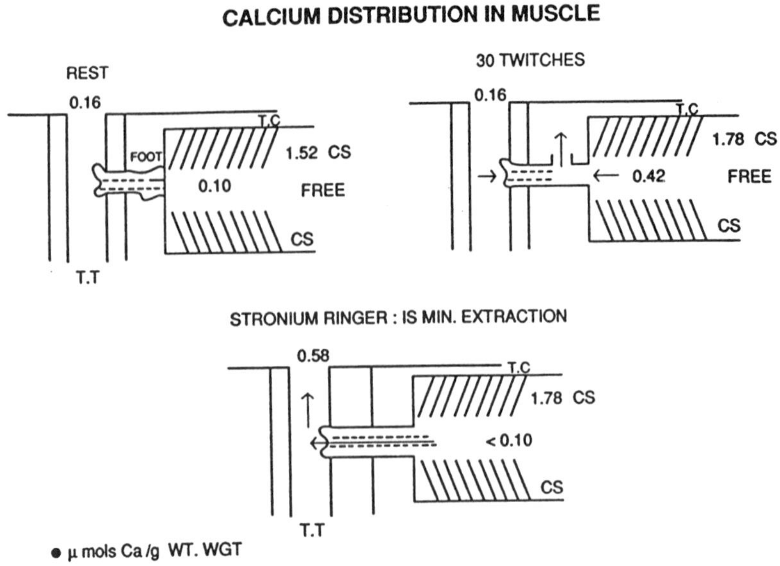

Figure 1

Schematic representation of calcium distribution between the extracellular volume and the transverse tubular network T.T. and the terminal cisternae (T.C.) at rest, following 30 twitches at 1 Hz, and after 15 min of extraction in strontium containing Ringer.

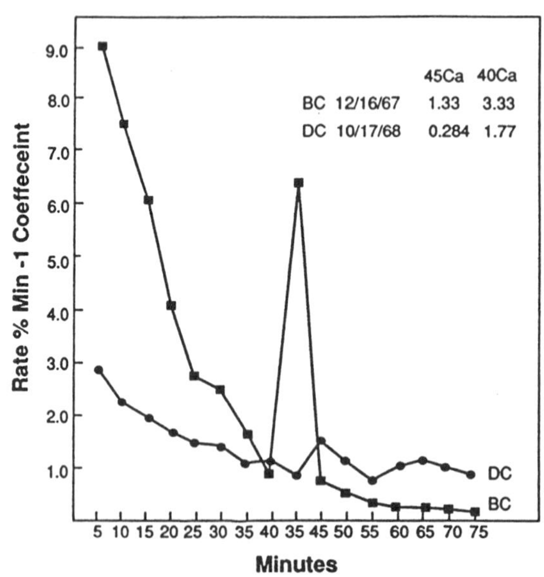

Figure 2

Spontaneous and oscillatory nature of calcium release from resting muscle equilibrated for 225 min with ^{45}Ca Ringer and washed out for 80 minutes. The rate coefficient is plotted for each 5 minute collection interval.

for calcium to be lost from the terminal cisternae it must be converted to a free form i.e. at 120 twitches the calsequestrin bound fraction is reduced to 1.31 µmol (g wet wgt. and the free is increased to 0.65 µmol/g wet wgt. Translocation occurs between bound and free forms of calcium within the T.C.

The junctionally operated feet channels appear to exist in a metastable state between open and closed state. At rest the most common state is a closed state, however, at aperiodic intervals these channels may open to allow an increased efflux to occur. In figure 2 are shown two examples of the oscillatory nature of these channels. The rate constant for ^{45}Ca efflux following 225 minutes of equilibration with ^{45}Ca Ringer is shown to have a large burst at 40 min for BC and to show oscillations for DC. The spontaneous burst in ^{45}Ca efflux in resting muscle while not common is observed in approximately 1 out of 10 washout experiments especially those muscle with a high sarcoplasmic reticulum calcium content.

Table 3. Effect of ryanodine on net calcium flux and oxygen uptake*

Condition	Time	Calcium uptake	Net	Calcium	Flux	0_2
K^+_o - log [Ryanodine]	min	mmol/g	P	nmol/g	up- min	take
2.5 m MK^+	10	10	+0.166	<.05	+16.6	++++
"	8	30	+0.330	<.05	+11.0	+++
20 m MK^+	10	10	-0.079	<.05	-7.9	+++
"	8	30	-0.010	N.S.	----	+++

*An N of 6 muscles used for each ryanodine concentration.

The complex nature of the ryanodine receptor operated foot channel is shown in table 3. At low concentrations of ryanodine 10^{-10}M and 10^{-8}M, calcium influx is increased to 16.6 and 11.0 nmol/g min over a 10 min period and oxygen uptake is markedly increased. When the membrane of the muscle fibers is depolorized to -50 mV by raising external potassium from 2.5 mM to 20 mM, ryanodine at 10^{-10}M causes a net efflux of 7.9 nmol/g min of calcium over a 10 min period. At 10^{-8}M ryanodine, depolarized muscles show no net calcium efflux but do show a marked increase in oxygen uptake.

The increased oxygen uptake is due to recycling of calcium and net uptake by the Ca-ATP'ase of the sarcoplasmic reticulum in polarized muscles. The consequence of calcium recycling is related to aerobic heat production. The time course of the increase in oxygen uptake at 10^{-10}M is shown in figure 3. Oxygen uptake peaks at 5 min and then slowly declines back to base level at 20 minutes. Since no contraction occurs at this concentration the ryanodine operated calcium channels must be open at a low conductance state insufficient to increased Ca^{2+}_i to the threshold for contraction i.e. $5x10^{-7}$M but sufficient to increase Ca^{2+}_i from $1.2x10^{-7}$M to $3.7x10^{-7}$M and cause calcium recycling and aerobic heat production (Lopez et al, 1983).

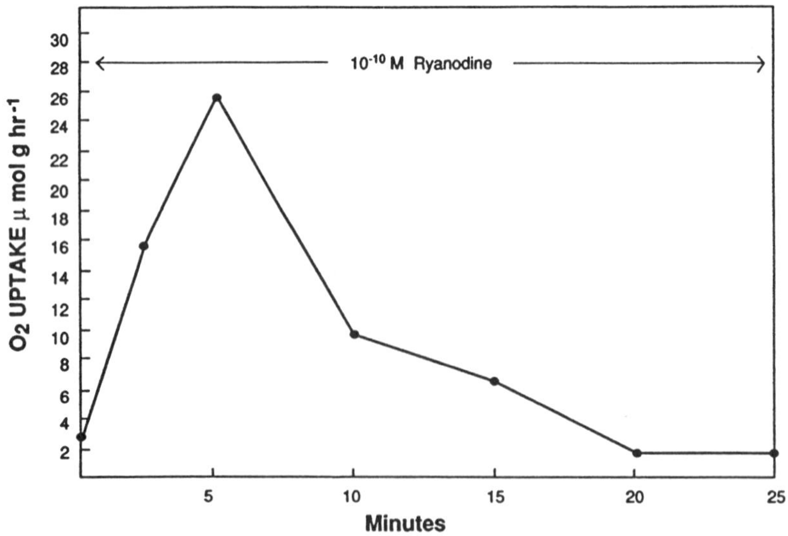

Figure 3

Time course of oxygen uptake by muscle exposed to 10^{-10}M ryanodine. The oxygen uptake mmol/g-1hr-1 peaks at 5 minutes and declines to basal levels at 20 minutes.

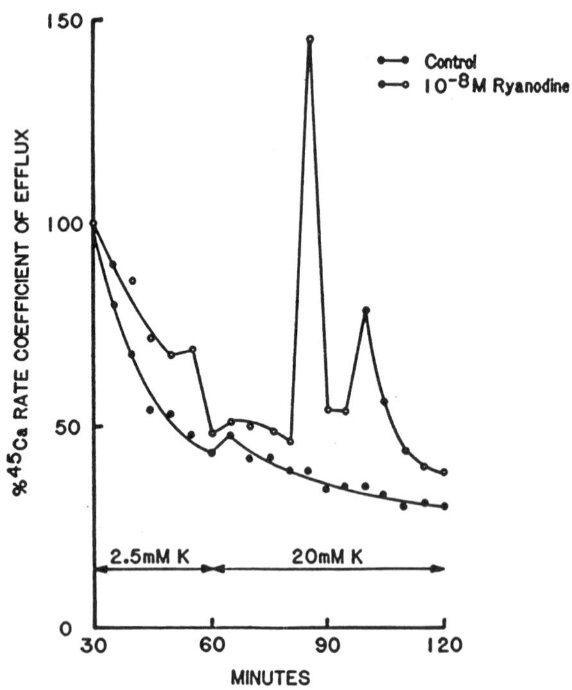

Figure 4

Oscillatory nature of calcium efflux following depolarization of muscle exposed to 10^{-8}M ryanodine. Efflux is expressed as the rate coefficient for each interval.

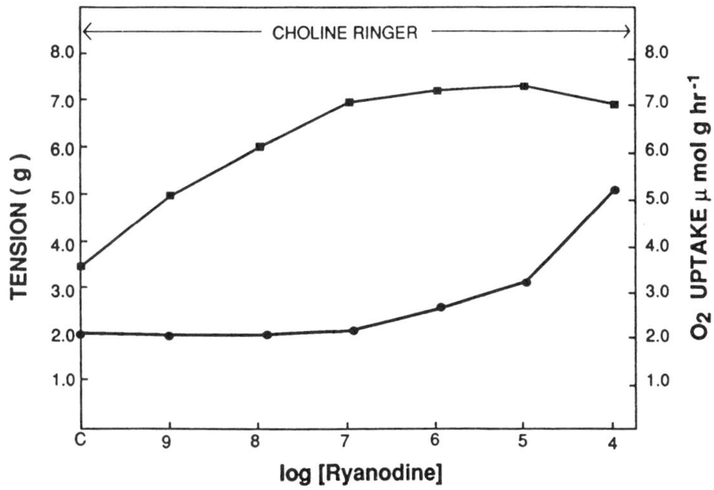

Figure 5

Ryanodine concentration-response plot of tension development and oxygen uptake in choline-Ringer.

The oscillatory nature of the calcium release channel of the sarcoplasmic reticulum is revealed by the action of 10^{-8}M ryanodine in depolarized muscles (fig. 4). In the presence of 20 $mK^+{}_o$, ryanodine at 10^{-8}M causes no net change in calcium content while causing oscillations in calcium efflux (fig. 4) indicating that the marked increase in O_2 uptake observed is due to a balance between increased calcium influx and efflux and associated calcium recycling by the sarcoplasmic reticulum.

The sodium calcium antiporter present in the sarcolemmal membrane plays a key role in helping to buffer the sarcoplasmic free calcium within physiological limits. When the sodium-calcium antiporter is blocked by replacing sodium of Ringer solution with choline, ryanodine causes a continual increase in oxygen uptake over the range of 10^{-9}M to 10^{-6}M (fig. 5). The threshold concentration for ryanodine contracture is reduced from 10^{-4}M (sodium Ringer) to 10^{-6}M. Inhibition of the sodium-calcium antiporter removes a powerful buffering system forcing the calcium released by ryanodine sensitive calcium channels to the sarcoplasm to be buffered solely by the Ca-ATP'ase of the sarcoplasmic reticulum. The increased activity the sarcoplasmic reticulum Ca-ATP'ase, causes greater splitting of ATP and its replacement by mitochondrial aerobic metabolism. As sarcoplasmic reticulum free calcium rises about $5x10^{-7}$M, the threshold for contraction, aerobic metabolism declines and is compensated for by an increased anaerobic metabolism.

In order to understand the integrative role that sarcolemmal L calcium channels, junctional sarcolemmal-terminal cisternae feet operated calcium channels, calcium release of calcium channels, and inositol triphosphate operated channels of the sarcoplasmic reticulum play in physiological functions of muscle (contraction, heat production and protein reserve) more physiological-biochemical experiments need to be performed. Pure biochemical studies of muscle contribute much valuable information, but only when related to physiological function do we begin to understand the integrative role that fine tuning of sarcoplasmic ionic calcium plays in various functional states of muscle. The ryanodine sensitive calcium channels have a remarkable plasticity in regard to various conductance states for calcium release from the terminal cisternae as well as allowing increased calcium conductance for calcium influx from the T-tubule. Ryanodine can cause the channels to either be open to varying degrees or to be maintained in a low conductance state.

The pharmacological action of ryanodine is dependent upon the functional state of the channel at the time of ryanodine-receptor interaction. The action of ryanodine to increase oxygen uptake in muscle at concentrations as low as 10^{-12}M suggests that there may be an endogenous agent similar to ryanodine that acts on the junctional feet channels to cause calcium cycling and heat production during hyperthermia (Bianchi, 1990).

References

Adams, B., Tanabe, T., Mikami, A., Numa, S., and Beam, K.G., 1990, Intramembrane charge movement restored in dysgenic skeletal muscle by injection of dihydropyridine receptor cDNAs, Nature, 346: 569-572.

Bianchi, C.P. and Shanes, A.M., 1959, Calcium influx in skeletal muscle at rest, during activity and during potasium contracture, J. Gen. Physiol., 42: 803-815.

Bianchi, C.P. and Narayan, S.R., 1982, Muscle fatigue and the role of Transverse Tubules, Science, 215: 295-296.

Bianchi, C.P. and Narayan, S.R., 1984, Effect of diazepam on calcium translocation during physiological muscle fatigue, J. Pharmacol. Exp. Ther., 231: 197-205.

Bianchi, C.P., 1990, Endogenous "Ryanodine agonist and hyperthermia", Life Sciences, 47: PL23-PL24.

Fabiato, A., 1985, Effect of ryanodine in skinned cardiac cells, Fed. Proc., 44: 2970-2976.

Hoffman, E.P., Knudson, C.M., Campbell, K.P. and Kunkel, L.M., 1987, Subcellular fractionation of dystrophin to the triads of skeletal muscle, Nature, 330: 754-758.

Ikemoto, N., Ronjat, M., Meszaros, L.G. and Koshita, M., 1989, Postulated role of calsequestrin in the regulation of calcium release from sarcoplasmic reticulum, Biochem., 28: 6764-6771.

Ikemoto, N., Antonia, B., Kang, J.J., Meszaros, L.G., and Ronjat, M., 1991, Intravesticular calcium transient during calcium release from sarcoplasmic reticulum, Biochemistry, 30: 5230-5237.

Imazawa, T., Smith, J.S., Coronado, R., and Campbell, K.P., 1987, Purified ryanodine receptor from skeletal muscle sarcoplasmic reticulum is the Ca^{2+}-permeable pore of the calcium release channel, J. Biol. Chem., 262: 16636-16643.

Lakatta, E.G. and Henderson, J.S., 1985, Spontaneous myocardial calcium oscillations: overview with emphasis on ryanodine and caffeine, Fed. Proc., 44: 2977-2983.

Lopez, L.R., Alamo, L., Caputo, C., DiPolo, R. and Vergara, J., 1983, Determination of ionic calcium in frog skeletal muscle fibers, Biophys. J., 453: 1-4.

Rios, E. and Pizarro, G., 1988, Voltage sensors and calcium channels of excitation-contraction coupling, New in Physiological Sciences, 3: 223-227.

Somlyo, A.V., Kitazawa, T., Gonzalez-Serratos, H., McClellan, G. and Somlyo, A.P., 1985, Ion movements associated with calcium release and uptake in the sarcoplasmic reticulum, In: Calcium in Biological Systems, R.P. Rubin, G.B. Weiss and J.W. Putney, ed., pp. 351-358, Plenum Press, New York.

Tanabe, T., Beam, K.G., Adams, B., Nudome, T. and Numa, S., 1990, Regions of the skeletal muscle dihyphopyridine receptor critical for excitation-contraction coupling, Nature, 346: 567-569.

TWO KINDS OF CALCIUM-INDUCED RELEASE OF CALCIUM

FROM THE SARCOPLASMIC RETICULUM OF SKINNED CARDIAC CELLS

Alexandre Fabiato

Department of Physiology and Biophysics
Medical College of Virginia
Richmond VA 23298, U.S.A.

ABSTRACT

This article compares the Ca^{2+}-induced release of Ca^{2+} that is triggered by a rapid increase of free Ca^{2+} concentration at the outer surface of the sarcoplasmic reticulum of a skinned cardiac cell to the spontaneous release of Ca^{2+} that is produced by a steady-state high free Ca^{2+} concentration which overloads the sarcoplasmic reticulum with Ca^{2+} in a skinned cardiac cell. The first process, that is triggered by a rapid increase of free Ca^{2+} concentration at the outer surface of the sarcoplasmic reticulum, has a time- and Ca^{2+}-dependent activation and inactivation, does not require any preload of the sarcoplasmic reticulum with Ca^{2+}, and is not affected by the addition of inositol(1,4,5)-trisphosphate. The second process, i.e. the spontaneous release of Ca^{2+} from the sarcoplasmic reticulum, is not inactivated by a high free Ca^{2+} concentration, requires an overload of the sarcoplasmic reticulum with Ca^{2+} and is enhanced by inositol(1,4,5)-trisphosphate. The filling inside the sarcoplasmic reticulum with Ca^{2+} is critical for the triggering of the spontaneous release of Ca^{2+}. On the other hand, the spontaneous release of Ca^{2+} has many similarities to the "Ca^{2+}-induced release of Ca^{2+}" that is observed for isolated sarcoplasmic reticulum vesicles incorporated into a lipid bilayer which is triggered by the increase of free Ca^{2+} concentration at the outer surface of the sarcoplasmic reticulum. Although the Ca^{2+}-induced release of Ca^{2+} with time- and Ca^{2+}-dependent activation and inactivation and the spontaneous release of Ca^{2+} are regulated by difference mechanisms, they are both inhibited by ryanodine, which suggests that they may take place through the same channel and may even have some partial common pathway.

INTRODUCTION

Ca^{2+}-induced release of Ca^{2+} from the sarcoplasmic reticulum was initially proposed as a hypothesis for excitation-contraction coupling of skeletal muscle (Bianchi & Shanes, 1959; Frank, 1960; see Frank, 1980 for a review). An increase of free Ca^{2+} concentration at the outer surface of the sarcoplasmic reticulum, resulting from transsarcolemmal Ca^{2+} influx from the extracellular space or Ca^{2+} displacement from a restricted intracellular compartment, would induce a Ca^{2+} release from the sarcoplasmic reticulum

Excitation-Contraction Coupling in Skeletal, Cardiac, and Smooth Muscle
Edited by G.B. Frank *et al.*, Plenum Press, New York, 1992

which, in turn, would activate the myofilaments. This hypothesis has received support from experiments in skinned (sarcolemma removed) fibers from skeletal (Endo et al., 1970; Ford & Podolsky, 1970, 1972a,b), cardiac (Fabiato & Fabiato, 1973, 1975a,b) amd smooth (Saida, 1982) muscle.

The physiological role of Ca^{2+}-induced release of Ca^{2+} from the sarcoplasmic reticulum is uncertain (Endo, 1977; Fabiato & Fabiato, 1977; Fabiato, 1981, 1983, 1985c; Stephenson, 1981a). In the intact mammalian cardiac muscle, the transsarcolemmal Ca^{2+} influx could be the trigger for this Ca^{2+}-induced release of Ca^{2+}. However, a modulatory role of the transsarcolemmal voltage, through a voltage-dependent charge movement, on the Ca^{2+} release is not yet excluded (Fabiato, 1989). In skeletal muscle, the Ca^{2+} trigger could be located in the clefts between the transverse tubules and the terminal cisternae (Frank, 1980, 1982) and the Ca^{2+}-induced release of Ca^{2+} could be an amplifying mechanism of a Ca^{2+} release initially triggered by a charge movement (Fabiato, 1983). Thus, the complexity of excitation-contraction coupling could increase when the diameter of the preparation increases. In the cells from the frog ventricle, which have a 3-5μm diameter, the transsarcollemal Ca^{2+} influx could suffice to activate the myofilament directly. In the mammalian cardiac cells, which have a diameter five times larger than that of the frog ventricular cells, Ca^{2+}-induced release of Ca^{2+} could be necessary to amplify the transsarcolemmal Ca^{2+} influx. In the skeletal muscle fibers, which have a diameter five times larger than that of the mammalian ventricular cells, the initial process may be Ca^{2+} independent and linked to charge movement in the transverse tubules; after this process has triggered a Ca^{2+} release from the sarcoplasmic reticulum, this Ca^{2+} release could be amplified by the Ca^{2+}-induced release of Ca^{2+} and thereby activate the myofilaments. However, there are valid arguments to propose that Ca^{2+}-induced release of Ca^{2+} may play no physiological role in either cardiac or skeletal muscle excitation-contraction coupling (Endo, 1977, 1985).

This article is limited to studies of cardiac muscle and its goal is not to fully answer the question of the physiological role of Ca^{2+}-induced release of Ca^{2+}, which I have always left unanswered (for instance, see footnote 5 in Fabiato, 1985c) but to describe comparatively two kinds of "Ca^{2+}-induced release of Ca^{2+} from the sarcoplasmic reticulum" for which the problem of physiological relevance should be differently assessed. The first of these processes, which I have always described as "Ca^{2+}-induced release of Ca^{2+} from the sarcoplasmic reticulum" presents a time- and Ca^{2+}-dependent activation and inactivation (Fabiato, 1985b). The second which I had previously also termed "Ca^{2+}-induced release of Ca^{2+} from the sarcoplasmic reticulum" (Fabiato and Fabiato, 1972) but I have more recently renamed "spontaneous release of Ca^{2+} from the sarcoplasmic reticulum" (Fabiato, 1983), occurs in the presence of a steady-state high concentration of Ca^{2+} in the sarcoplasmic reticulum, has no time dependence and is not inactivated by high concentrations of Ca^{2+}.

I. Ca^{2+}-INDUCED RELEASE OF Ca^{2+} FROM THE SARCOPLASMIC RETICULUM TRIGGERED BY A RAPID INCREASE OF FREE Ca^{2+} CONCENTRATION AT THE OUTER SURFACE OF THE SARCOPLASMIC RETICULUM OF A SKINNED CARDIAC CELL

The preparations used for the experiments describing this phenomenon are obtained by homogenization followed by microdissection into small bundles of myofibrils which are individually packed and surrounded by the the sarcoplasmic reticulum. The dimensions are 6-10μm in width, 4-7μm in thickness and 20-60μm in length. The volume ranges from 15% of that of an intact cardiac cell from the adult mammalian ventricle (Fabiato, 1981) to

2% of that of an intact canine cardiac Purkinje cell (Fabiato, 1985a). The removal by microdissection of the cellular structures that keep the myofibrils packed together in an intact cell permits a wide separation of individual myofibrils with sarcoplasmic reticulum wrapped around them (Fabiato, 1985a). Extensive tests have demonstrated that the externally applied solutions have direct access by hydraulic bulk flow to the outer surface of the sarcoplasmic reticulum which is wrapped around individual myofibrils. The slow diffusion of Ca^{2+} (Kushmerick & Podolsky, 1969) takes place only through each myofibril (0.3-0.4 μm radius) after Ca^{2+} has been released from the sarcoplasmic reticulum.

Rapid (millisecond range) changes of solutions are effected by a microprocessor-controlled system of microinjections and aspirations with microsyringes controlled by stepping motors. The transmission fluid contained in the microsyringes is less than 9 μl water that has been degassed to make it incompressible. The water is separated by ion-free mineral oil from the very small volume (2 nl) of physiological solution generally containing aequorin. Part of this volume is injected around the skinned cell that is bathed in oil (Fabiato, 1981). Hence, an artificial 'chamber' of less than 2 nl of bathing aqueous saline solution is constituted around the skinned cell whose volume (1-3 nl) is till 500 to 1000 times less.

Solution changes are done by a microaspiration of the previous solution followed, after a few ms, by the microinjection of the new solution. Thus, there is no mixing of the aequorin-containing solutions at various free Ca^{2+} concentration levels. Each solution is reaspirated in the micropipette in which it was originally contained. The photomultiplier tube permanently monitors the bioluminescence emitted by each of the aequorin-containing solutions, whether this aequorin is still in the micropipette tip, injected in the experimental chamber or reaspirated in the micropipette. This results in a very large steady light level which is balanced by a counter-bias voltage and on top of which the light transients caused by Ca^{2+} release from the sarcoplasmic reticulum are recorded. These are the major reasons for which a solution change caused no artefact on the aequorin light recording (Fig. 1). Additional details on the methodology that permits the complete elimination of any artefact on the light recording are explained in Fabiato (1985a).

Despite the small volume of microinjected solution, the hydraulic flow is very rapid (135-270 $cm \cdot s^{-1}$). The micropipettes are oriented at an angle with respect to the skinned cell so that the flow does not strike it directly, but instead strikes the oil-water interface. A rebounding hydraulic flow originates presumably from the entire internal surface area of the ellipsoid constituted by the drop of aqueous solution in oil. The compression of the oil reduces the flow rate. The rebounding flow reaches the skinned cell from all directions, symmetrically in the three axes. For this reason, a microinjection does not cause an artefact on the tension recording.

Turbulence caused by the hydraulic flow between the myofibrils further reduces the flow rate. It remains, however, very high relative to the small dimensions of the skinned cell. After its aspiration, some of the previous solution remains in the skinned cell. One half-step (the stepping motor is used in this mode) of microinjection of the new solution should be sufficient to dilute the remaining previous solution by a factor of more than 50. Thus, the change of solution at the outer surface of the sarcoplasmic reticulum could theoretically be achieved in one half-step (1 ms for Fig. 1). In fact, measurements indicated that the change of solution composition at the outer surface of the sarcoplasmic reticulum takes about twice as long (Fabiato, 1985a).

The microprocessor controlling the microinjection-aspiration system displays signals on the recorder simultaneously with the signals of the tension and aequorin light recordings (Fig. 1). At the end of the experiment, the signals are redisplayed at a speed 20 times higher (lower trace in Fig. 1) to facilitate their analysis. The signal is upward for an injection, downward for an aspiration. The timing and duration of the signals are accurate. Their amplitudes differ only to distinguish the three channels of microinjection-aspiration; the volumes microinjected and aspirated are always the same. The photographic enlargement of the right-hand panels of Fig. 1 shows the time relationship between the microinjection of each solution and its effect on the aequorin light signal for each of the two Ca^{2+} transients.

An experiment ends by a determination of maximum light, by discharging all of the aequorin contained in the skinned cell with a saturating free Ca^{2+} concentration. The myoplasmic free Ca^{2+} concentration reached at the peak of the Ca^{2+} transient resulting from Ca^{2+} release from the sarcoplasmic reticulum is estimated from (1) the ratio of peak light to maximum light and a calibration curve of the Ca^{2+} sensitivity of the aequorin batch used for the experiment, under the same ionic conditions, (2) the ratio of peak tension to maximum tension and the steady-state tension free Ca^{2+} concentration curve (Fabiato, 1985a).

These technical details are needed to understand Fig. 1 which shows some of the major properties of the Ca^{2+}-induced release of Ca^{2+} from the cardiac sarcoplasmic reticulum. They will not be repeated for the other figures using the same techniques and display.

A skinned cardiac cell from the adult rat ventricle (Fig. 1) was bathed in a solution at a resting free Ca^{2+} concentration low enough to prevent the occurrence of a cyclic release of Ca^{2+} from the sarcoplasmic reticulum: $-\log_{10}$ free Ca^{2+} concentration (pCa) 7.4 in a rat ventricular skinned cardiac cell. The absence of a cyclic release of Ca^{2+} from the sarcoplasmic reticulum was taken as evidence that it was not overloaded with Ca^{2+} in the presence of the resting free Ca^{2+} concentration (Fabiato, 1983). Ca^{2+}-induced release of Ca^{2+} from the sarcoplasmic reticulum was triggered by a rapid increase of free Ca^{2+} concentration at its outer surface by aspiration of the solution at pCa 7.4 and injection of a solution at pCa 6.25. This was rapidly followed by a return to the initial low free Ca^{2+} concentration by an opposite aspiration and injection (Fig. 1), which prevented the overload the sarcoplasmic reticulum.

This Ca^{2+}-induced release of Ca^{2+} can be triggered with a much lower increase of free Ca^{2+} concentration than that needed to activate the myofilaments directly. The amplitude of the Ca^{2+} release, estimated from aequorin and tension calibrations, increases when the free Ca^{2+} concentration trigger is increased up to an optimum, pCa 6.25 in a rat ventricular skinned cell (Fabiato, 1981, 1983). Above the optimum, further increase of free Ca^{2+} concentration inhibits Ca^{2+}-induced release of Ca^{2+} to an extent that increases with the free Ca^{2+} concentration trigger. Eventually, no detectable Ca^{2+} release is obtained with a free Ca^{2+} concentration trigger higher than pCa 5.15 in the rat ventricular skinned cell (see the first series of aspiration-injection signals in Fig. 5).

The inhibition of Ca^{2+}-induced release of Ca^{2+} by a supraoptimum free Ca^{2+} concentration occurs at the free Ca^{2+} concentration outside the sarcoplasmic reticulum reached at the peak of Ca^{2+} release. This negative feedback helps to explain why Ca^{2+}-induced release of Ca^{2+} from the sarcoplasmic reticulum is graded rather than all or none as previously assumed (Costantin & Taylor, 1973). The inactivation of Ca^{2+} release at the free Ca^{2+} concentration level reached at the peak of Ca^{2+} release is demonstrated by the experiment shown in Fig. 1. During the ascending phase of the Ca^{2+} transient, the premature exposure of the outer surface of the sarcoplasmic reticulum to a free Ca^{2+} concentration that would have been reached only at the peak of the Ca^{2+} transient resulted in a curtailment of this Ca^{2+} transient after a change of slope of the ascending phase. This is visible on the photographic enlargement of the right-hand panel of Fig. 1.

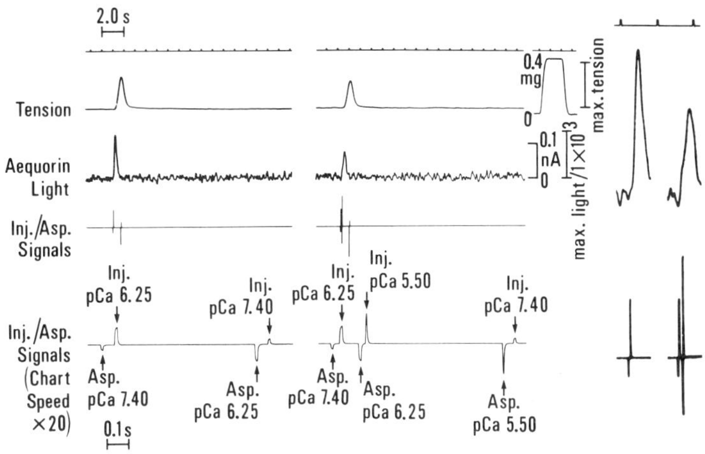

Fig. 1 Triggering of Ca^{2+}-inducea release of Ca^{2+} sarcoplasmic reticulum with an optimum increase of bulk solution free Ca^{2+} concentration and premature inactivation of the Ca^{2+} release by a supraoptimum additional increase of free Ca^{2+} concentration. The experiment was done in a skinned cardiac cell from the adult rat ventricle which was 8 μm wide, 6 μm thick and 54 μm long. The solutions contained 0.05 mM total EGTA, pMg 2.5, pMgATP 2.5, 12 mM phosphocreatine, 15 U ml^{-1} creatine phosphokinase, 0.16 M ionic strength with K^+ and Cl^- as major ionic species and 30mM N,N-bis(2-hydroxyethyl)-2-aminoethanesulphonic acid (BES) at pH 7.1 and 22° C. The filtration was 30 Hz for the aequorin light signal ad 5Hz for the tension signal. The injection of the solution at pCa 5.5 exposed the outer surface of the sarcoplasmic reticulum to a free Ca^{2+} transient if the release induced by pCa 6.15 had been allowed to follow its spontaneous course. This figure was modified from Fig. 1 of Fabiato (1983). The maximum light and the calibration of aequorin light as a function of the free Ca^{2+} concentration were corrected to take into account recent data and calibration curves with the same aequorin batch (Fabiato, 1985a). From this it was inferred that a pCa of 5.25 was reached at the peak of the first Ca^{2+} transient. A consistent result (pCa 5.3) was inferred from the ratio of the amplitude of the tension transient to the maximum tension and the tension-pCa curve (Fabiato, 1981). The reproduction, with modification, was authorized by the American Physiological Society.

The threshold of free Ca^{2+} concentration change at the outer surface of the sarcoplasmic reticulum necessary to trigger Ca^{2+} release from the sarcoplasmic reticulum and the amplitude of the Ca^{2+}-release elicited by a given increase in free Ca^{2+} concentration are dependent upon the rate of this increase. A high rate induces release whereas a low rate inhibits and causes Ca^{2+} accumulation (Fabiato 1982a, b, 1983, 1985b, c). This explains the failure to observe this type of release in large 'chemically skinned' multicellular cardiac preparations (Fabiato, 1982a).

Evidence has been presented that Ca^{2+}-induced release of Ca^{2+} occurs through a channel with time- and Ca^{2+}-dependent activation and inactivation

(Fabiato, 1983, 1985b). This evidence was reported before any of the single channel recording from sarcoplasmic reticulum incorporated into a lipid bilayer had been done. The inactivation appears to be caused by the binding of Ca^{2+} to a single site at the outer surface of the sarcoplasmic reticulum because it follows first-order kinetics and has a low Q_{10} (Fabiato, 1985b). The inactivating site has a higher affinity for Ca^{2+} but a lower rate constant than the activating site (Fabiato, 1985b). The latter is hypothesized to present a high degree of cooperativity. Ca^{2+} release would occur during the lapse of time when the gating mechanism of the channel is already activated and not yet inactivated. The channel should go through at least four states (Fabiato, 1985b; Fig. 2): (1) open as a result of the increase in concentration of free Ca^{2+} used as a trigger, which results in Ca^{2+} binding to the activating site; (2) closed as a consequence of the large increase of free Ca^{2+} concentration resulting from the Ca^{2+} release from the sarcoplasmic reticulum, which caused Ca^{2+} binding to the inactivating site; (3) refractory; and (4) activatable. A computer model based on this concept has been shown to fit the experimental data provided that the Ca^{2+} triggering site and the Ca^{2+} release site at the outer surface of the sarcoplasmic reticulum are functionally separated (Wong et al 1987).

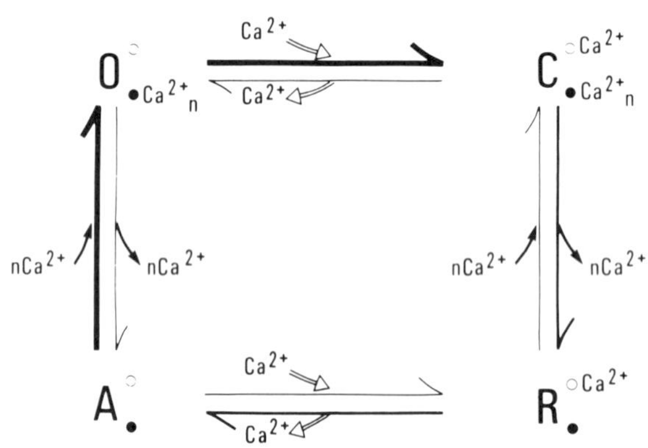

Fig. 2 - Model for the Ca^{2+}-induced release of Ca^{2+} from the sarcoplasmic reticulum with time- and Ca^{2+}-dependent activation and inactivation. The activating site is represented by the solid circle, the inactivating Ca^{2+} binding site is represented by the clear circle. O = open, C = closed, R = refractory, and A = activatable. The letter n before Ca^{2+} or in subscript after Ca^{2+} corresponds to the cooperative binding of several calcium ions.

There is no evidence for the participation of any of the partial reactions of the Ca^{2+} pump of the sarcoplasmic reticulum, which have been described with fragmented preparations of sarcoplasmic reticulum, in the mechanism of Ca^{2+}-induced release of Ca^{2+} from it (Fabiato, 1985b). This time-dependent Ca^{2+}-induced release of Ca^{2+} and the caffeine-induced release of Ca^{2+} do not share a common mechanism, because caffeine-induced release of Ca^{2+} can be obtained under conditions inhibiting Ca^{2+}-induced Ca^{2+} release. This Ca^{2+}-induced release of Ca^{2+} presents a refractory period whereas the caffeine-induced release of Ca^{2+} does not (Fabiato, 1985b). On the other hand both are inhibited by ryanodine (Fabiato, 1985f), which suggest that they occur through a common Ca^{2+} channel across the sarcoplasmic reticulum membrane.

This Ca^{2+}-induced release of Ca^{2+} does not require any preload of the sarcoplasmic reticulum with Ca^{2+} other than the bathing in the low resting free Ca^{2+} concentration; pCa 7.4 in Fig. 1. Its amplitude is optimum for a free Mg^{2+} concentration at about physiological intracellular level (Fabiato & Fabiato, 1979). Nothing in this demonstration is based on a precise definition of the free Ca^{2+} concentration at the outer surface of the sarcoplasmic reticulum necessary to induce Ca^{2+} release, which is unknown as stated in all articles on this subject (e.g. p. 492 in Fabiato, 1981).

This Ca^{2+}-induced release of Ca^{2+} from the sarcoplasmic reticulum can liberate enough Ca^{2+} to account for the twitch obtained under optimum inotropic conditions in intact mammalian cardiac cells (Fabiato, 1981). The degree of prominence of this Ca^{2+}-induced release of Ca^{2+} from the sarcoplasmic reticulum varies among animal species and developmental stages in a way opposite to that of the transsarcolemmal Ca^{2+} influx (Fabiato & Fabiato, 1978; Sutko & Willerson, 1980; Bers *et al.*, 1981, Fabiato, 1982a,b; Bers, 1985). This Ca^{2+}-induced release of Ca^{2+} from the sarcoplasmic reticulum was not observed in the cardiac cells with the smallest diameter: those from the frog ventricle and the fetal mammalian ventricle.

The Ca^{2+}-induced release of Ca^{2+} with time- and Ca^{2+}-dependent activation and inactivation is much faster than the Ca^{2+} release induced by inositol(1,4,5)-trisphosphate [Ins(1,4,5)P_3] and is not effected by inositol(1,4,5)-trisphosphate (Fabiato, 1990). Thus, comparing Figs. 3a and 3b, taking into account the different tension scales (five times higher in Fig. 3a) shows that there is a great difference in the rate of tension development, thus in the rate of total calcium release from the sarcoplasmic reticulum. This rate of total calcium release has been measured directly and has been found to be much higher for the Ca^{2+}-induced release of Ca^{2+} than for the Ins(1,4,5)P_3-induced release of Ca2+ (Fabiato, 1990). In the experiment shown in Fig. 3c the two stimuli were combined. At the same time as the free Ca^{2+} concentration was increased from pCa 7.75 to pCa 6.25, 10 μM Ins(1,4,5)P_3 was applied. The initial tension transient was not modified in amplitude as compared to that observed in Fig. 3b. The only effect of this inclusion of Ins(1,4,5)P_3 in the solution was to produce a distinct second Ca^{2+} transient of much smaller amplitude than the first one. Note again that the amplification for the tension recording was 5 times greater in Fig. 3a than in Fig. 3b and 3c.

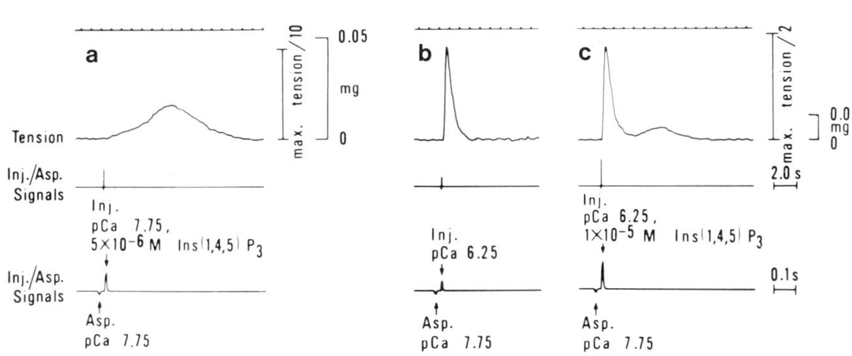

Fig. 3 - Ins(1,4,5)P_3-induced release of Ca^{2+} (a) Ca^{2+}-induced release of Ca^{2+} (b) and combination of the two stimuli (c) in a skinned cardiac cell from the rat ventricle. Reproduced from Fig. 1 of Fabiato (1986) and Fig. 1 of Fabiato (1990) with permission of the publishers of the Proceedings of the International Union of Physiological Sciences and of Pergamon Press.

The hypothesis of Ca^{2+}-induced release of Ca^{2+} from the sarcoplasmic reticulum is consistent with several aspects of excitation-contraction coupling of the intact mammalian heart (Chapman, 1983), including the well-established observation that the transsarcolemmal Ca^{2+} influx also serves to load the sarcoplasmic reticulum with Ca^{2+} (Fabiato, 1985c). However, these observations are only circumstantial evidence for a possible physiological role of Ca^{2+}-induced release of Ca^{2+} from the sarcoplasmic reticulum in cardiac excitation-contraction coupling. In addition, the suggestion by Endo (1985) that this Ca^{2+} release may be at least in part a non-physiolical process caused by the rapid solution changes cannot be completely discounted.

II SPONTANEOUS CYCLIC RELEASE OF Ca^{2+} FROM THE SARCOPLASMIC RETICULUM OF A SKINNED CARDIAC CELL IN THE CONTINUOUS PRESENCE OF A HIGH FREE Ca^{2+} CONCENTRATION OVERLOADING THE SARCOPLASMIC RETICULUM

When a skinned cardiac cell (Fabiato & Fabiato, 1973, 1975a,b) or a partly skinned group of cardiac cells (Fabiato & Fabiato, 1972) is continuously exposed to a sufficiently high free Ca^{2+} concentration it presents a spontaneous cyclic release of Ca^{2+} from the sarcoplasmic reticulum which generates cyclic contractions. If the preparation is large enough and contains remaining pieces of sarcolemma, the contraction propagates as a wave at 50-175 $\mu m\ s^{-1}$ at 22°C, a speed that increases with increased free Ca^{2+} concentration (Fabiato & Fabiato, 1972; Fabiato, 1985d). Similarly, the frequency and amplitude of the cyclic concentrations increase when the free Ca^{2+} concentration is increased. They are also increased by a lower concentration of caffeine but inhibited by a continuous exposure to a higher concentration (Fabiato & Fabiato, 1972, 1973).

Since our first report on this subject, we emphasized that only the first contraction induced by an increase of free Ca^{2+} concentration in a previously quiescent skinned or disrupted cell demonstrates a perhaps physiologically relevant Ca^{2+} release (see Fig. 6 in Fabiato & Fabiato, 1972). In contrast, the cyclic contractions were deemed to be related to Ca^{2+}-overload of the sarcoplasmic reticulum. This distinction between the two processes has been repeated in all subsequent articles (see p. 471 in Fabiato, 1981)

Each of the spontaneous Ca^{2+} release transients in the cyclic sequence appears to occur when the loading of the sarcoplasmic reticulum with Ca^{2+} has reached a critical level. This is suggested by the observation that the amplitude of the caffeine-induced contraction increases progressively as the delay between the previous spontaneous Ca^{2+} release and the application of caffeine during a cycle of spontaneous Ca^{2+} release (see Fig. 6 in Fabiato and Fabiato, 1975a). The Ca^{2+} reaccumulation into the sarcoplasmic reticulum after a Ca^{2+} release is rapid during the first second, but it continues more slowly during many more seconds (see Figs. 15 and 16 in Fabiato, 1985b). The total duration of the rapid and slow phases of the Ca^{2+} reaccumulation into the sarcoplasmic reticulum at a given bathing concentration of free Ca^{2+} appears to be about equal to the interval between spontaneous cyclic releases and varies in the same direction with variation of the temperature, the ionic composition of the bathing solution and the nature of the cardiac tissues studied. A progressive filling of the sarcoplasmic reticulum between cyclic releases can be demonstrated by monitoring the fluorescence of chlorotetracycline (Fig. 4) which increases when the free Ca^{2+} concentration inside the sarcoplasmic reticulum increases (Fabiato, 1982b). This free Ca^{2+} is in rapid equilibrium with the Ca^{2+} bound to the membrane of the sarcoplasmic reticulum, the variations in concentration of which are detected by changes in the fluorescence of the Ca^{2+}-chlorotetracycline chelate.

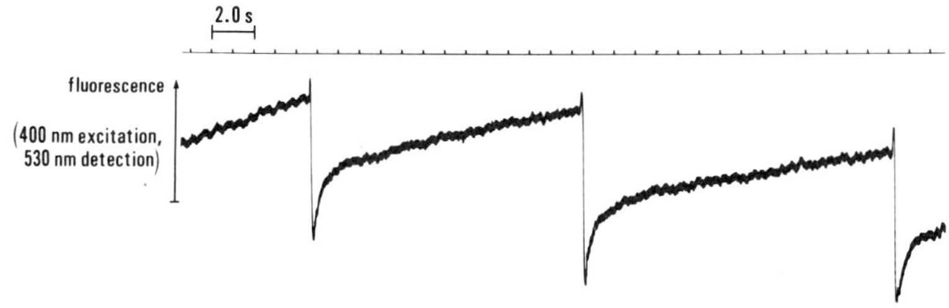

Fig. 4. *Demonstration of the continuous filling of the sarcoplasmic reticulum with Ca^{2+} during the interval between spontaneous cyclic release of Ca^{2+} through the monitoring of the fluorescence of the Ca^{2+}-chlorotetracycline chelate. The experiment was done in a skinned cardiac cell from the rat ventricle that was 11 µm wide, 8.5 µm thick and 61 µm long and had its myosin extracted to eliminate the contraction (see Fabiato 1982b, for technique and explanation of the rapid and small increase preceding the large decrease of fluorescence). The drift was caused by the progressive leak of chlorotetracycline from the sarcoplasmic reticulum. Hence the spontaneous Ca^{2+} release may have started at about the same free Ca^{2+} concentration inside the sarcoplasmic reticulum even though this corresponded to a progressively lower level of fluorescence. The bulk solution pCa was 6.0, the pMg was 2.5, the pMgATP was 2.5, the pH was 7.1, the total EGTA was 0.05 mM, the ionic strength was 0.16M with K^+ and Cl^- as major ionic species, and the temperature was 22° C. The solutions contained 12 mM and 15 U ml^{-1} creatine phosphokinase.*

The spontaneous cyclic Ca^{2+} release occurs under conditions that inhibit the Ca^{2+}-induced release of Ca^{2+} that has time- and Ca^{2+}-dependent activation and inactivation (Fig. 5). An increase of free Ca^{2+} concentration from pCa 7.4 to pCa 5.15 by aspiration of a solution at pCa 7.4 and injection of a solution at pCa 5.15, rapidly followed by a return to pCa 7.4 through the reverse procedure, failed to induce a significant tension transient (first series of aspiration-injection signals in Fig. 5), because this high free Ca^{2+} concentration inhibited the Ca^{2+}-induced release of Ca^{2+} from the sarcoplasmic reticulum. When the same increase of free Ca^{2+} concentation was applied and maintained, by omitting the return to pCa 7.4, cyclic contractions appeared after a delay of 1.0s (second injection of pCa 5.15 not followed by an aspiration in Fig. 5). The powerful Ca^{2+} accumulation into the sarcoplasmic reticulum, which is wrapped around individual myofibrils, prevented the externally applied Ca^{2+} to activate the myofilaments directly (see Fabiato, 1985a, b for additional evidence for such a 'screening' effect of the sarcoplasmic reticulum).

In another experimental run (not shown) with this same skinned cell used for Fig. 5, the application of the solution of pCa 5.15 during 0.5s (followed by its reaspiration and the reinjection of pCa 7.4) did not result in a spontaneous Ca^{2+} release but shortened to 0.6s the delay between a new application of pCa 5.15 (after 1.0s in pCa 7.4) and the occurrence of a spontaneous release of Ca^{2+}.[1] This suggest that the delay observed in Fig. 5 corresponds to the time taken by over-filling the sarcoplasmic reticulum. Thus, a prior exposure to pCa 5.15 shortened the delay by causing a partial

[1] See Fig. 5, p. 479, in Fabiato & Fabiato, 1975a, for another example of the decrease of the latency between the increase of free Ca^{2+} in the solution and the initiation of the spontaneous release of Ca^{2+} from the sarcoplasmic reticulum.

refilling of the sarcoplasmic reticulum. In contrast, the prior exposure to pCa 5.15 should have further delayed the Ca^{2+} release if the delay had corresponded to the time required fora rapid Ca^{2+} accumulation into the sarcoplasmic reticulum to decrease the myoplasmic free Ca^{2+} concentration at its outer surface below that set in the bulk solution. A prefilling of the sarcoplasmic reticulum should reduce the rate of Ca^{2+} reaccumulation into it.

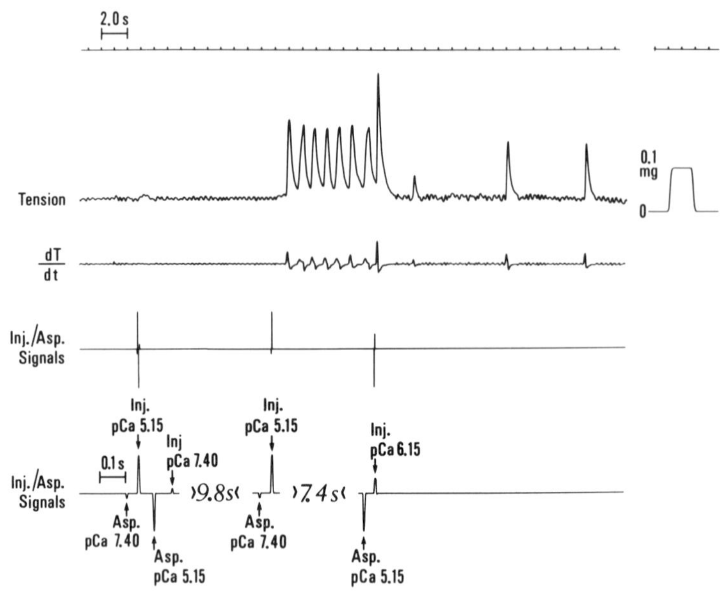

Fig. 5. *Spontaneous cyclic release of Ca^{2+} from the sarcoplasmic reticulum of a skinned cardiac cell from the rat ventricle under conditions inhibiting Ca^{2+}-induced release of Ca^{2+} that has time- and Ca^{2+}-dependent activation and inactivation. The skinned cell was 9 μm wide, 7 μm thick and 47 μm long. The pMg was 2.5, the pMgATP was 2.5, the pH was 7.1, the ionic strength was 0.16 M with K^+ and Cl^- as major ionic species, and the temperature was 22°C. The solutions contained 12 mM phosphocreatine and 15 U ml^{-1} creatine phosphokinase. The first aspiration-injection series failed to trigger a Ca^{2+}-induced release of Ca^{2+}. Consequently a spontaneous Ca^{2+} release was observed. The third aspiration-injection series decreased the free Ca^{2+} concentration, which resulted in a 'paradoxical' enhancement of the spontaneous Ca^{2+} release. Reproduced from Fig. 2 of Fabiato (1985d) with authorization of Steinkopff Verlag.*

After a few seconds of cyclic contractions, the free Ca^{2+} concentration was rapidly decreased from pCa 5.15 to pCa 6.15 (third aspiration-injection series in Fig. 5). This decreased the frequency and amplitude of the cyclic contractions (Fig 5). But, the first contraction transient just following the decrease in free Ca^{2+} concentration was considerably increased in amplitude (Fig. 5). This observation could suggest that the spontaneous Ca^{2+} release is facilitated by an increase of the gradient of free Ca^{2+} concentration across the sarcoplasmic reticulum membrane. Alternatively, the transient decrease of free Ca^{2+} concentration to pCa 6.15 may have removed the inhibition of the Ca^{2+}-induced release of Ca^{2+}, which could permit early part of the spontaneous Ca^{2+} release just following the decrease of free Ca^{2+} to trigger a Ca^{2+}-induced release of Ca^{2+}. Thus, one cannot exclude the possibility that, even if the spontaneous release is initiated from inside the sarcoplasmic reticulum membrane rather than from outside, it may be subsequently amplified by a Ca^{2+}-induced release of Ca^{2+}

triggered from outside the sarcoplasmic reticulum membrane (Fabiato 1985a). In addition, as will be explained later, the spontaneous release of Ca^{2+} shares some properties with the Ca^{2+}-induced release of Ca^{2+} detected by single channel recording in sarcoplasmic reticulum incorporated in lipid bilayers (Meissner and Henderson, 1987, Rousseau and Meissner, 1989) which is triggered from the trans side of the bilayer that corresponds to the outer surface of the sarcoplasmic reticulum. Finally, both the time- and Ca^{2+}-dependent Ca^{2+}-induced release of Ca^{2+} and the spontaneous cyclic release of Ca^{2+} from the sarcoplasmic reticulum are inhibited by ryanodine (Fabiato, 1985f). This suggests that they may occur through a common Ca^{2+} channel across the sarcoplasmic reticulum membrane.

The following additional pieces of evidence for different control mechanisms of the Ca^{2+}-induced release of Ca^{2+} that has a time- and Ca^{2+}-dependent activation and inactivation and the spontaneous cyclic release of Ca^{2+} have been reported (see pp. 230-32 in Fabiato, 1985a). (1) The spontaneous cyclic release of Ca^{2+} is not affected by the rate of free Ca^{2+} concentration change at the outer surface of the sarcoplasmic reticulum. Once a spontaneous release has been initiated in the presence of a solution at a sufficiently high free Ca^{2+} concentration, the aspiration of this solution followed about 50ms later by the reinjection of the same solution had no effect at any time during the interval between spontaneous Ca^{2+} releases. (2) A free Ca^{2+} concentration sufficiently high to permit spontaneous cyclic release partly inactivates the Ca^{2+}-induced release of Ca^{2+}. (3) When an increase of free Ca^{2+} concentration supraoptimum for the time- and Ca^{2+}- dependent Ca^{2+}-induced release of Ca^{2+} is applied during the ascending phase of a spontaneous cyclic release, this does not affect the spontaneous release of Ca^{2+}. Therefore, the Ca^{2+}-induced release of Ca^{2+} that has a time- and Ca^{2+}-dependent activation and inactivation and the spontaneous release of Ca^{2+} are triggered by at least partly different mechanisms and should, accordingly, be given different names.

The lack of inhibition of Ca^{2+}-induced release of Ca^{2+} by a high concentration of Ca^{2+} resembles the lack of inhibition by a high Ca^{2+} concentration that is observed during single channel recording from sarcoplasmic reticulum incorporated into a lipid bilayer (Meissner and Henderson, 1987; Rousseau and Meissner, 1989). However, in this preparation of isolated sarcoplasmic reticulum incorporated into a lipid bilayer a high-Ca^{2+}-induced inactivation can be demonstrated in the absence of MgATP (Kawano and Coronado, 1991; Meissner and Henderson, 1987). In addition, the Ca^{2+}-induced release of Ca^{2+} observed for the sarcoplasmic reticulum incorporated into a lipid bilayer is triggered by a slow increase of free Ca^{2+} concentration and is not time dependent, which is an additional similarity with the spontaneous release of Ca^{2+} observed in skinned cardiac cells. The spontaneous release of Ca^{2+} from the sarcoplasmic reticulum in skinned cardiac cell responds to an increase of free Ca^{2+} concentration by an increase of the frequency of the cyclic contractions (Fabiato and Fabiato, 1975a). This may have some relationshiop to the increase of frequency of opening of the channel observed when the free Ca2+ concentration is elevated in the cis-chamber of preparation of isolated sarcoplasmic reticulum incorporated into a lipid bilayer. (Meissner and Henderson, 1987; Rousseau and Meissner, 1989). Finally, both spontaneous release of Ca^{2+} in skinned cardiac cells and the Ca^{2+}-induced release of Ca^{2+} detected in sarcoplasmic reticulum vesicles incorporated into a lipid bilayer are enhanced by caffeine (Fabiato and Fabiato 1972, 1975a and Rousseau and Meissner, 1989). Therefore, the Ca^{2+}-induced release of Ca^{2+} from sarcoplasmic reticulum incorporated into a lipid bilayer is much more similar to the spontaneous release of Ca^{2+} in skinned cardiac cells than to the phenomenon I have described as Ca^{2+}-induced release of Ca^{2+} from the sarcoplasmic reticulum in skinned cardiac cells and which has a time and calcium dependence (Fabiato, 1983 and 1985b). For this reason, I am currently attempting to induce a Ca^{2+} release into lipid bilayer by flash photolysis of a caged Ca^{2+} compound to permit a rapid increase of free Ca^{2+} concentration in the *cis* chamber; i.e. at the outer surface of the sarcoplasmic reticulum. The goal is to check whether one caon observe in this preparation a Ca^{2+} release with time- and Ca^{2+}-dependent activation and inactivation similar to that observed in skinned cardiac cells.

Another characteristic in which the spontaneous release of Ca^{2+} and Ca^{2+}-induced release of Ca^{2+} differs is that the spontaneous release of Ca^{2+} is affected by Ins(1,4,5)P_3. An application of Ins(1,4,5)P_3 increases, the frequency and amplitude of the cyclic contractions (Fabiato, 1986, 1990; and Fig. 6). This result is similar to that reported by Nosek et al (186) in chemically-skinned cardiac muscle.

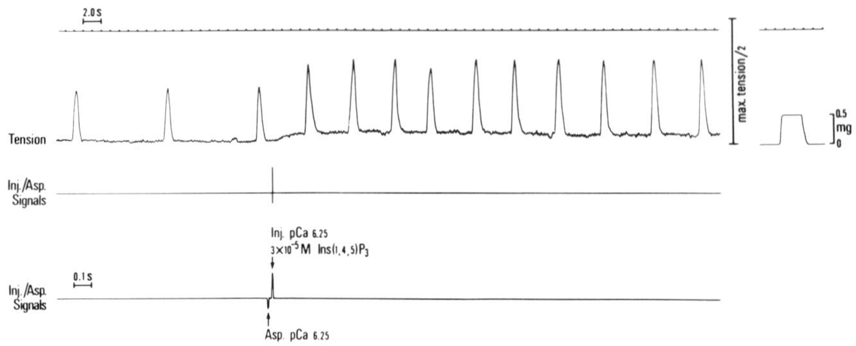

Fig. 6 Tension recording in rat ventricular skinned cell. Cyclic contractions caused by spontanous release of Ca^{2+} from the sarcoplasmic reticulum observed in the continous presence of a free Ca^{2+} concentration of 6.25. The inclusion of 10^{-5} M Ins(1,4,5)P_3 in the solution enhanced the frequency and amplitude of the cyclic contractions.

As previously mentioned at the beginning of this second major section of the article, the spontaneous release of Ca^{2+} from the sarcoplasmic reticulum of skinned cardiac cells is often associated to the propagation of contractile waves (Fabiato and Fabiato, 1972). The occurrence of propagated contractile waves also requires a Ca^{2+} overload of the sarcoplasmic reticulum. The spontaneous Ca^{2+} release from one area of the sarcoplasmic reticulum could induce an overload of adjacent areas causing them to release Ca^{2+} which would in turn overload other areas (Fabiato & Fabiato, 1972). This 'autocatalytic' spontaneous Ca^{2+} release is not, however, the only factor controlling the propagation of contractile waves. Remaining pieces of sarcolemma also play a role, either by propagating slow Ca^{2+} action potential or by cyclically releasing rebinding Ca^{2+}. Propagated contractile waves are never observed in a skinned single cardiac cell (Fabiato & Fabiato, 1973). This could have been explained by the rapid and homogenous diffusion of the bathing solution into all areas of these small preparations. It is, however, possible to obtain skinned cells with a large cross section (up to 40 μm in width) from the canine cardiac Purkinje tissue, which also presents the advantage to be devoid of transverse tubules. A localized application of Ca^{2+} by microiontophoreseis at one end of such a large skinned single cell from the canine Purkinje tissue causes repeated localized contractions related to spontaneous Ca^{2+} release from the sarcoplasmic reticulum but no propagated contractile waves (see p. 491 in Fabiato, 1985e). In contrast, propagated contractile waves are readily obtained when the same localized Ca^{2+} iontophoresis is applied at one end of a single canine cardiac Purkinje cell with sarcolemma disrupted at the ends but present laterally.

Neither spontaneous cyclic contractions nor propagated contractile waves were observed in intact, enzymatically isolated adult cardiac cells (Fabiato & Fabiato, 1972). Therefore the spontaneous cyclic release of Ca^{2+} was deemed unphysiological. Quiescence in the absence of electrical stimulation was one of our major criteria for intactness of isolated adult cells (Fabiato & Fabiato, 1972). This criterion has now been adopted by others (Brady *et al.*, 1979; Altschuld *et al.*, 1980; Krueger *et al.*, 1980). In contrast, damaged, enzymatically separated, adult cardiac cells presented spontaneous cyclic contractions and contractile waves similar to those observed in Ca^{2+}-overload groups of cells with disrupted sarcolemma (Fabiato & Fabiato, 1972). These observations were confirmed and extended by more recent experiments with enzymatically separated adult cardiac cells (Fabiato, 1981, 1985d) and by a series of studies from Professor Inesi's group on isolated adult cardiac cells with leaky sarcolemma, which included elegant computer image analysis of the contractile waves (Rieser *et al.*, 1979), correlation with Ca^{2+} transport measurements (Dani *et al.*, 1979) and electron probe analysis (Chiesi *et al.*, 1981). The spontaneous cyclic release of Ca^{2+} from the sarcoplasmic reticulum in multicellular cardiac preparations overloaded with Ca^{2+} by digitalis intoxication strengthens the hypothesis proposed for this spontaneous cyclic release (Kass *et al.*, Tsien *et al.*, 1978, 1979; Kass & Tsien, 1982; Orchard *et al*, 1983). This phenomenon may play a major role in the generation of arrhythmias during digitalis intoxication and under other pathological conditions increasing the intracellular free Ca^{2+} concentration (see Discussion in Fabiato & Fabiato, 1972, and Fig. 15 in Fabiato & Fabiato, 1978)

A different appraisal of the spontaneous cyclic release of Ca^{2+} from the cardiac sarcoplasmic reticulum has been given in a series of elegant studies from the group of Dr. Lakatta. They have more recently observed cyclic contractions and contractile waves in isolated adult cardiac myocytes (Capogrossi *et al.*, 1984a, b). They demonstrated that these cyclic contractions have the same characteristics as the fluctuations of scattered light intensity that they had previously described in nonstimulated papillary muscles with a powerful methodology (Lakatta & Lappé, 1981; Stern *et al.*, 1981, 1983). Because the scattered light intensity fluctuations were observed in papillary muscles that would be termed 'intact' by most of the usual criteria, Dr. Lakatta and his collaborators have suggested that the spontaneous cyclic release of Ca^{2+} from the sarcoplasmic reticulum may be physiological (Stern *et al.*, 1981, 1983). Thoir attractive hypothesis that the diastolic myoplasmic free Ca^{2+} concentration is continuously variable has important implications for interpreting some physiological measurement, phenomena and properties of cardiac muscle including the diastolic compliance (Stern *et al.*, 1983).

It may be suggested, however, that multicellular preparations are considered as 'intact' only insofar as the great majority of their cells are indeed intact. A few damaged, and hence Ca^{2+}-overloaded, cells may cause misleading phenomena. Cells are often damaged in the area near each end of a papillary muscle. A speckle movement in the light diffraction pattern (Krueger & Pollack, 1975), which is likely to be the same phenomenon as the scattered light intensity fluctuations described by Lakatta & Lappé (1981) and Stern *et al.*, (1981, 1983), is observed preferentially in this area near each end (see p. 631 in Krueger & Pollack, 1975)). The core of a papillary muscle will be hypoxic if it is too thick. The surface of papillary muscle bleaches during superfusion with 'physiological' saline solutions, which indicates that at least the superficial layer(s) of cells are damaged. Under such superfusion, the action potential loses part of the initial spike seen under blood perfusion (Guilbault *et al.*, 1964) and its conduction velocity decreased (Rosen et al., 1972) Early efforts to improved 'physiological' solutions, for instance by including serum albumin (Page, 1962), are seldom taken into consideration. The same bleaching and electrophysiological alterations are observed at the epicardial surface of the whole heart perfused by the Langendorff method and immersed in a usual 'physiological' saline solution (Guilbault *et al.*, 1964).

Even if the sarcomere length intensity fluctuations were not a physiological phenomenon, the hypothesis of Stern *et al.*, (1983) may very well be correct. The previous observations by others of changes of diastolic compliance during positive inotropic interventions, which Stern *et al.* (1983) explain by sarcomere length intensity fluctuation, may have been made, in fact, under nonphysiological conditions causing a Ca^{2+} overload of the sarcoplasmic reticulum. High concentrations of isoproterenol also caused a spontaneous cyclic release of Ca^{2+} in stimulated intact cardiac cells, which were quiescent by electrical stimulation in the presence of a physiological extracellular free Ca^{2+} concentration (see Fig. 6c in Fabiato, 1981).

From this review of its characteristics, it appears that the spontaneous release of Ca^{2+} from the sarcoplasmic reticulum, which is time-independent and not inhibited by high Ca^{2+} concentration, is probably not a physiological process but may be of great pathophysiological interest.

ACKNOWLEDGEMENT

This work is supported by Grant No. 90-1207 from the American Heart Association.

REFERENCES

Altschuld, R., Gibb, L., Ansel, A., Hohl, C., Kruger, F.A. and Brierley, G.P. (1980) Calcium tolerance of isolated rat heart cells. *J. Mol. Cell. Cardiol.* 12, 1383-95.

Bers, D.M. (1985) Ca influx and sarcoplasmic reticulum Ca release in cardiac muscle activation during postrest recovery. *Am. J. Physiol.* 248, H366-81.

Bers, D.M., Philipson, K.D. and Langer, G.A. (1981) Cardiac contractility and sarcolemmal calcium binding in several cardiac muscle preparations. *Am. J. Physiol.* 240, H576-83.

Bianchi, C.P. & Shanes, A.M. (1959) Calcium influx in skeletal muscle at rest, during activity, and during potassium contracture. *J. Gen. Physiol.* 42, 803-15.

Brady, A.J., Tan, S.T. & Ricchiuti, N.V. (1979) Contractile force measured in unskinned isolated adult rat heart fibres. *Nature, Lond.* 282, 728-9.

Capogrossi, M.C., Fraticelli, A and Lakatta, E.G. (1984a) Ca^{2+} release from the sarcoplasmic reticulum: Different effects of ryanodine and caffeine. *Fedn. Proc. Fed. Am. Socs. Exp. Biol.* 43, 820.

Capogrossi, M.C., Kort, A. A., Spurgeon, H.A., Suarez-Isla, B.A. and Lakatta, E.G. (1984b) Spontaneous contractile waves and simulated contractions exhibit the same Ca^{2+} and species dependence in isolated cardiac myocytes and papillar muscles. *Biophys.* 45, 94a.

Chapman, R.A. (1983) Control of cardiac contractility at the cellular level. *Am. J. Physiol.* 245, H535-52.

Chiesi, M., Ho., M.M., Inesi, G., Somlyo, A.V. and Somlyo, A.P. (1981) Primary role of sarcoplasmic reticulum in phasic contractile activation of cardiac myocytes with shunted myolemma. *J. Cell Biol.* 91, 728-42.

Costantin, L.L and Taylor, S.R. (1973) Graded activation in frog muscle fibers. *J. Gen. Physiol.* 61, 424-43.

Dani, A.M., Cittadini, A. and Inesi, G. (1979) Calcium transport and contractile activity in dissociated mammalian heart cells. *Am. J. Physiol.* 237, C147-55.

Endo, M. (1977) Calcium release from the sarcoplasmic reticulum. *Physiol. Rev.* 57, 71-108.

Endo,M. (1985) Calcium release from sarcoplasmic reticulum. In Current Topics in Membranes and Transport, Vol 25, 181-230.

Endo, M., Tanaka, M. and Ogawa, Y. (1970) Calcium induced release of calcium from the sarcoplasmic reticulum of skinned skeletal muscle fibres. *Nature, Lond.* 228, 34-6.

Fabiato, A. (1981) Myoplasmic free calcium concentration reached during the twitch of an intact isolated cardiac cell and during calcium-induced release of calcium from the sarcoplasmic reticulum of a skinned cardiac cell from the adult rat or rabbit ventricle. *J. Gen. Physiol.* 78, 457-97.

Fabiato, A. (1982a) Calcium release in skinned cardiac cells: Variations with species, tissues and development. *Fedn Am. Socs Exp. Biol.* 41, 2238-44.

Fabiato. A. (1982b) Fluorescence and differential light absorption recordings with calcium probes and potential-sensitive dyes in skinned cardiac cell. Can. J. Physiol. Pharmac. 60, 556-67.

Fabiato, A. (1983) Calcium-induced release of calcium from the cardiac sarcoplasmic reticulum. *Am. J. Physiol.* 245, C1-14.

Fabiato, A. (1985a) Rapid ionic modifications during the aequorin-detected calcium transient in a skinned canine cardiac Purkinje cell. *J. Gen. Physiol.* 85, 189-246.

Fabiato, A. (1985b) Time and calcium dependence of activation and inactivation of calcium-induced release of calcium from the sarcoplasmic reticulum of a skinned canine cardiac Purkinje cell. *J. Gen. Physiol.* 85, 247-89.

Fabiato, A. (1985c) Simulated calcium current can both cause calcium loading in and trigger calcium release from the sarcoplasmic reticulum of a skinned canine Purkinje cell. *J. Gen. Physiol.* 85, 291-320.

Fabiato, A. (1985d) Spontaneous versus triggered contractions of 'calcium-tolerant' cardiac cells from the adult rat ventricle. *Basic Res. Cardiol.* 80, suppl. 2, 178-82.

Fabiato, A. (1985e) Appraisal of the hypothesis of the 'depolarization-induced' release of calcium from the sarcoplasmic reticulum in skinned cardiac cells from the rat or pigeon ventricle. In Structure and Function of the Sarcoplasmic Reticulum, ed. by Fleischer, S. and Tonomura, Y., pp. 479-519. New York: Academic Press.

Fabiato, A. (1985f) Effects of ryanodine on skinned cardiac cells *Fedn. Am. Socs. Exp. Biol.* 44, 2970-2976.

Fabiato, A. (1986) Inositol(1,4,5)-trisphosphate-induced versus Ca^{2+}-induced release of Ca^{2+} from the cardiac sarcoplasmic reticulum. *Proc. Int. Union Physiol. Sci.* 16, 350.

Fabiato, A. (1989) Appraisal of the physiological relevance of two hypotheses for the mechanisms of calcium release from the mammalian cardiac sarcoplasmic reticulum; calcium-induced release versus charge-coupled release. *Mol. Cell. Biochem.* 89, 135-140.

Fabiato, A. (1990) Comparison and relation between inositol(1,4,5)-trisphosphate-induced release and calcium-induced release of calcium from the cardiac sarcoplasmic reticulum. In Recent Advances in Calcium Channels and Calcium Antagonists, ed. by K. Yamada and S. Shibata, New York: Pergamon Press.

Fabiato, A. and Fabiato, F. (1972). Excitation-contraction coupling of isolated cardiac fibers with disrupted or closed sarcolemmas. Calcium-dependent cyclic and tonic contractions. *Circ. Res.* 31, 293-307.

Fabiato, A. and Fabiato F. (1973) Activation of skinned cardiac cells: subcellular effects of cardioactive drugs. *Eur. J. Cardiol.* 1/2, 143-55.

Fabiato, A. and Fabiato F. (1975a) Contractions induced by a calcium-triggered release of calcium from the sarcoplasmic reticulum of single skinned cardiac cells. *J. Physiol., Lond.* 249. 469-95.

Fabiato, A. and Fabiato F. (1975b) Effects of magnesium on contractile activation of skinned cardiac cells. *J. Physiol., Lond.* 249. 469-95.

Fabiato, A. and Fabiato F. (1977) Calcium release from the sarcoplasmic reticulum. *Circ. Res.* 40, 119-29.

Fabiato, A. and Fabiato F. (1978) Calcium-induced release of calcium from the sarcoplasmic reticulum of skinned cells from adult human, dog, cat, rabbit, rat, and frog hearts and from fetal and new-born rat ventricles. *Ann. N.Y. Acad. Sci.* 307, 491-522.

Fabiato, A. and Fabiato, F. (1979) Calculator programs for computing the composition of the solutions containing multiple metals and ligands used for experiments is skinned muscle cells. *J. Physiology, Paris* 75, 463-505.

Ford, L.E. and Podolsky, R.J. (1970) Regenerative calcium release within muscle cells. *Science, N.Y.* 167, 58-9.

Ford, L.E. and Podolsky, R.J. (1972a) Calcium uptake and force development by skinned muscle fibres in EGTA buffered solutions. *J. Physiol., Lond.* 223, 1-19.

Ford, L.E. and Podolsky, R.J. (1972b) Intracellular calcium movements in skinned muscle fibres. *J. Physiol., Lond.* 223, 21-33.

Frank, G.B. (1960) Effects of changes in extracellular calcium concentration on the potassium-induced contracture of frog's skeletal muscle. *J. Physiol., Lond.* 151, 518-38.

Frank, G.B. (1980) The current view of the source of trigger calcium in excitation-contraction coupling in vertebrate skeletal muscle. *Biochem. Pharmac.* 29, 2399-406.

Frank, G.B. (1982) Roles of extracellular and 'trigger' calcium ions in excitation-contraction coupling in skeletal muscle. *Can. J. Physiol. Pharmac.* 60, 427-39.

Guilbault, P., Coraboeuf, E. and Loulergue, F. (1964): Possibilités d'étude de l'activité électrique et mécanique du coeur isolé de lapid dans des conditions de survie de longue durée. *C.R. Soc. Biol.* 158, 978-82.

Kass, R.S. and Tsien, R.W. (1982) Fluctuations in membrane current driven by intracellular calcium in cardiac Purkinje fibers. *Biophys. J.* 38, 259-69.

Kass, R.S., Lederer, W.J., Tsien, R.W. and Weingart, R. (1978): Role of calcium ions in transient inward currents and aftercontractions induced by strophanthidin in cardiac Purkinje fibres. *J. Physiol., Lond.* 281, 187-208.

Kawano, S. and Coronado, R. (1991) Ca^{2+} dependence of Ca^{2+} relase channel activity in the sarcoplasmic reticulum of cardiac and skeletal muscle. *Biophysical Journal* 59:600a.

Kruger, J.W., Forletti, D. and Wittenberg, B.A, (1980) Uniform sarcomere shortening behavior in isolated cardiac muscle cells. *J. Gen. Physiol.* 76, 587-607.

Krueger, J.W. and Pollack, G.H. (1975) Myocardial sarcomere dynamics during isometric contraction. *J. Physiol., Lond.* 251, 627-43.

Kushmerick, M.J. and Podolsky, R.J. (1969) Ionic mobility in muscle cells. *Science, N.Y.* 166, 1297-8.

Lakatta, E.G. and Lappé, D.L. (1981) Diastolic scattered light fluctuation, resting force and twitch force in mammalian cardiac muscle. *J. Physiol., Lond.* 315, 369-94.

Meissner, G. and Henderson, J.S. (1987): Rapid calcium release from cardiac sarcoplasmic reticulum vesicles is dependent on Ca^{2+} and is modulated by Mg^{2+}, adenine nucleotide, and calmodulin. *J. Biol. Chem.* 262, 3065-73.

Nosek, T.M., Williams, M.F., Ziegler, S.T. and Godt, R.E. (1986) Inositol trisphosphate enhances calcium release in skinned cardiac and skeletal muscle. *Am. J. Physiol.* 250: C807-11.

Orchard, C.H,, Eisner, D.A. and Allen, D.G. (1983) Oscillations of intracellular Ca^{2+} in mammalian cardiac muscle. *Nature*, Lond. 304, 735-8.

Page, E. (1962) Cat heart muscle *in vitro*. II. The steady state resting potential in quiescent papillary muscles. *J. Gen. Physiol.* 46, 189-99.

Reiser, G., Sabbadini, R., Paolini, P., Fry, M. and Inesi, G. (1979) Sarcomere motion in isolated cardiac cells. *Am. J. Physiol.* 236, C70-7.

Rosen, M.R., Gelband, H. and Hoffman, B.F. (1972) Effects of blood perfusion on electrophysiological properties of isolated canine Purkinje fibers, *Circ. Res.* 30: 575-87.

Rousseau, E. and Meissner, G. (1989) Single cardiac sarcoplasmic reticulum Ca^{2+}-release channel: activation by caffeine, *Am. J. Physiol.* 256, H328-33.

Saida, K. (1982) Intracellular Ca release in skinned smooth muscle. *J. Gen. Physiol.* 80, 191-202.

Stephenson, E.W. (1981a) Activation of fast skeletal muscle: Contributions of studies on skinned fibers. *Am. J. Physiol.* 240, C1-19.

Stern, M.D., Kort, A.A., Bhatnagar, G.M. and Lakatta, E.G. (1981) Laser scattering fluctuations in non-beating rat myocardium related to spontaneous myoplasmic calcium oscillations. *Biophys. J.* 33, 284a.

Stern, M.uD., Kort, A.A., Bhatnagar, G.M. and Lakatta, E.G. (1983) Scattered-light intensity fluctuations in diastolic rat cardiac muscle caused by spontaneous Ca^{2+}-dependent cellular mechanical oscillations. *J. Gen. Physiol.* 82, 199-53.

Sutko, J.L. and Willerson J.T. (1980) Ryanodine alteration of the contractile state of rat ventricular myocardium. Comparison with dog, cat, and rabbit ventricular tissues. *Circ. Res.* 46, 332-43.

Tsien, R.W., Kass, R.S. and Weingate, R. (1978) Calcium ions and membrane current changes induced by digitalis in cardiac Purkinje fibers. *Ann. N.Y. Acad. Sci.* 307: 483-90.

Wong, A.Y.K., Fabiato, A., Bassingthwaighte, J.B., (1987): Model of Ca release mechanisms from the sarcoplasmic reticulum: Ca-mediated activation, inactivation and reactivation. In Electromechanical Activation, Metabolism and Perfusion of the Heart, Simulation and Experimental Models, ed. by S. Sideman and R. Beyer, pp. 281-295. Dordrecht: Martinus Nijhoff Press.

THE SARCOPLASMIC RETICULUM OF SKELETAL MUSCLE: A LOOK FROM INSIDE

Pompeo Volpe, Adelina Martini and Alessandra Nori

Centro di Studio per la Biologia e Fisiopatologia Muscolare del CNR, Istituto di Patologia Generale, Universita' degli Studi di Padova, via Trieste 75, 35121 PADOVA, Italy

INTRODUCTION

The sarcoplasmic reticulum (SR), the membrane-bound calcium store of skeletal muscle, controls the contraction-relaxation cycle by raising and lowering the myoplasmic free calcium concentration, and consists of two continuous yet distinct regions, the free or non-junctional SR and the junctional SR, i.e., the area of terminal cisternae (TC) directly facing the transverse tubules (Costello et al., 1986; Fleischer and Inui, 1989).

Skeletal muscle contraction is elicited by the release of calcium via calcium release channels confined to the junctional area of TC (Fleischer and Inui, 1989). After calcium-pump mediated active uptake, calcium is largely stored in the TC lumen where it is believed to be bound, at least in part, to calsequestrin (CS), a low affinity, high capacity calcium binding protein (MacLennan and Wong, 1971; Fliegel et al., 1987). Despite some key questions, which are still being investigated, the basic phenomenology of both calcium uptake and calcium release has been studied in depth in the last twenty years (for review see Fleischer and Inui, 1989).

The SR lumenal content has been observed by electron microscopy, yet its nature has remained elusive: homogeneous, nodular material has been described in the non-junctional region (see Franzini-Armstrong et al., 1987), and cristalline arrays in the junctional region (Saito et al., 1984; Costello et al., 1986). Calsequestrin, the "major" intralumenal SR protein, plays an important role in buffering intralumenal calcium, although the intimate mechanisms involved in calcium binding and unbinding and their relationship to calcium release to the myoplasm are yet to be clarified (cfr. Volpe and

Excitation-Contraction Coupling in Skeletal, Cardiac, and Smooth Muscle
Edited by G.B. Frank *et al.*, Plenum Press, New York, 1992

Simon, 1991). The function of several, "minor" intralumenal SR proteins, which have been discovered and partially characterized in the last fifteen years, is largely unknown (cfr. Cala et al., 1990). In this article, we shall review the topic of intralumenal SR proteins with emphasis on the control of intralumenal [Ca^{2+}].

INTRALUMENAL SR PROTEINS

There are several intralumenal SR proteins, either extrinsic or internal loops of intrinsic (integral membrane) proteins (Table 1). Accepted criteria to define intralumenal, extrinsic proteins are: a) extractability from SR vesicles by EGTA, EDTA or Na-carbonate at alkaline pH and hypotonic conditions (Cala and Jones, 1983; Cala et al., 1990); b) inaccessibility to trypsin in intact SR vesicles; c) lack of labeling with photoactivatable phospholipid analogues (Volpe et al., 1987); d) lack of stretches of hydrophobic amino acid sequences, i.e., putative transmembrane sequences.

Table 1. Intralumenal Proteins of Skeletal Muscle SR

	Mr deduced	Junctional	Non-junctional	KDEL	Glycosylated
Extrinsic Proteins					
CS (skeletal)	41,630	++	--	no	yes
CS (cardiac)	45,269	++	--	no	yes
Sarcalumenin	95,710	--	++	no	yes
53 KDa glycoprotein	52,241	--	++	no	yes
HCP [a]	92,933	+ [a]	+	no	no
CR	46,567	++ [b]	--	yes	yes
55 KDa T_3BP [c]	56,681	?	?	yes	yes
BiP	78,000 [d]	?	?	yes	yes
Intrinsic Proteins					
Calcium pump	109,763	--	++		
Ryanodine receptor	563,584	++	--		
CS-binding proteins	26,000 [e]	++	--		
	30,000 [e]	++	--		

a Damiani and Margreth (1991), on the contrary, provided evidence that HCP is not intralumenal and is preferentially located in the junctional SR

b Based on immunofluorescence staining of frozen sections (Smith and Koch, 1989; Fliegel et al., 1989a)

c RDEL instead of KDEL (Fliegel et al., 1990)

d Apparent molecular weight (see Fig. 1)

e Apparent molecular weight (cfr. Mitchell et al., 1988; Damiani and Margreth, 1990)

Intralumenal extrinsic SR proteins are listed in Table 1. There are two isoforms of CS, a skeletal type expressed in both fast- and slow-twitch muscles (Fliegel et al., 1989b; Damiani et al., 1990) and a cardiac type exclusively expressed in slow-twitch muscle (Scott et al., 1988; Fliegel et al., 1989b; Damiani et al., 1990), which are the products of two different genes. CS is restricted to the junctional SR and is a major component of the electron-dense material observed in thin-section electron microscopy (Saito et al., 1984; Costello et al., 1986). Both CS isoforms are highly acidic proteins due to the presence of 29% acidic amino acid residues (Fliegel et al., 1987; Scott et al., 1988), and undergo substantial conformational changes upon calcium binding. Both CS isoforms are glycoproteins whose complex carbohydrate chains (MacLennan et al., 1983) appear to require processing through the Golgi apparatus.

Sarcalumenin (Leberer et al., 1989b) is a glycoprotein which contains a highly acidic NH_2-terminus (25% of acidic residues) and is localized in the non-junctional SR (Leberer et al., 1990). By alternative splicing, the sarcalumenin gene yields a second gene product, the 53 KDa glycoprotein, a truncated protein of Mr 52,421 which lacks the tail of acidic amino acid residues (Leberer et al., 1989a, 1990). Both proteins are localized in the non-junctional SR (Leberer et al., 1990).

The histidine-rich calcium binding protein (HCP) has been identified on the account of its ability to bind, in vitro, low density lipoproteins (LDL) with high affinity. HCP appears to be the only intralumenal SR proteins not to be glycosylated. HCP has been reported to be localized intralumenally, both in the junctional and free SR (Hofmann et al., 1989a), and to have no putative transmembrane sequences (Hofmann et al., 1989b). A different conclusion was reached by Damiani and Margreth (1991) who showed that the LDL-binding protein was susceptible to trypsin digestion, i.e., it appeared to be an extralumenal protein, and was subfractionated to the junctional SR.

Calsequestrin, sarcalumenin and HCP are selectively phosphorylated by casein kinase II (Cala et al., 1990; Cala and Jones, 1991). It has been suggested that phosphorylation occur in the Golgi complex and may play a role in protein targeting to the SR (Cala et al., 1990; Cala and Jones, 1991).

Calreticulin [CR, previously referred to as CRP55, high-affinity calcium binding protein (HACBP) or CAB-63; Ostwald and MacLennan, 1974, Waisman et al., 1985, Macer and Koch, 1988, Smith and Koch, 1989, Fliegel et al., 1989c] is an acidic protein (26% of acidic residues). Analysis of primary sequence reveals a higly acidic COOH-terminus (37 out of 52 residues) following a proline-rich segment. Primary structure homology between CS and CR exists only within two short amino acid stretches (Smith and Koch, 1989).

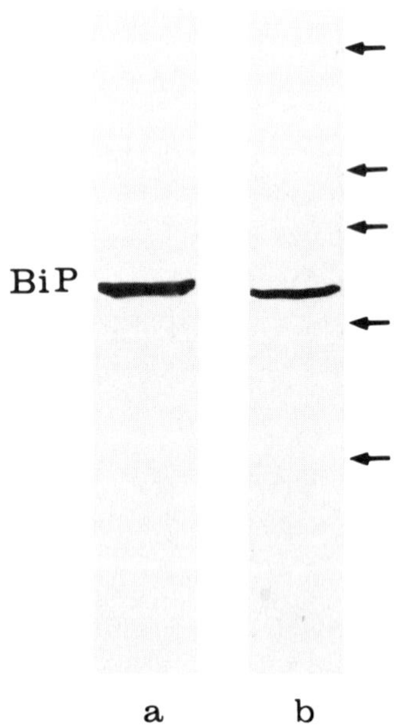

Fig. 1. Immunological evidence for the presence of BiP in SR-enriched membrane fractions of rabbit fast-twitch skeletal muscle. Chicken cerebellum microsomes (lane a) and rabbit skeletal muscle SR (lane b) were prepared as described by Volpe et al. (1990) and Saito et al. (1984), respectively. One hundred ug of protein was loaded on a 5-10% linear polyacrylamide gradient gel, transferred to nitrocellulose and decorated with anti-BiP antibodies (Bole et al., 1986), as previously described (Volpe et al., 1990). Arrows indicate the position of Bio-Rad high molecular weight standards (in daltons): myosin 200,000; galactosidase 116,250; phosphorylase b 97,400; bovine serum albumin 66,200; ovalbumin 42,700.

The last additions to an ever-growing list, are the 55 KDa multifunctional T3BP [also referred to as ERp 59 or PDI (Protein Disulfide Isomerase); Fliegel et al., 1990] and BiP (Fig. 1). BiP (*Bi*nding *P*rotein), a resident intralumenal endoplasmic reticulum (ER) protein (Bole et al., 1986), has been identified in SR-enriched vesicles isolated from rabbit fast-twitch skeletal muscle. Fig. 1 shows the Western blot of cerebellum microsomal fraction (lane a) and SR vesicles (lane b) decorated with monoclonal anti-BiP antibodies (Bole et al., 1986). Both membrane fractions contain a polypeptide having an apparent molecular weight of about 78,000 daltons. In skeletal muscle fiber, it remains to be ascertained whether BiP is located in the lumen of the SR or within discrete (ER ?) compartments. In other cell types and cell lines, BiP appears to be localized not only in the lumen of rough-surfaced ER but also in that of smooth-surfaced ER (Tooze et al., 1989; Bole et al., 1990; Villa et al., 1991).

By far the most abundant of intralumenal extrinsic proteins is CS, and with the exception of the 53 KDa glycoprotein, all of them are calcium binding proteins or bear calcium binding sites (see below). The function of the 53 KDa glycoprotein is unknown; the suggested stimulatory effect on the coupling ratio of the calcium pump (Leonards and Kutchai, 1985) and on the regulation of calcium transport (Chiesi and Carafoli, 1982) has been questioned by Leberer et al. (1989b) on the account of experiments of cotransfection of COS-1 cells with cDNAs of both the calcium pump and 53 KDa glycoprotein.

All other extrinsic proteins, besides playing a role in intralumenal calcium homeostasis (see below), may have additional functions: for instance the 55 KDa multifunctional T3BP could be endowed with disparate activities such as disulfide isomerase, prolyl 4-hydroxylase, oligosaccharyl tranferase [all enzymatic activities involved in the post-translational modification of newly-translocated polypeptides in the ER (Gething and Sambrook, 1990)], and thyroid hormone binding. Thus, as for the homologous ER proteins of non-muscle cells, skeletal muscle 55 KDa T3BP and BiP may facilitate oligomerization and proper folding/refolding of both membrane proteins and extrinsic proteins (Gething and Sambrook, 1990; Sambrook, 1990) in the SR lumen. For instance, BiP has been shown to play "an important role in conformational maturation and/or editing of unassembled acetylcholine receptors subunits in vivo" in the muscle-like cell line BC3H-1 (Blount and Merlie, 1991).

Internal loops of integral membrane proteins can be considered as additional intralumenal proteins. Among them, 4-10 amino acid stretches of the calcium release channel ("ryanodine receptor") in the junctional SR (Takeshima et al., 1989; Zorzato et al., 1990); 6 stretches of the calcium pump in the non-junctional SR (MacLennan et al., 1985). Internal loops of both the calcium pump (MacLennan et al., 1985) and ryanodine receptor (Zorzato et al., 1990) contain acidic amino acid residues which might account for the existence of low affinity calcium binding sites (see Table 2). Polypeptides having apparent MW of 26,000-30,000 daltons (cfr. Volpe et al., 1987) have been described as putative CS-binding proteins by Mitchell et al. (1988) and Damiani and Margreth (1990), and might correspond to subunits of the "strands" connecting CS to the internal layer of the junctional SR membrane (cfr. Franzini-Armstrong et al., 1987).

SPECIFIC SR AND "SPECIFIC" ER PROTEINS

Porter and Palade (1957) first described the SR of skeletal muscle as a specialized form of ER. On the basis of several, elegant studies, a plausible scheme for the biogenesis of SR would be as follows: the non-junctional SR forms as tubular extension (vesicle-budding) of rough ER (Ezerman and Ishikawa, 1967), whereas clathrin-coated CS-containing vesicles (junctional SR ?) originate from the cis/medial Golgi complex (Thomas et al., 1989); discrete ryanodine-receptor containing vesicles (junctional SR) appear to fuse with forming free SR at times coincident with transverse tubules formation (cfr. Fig. 8 in Yuan et al., 1991). During early stages of (rat) muscle ontogenesis (Schiaffino and Margreth, 1969) and during differentiation of (chick) muscle cells in vitro (Ezerman and Ishikawa, 1967), continuity of smooth-surfaced and rough-surfaced membranes was observed. In mature skeletal muscle fibers, continuity between SR and the scanty, perinuclear rough ER (Fambrough and Devreotes, 1978) was never (seldom) observed. Thus, the SR has long been portrayed as a separate cytological entity, different from the ER in structure, function and composition.

The SR, however, appears to express CR (Fliegel et al., 1989a, c; Smith and Koch, 1989), the 55 KDa multifunctional T_3BP (Fliegel et al., 1990) and BiP (Fig. 1), proteins which contain at the COOH terminus the K(R)DEL sequence (see Table 1), i.e., the sequence that serves as a signal for the selective retention of ER proteins, possibly through recycling via a pre-Golgi compartment (Munro and Pelham, 1987). CR, PDI, BiP along with endoplasmin constitute the specific complement of ER resident proteins in non-muscle cells, and have been collectively referred to as reticuloplasmins (Koch, 1987). The occurrence of, at least, three "specific" ER proteins in SR-derived vesicles forces to reconsider the notion of two separate cytological entities, pending the results of immunolabeling experiments, which are in progress: the subcellular localization of "ER" proteins in relation to that of "typical" SR proteins, i.e., CS, calcium pump and ryanodine-sensitive calcium release channel, should clarify this issue.

CALCIUM BINDING PROTEINS

Table 2 lists known intralumenal SR calcium binding proteins. Some of them, CS, CR, sarcalumenin and internal loops of calcium pump, have several, low affinity calcium binding sites and play an obvious role in buffering intralumenal [Ca^{2+}], especially CS and the calcium pump given their abundance, i.e., 10% and 50% of total SR protein. The physiological relevance of low affinity calcium binding proteins present in low amounts is not known.

The high affinity binding site of CR (Table 2) should always be saturated with calcium and is, likely, relevant to the tertiary configuration of the protein. Calcium binding characteristics of BiP, 55 KDa multifunctional T_3BP and HCP are not known, but are expected to be of the low affinity type because of the presence of clusters of asp and glu residues interdispersed among hydrophobic tracts.

Table 2. Calcium Binding Properties of Intralumenal SR Proteins

	^{45}Ca overlay [a]	K_d (mM)	B_{max} (mol/mol)
CS	yes	0.6-1	50
CR	?	5	25
		0.002	1
Sarcalumenin	yes	0.6	35
HCP	yes [b]	?	?
55 KDa T_3BP [c]	?	?	?
BiP [d]			
Calcium pump	yes	1-7 [e]	1-12 [e]
Ryanodine receptor	yes	?	?

[a] Zorzato and Volpe (1988)
[b] Hoffman et al. (1989a)
[c] It may be a calcium binding protein given the acidic COOH-terminal amino acid sequence (Fliegel et al., 1990)
[d] In non-muscle cells, it is a calcium binding protein (Macer and Koch, 1988)
[e] See Fig. 1 in Volpe and Simon (1991)

As shown in Table 1, some calcium binding proteins are preferentially located in the junctional SR, e.g., CS, some are in the non-junctional SR, e.g., calcium pump and sarcalumenin. Leberer et al. (1989a, b; 1990) have suggested the possible formation of calcium binding complexes in the lumen of the non-junctional SR which, in our view, could be in continuity with the CS network present in the TC lumen (Saito et al., 1984; Costello et al., 1986; Franzini-Armstrong et al., 1987). The "calcium binding complex" hypothesis is based on the presence of the motif ser-glu-glu in sarcalumenin, 53 KDa glycoprotein, HCP, CS and 55 KDa multifunctional T_3BP [in a modified form ser-asp-glu (Fliegel et al., 1990)]. Such a sequence appears to play an important role in protein-protein interaction (Earnshaw, 1987) and should be responsible for the calcium-dependent self-association and interaction of intralumenal SR proteins. Calcium bridging may also intervene to connect extrinsic proteins with intrinsic proteins, e.g., via the intralumenal acidic sequence glu-glu-gly-glu-glu of the calcium pump (MacLennan et al., 1985). Complexes and networks of intralumenal proteins may also perform a scaffolding or structural role.

CALCIUM BUFFERING AND INTRALUMENAL $[Ca^{2+}]_i$

The type, relative amount (low versus high) and localization (junctional versus non-junctional) of calcium binding proteins are relevant not only to calcium buffering, but also to regulation of calcium release. Complexes and networks of intralumenal proteins may constitute ionic diffusional barriers and define, although transiently, concentration gradients. For instance, Leberer et al. (1989a) have suggested that "a Ca^{2+} binding complex in the lumen of longitudinal SR could act to reduce the Ca^{2+} gradient built up by the Ca^{2+} pump and consequently accelerate Ca^{2+} transport". Moreover, intralumenal $[Ca^{2+}]_i$ and its changes are known to gate calcium release channels (Valdivia et al., 1989; Nelson and Nelson, 1990) and localized variations of calcium concentrations within a putative sub-junctional compartment might exercise an even finer control on channel gating.

The kinetics of calcium release from TC are determined not only by the gating properties of the calcium release channels but also by the calcium gradient ($[Ca^{2+}]_i/[Ca^{2+}]_o$) across the SR membrane. Such a gradient is established by the calcium pump and thought to depend mainly upon: 1) the affinity and capacity of CS and other intralumenal calcium binding proteins, and 2) rate constants of calcium unbinding which are expected, however, to be fast and not rate-limiting. Direct measurements of intralumenal $[Ca^{2+}]_i$ are not available; estimates of intralumenal $[Ca^{2+}]_i$ have provided values between 2.3 and 3.6 mM (Maylie et al., 1987; Volpe and Simon, 1991, and references therein). If these estimates were to be correct, the $[Ca^{2+}]_i/[Ca^{2+}]_o$ ratio would be around $5x10^4$, a value which is near the theoretical limits achieved by the calcium pump.

Taking into account calcium binding sites on CS and other intralumenal binding sites (Table 2, and Volpe and Simon, 1991), at least 75% of calcium binding sites on CS would be saturated with calcium, under resting conditions. Computer simulations of calcium release have indicated that the bulk of calcium released to the myoplasm, during a twitch, is free in the SR lumen and does not unbind (cfr. Volpe and Simon, 1991). It has been observed that a calcium-dependent conformational change of CS precedes calcium release in vitro (Ikemoto et al., 1989). If 75% of released calcium is not bound to CS, a calcium-dependent conformational change may follow but not precede the opening of calcium release channels (Volpe and Simon, 1991). Only during a tetanus a sizeable amount of calcium released to the myoplasm would come off CS and other calcium binding proteins (Volpe and Simon, 1991).

REFERENCES

Blount, P. and L.J.P. Merlie, 1991, BiP associates with newly synthesized subunits of the mouse nicotinic receptor, J. Cell Biol.,113: 1125.

Bole, D.G., L.M. Hendershot and J.F. Kenney, 1986, Posttranslational association of immunoglobulin heavy chain binding proteins with nascent heavy chains in non secreting and secreting hybridomas. J. Cell Biol., 102: 1558.

Bole, D.G., R. Down, M. Doriaux and J.J. Jamieson, 1990, Immunocytochemical localization of BiP to the rough endoplasmic reticulum: evidence for protein sorting by selective retention. J. Histochem. Cytochem., 37: 1817.

Cala, S.E. and L.R. Jones, 1983, Rapid purification of calsequestrin from cardiac and skeletal muscle sarcoplasmic reticulum vesicles by Ca^{2+}-dependent elution from phenyl-Sepharose. J. Biol. Chem., 258: 11932.

Cala, S.E. and L.R. Jones, 1991, Phosphorylation of cardiac and skeletal muscle calsequestrin isoforms by casein kinase II. J. Biol. Chem., 266: 391.

Cala, S.E., B.T. Scott and L.R. Jones, 1990, Intraluminal sarcoplasmic reticulum Ca^{2+}-binding proteins. Sem. Cell Biol., 1: 265.

Chiesi, M. and E. Carafoli, 1982, The regulation of Ca^{2+} transport by rat skeletal muscle sarcoplasmic reticulum. Role of calmodulin and the 53,000 dalton glycoprotein. J. Biol. Chem., 257: 984.

Costello B., C. Chadwick, A. Saito, A. Chu, A. Maurer and S. Fleischer, 1986, Characterization of the junctional face membrane of terminal cisternae of sarcoplasmic reticulum. J. Cell Biol., 103: 741.

Damiani, E., and A. Margreth, 1990, Specific protein-protein interactions of calsequestrin with junctional sarcoplasmic reticulum of skeletal muscle. Biochem. Biophys. Res. Commun., 172: 1253.

Damiani, E. and A. Margreth, 1991, Subcellular fractionation to junctional sarcoplasmic reticulum and biochemical characterization of the 170 kDa Ca^{2+}- and LDL-binding protein in rabbit skeletal muscle. Biochem. J., in press.

Damiani, E., P. Volpe and A. Margreth, 1990, Coexpression of two isoforms of calsequestrin in rabbit slow-twitch muscle. J. Muscle Res. Cell Motil., 11: 522.

Earnshaw, W.C., 1987, Anionic regions in nuclear proteins. J. Cell Biol., 105:1479.

Ezerman, E.B. and H. Ishikawa, 1967, Differentiation of the sarcoplasmic reticulum and T-system in developing chick skeletal muscle in vitro. J. Cell Biol., 35: 405.

Fambrough, D.M. and P.N. Devreotes, 1978, Newly synthesized acetylcholine receptors are located in the Golgi apparatus. J. Cell Biol., 76: 237.

Fleischer, S. and M. Inui, 1989, Biochemistry and biophysics of excitation-contraction coupling. Annu. Rev. Biophys. Biophys. Chem., 18: 333.

Fliegel, L., K. Burns, M. Opas and M. Michalak, 1989a, The high-affinity calcium binding protein of sarcoplasmic reticulum. Tissue distribution and homology with calregulin. Biochim. Biophys. Acta, 982: 1.

Fliegel, L., E. Leberer, N.M. Green and D.H. MacLennan, 1989b, The fast-twitch muscle calsequestrin isoform predominates in rabbit slow-twitch soleus muscle. FEBS Lett., 242: 297.

Fliegel, L., K. Burns, D.H. MacLennan, R.A.F. Reithmeier and M. Michalak, 1989c, Molecular cloning of the high affinity calcium binding protein (calreticulin) of skeletal muscle sarcoplasmic reticulum. J. Biol. Chem., 264: 21522.

Fliegel, L., E. Newton, K. Burns and M. Michalak, 1990, Molecular cloning of cDNA encoding a 55-KDa multifunctional thyroid hormone binding protein of skeletal muscle sarcoplasmic reticulum. J. Biol. Chem., 265: 15496.

Fliegel, L., M. Ohnishi, M.R. Carpenter, V.H. Khanna, A.F. Reinhart, R.A.F Reithmeier and D.H. MacLennan, 1987, Amino acid sequence of rabbit fast-twitch skeletal muscle calsequestrin deduced from cDNA and peptide sequencing. Proc. Natl. Acad. Sci. (USA), 84: 1167.

Franzini-Armstrong, C., L.J. Kenney and E. Varriano-Marston, 1987, The structure of calsequestrin in triads of vertebrate skeletal muscle: a deep-etch study. J. Cell Biol., 105: 49.

Gething, M.-J. and J. Sambrook, 1990, Transport and assembly processes in the endoplasmic reticulum. Sem. Cell Biol., 1: 65.

Hofmann, S.L., M.S. Brown, E. Lee, R.K. Pathak, R.G.W. Anderson and J.L. Goldstein, 1989a, Purification of a sarcoplasmic reticulum protein that binds calcium and plasma lipoproteins. J. Biol. Chem., 264: 8260.

Hofmann, S.L., J.L. Goldstein, K. Orth, C.R. Moorman, C.A. Slaughter and M.S. Brown, 1989b, Molecular cloning of histidine-rich calcium binding protein of sarcoplasmic reticulum that contains highly conserved repeated elements. J. Biol. Chem., 264: 18083.

Ikemoto, N., M. Ronjat, L.G. Meszaros and M. Koshita, 1989, Postulated role of calsequestrin in the regulation of calcium release from sarcoplasmic reticulum. Biochemistry, 28: 6764.

Koch, G.L.E., 1987, Reticuloplasmins: a novel group of proteins in the endoplasmic reticulum. J. Cell Sci., 87: 491.

Leberer, E., J.M.H Charuk, N.M. Green and D.H. MacLennan, 1989a, Molecular cloning and expression of cDNA encoding a lumenal calcium binding glycoprotein from sarcoplasmic reticulum. Proc. Natl. Acad. Sci. (USA), 86: 6047.

Leberer, E., J.M.H. Charuk, D.M. Clarke, N.M. Green, E. Zubrzycka-Gaarn and D.H. MacLennan, 1989b, Molecular cloning and expression of cDNA encoding the 53,000-dalton glycoprotein of rabbit skeletal muscle sarcoplasmic reticulum. J. Biol. Chem., 264: 3484.

Leberer, E., B.G. Timms, K.P. Campbell and D.H. MacLennan, 1990, Purification, calcium binding properties, and ultrastructural localization of the 53,000 and 160,000 (sarcalumenin) dalton glycoproteins of sarcoplasmic reticulum. J. Biol. Chem., 265: 10118.

Leonards, K.S. and H. Kutchai, 1985, Coupling of Ca^{2+} transport to ATP hydrolysis by the Ca^{2+}-ATPase of sarcoplasmic reticulum. Potential role of the 53-Kilodalton glycoprotein. Biochemistry, 24: 4876.

Macer, D.R.J. and G.L.E. Koch, 1988, Identification of a set of calcium-binding proteins in reticuloplasm, the luminal content of the endoplasmic reticulum. J. Cell Sci., 91: 61.

MacLennan, D.H. and P.T.S. Wong, 1971, Isolation of a calcium-sequestering protein from sarcoplasmic reticulum. Proc. Natl. Acad. Sci. (USA), 68: 1231.

MacLennan, D.H., C.J. Brandl, B. Korczak and N.M. Green, 1985, Amino acid sequence of a Ca^{2+}+ Mg^{2+}-dependent ATPase from rabbit muscle sarcoplasmic reticulum, deduced from its complementary DNA sequence. Nature (London), 316: 699.

MacLennan, D.H., K.P. Campbell and R.A.F Reithmeier, 1983, in: "Calcium and Cell Function:, Chung, W.Y. ed., Academic Press, New York, pp. 151-173.

Maylie, J., M. Irving, N. Leung Sizto, G. Boyarsky and W.K Chandler, 1987, Calcium signals recorded from cut frog twitch fibers containing tetramethylmurexide. J. Gen. Physiol., 89: 145.

Mitchell, R.D., H.K.B. Simmerman and L.R. Jones, 1988, Calcium binding effects on protein conformation and protein interactions of canine cardiac calsequestrin. J. Biol. Chem., 263: 1376.

Munro, S. and H.R.B. Pelham, 1987, A C-terminal signal prevents secretion of luminal ER proteins. Cell, 48: 899.

Nelson, T.E. and K. Nelson, 1990, Intra- and extraluminal sarcoplasmic reticulum membrane regulatory sites for Ca^{2+}-induced Ca^{2+} release. FEBS Lett., 263: 292.

Ostwald, T.J. and D.H MacLennan, 1974, Isolation of a high affinity calcium-binding protein from sarcoplasmic reticulum. J. Biol. Chem., 249: 974.

Porter, K.R. and G.E. Palade, 1957, Studies on the sarcoplasmic reticulum. III. Its form and distribution in striated muscle cells. J. Biophys. Biochem. Cytol., 3: 269.

Saito, A., S. Seiler, A. Chu and S. Fleischer, 1984, Preparation and morphology of terminal cisternae from rabbit skeletal muscle. J. Cell Biol., 99: 875.

Sambrook, J.F., 1990, The involvement of calcium in transport of secretory proteins from the endoplasmic reticulum. Cell, 61: 197.

Schiaffino, S. and A. Margreth, 1969, Coordinated development of the sarcoplasmic reticulum and T system during postnatal differentiation of rat skeletal muscle. J. Cell Biol., 41: 855.

Scott, B.T., H.K.B. Simmerman, J.H. Collins, B. Nadal-Ginard and L.R. Jones, 1988, Complete amino acid sequence of canine cardiac calsequestrin deduced by cDNA cloning. J. Biol. Chem., 263: 8958.

Smith, M.J. and Koch, G.L.E., 1989, Multiple zones in the sequence of calreticulin (CRP 55, calregulin, HACBP) a major calcium binding ER/SR protein. EMBO J., 8: 3581.

Takeshima, H., S. Nishimura, T. Matsumoto, H. Ishida, K. Kangawa, N. Minamino, H. Matsuo, M. Ueda, M. Hanaoka, T. Hirose and S. Numa, 1989, Primary structure and expression from complementary DNA of skeletal muscle ryanodine receptor. Nature (London), 339: 439.

Thomas, K., J. Navarro, R.J.J. Benson, K.P. Campbell, R.L. Rotundo and R.E. Fine, 1989, Newly synthesized calsequestrin, destined for the sarcoplasmic reticulum, is contained in early/intermediate Golgi-derived clathrin-coated vesicles. J. Biol. Chem., 264: 3140.

Tooze, J., M. Hollinshead, S.D. Fuller, S.A. Tooze and W.B. Huttner, 1989, Morphological and biochemical properties in AtT-20 cells and their growth cones. Eur. J. Cell Biol., 49: 259.

Valdivia, C., H. Valdivia, J. Vilven and R. Coronado, 1989, Proton gating of calcium release channels in vesicles derived from junctional sarcoplasmic reticulum. Biophys. J., 55: 88a.

Villa, A., P. Podini, D.O. Clegg, T. Pozzan and J. Meldolesi, 1991, Intracellular Ca^{2+} stores in chicken Purkinje neurons: differential distribution of the low affinity-high capacity calcium binding protein, calsequestrin, of Ca^{2+} ATPase and of the ER lumenal protein, BiP. J. Cell Biol., 113: 779.

Volpe, P., 1989, The unraveling architecture of the junctional sarcoplasmic reticulum. J. Bioenerg. Biomem., 21: 215.

Volpe, P. and B.J. Simon, 1991, The bulk of calcium released to the myoplasm is free in the sarcoplasmic reticulum and does not unbind from calsequestrin. FEBS Lett., 278: 274.

Volpe, P., H.E. Gutweniger and C. Montecucco, 1987, Photolabeling of integral proteins of skeletal muscle sarcoplasmic reticulum. Comparison of junctional and non-junctional membrane fractions. Arch. Biochem. Biophys., 253: 138.

Volpe, P., B.H. Alderson-Lang, L. Madeddu, E. Damiani, J.H. Collins and A. Margreth, 1990, Calsequestrin a component of the inositol 1,4,5-trisphosphate-sensitive Ca^{2+} store of chicken cerebellum. Neuron, 5: 713.

Waisman, D.M., B.P. Salimath and M.J. Anderson, 1985, Isolation and characterization of CAB-63, a novel calcium-binding protein. J. Biol. Chem., 260: 1652.

Yuan, S., W. Arnold and A.O. Jorgensen, 1991, Biogenesis of transverse tubules and triads: immunolocalization of the 1,4-dihydropyridine receptor, TS28, and the ryanodine receptor in rabbit skeletal muscle developing in situ. J. Cell Biol., 112: 289.

Zorzato, F. and P. Volpe, 1988, Calcium binding proteins of junctional sarcoplasmic reticulum. Detection by Ca^{2+} ligand overlay. Arch. Biochem Biophys., 261: 324.

Zorzato, F., J. Fujii, K. Otsu, M. Phillips, N.M. Green, F.A. Lai, G. Meissner and D.H. MacLennan, 1990, Molecular cloning of cDNA encoding human and rabbit forms of the Ca^{2+} release channel (ryanodine receptor) of skeletal muscle sarcoplasmic reticulum. J. Biol. Chem., 265: 2244.

LIGAND BINDING AND COOPERATIVE INTERACTIONS AMONG THE SUBUNITS OF THE TETRAMERIC Ca^{2+} RELEASE CHANNEL COMPLEX OF SARCOPLASMIC RETICULUM

Gerhard Meissner

Departments of Biochemistry and Biophysics
and Physiology
University of North Carolina
Chapel Hill, NC 27599-7260

INTRODUCTION

In skeletal muscle, rapid release of Ca^{2+} ions from the intracellular membrane compartment, sarcoplasmic reticulum, (SR) is triggered by a surface membrane action potential that is communicated to the SR at specialized areas where the "junctional" SR comes in close contact with the surface membrane and tubular infoldings of the surface membane (T-tubule); at these areas protein bridges are present which span the gap between the two membrane systems (Peachey and Franzini-Armstrong, 1983). These structures have been variously termed "feet" (Franzini-Armstrong, 1970), "bridges" (Somlyo, 1979), "pillars" (Eisenberg and Eisenberg, 1982) or "spanning proteins" (Caswell and Brandt, 1989) and are now believed to be identical with the ryanodine receptor or SR Ca^{2+} release channel (for reviews see Lai and Meissner, 1989; Fleischer and Inui, 1989).

The specific mechanism of signal transmission from the T-tubule to SR remains to be fully defined. In skeletal muscle, a unique mechanism, referred to as the mechanical coupling mechanism, has been formulated which suggests that small transient charge movements observed during T-tubule depolarization lead to the opening of a SR Ca^{2+} release channel (Schneider *et al.*, 1973). An important implication of the mechanical coupling model is that the SR Ca^{2+} release channel is in direct contact with a voltage-sensing molecule in the T-tubule membrane. Recent biophysical, pharmacological and molecular biological evidence has suggested that the dihydropyridine (DHP) receptor for calcium channel antagonists is the voltage-sensor in excitation-contraction (E-C) coupling in skeletal muscle (Rios and Pizarro, 1988; Tanabe *et al.*, 1988). The involvement of a M_r 90,000 protein (triadin) in forming a direct physical link between DHP receptor and SR Ca^{2+} release channel has been postulated (Kim *et al.*, 1990).

In this chapter, we describe single channel, vesicle ion flux and ligand binding measurements that we have carried out

Excitation-Contraction Coupling in Skeletal, Cardiac, and Smooth Muscle
Edited by G.B. Frank *et al.*, Plenum Press, New York, 1992

to determine several of the structural and functional features of the Ca^{2+} release channel of mammalian skeletal muscle. Our studies have indicated that the skeletal muscle Ca^{2+} release channel is a large protein complex comprised of four cooperatively coupled M_r 565,000 subunits.

RESULTS AND DISCUSSION

Table 1 summarizes several of the presently known properties of the mammalian skeletal muscle Ca^{2+} release channel as determined in our laboratory. The skeletal channel complex has been purified from junctionally derived SR vesicles by sucrose gradient centrifugation, following its solubilization in the presence of the zwitterionic detergent Chaps and radiolabeling with the channel-specific ligand [^{3}H]ryanodine. The ryanodine receptor migrated on the gradients as a 30S complex comprised of four negatively charged, electrophoretically identical polypeptides of apparent relative molecular mass (M_r) of 400,000 on SDS

Table 1. Properties of skeletal muscle Ca^{2+} release channel

1. Composition:	30S homotetramer of M_r ~2.3 x 10^6 (four ~565,000 subunits)
2. Structure:	four-leaf clover (quatrefoil)
3. Ryanodine binding:	one high-affinity and 2-3 low-affinity sites per 30S complex
4. Conductance:	100 pS with 50 mM Ca^{2+} 600 pS with 500 mM Na^+
5. Regulation:	
Endogenous ligands:	Activation by μM Ca^{2+} and mM ATP, inhibition by mM Mg^{2+}
Exogenous ligands:	Activation by nM ryanodine and mM caffeine, inhibition by mM ryanodine and μM ruthenium red

polyacrylamide gels (Lai *et al.*, 1988; 1989), in reasonable agreement with a calculated M_r ~565,000 deduced from cDNA sequence analysis (Takeshima *et al.*, 1988; Zorzato *et al.*, 1989). The purified complex bound approximately 500 pmol [^{3}H]ryanodine per mg protein, which, taking a calculated M_r 565,000 into account, corresponds to one high-affinity site per 30S tetramer. Upon reconstitution of the purified receptor complex into planar lipid bilayers, monovalent and divalent cation conducting channels were evident. These were activated by Ca^{2+}, ATP and caffeine, inhibited by Mg^{2+} and ruthenium red, and modified by ryanodine, as observed for the native Ca^{2+} conducting channel. Electron microscopic exami-

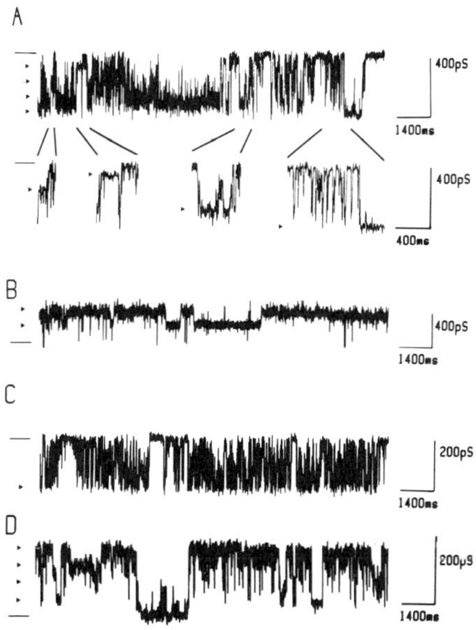

Fig. 1. Multiple conductance levels of the Na^+-conducting Ca^{2+} release channel. Single channel currents of four separate experiments were recorded in symmetrical 100 mM NaCl, 100 µM $CaCl_2$, 100 µM EGTA, 20 mM NaPipes, pH 7 buffer. Baseline currents are denoted by bar and conductance levels by arrows on left of traces. Holding potentials were +18 mV (A), -8 mV (B), +30 mV (C) and -40 mV (D) (in modified form from Liu *et al.*, 1989).

nation further revealed the quatrefoil structure described for the feet. These results have suggested that skeletal muscle contains a ryanodine-sensitive SR Ca^{2+} release pathway which is identical with the large protein structures (feet) that span the T-SR junctional gap.

A Ca^{2+}-gated, ryanodine- and caffeine-sensitive 30S Ca^{2+} release channel complex has been also isolated from cardiac muscle (Anderson *et al.*, 1989). More recently, we have also obtained evidence for the presence of a 30S Ca^{2+}-conducting channel complex in smooth muscle (Herrmann-Frank *et al.*, 1991; Xu *et al.*, 1991) and brain (Lai *et al.*, 1990; 1991). A similar high molecular weight, intracellular Ca^{2+} release channel complex may be present therefore, in most, if not all excitable tissues.

In Fig. 1, the rabbit skeletal muscle Ca^{2+} release channel was purified in the absence of [^{3}H]ryanodine and then directly incorporated into a planar lipid bilayer in the presence of a symmetrical NaCl buffer. The reconstituted SR Ca^{2+} release channel does not conduct Cl^- anions, and it conducts monovalent cations more efficiently than it does Ca^{2+} (Lai *et al.*, 1988; Liu *et al.*, 1989). Accordingly, the channel can be conveniently monitored in a monovalent cation medium such as 100 mM NaCl. Fig. 1 shows representive current traces of four separate experiments following the apparent incorporation of a single channel into the bilayer. The upper trace of Fig. 1 shows a major conductance of 400 picoSiemens (pS) and additional conductance levels of 300, 200 and 100 pS. In Fig. 1B, a 400 pS main conductance and a subconductance of 200 pS were recorded. In many of the experiments, the main conductance observed was 200 pS (Fig. 1C), which frequently entered subconductance levels of 150, 100 and 50 pS (Fig. 1D). Appearance of up to four nearly equiva-lent conductances could be due either to the presence of a channel comprised of four subunits and each containing an individually conducting pore or a single conducting pore pre-

Table 2. Ca^{2+} release properties of skeletal Ca^{2+} release vesicles

Additions to release medium	$^{45}Ca^{2+}$ efflux
	k_1 (s^{-1})
10^{-9} M Ca^{2+}	0.1
10^{-9} M Ca^{2+}, 5 x 10^{-3} M ATP	12
4 x 10^{-6} M Ca^{2+}	1.5
2 x 10^{-6} M Ca^{2+}, 5 x 10^{-3} M AMP-PCP	90
4 x 10^{-6} M Ca^{2+}, 5 x 10^{-3} M Mg^{2+}, 5 x 10^{-3} M AMP-PCP	8
4 x 10^{-6} M Ca^{2+}, 5 x 10^{-3} M Mg^{2+}	0.004
5 x 10^{-3} M Ca^{2+}	0.01

First order rate constants of Ca^{2+} release were determined by passive loading of SR vesicles with 5 mM $^{45}Ca^{2+}$ and measurement of $^{45}Ca^{2+}$ efflux rates with the use of a rapid-mixing quench-filtration protocol (Meissner *et al.*, 1986).

sent within the tetrameric assembly of the M_r 565,000 subunits (Liu *et al.*, 1989). In both cases, cooperative interactions among the subunits must be postulated to take place to account for the multiple conductance behavior of the channel.

The existence of cooperatively interacting domains within the tetrameric channel complex has been also shown by studying the interaction of the channel with its three endogenous regulatory ligands Ca^{2+}, Mg^{2+} and ATP. Table 2 summarizes the results of rapid-mixing and quench vesicle - $^{45}Ca^{2+}$ flux measurements which have indicated that the three ligands dramatically affect the Ca^{2+} efflux rate from skeletal SR vesicles. In the presence of a low extravesicular Ca^{2+} concentration (10^{-9} M free Ca^{2+}), channel-mediated Ca^{2+} release is slow, occurring with a first order rate constant of about 0.1 s^{-1}. Increase in extravesicular free Ca^{2+} concentration to 2 μM and addition of the nonhydrolyzable ATP analog AMP-PCP increased the release rate by a factor of about 1000 fold. Intermediate release rates were observed when the skeletal muscle Ca^{2+} release channel was activated at 10^{-9} M free Ca^{2+} by 5 mM ATP, or by 4 μM free Ca^{2+} alone. Mg^{2+} had an inhibitory effect when added to the micromolar Ca^{2+} release media containing or lacking AMP-PCP. Similarly, $^{45}Ca^{2+}$ release was slow when vesicles were transferred to a medium containing 5 mM free Ca^{2+}.

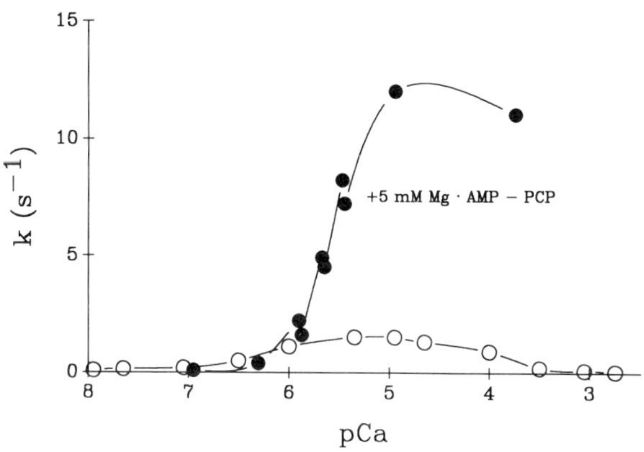

Fig. 2. Dependence of first order rate constants of $^{45}Ca^{2+}$ efflux on Ca^{2+} concentration in the presence and absence of Mg^{2+} and AMP-PCP. SR vesicles were passively loaded with 5 mM $^{45}Ca^{2+}$ and diluted into media containing the indicated concentrations of free Ca^{2+} and zero (o) or 5 mM AMP-PCP plus 5 mM Mg^{2+} (•) (in modified form from Meissner *et al.*, 1986).

Fig. 2 compares the rate constants of Ca^{2+} release on free Ca^{2+} concentration in the presence and absence of 5 mM Mg^{2+} plus 5 mM AMP-PCP. In the absence of Mg^{2+} and adenine nucleotide, Ca^{2+} release from skeletal SR vesicles was maximal in the micromolar Ca^{2+} concentration range. A biphasic curve like the one shown in Fig. 2 can be most easily explained by assuming that the channel possesses high affinity, activating and low-affinity, inhibitory Ca^{2+} binding sites. Addition of 5 mM Mg^{2+} and nucleotide, which results in the formation of a Mg•nucleotide complex, increased the rate constant of Ca^{2+} release about tenfold at 10^{-5} - 10^{-4} M free Ca^{2+}. A second effect of these two ligands was to increase the cooperativity of Ca^{2+} activation of Ca^{2+} release (n_H increased from 1.0 to 2.1, Meissner *et al.*, 1986), thereby rendering the channel sensitive to Ca^{2+} in a narrower concentration range.

We have formulated a minimum model to explain the effects of Ca^{2+}, Mg^{2+} and adenine nucleotide on the Ca^{2+} release

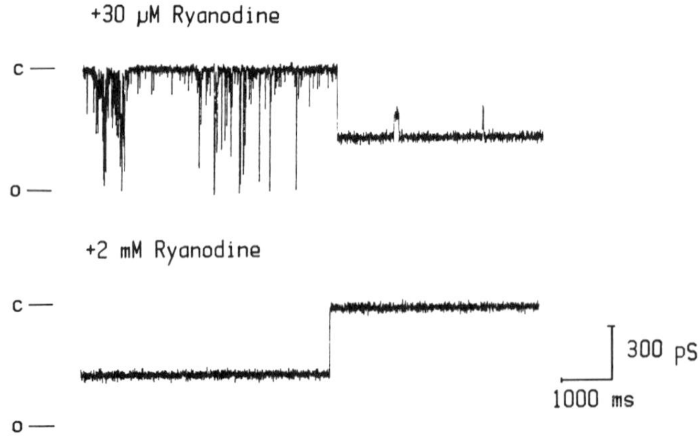

Fig. 3. Effect of ryanodine on a single reconstituted, purified Ca^{2+} release channel. Shown are single channel recordings of K^+ currents of the purified 30S channel complex incorporated into a planar lipid bilayer in symmetric 250 mM KCl buffer with 50 μM free Ca^{2+}. Unitary conductance = 700 pS. Holding potential = 20 mV. Upper trace shows appearance of subconducting state with open probability (P_o) ~1 following several minutes after cis addition of 30 μM ryanodine. An additional, infrequently observed substate can also be noticed. Lower trace illustrates the suddent transition from the subconductance state to a fully closed state within one minute after cis addition of 2 mM ryanodine. Bars on left represent the open (o) and closed (c) channel (from Lai *et al.*, 1989).

behavior of junctionally derived SR vesicles (Meissner *et al.*, 1986). The model proposes that the SR Ca^{2+} release channel is an allosteric enzyme which has n binding sites each for Ca^{2+}, Mg^{2+} and ATP. The ligand-free form of the channel represents a low Ca^{2+} releasing state of the skeletal channel (Table 2, Fig. 2). Distinct channel intermediates are formed by the binding of ligands such as Ca^{2+} to high affinity sites (K_{Ca} ~1 μM), which partially activates the channel. Binding of adenine nucleotide to the Ca^{2+}-activated channel further increases channel activity. The activating action of Ca^{2+} and adenine nucleotide is partially reversed by binding of millimolar Ca^{2+} or Mg^{2+} to low affinity, cooperatively interacting, inhibitory channel sites.

The presence of cooperatively coupled domains within the 30S SR Ca^{2+} release channel complex has been also revealed in studies with the channel-specific probe ryanodine. In single channel measurements, ryanodine modified the gating and conductance of the channel in a highly characteristic manner (Rousseau *et al.* 1987; Lai *et al.*, 1989). Micromolar concentrations of ryanodine induced the formation of a channel state characterized by long open intervals and a reduced conductance, whereas millimolar concentrations of ryanodine completely closed the channel (Fig. 3). Furthermore, the ryanodine-modified, subconducting channel state became insensitive to regulation by Ca^{2+}, Mg^{2+} and ATP, which all greatly affect the gating of the ryanodine-unmodified channel. [^{3}H]Ryanodine binding studies have indicated that ryanodine exerts its diverse actions by binding in an apparently cooperative or noncooperative manner to the channel, depending on the experimental conditions. Fig. 4 (upper panel) shows that ryanodine binding was dependent on free Ca^{2+} concentration at nanomolar ryanodine, whereas at elevated concentrations no significant dependence of binding on Ca^{2+} concentration could be discerned. Scatchard plot analysis of ryanodine binding data indicated the presence of a high-affinity site with a B_{max} value of about 7 pmol/mg protein and K_D of 6 and 8 nM at 1 mM and 20 μM free Ca^{2+}, respectively. Reduction of free Ca^{2+} to 0.1 μM and less resulted in loss of high-affinity ryanodine binding. The binding data of the upper panel of Fig. 4 also revealed the presence of two to three additional low-affinity sites per high-affinity site, suggesting the presence of up to four ryanodine binding sites per tetrameric 30S channel complex. In the lower panel of Fig. 4, ryanodine binding characteristics of the channel were further analyzed by rearranging data in form of Hill plots. At low Ca^{2+} concentration (0.1 μM), data could be fitted by a straight line with a Hill coefficient (n_H) of 0.8, suggesting no or only little interaction between the binding sites. At elevated Ca^{2+} concentrations, Hill plots showed a nonlinear behavior over the nanomolar to micromolar ryanodine concentration range. At nanomolar ryanodine, n_H was 0.9 whereas in the lower affinity binding range, n_H decreased to 0.45, suggesting a negative cooperativity of ryanodine binding among the lower affinity sites. An apparently contradictory result, indicating a positive cooperative interaction between high and low affinity sites of ryanodine binding, was obtain-ed when the effects of elevated concentrations of ryanodine on the rate of [^{3}H]ryanodine dissociation from the high-affinity site were examined. In Fig. 5, SR membranes were labeled in the presence of 100 μM free Ca^{2+} with a relatively low (10 nM) concentration of [^{3}H]ryanodine and then diluted into four media containing either (i) 0.05 μM free Ca^{2+}, (ii) 0.05 μM free Ca^{2+} plus unlabeled ryanodine, (iii) 100 μM free Ca^{2+}, or (iv) 100 μM free Ca^{2+} plus unlabeled ryanodine. A relatively high concentration of unlabeled ryanodine (200 μM) was used to achieve rapid binding of ryanodine to low-affinity sites. Dilution of membranes into the two media containing 200 μM ryanodine resulted in a much slower decrease in [^{3}H]ryanodine binding with time. In the 100 μM Ca^{2+} medium, which favors high-affinity binding, incomplete dissociation of [^{3}H]ryanodine was observed, presumably because a new equilibrium between bound and free [^{3}H]ryanodine was established. These observations provided direct evidence for an allosteric behavior of the channel complex and further suggested that

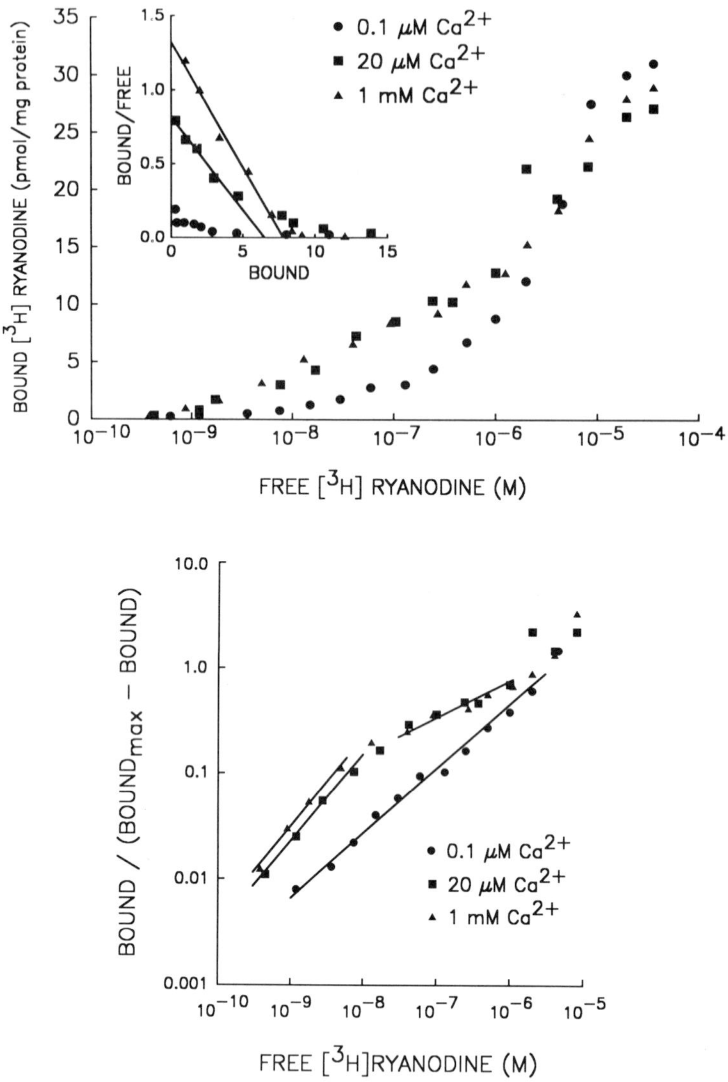

Fig. 4. Dependence of [^{3}H]ryanodine binding on Ca^{2+} concentration. SR vesicles were incubated for 15 h at 22°C in a medium containing 1.0 M NaCl, 20 mM NaPipes, pH 7, 1 mM DIFP, 5 mM AMP, 2 nM-3 mM [^{3}H]ryanodine and indicated concentrations of free Ca^{2+}. Specific binding is shown in upper panel, and in the form of Scatchard (inset in upper panel) and Hill (lower panel) plots (from Lai *et al.*, 1989).

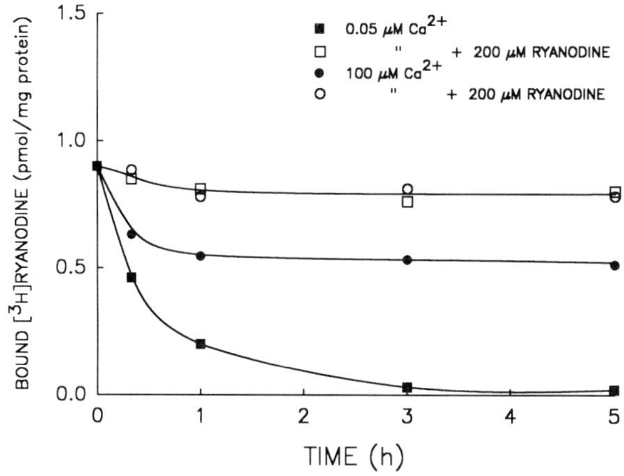

Fig. 5. Rate of [^{3}H]ryanodine dissociation in the presence and absence of unlabeled ryanodine. High-affinity site of SR Ca^{2+} release channel was labeled with [^{3}H]ryanodine by incubation of SR vesicles in a medium containing 100 μM free Ca^{2+} and 10 nM [^{3}H]ryanodine. [^{3}H]ryanodine dissociation was initiated by diluting vesicles at 37°C 20-fold into media containing the indicated concentrations of free Ca^{2+} and unlabeled ryanodine (from Lai *et al.*, 1989).

occupation of the low-affinity sites induced protein conformational changes which resulted in a partial occlusion of the ligand-occupied, high-affinity site (Lai *et al.*, 1989). In addition to Ca^{2+}, the rates and cooperativity of ryanodine binding to the skeletal muscle receptor are affected by Mg^{2+}, ATP, ionic strength, proteolytic digestion, as well as other ligands that have been shown to modulate Ca^{2+} release channel activity (for review see Lai and Meissner, 1989).

In conclusion, our studies have indicated that various endogenous and exogenous ligands regulate SR Ca^{2+} release channel activity in a noncooperative or cooperative manner, depending on the experimental conditions. Presence of cooperatively interacting sites suggests that occupation of these sites by their respective ligands results in protein conformational changes among the four subunits of the tetrameric SR Ca^{2+} release channel complex. Futher studies are needed to more fully characterize the ligand binding sites and their interactions within the large SR Ca^{2+} release channel complex.

ACKNOWLEDGMENT

Research reported here was supported by NIH grant AR18687.

REFERENCES

Anderson, K., Lai, F. A., Liu, Q. Y., Rousseau, E., Erickson, H. P., and Meissner, G., 1989, Structural and functional characterization of the purified cardiac ryanodine receptor - Ca^{2+} release channel complex. J Biol Chem, 264:1329-1335.

Caswell, A. H., and Brandt, N. R., 1989, Triadic proteins of skeletal muscle. J Bioenerg Biomembr, 21:149-162.

Eisenberg, B. R., and Eisenberg, R. S., 1982, The T-SR junction in contracting single skeletal muscle fibers. J Gen Physiol 79:1-19.

Fleischer, S., and Inui, M., 1989, Biochemistry and biophysics of excitation-contraction coupling. Annu Rev Biophys Chem 18:333-364.

Franzini-Armstrong, C., 1970, Studies of the triad. I. Structure of the junction in frog twitch fibers. J Cell Biol 47:488-499.

Herrmann-Frank, A., Darling, E., and Meissner, G., Functional characterization of the Ca^{2+}-gated Ca^{2+} release channel of vascular smooth muscle sarcoplasmic reticulum. Eur J Physiol, in press.

Kim, K. C., Caswell, A. H., Talvenheimo, J. A., and Brandt, N. R., 1990, Isolation of a terminal cisternae protein which may link the dihydropyridine receptor to the junctional foot protein in skeletal muscle. Biochemistry 29:9281-9289.

Lai, F. A., Erickson, H. P., Rousseau, E., Liu, Q. Y., and Meissner, G., 1988, Purification and reconstitution of the calcium release channel from skeletal muscle. Nature 331:315-319.

Lai, F. A., and Meissner, G., 1989, The muscle ryanodine receptor and its intrinsic Ca^{2+} channel activity. J Bioenerg Biomembr 21:227-246.

Lai, F. A., Misra, M., Xu, L., Smith, H. A., and Meissner, G., 1989, The ryanodine receptor-Ca^{2+} release channel complex of skeletal muscle sarcoplasmic reticulum. Evidence for a cooperatively coupled, negatively charged homotetramer. J Biol Chem 264:16776-16785.

Lai, F. A., Xu, L., and Meissner, G., 1990, Identification of a ryanodine receptor in rat and bovine brain. Biophys J 57:529a.

Lai, F. A., Xu, L., Kumari, G., Lee, H. B., and Meissner, G., 1991, Purification and functional reconstitution of a bovine brain ryanodine receptor. Biophys J 59:272a.

Liu, Q. Y., Lai, F. A., Rousseau, E., Jones, R. V., and Meissner, G., 1989, Multiple conductance states of the purified calcium release channel complex from skeletal sarcoplasmic reticulum. Biophys J 55:415-424.

Meissner, G., Darling, E., and Eveleth, J., 1986, Kinetics of rapid Ca^{2+} release by sarcoplasmic reticulum. Effects of Ca^{2+}, Mg^{2+} and adenine nucleotides. Biochemistry 25:236-244.

Peachey, L. D., and Franzini-Armstrong, C., 1983, Structure and function of membrane systems of skeletal muscle cells. in "Skeletal muscle" (Peachey LD, Adrian RH, Geiger SR, eds) Handbook of Physiology, Sect. 10. American Physiological Society, Bethesda, MD, pp. 23-71.

Rios, E., and Pizarro, G., 1988, Voltage sensors and calcium channels of excitation-contraction coupling. NIPS 3:223-227.
Rousseau, E., Smith, J. S., and Meissner, G., 1987, Ryanodine modifies conductance and gating behavior of single Ca^{2+} release channel. Am J Physiol 253:C364-C368.
Schneider, M. F., and Chandler, W. K., 1973, Voltage dependent charge movement in skeletal muscle: a possible step in excitation-contaction coupling. Nature 242:244-246.
Somlyo, A. V., 1979, Bridging structures spanning the junctional gap at the triad of skeletal muscle. J Cell Biol 80:743-750.
Takeshima, H., Nishimura, S., Matsumoto, T., Ishida, H., Kangawa, K., Minamino, N., Matsuo, H., Ueda, M., Hanaoka, M., Hirose, T., and Numa, S., 1989, Primary structure and expression from complementary DNA of skeletal muscle ryanodine receptor. Nature 339:439-445.
Tanabe, T., Beam, K. G., Powell, J. A., and Numa, S., 1988, Restoration of excitation-contaction coupling and slow calcium current in dysgenic muscle by dihydropyridine receptor complementary DNA. Nature 336:134-139.
Xu, L., Lai, F. A., Ettner, E., Fay, F. S., and Meissner, G., 1991, Evidence for a 30S ryanodine receptor-Ca^{2+} release channel complex in stomach smooth muscle of toad. Biophys J 59:67a.
Zorzato, F., Fujii, J., Otsu, K., Phillips, M., Green, N. M., Lai, F. A., Meissner, G., and MacLennan, D. H., 1990, Molecular cloning of cDNA encoding human and rabbit forms of the Ca^{2+} release channel (ryanodine receptor) of skeletal muscle sarcoplasmic reticulum. J Biol Chem 265:2244-2256.

CONTROL OF CALCIUM RELEASE FROM THE SARCOPLASMIC RETICULUM

G.D. Lamb and D.G. Stephenson

Department of Zoology
La Trobe University
Bundoora, Melbourne, 3083
Australia

INTRODUCTION

In order to study many of the important events in excitation-contraction (E-C) coupling in skeletal muscle, it is necessary to have access to the intracellular environment while still retaining the normal voltage-sensor control over Ca^{2+} release from the sarcoplasmic reticulum (SR). We have achieved this using a skinned fibre preparation similar to that described previously (Donaldson, 1985; Stephenson, 1985; Fill & Best, 1988), with modifications that improve the viability of the preparation and enable normal E-C coupling to be distinguished from non-physiological Ca^{2+} release that can occur. We review here some of the more important findings obtained with this preparation.

The T-system Reseals after Skinning a Fibre

A segment of a single fibre, from either the iliofibularis muscle of the toad or the extensor digitorum longus (e.d.l.) muscle of the rat, was mechanically 'skinned' under paraffin oil by peeling back the sarcolemma. The responses of the fibre segment to the stimulation described previously (Donaldson, 1985; Stephenson, 1985; Fill & Best, 1988) and below, conclusively showed that the transverse-tubular (t-) system resealed and repolarized after such skinning. This was further verified by experiments with a membrane impermeable salt of the Ca^{2+}-sensitive dye, Fluo-3. A droplet of Ringer extracellular solution (2mM Ca^{2+}) and 1mM Fluo-3, at the end of a capillary tube, was moved back and forth along an intact fibre for several minutes to allow diffusion of the solution into the t-system. The fibre was then skinned and placed in a potassium solution with low $[Ca^{2+}]$ (pCa > 8, 1mM EGTA) and examined with a confocal laser scanning microscope. Under the 488nm laser illumination, bright spots were observed in a regular arrangement correlating to the sarcomeric spacing. These bright spots were present for more than an hour with the fibre in the low $[Ca^{2+}]$ solution, but were absent in control fibres which either had not been exposed to the Fluo-3 or had not been skinned after exposure to Fluo-3, allowing the Fluo-3 to diffuse out again and the EGTA to chelate the free Ca^{2+}. Thus, this data indicates that the t-system resealed after skinning a fibre, trapping the Fluo-3 in the t-system.

Excitation Contraction Coupling in Skeletal, Cardiac, and Smooth Muscle
Edited by G.B. Frank *et al.*, Plenum Press, New York, 1992

A segment of a skinned fibre was mounted on a force transducer and bathed in a K_2HDTA solution (Hexamethylene-diamine-tetraacetate) with 1mM free Mg^{2+}, 8mM total ATP, 50μM EGTA to give pCa 6.7, and pH 7.1 (Lamb & Stephenson, 1990a). Na^+ was also present in the solution to enable Na/K pump in the t-system membrane to keep a high $[Na^+]$ and low $[K^+]$ in the t-system, thus polarizing the t-system membrane. Previous studies have 'depolarized' the fibre by rapidly replacing a similar potassium solution with a choline chloride (ChCl) solution. Lowering the $[K^+]$ and raising the $[Cl^-]$ in the bathing solution in this way should indeed depolarize the t-system, but the high $[Cl^-]$ will also make the inside of the SR more negative and may have osmotic effects. Instead, noting that the SR is almost equally permeable to Na^+ and K^+, we chose to substitute all the K^+ in the bathing solution with Na^+ with no change of anion, which should still depolarize the t-system with little or no direct effect on the SR. Such substitution elicited a large, rapidly-rising response in virtually every toad and rat fibre examined (eg Fig. 1), indicating that the stimulus rapidly induced massive Ca^{2+} release from the SR. The response spontaneously declined completely within about 2 to 4s. Experiments in which the fibre was moved rapidly back into paraffin oil after initiating a response, indicated that the decline in force was determined primarily by the re-uptake of Ca^{2+} within the fibre, rather than by diffusion of Ca^{2+} into the bathing solution (Lamb & Stephenson, 1990a).

Another, virtually identical, response could be elicited after repolarizing the t-system and 'repriming' the voltage-sensors, by returning the fibre to the K^+ solution for 30s or more before substituting the K^+ with Na^+ again (Fig. 1). This procedure could be repeated between 10 to 30 times in iliofibularis fibres from the toad before the response 'ran-down' and only small responses or no response could be elicited (Lamb & Stephenson, 1990a). In rat e.d.l. fibres between 6 and 16 responses could be elicited before the fibre ran-down (Lamb & Stephenson, 1991b).

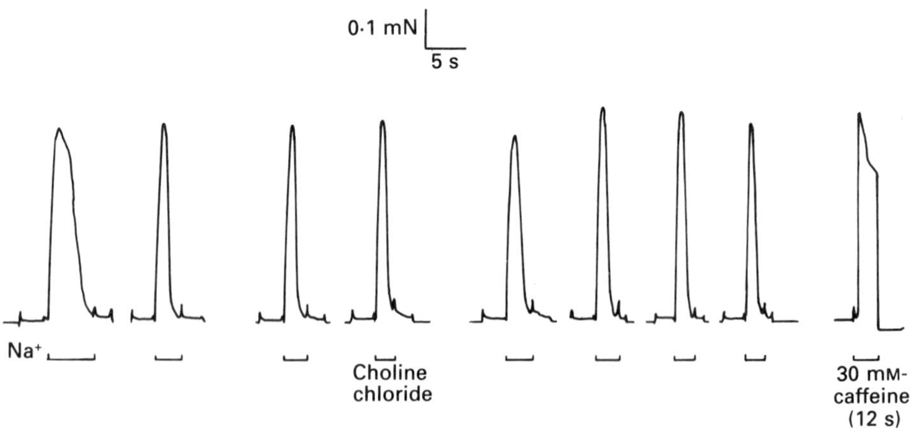

Fig. 1. Isometric contractions, in a mechanically skinned fibre from toad muscle, elicited by rapid substitution of K_2HDTA with Na_2HDTA (or with ChCl in one case) over the periods indicated by the bars. The fibre was 'reprimed' in the K_2HDTA solution for 30s between successive stimulations. (From Lamb & Stephenson (1990a) with permission).

These responses, induced by replacing the 'internal' K^+ with Na^+, are highly analagous to potassium contractures in intact fibres (Hodgkin & Horowicz, 1960), where the external Na^+ is replaced with K^+. Thus, it is not surprising that both activation and steady-state inactivation curves, like those obtained by raising the external $[K^+]$ to various levels in intact fibres, can also be obtained in skinned fibres by lowering the $[K^+]$ to intermediate levels (Lamb & Stephenson, 1990a).

As expected, all treatment designed to 'inactivate' the voltage-sensors in the t-system invariably abolished this fast depolarization-induced response (Lamb and Stephenson, 1990a). Thus, permanently depolarizing the t-system by a) perforating the t-system with saponin, b) 'splitting' the fibre rather than skinning it, and c) blocking Na/K exchange with ouabain, invariably abolished the responses to Na^+ substitution. Similarly, bathing a fibre in an external solution with very low $[Ca^{2+}]$ and $[Mg^{2+}]$ before skinning it, which should also inactivate the voltage-sensors, also prevented the response. Ruthenium red (RR, 2μM), a well-known blocker of the ryanodine receptor/Ca^{2+} release channel, completely abolished the depolarization-induced response produced by Na^+ substitution, even when the fibre was heavily loaded with Ca^{2+} (Fig. 2B). The concentration of RR producing half-maximal block of the response was approximately 0.3μM in both toad and rat muscle (Lamb & Stephenson, 1990a, 1991b).

Responses to ChCl Substitution

When a fibre was stimulated soon after skinning, without additional Ca^{2+} loading (ie without prolonged exposure to solutions either weakly Ca^{2+}-buffered to pCa < 6.7 or strongly Ca^{2+}-EGTA buffered to pCa < 7.0), ChCl substitution elicited a similar response to that observed with Na^+ substitution; the response rose rapidly (<1s) to a near maximal level and then spontaneously declined within several seconds (eg Fig. 1). However, if a fibre had been moderately to heavily loaded with Ca^{2+}, in addition to the fast depolarization-induced response, ChCl substitution could also evoke a slowly-rising, prolonged response in both toad (eg Fig. 2A) and rat muscle (Lamb & Stephenson, 1991b). The precise kinetic division between the fast depolarization-induced response and this prolonged response, named the 'second component', was often ill-defined, particularly in rat muscle fibres. Nevertheless, a number of other properties of the 'second component' clearly distinguished it from the depolarization-induced response (Lamb & Stephenson, 1990a). Firstly, ChCl substitution still elicited this response when the fast response had run-down or been abolished by permanently depolarizing the t-system by saponin treatment, ouabain in the t-system or splitting rather than skinning the fibre. Secondly, temporarily depolarizing the t-system by prolonged exposure to the Na^+ solution, abolished the fast depolarization-induced response to ChCl substitution but not the 'second component'. Thirdly, D600, which produced a use-dependent paralysis of the voltage-sensors (see below), did not affect the second component. Finally, 2μM ruthenium red, while completely blocking the response to Na^+ substitution and the rapid response to ChCl substitution, did not block the 'second component' response to ChCl substitution in heavily loaded fibres. The first three of these points show that the 'second component' is not produced by, and does not depend on, activation of the voltage-sensors in the t-system, and the last point indicates that 'second component' involves Ca^{2+} release by some pathway not involved in normal E-C coupling. It was often noted that after a second component response, the depolarization-induced response was abolished (eg Fig. 2A) and that fibres broke when stimulate with caffeine or high $[Ca^{2+}]$; this suggests that, unlike the depolarization-induced response, the second component response in some way damaged the fibres.

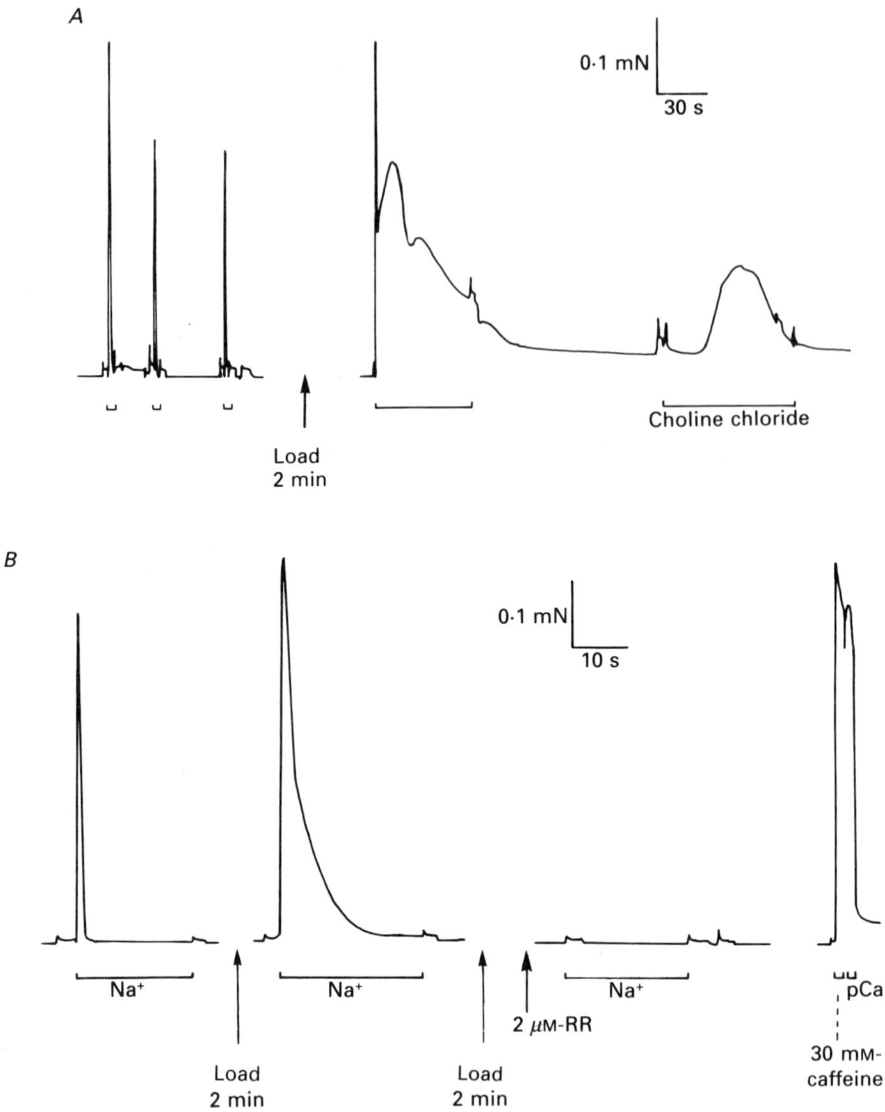

Fig. 2 A. After a fibre had been heavily loaded with Ca^{2+}, ChCl substitution induced a slow, prolonged 'second component' in addition to the fast depolarization-induced response. B. Na^+ substitution only induced the fast depolarization-induced response, even after heavily loading a fibre with Ca^{2+} and this response was completely blocked by 2μM ruthenium red. (From Lamb & Stephenson (1990a) with permission).

Effects of Caffeine and D600

Addition of 1mM caffeine induced a large response in toad fibres (Fig. 3), particularly if the [Ca^{2+}] was slightly elevated (eg to pCa 6.3). Similar responses in rat e.d.l. fibres required higher caffeine concentrations (2 to 5mM). The responses to these low concentrations of caffeine were completely blocked by 2μM ruthenium red in both types of muscle. D600 produces a use-dependent paralysis in intact fibres (Eisenberg et al, 1983) by locking the voltage-sensors into an inactivated state. Similarly here, a single depolarization (by either ChCl or Na^+ substitution) abolished all subsequent depolarization-induced responses

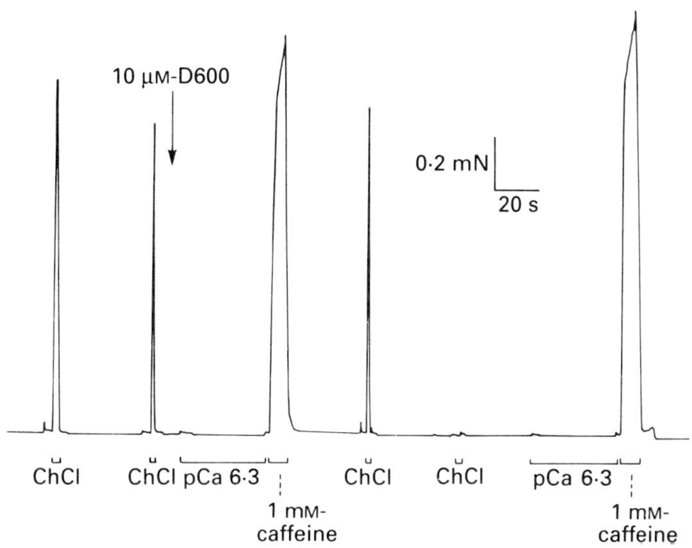

Fig. 3. After addition of 10μM D600 (in all solutions after arrow) a large response to caffeine had no effect on the response to depolarization with ChCl, but after that single ChCl response a subsequent ChCl substitution gave no response. Subsequent responses to 1mM caffeine appeared to be unaffected. Before each caffeine response each fibre was bathed for 30s in a solution at pCa 6.3 (50μM EGTA), and after each response had peaked the caffeine was washed out. (From Lamb & Stephenson (1990b) with permission).

for several minutes or more (at 22°C) (Fig. 3). A large prolonged response to caffeine in the presence of D600 did not affect the response to a subsequent depolarization, and conversely locking the voltage-sensors into an inactivated state by a depolarization in D600 did not affect the response to caffeine (Fig. 3). These results are consistent with the current view that caffeine does not activate the voltage-sensors but instead directly activates the Ca^{2+} release channels in the SR which are involved in normal E-C coupling.

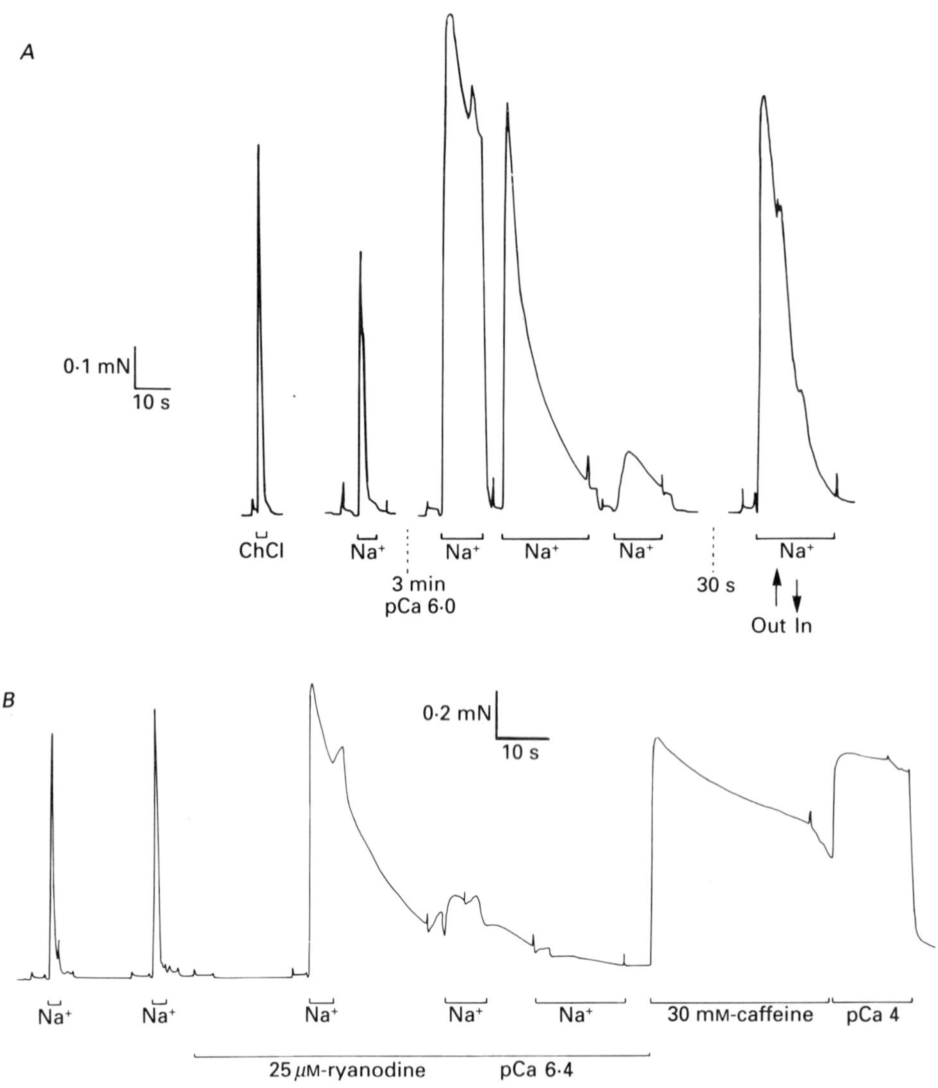

Fig. 4. Ryanodine prevents repolarization from terminating Ca^{2+} release. A. Depolarization by Na^{+} substitution (horizontal bars) induced a prolonged response after heavily loading the fibre (by 3 minutes in pCa 6.0 solution, 50μM EGTA). This response was rapidly terminated by repolarizing the fibre in the K^{+} solution (absence of bar) and then was re-initiated by depolarizing the fibre in Na^{+} again. Moving the fibre out of and back into the Na^{+} solution did not terminate the response. B. In the presence of 25μM ryanodine, repolarization could not terminate the response, presumably because some release channels were initially blocked open (and then shut) by the ryanodine. (From Lamb & Stephenson (1990b) with permission).

Ryanodine Prevents Repolarization from Terminating Ca^{2+} Release

A distinguishing feature of Ca^{2+} release in skeletal muscle is that it remains under control of the voltage-sensors in the t-system; voltage-clamp experiment with cut muscle fibres from frog show that repolarization rapidly terminates Ca^{2+} release from the SR (Melzer et al, 1986). Similarly, in the experiments described here, repolarizing a skinned fibre by returning to the K^+ solution, rapidly and completely terminated the response (Fig. 4A). In contrast, following a depolarization in the presence of ryanodine (25μM), repolarization did not rapidly terminate the response (Fig. 4B). This is consistent with reports on single channels, which describe how ryanodine can lock the release channels into a partially open state (Smith et al, 1988), and further shows that not even deactivation of the voltage-sensors can close the channels again. In many fibres in these experiments, the SR remained depleted of Ca^{2+} even after attempts at Ca^{2+} loading, implying that at least some release channels remained locked open. Nevertheless, the absence of a depolarization-induced response in some other fibres in which 30mM caffeine could still elicit a maximal response (eg Fig. 4B), seemed to imply that the SR was not depleted of Ca^{2+} and that most of the release channels eventually must be blocked shut and not open, as reported for high concentrations of ryanodine (see references in Lamb & Stephenson, 1990b). If the free $[Ca^{2+}]$ was high enough (< pCa 6.7), fibres usually gave a spontaneous, transient response within one minute of applying 25μM ryanodine. This spontaneous reponse is consistent with ryanodine potentiating Ca^{2+}-induced Ca^{2+} release (see below) by initially locking some release channels open, giving large Ca^{2+} efflux and opening other channels and so on.

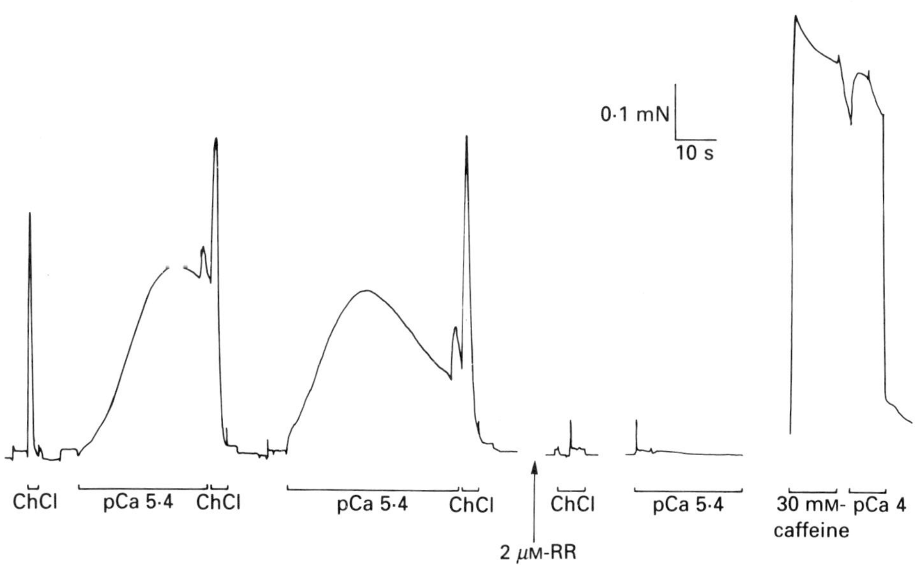

Fig. 5. Weak Ca^{2+}-induced Ca^{2+} release. Bathing a fibre in a K^+ solution weakly buffered (50μM EGTA) to a high $[Ca^{2+}]$ (pCa 5.4) could induce a slowly rising, sub-maximal response. It seems likely that this response was due to the exogenous Ca^{2+} inducing Ca^{2+} release from the SR, as the response was completely blocked by 2μM ruthenium red. This weak 'Ca^{2+}-induced Ca^{2+} release' was much slower and smaller than the depolarization-induced responses which could still be elicited at any time by ChCl substitution. (From Lamb & Stephenson (1990b) with permission).

EFFECT OF MYOPLASMIC Ca^{2+} ON CALCIUM RELEASE

Under certain circumstances, myoplasmic Ca^{2+} can stimulate Ca^{2+} release in skinned fibres (see reviews by Stephenson, 1981; Endo, 1985) and can directly activate the Ca^{2+} release channels (Smith et al, 1986; Meissner et al., 1988). This might suggest that some simple version of Ca^{2+}-induced Ca^{2+} release (CICR) is important in normal E-C coupling. However, in agreement with Endo (1985) the following experiments suggest that this is not the case.

Firstly, the response to depolarization following two or three depolarizations (by Na^{+} substitution) in the presence of 10mM EGTA (pCa >9), depleted the SR of releasable Ca^{2+}, whereas a similar duration exposure to 10mM EGTA, without any depolarizations, caused virtually no depletion (Lamb & Stephenson, 1990b). This implies that activating the voltage-sensors by depolarization could still trigger massive Ca^{2+} release even though the 'myoplasmic' [Ca^{2+}] was buffered to very low levels, at least initially. Secondly, exposing a skinned fibre to a solution very weakly buffered to a high [Ca^{2+}] (pCa 5.4, 50μ EGTA), induced a slowly rising, moderate response (Fig. 5). This response was completely blocked by 2μM RR, indicating that it involved Ca^{2+} release from the SR and not just diffusion of Ca^{2+} into the fibre from the weakly-buffered bathing solution. The Ca^{2+} diffusing into the fibre from the effectively unbuffered bathing solution apparently was taken up rapidly by the SR before it affected the contractile apparatus. Thus, in these solutions with 1mM free Mg^{2+}, Ca^{2+} could induce some Ca^{2+} release, but this release was smaller and much slower than depolarization-induced release of Ca^{2+} in the same fibres (eg Fig. 5). The slow rise and eventual decline of the Ca^{2+}-induced response was not due to the slow diffusion of Ca^{2+} into the fibre inactivating the channels before it fully activated them, because depolarization could still elicit a large, fast response at any time (Fig. 5). Thirdly, as mentioned above, repolarization can stop Ca^{2+} release at any time, indicating that the raised myoplasmic [Ca^{2+}] is not sufficient to keep the release channels open.

Thus, these experiments make it clear that, under the 'physiological' conditions used here (1mM free Mg^{2+}, 8mM total ATP, pH 7.0), E-C coupling does not involve simple CICR, but instead Ca^{2+} release stays under the control of the voltage-sensors. Nevertheless, when the voltage-sensors have been activated, Ca^{2+} may well help potentiate Ca^{2+} release (see below). The importance of Mg^{2+} in allowing the voltage-sensor to maintain control of Ca^{2+} release is described below.

EFFECT OF Mg^{2+} ON CALCIUM RELEASE

Meissner et al (1988) have shown that in the presence of physiological [ATP], Ca^{2+} release channels in lipid bilayers and in SR vesicles spontaneously open in the absence of Mg^{2+}, even at nanomolar free [Ca^{2+}]. We first investigated whether such an inhibitory effect of Mg^{2+} was important in functional muscle fibres where the release channels were still controlled by the voltage-sensors in the t-system.

Low [Mg^{2+}]

Lowering the free [Mg^{2+}] to 0.05mM, without any depolarization or change in the total [ATP], free [Ca^{2+}] or pH, invariably induced a large response in every rat or toad fibre loaded to normal levels with Ca^{2+} (eg Fig. 6). After such a response, the SR became substantially depleted of Ca^{2+} and neither depolarization nor caffeine (2mM) could elicit any response, unless the SR was reloaded with Ca^{2+} (in the presence of 1mM Mg^{2+}). This response to low [Mg^{2+}] was unaffected by inactivation of the

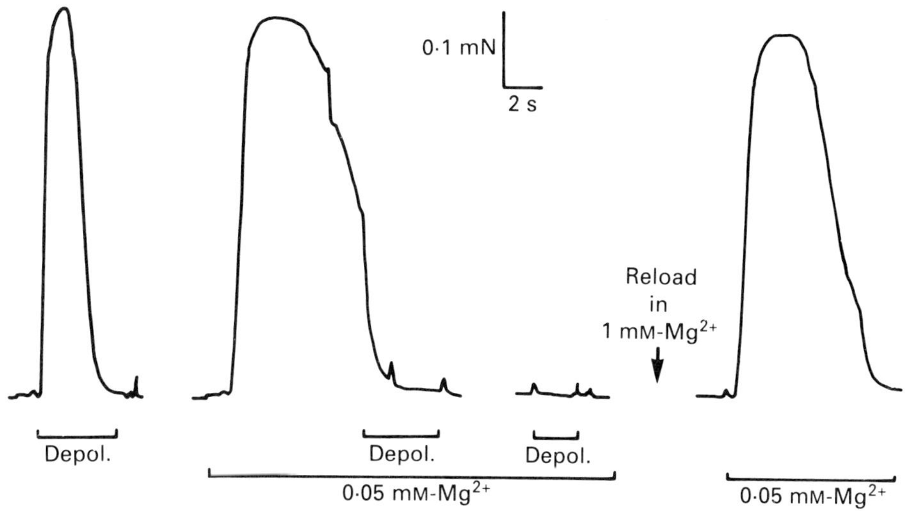

Fig. 6. Lowering $[Mg^{2+}]$ induces a response in a skinned muscle fibre. In 1mM free Mg^{2+}, depolarization of the t-system (by Na^+ substitution, labelled 'Depol.') induced a rapid contraction (left part of trace). Lowering the $[Mg^{2+}]$ to 0.05mM, without any depolarization, also induced a large response (middle trace) after a short diffusional delay, but subsequent depolarization-induced contractions could only be elicited after returning to 1mM Mg^{2+} and reloading the SR with Ca^{2+}. (From Lamb & Stephenson (1991a) with permission).

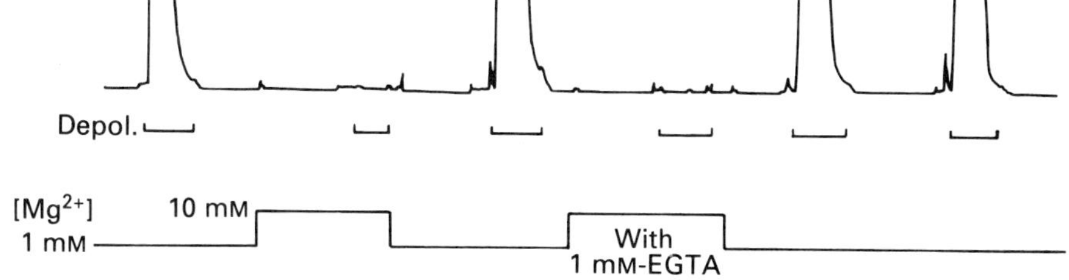

Fig. 7. Effect of 10mM Mg^{2+} on depolarization-induced release of Ca^{2+}. Depolarization in 10mM Mg^{2+} did not produce any contraction, and depolarization in 10mM Mg^{2+} plus 1 mM EGTA had little effect on the following response in 1mM Mg^{2+}, indicating that virtually no Ca^{2+} was released by depolarization in 10mM Mg^{2+}. Time scale: 2s during depolarizations (ChCl), 30s otherwise. (From Lamb & Stephenson (1991a) with permission).

voltage-sensors (by prolonged exposure to the Na^+ solution or by perforation of the t-system with saponin), but was completely blocked by 2μM RR. Lowering the [Mg^{2+}] to 0.05mM even induced massive Ca^{2+} release when the [Ca^{2+}] in the bathing solution was buffered to very low levels (pCa > 8.7, 2mM free EGTA). In the absence of ATP (with the fibre in rigor), neither the complete removal of Mg^{2+} (no Mg added and 1mM EDTA) nor depolarization could induce measurable Ca^{2+} release (Lamb & Stephenson, 1991a). If the free [Mg^{2+}] was kept constant at 1mM and the [MgATP] was lowered to the same level as in the 0.05mM free Mg^{2+} solutions (2mM MgATP), there was no spontaneous release of Ca^{2+}. Furthermore, depolarization could still elicit similar reponses as under the standard conditions (with 7mM MgATP), and the pump in the SR rapidly retrieved the released Ca^{2+}. Altogether, these results are totally consistent with the single channel and vesicle experiments of Meissner et al (1988) and furthermore show that Mg^{2+} plays an important role in functional muscle fibres by inhibiting the Ca^{2+} release channels when the voltage-sensors are not activated, ie when the voltage-sensors are in their resting or inactivated states. Nevertheless, it is clear that activating the voltage-sensors overcomes the inhibitory effect of physiological concentrations of Mg^{2+} (1mM) and elicits massive Ca^{2+} release.

High [Mg^{2+}]

In the presence of 10mM Mg^{2+} (pCa 6.7), depolarization did not induce any response in the skinned fibres (Fig. 7). Furthermore, depolarization in the presence of 10mM Mg^{2+} and 1mM EGTA (to chelate released Ca^{2+}) did not cause noticeable depletion of the Ca^{2+} in the SR, as indicated by the large response to a subsequent response in 1mM Mg^{2+} (Fig. 7). In contrast, a single depolarization in 1mM Mg^{2+} and 1mM EGTA reduced the following response under standard conditions to about 10% of the preceding control response, indicating substantial depletion of releasable Ca^{2+} (Lamb & Stephenson, 1991a). Thus, in 10mM Mg^{2+}, depolarization was unable to elicit much Ca^{2+} release from the SR. This blocking effect of Mg^{2+}, which was half-maximal at about 2.5mM free Mg^{2+}, was not due to Mg^{2+} interfering with the ability of depolarization to activate the voltage-sensors, as D600 could still produce a use-dependent 'paralysis' of the voltage-sensors in 10mM Mg^{2+} (Lamb & Stephenson, 1991a).

Now, as the release channels are permeable to Mg^{2+} as well as Ca^{2+}, it might be supposed that this inhibitory effect of Mg^{2+} is solely the result of myoplasmic Mg^{2+} competing with Ca^{2+} from the SR for common binding sites within the release channel. Nevertheless, this does seem to be the case. Firstly, in contrast to the results with 1mM Mg^{2+}, ryanodine (25μM) appeared to have little or no effect on the release channels during and following a depolarization in 10mM Mg^{2+}; the large response to depolarization one minute after returning to 1mM Mg^{2+}, indicated that the channels neither had been blocked open nor shut by ryanodine in 10mM Mg^{2+} (Fig. 8). This suggest that the depolarization in 10mM Mg^{2+} did not open the release channels. Secondly, it was possible to elicit a large response in 10mM Mg^{2+}, if the free [Ca^{2+}] in the bathing solution was raised (to pCa 5.8) and the fibre was depolarized in the presence of the potentiating agent caffeine (2mM) (Fig. 9). No such response could be elicited in the presence of 2μM ruthenium red. Thus, it seems that the release channels can conduct very large efflux of Ca^{2+} in the presence of 10mM myoplasmic Mg^{2+}. Together, these data suggest that elevated myoplasmic Mg^{2+} directly interferes with the ability of the voltage-sensors to activate the Ca^{2+} release channels. This may well be an important factor in the decreased Ca^{2+} release observed upon depolarization in fatigued muscle fibres, where the hydrolysis of ATP to ADP should have led to an increase in the free [Mg^{2+}].

The Voltage-sensors and the Role of Mg^{2+} Inhibition

Consideration of the responses to both low and high [Mg^{2+}] gives an interesting insight into the control of Ca^{2+} release in skeletal muscle. It appears that the stronger inhibitory effect of Mg^{2+} on the release channels in skeletal muscle compared with cardiac muscle (Meissner et al, 1988) is necessary to sufficiently dampen CICR at physiological [Mg^{2+}] (1mM) to enable the voltage-sensors to keep control of Ca^{2+} release. Activation of the voltage-sensors appears to have the same effect on the release channels in skeletal muscle as does lowering the [Mg^{2+}]. As activation of the voltage-sensors can overcome the inhibitory effect of 1mM Mg^{2+} but not 10mM Mg^{2+}, activating the voltage-sensors seems equivalent to lowering the affinity of the release channels to Mg^{2+} (by about 10 to 20 fold). If activating the voltage-sensors does indeed exert its

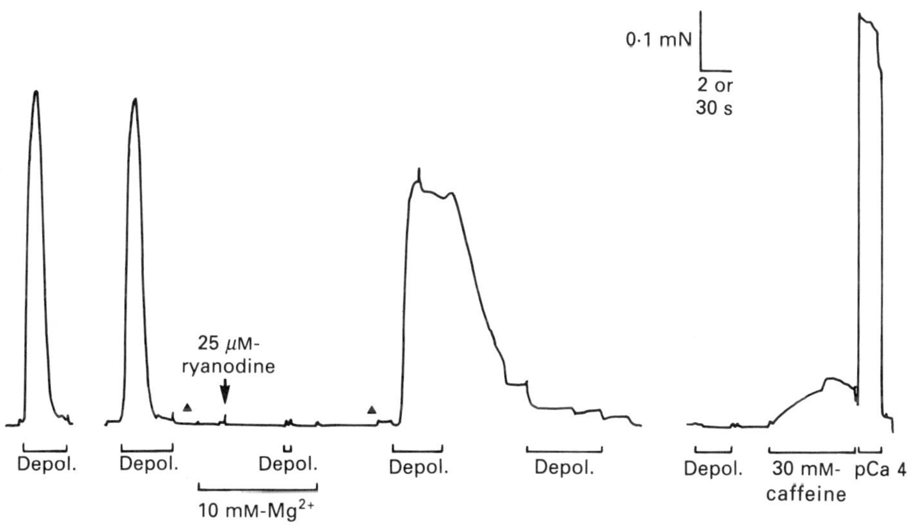

Fig. 8. Ryanodine (25µM) had little if any effect when the fibre was depolarized in 10mM Mg^{2+}, but with a single depolarization in 1mM Mg^{2+} ryanodine blocked open many Ca^{2+} channels, depleting the fibre of Ca^{2+}. Ryanodine was present in all solutions from arrow until 30mM caffeine. Time scale 30s between (▲) and 2s elsewhere. (From Lamb & Stephenson (1991a) with permission).

effect on the release channels by lowering their affinity for Mg^{2+}, this would reconcile and unify the apparently conflicting mechanisms of CICR and voltage-sensor control of Ca^{2+} release. Such a proposal would also explain why myoplasmic Ca^{2+} cannot initiate Ca^{2+} release when the voltage-sensor is in its resting state, but might potentiate Ca^{2+} release when the sensor is activated (Fig.10). In this hypothesis, activating the voltage-sensor exerts an effect on the release channels (by removing the inhibitory effect of Mg^{2+}), but inactivation of the sensors has no effect. This is consistent with the observations that the response of the release channels to direct activation by caffeine (Fig. 3), exogenous Ca^{2+} or ryanodine, was similar regardless of whether the voltage-sensors were in their resting or inactivated states (Lamb & Stephenson, 1991a).

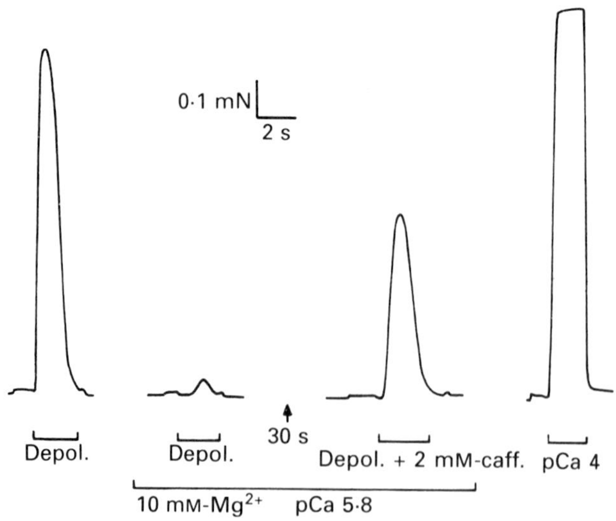

Fig. 9. Depolarization in 10mM Mg^{2+} could elicit a small response if the $[Ca^{2+}]$ in the bathing solution was raised 60s before depolarization (weakly buffered by 50μM EGTA at pCa 5.8). Depolarization in the same solution with 2mM caffeine produced a much larger response, indicating that channels were not simply blocked in 10mM Mg^{2+}, but rather that the depolarization alone could not open them. (From Lamb & Stephenson (1991a) with permission).

ROLE OF G-PROTEINS

It has been proposed that a GTP binding (G-) protein may be involved in linking the voltage-sensors to the release channels, possibly by facilitating the production of inositol trisphosphate (IP_3) (Di Virgilio et al, 1986). However, our experiments indicate that this is not the case. G-proteins become activated after exchanging bound GDP with GTP. Activation of the associated membrane receptor catalyses this exchange, thus activating the G-protein until it uses its inherent ability to hydrolyse the bound GTP back to GDP. Binding of GTPγS, a non-hydrolysable analogue of GTP, should effectively irreversibly activate any G-protein. Nevertheless, the presence of GTPγS (50μM to 1mM) had no effect whatsoever on the size, duration or number of depolarization-induced responses in both rat and toad muscle (Lamb & Stephenson, 1991b), even though such repeated activation should have required the repeated binding and hydrolysis of GTP if a G-protein did indeed link the sensors to the release channels.

As production of IP_3 is dependent on G-protein activation in other systems, this data also suggests that IP_3 does not link the voltage-sensors to the release channels. The fact that depolarization can continue to trigger Ca^{2+} release for many seconds, and that repolarization can stop this release at any time (Fig. 4), also makes it unlikely that any diffusable second messenger is necessary for linking the voltage-sensors to the release channels. Of course, it is still possible that IP_3 is produced by the activation of some surface receptor and its related G-protein, and that the IP_3 then triggers Ca^{2+} release from some type of Ca^{2+} release channel in the SR; nevertheless, this should be distinguised from normal E-C coupling.

G-proteins may still play a role in E-C coupling by modulating the response of the voltage-sensors to depolarization. However, our experiments did not provide any data to indicate this: GTPγS (250μM) had no effect on the contraction 'threshold' to sub-maximal depolarizations in skinned fibres, and GTPγS (1mM) had no effect on the amount or voltage-dependence of the asymmetric charge movement in rat e.d.l. fibres voltage-clamped with the vaseline gap technique (Lamb & Stephenson, 1991b). Nevertheless, it should be pointed out that the modulating G-protein or some related compound may be been 'washed out' in these experiments or alternatively the exchange of GDP with GTPγS on the G-protein may have been extremely slow without hormonal activation of the associated (unknown) surface receptor. Consequently, these experiments do not preclude a role of a G-protein in modulating the voltage-sensors.

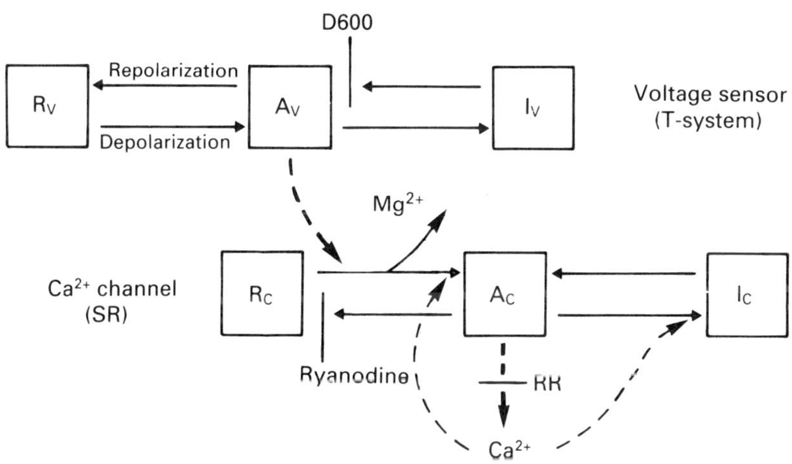

Fig. 10. Schematic diagram showing the effects of t-system potential and various agents on the states of the voltage-sensor in the t-system and the Ca^{2+} release channel in the SR. The voltage-sensor (subscript V) and release channel (subscript C) move between their respective resting (R), activated (A) and inactivated (I) states. D600 and ryanodine (Ry) hinder certain transitions as indicated by a bar blocking an arrow, and ruthenium red (RR) blocks Ca^{2+} release. Activation of the voltage-sensor strongly favours activation of the release channel, possibly by inducing the dissociation of Mg^{2+} from an inhibitory site on the release channel. By itself, myoplasmic Ca^{2+} only weakly activates the release channel, but in conjunction with activation of the voltage-sensor Ca^{2+} may further potentiate release. High myoplasmic $[Ca^{2+}]$ may also inhibit Ca^{2+} release by promoting inactivation of the release channel. (From Lamb & Stephenson (1990b) with permission).

CONCLUSION

These experiments have shown that activating the voltage-sensors in the t-system activates the Ca^{2+} release channels in the SR in a relatively direct manner, not involving a G-protein. Activating the voltage-sensors appears to have a similar effect to lowering the myoplasmic [Mg^{2+}], suggesting that the voltage-sensor may activate the release channels by lowering their affinity for Mg^{2+}, thereby removing the associated inhibition. Ca^{2+} does not readily activate Ca^{2+} release at physiological [Mg^{2+}] when the voltage-sensors have not been activated or have been inactivated, but Ca^{2+} may well potentiate further release when the activation of the voltage-sensors has overcome the resting inhibition by Mg^{2+}.

REFERENCES

Di Virgilio, F., Salviati, G., Pozzan, T., and Volpe, P., 1986, Is a guanine neucleotide-binding protein involved in excitation-contraction coupling in skeletal muscle?, EMBO Journal 5, 259-262.

Donaldson, S.K.B., 1985, Peeled mammalian skeletal muscle fibres. Possible stimulation of Ca^{2+} release via a transverse-tubule-sarcoplasmic reticulum mechanism, Journal of General Physiology **86**, 501-525.

Eisenberg, R.S., McCarthy, R.T., and Milton, R. L., 1983, Paralysis of frog skeletal muscle fibres by the calcium channel antagonist D600, Journal of Physiology **341**, 495-505.

Endo, M., 1985, Calcium release from sarcoplasmic reticulum, Current Topics in Membranes and Transport **25**, 181-230.

Fill, M.D., and Best, P.M., 1988, Contractile activation and recovery in skinned frog muscle stimulated by ionic substitution, American Journal of Physiology **254**, C107-C114.

Lamb, G.D., and Stephenson, D.G., 1990a, Calcium release in skinned muscle fibres of the toad by transverse tubular depolarization or by direct stimulation, Journal of Physiology **423**, 495-517.

Lamb, G.D., and Stephenson, D.G., 1990b, Control of calcium release and the effect of ryanodine in skinned muscle fibres of the toad, Journal of Physiology **423**, 519-542.

Lamb, G.D., and Stephenson, D.G., 1991a, Effect of magnesium ions on the control of calcium release in skeletal muscle fibres of the toad, Journal of Physiology **434**, 507-528.

Lamb, G.D. and Stephenson, D.G., 1991b, Excitation-contraction coupling in skeletal muscle fibres from rat and toad in the presence of GTPγS, Journal of Physiology (In Press).

Meissner, G., Rousseau, E., Lai, F.A., Liu, Q-Y., and Anderson, K.A, 1988, Biochemical characterization of the Ca^{2+} release channel of skeletal and cardiac sarcoplasmic reticulum, Molecular and Cellular Biochemistry **82**, 59-65.

Melzer, W., Schneider, M.F., Simon, B. and Szücs, G., 1986, Intramembrane charge movement and calcium release in frog skeletal muscle fibres. Journal of Physiology **373**, 481-511.

Smith, J.S., Coronado, R. and Meissner, G., 1986, Single channel measurements of the calcium release channel from skeletal muscle sarcoplas-mic reticulum, Journal of General Physiology **88**. 573-588.

Stephenson, E.W., 1981, Activation of fast skeletal muscle: contributions of studies on skinned fibers, American Journal of Physiology **240**, C1-C9.

Stephenson, E.W., 1985, Excitation of skinned muscle fibers by imposed ion gradients. I. Stimulation of ^{45}Ca efflux at constant [K][Cl] product, Journal of General Physiology **86**, 813-832.

MECHANISMS OF CALCIUM RELEASE FROM TERMINAL CISTERNAE IN CRUSTACEAN MUSCLE

Eduardo Rojas, Verónica Nassar-Gentina,
Michael E. Pollard and Mario Luxoro

Laboratory of Cell Biology and Genetics, NIDDK
National Institutes of Health, Bethesda, USA
and Laboratory of Cell Physiology, Faculty of
Sciences, University of Chile, Viña del Mar, Chile

INTRODUCTION

The chain of events leading to the release of Ca^{2+} from the terminal cisternae of the sarcoplasmic reticulum (SR) in vertebrate twitch skeletal muscle fibers begins with the action potential, which is propagated along the transverse tubular membrane (T tubules). The exact link between membrane potential changes in the T tubules (Heiny & Vergara, 1984) and Ca^{2+} release from the SR remains unresolved (Frank, 1980) despite the fact that the inward spread of mechanical activation has been measured (Costantin, 1975). Until recently, most of the experimental work on E-C coupling has been interpreted in terms of the direct electro-mechanical coupling hypothesis. In this model charged gating particles (Chandler, Rakowski & Schneider, 1976) control directly the release of Ca^{2+} from the SR (Hui & Gilly, 1979; Franzini-Armstrong, Block & Ferguson, 1990).

In the late sixties Bianchi and Bolton (1967) adopted the idea of chemical mediators in excitation-contraction coupling (E-C coupling). In this model depolarization of the plasmalemma of the twitch muscle fiber was proposed to induce the release of trigger substances from the T-tubules. Twenty years later, Vergara, Tsien and Delay (1985) produced, for the first time, compelling evidence in support of the involvement of chemical mediators in E-C coupling. The trigger substance was identified as inositol 1,4,5-trisphosphate ($InsP_3$). They demonstrated that $InsP_3$ was rapidly produced in stimulated muscle fibers and that intracellular application of exogenous $InsP_3$ caused the release of Ca^{2+} and contracture of frog twitch fibers. Support for this hypothesis in skeletal muscle from crustacea was also obtained recently in our laboratory (Rojas <u>et al</u>., 1987). We found that exogenous $InsP_3$ and caffeine elicited distinct but nonetheless multiphasic Ca^{2+} release from skinned muscle fibers from barnacle (or lobster) exposed to aequorin. We also noted that, provided the stores had been previously loaded with Ca^{2+}, while the Ca^{2+} release following applications of $InsP_3$ were transitory, the further addition of caffeine nevertheless always resulted in a maintained Ca^{2+} release signal. In addition, we observed that the application of $InsP_3$ in presence of caffeine (> 6 mM) continued to induce transitory Ca^{2+} release responses. In this situation, $[Ca^{2+}]$ always returned to the level obtained with caffeine alone. If $InsP_3$ and caffeine elicited Ca^{2+} release from a common pool, pretreatment with caffeine should limit the Ca^{2+} release induced by a subsequent treatment with $InsP_3$. However, maximal doses $InsP_3$ evoked similar Ca^{2+} release responses regardless of the presence or absence of caffeine (Rojas et al., 1987).

These results suggest that Ca^{2+} released by $InsP_3$ can be taken up by the $InsP_3$-sensitive pool and thus, Ca^{2+} released by $InsP_3$ and caffeine originated from different intracellular pools. To further test this hypothesis, we have used protocols consisting of applications of $InsP_3$ to skinned fibers between treatments with either caffeine or ryanodine. The advantage of compounds like caffeine and ryanodine is that they are non-hydrolyzable, and therefore might be expected to exert their effects until their specific compartments were exhausted. Furthermore, we examined the possibility that tetracaine, known to block E-C coupling in vertebrate skeletal muscle (Vergara & Caputo, 1983) might interfere only with the specific Ca^{2+} release mechanism. By contrast, the action of $InsP_3$ might be expected to be transitory, since $InsP_3$ molecules are rapidly destroyed by $InsP_3$-ases. We therefore hypothesized that if ryanodine and caffeine accessed the same compartment, treatment with one of the drugs should deplete of Ca^{2+} the potentially common compartment for sensitivity to the other. Alternatively, if the two drugs accessed different compartments, Ca^{2+} release elicited by one drug would not exhaust the other pool and sequential application of the two drugs should be additive. Similarly, one could ask the same question of $InsP_3$ and ryanodine by the same type of experiment.

In the present paper we show that Ca^{2+} responses to consecutive applications of caffeine and ryanodine are always additive. In addition, we observed that while teracaine blocks mechanical activation induced by caffeine, the anesthetic has no noticeable effects on $InsP_3$-induced Ca^{2+} release from the skinned fiber and development of tension in the intact fiber. We also show here that the complex Ca^{2+} release response induced by $InsP_3$ can be described by a minimal empirical model involving two components with different kinetics. Taken together our results suggest that in crustacean skeletal muscle $InsP_3$, ryanodine and caffeine act on different intracellular Ca^{2+} stores.

METHODS

Exogenous $InsP_3$-induced contractures and Ca^{2+} release from intracellular stores were measured in skinned fibers from the depressor muscle of the scutum of the giant barnacle Megabalanus psittacus. Minced muscle fibers from meropodites of the walking legs of the lobster <u>Homarus americanus</u> were also used to measure $InsP_3$-evoked Ca^{2+} transients.

Intracellular perfusion and tension measurements. The methods used for intracellular perfusion, for membrane potential control and, for the simultaneous recordings of calcium currents and mechanical activation have been described elsewhere (Keynes, Rojas, Taylor & Vergara, 1973). Intracellular perfusion was carried out with the muscle fiber mounted horizontally and held in position by the tendon end, which was attached to the shaft of a tension transducer and to a glass cannula inserted at the opposite end. With the fiber bathed in artificial sea water free of calcium (Ca^{2+}-free ASW in mM: 450 NaCl, 10 KCl, 60 MgC_2, 10 TrisCl, pH=7.6), a platinum capillary coated with black platinum was used as a current electrode and was inserted from the cannulated end along the axis of the fiber. During this operation, intracellular medium (in mM: 300 Cs-aspartate, 2 MgATP, 10 NaHepes, 600 sucrose, pH = 7.2) was injected continuously so as to separate the myofibrils in front of the tip of the platinum capillary, which was continuously advanced along the axis of the fiber.

Calcium measurements. The light resulting from the reaction of aequorin and ionized calcium was measured as described in detail elsewhere (Rojas, Forsberg & Pollard, 1986). The reaction chamber consisted of a plastic tube (1000 uL), with hydrophobic surfaces so as to keep 200 uL of solution used in the assays confined to the bottom. The tube was mounted vertically between two photomultipliers. Solutions containing either $InsP_3$, caffeine, or ryanodine were added in small volumes (20 uL) to the reaction medium using a microlitre pipette (Gilson) through a light-tight holder. Skinned fibers (wet weight <u>ca</u>. 50 mg), which had been incubated for 10 min in the reaction chamber in 50 uL of the internal medium (in mM: 180 K-aspartate, 2 MgATP, 10 NaHepes, 600 sucrose, pH=7.2) containing aequorin, were stimulated by rapidly injecting 20 uL of internal medium containing $InsP_3$, caffeine or ryanodine. Emitted light from the reaction chamber was detected using both photomultipliers. The anode of each photomultiplier was held at virtual ground voltage by means of an operational amplifier with negative feedback resistance (10^{10} Ohm).

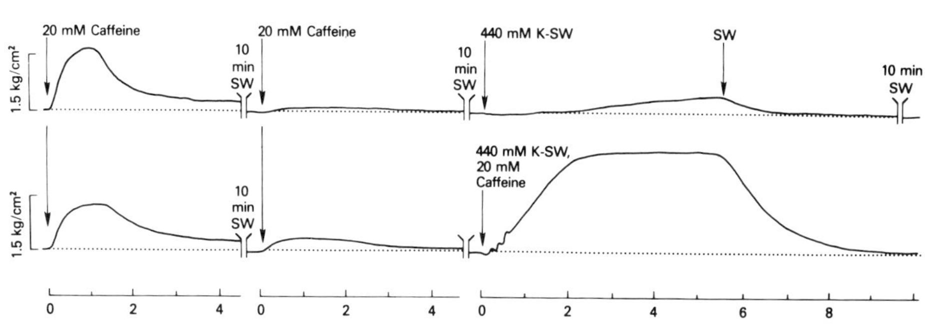

Fig. 1 Depletion of intracellular calcium pool sensitive to caffeine in barnacle muscle. $[Ca^{2+}]_o$ = 10 mM.

RESULTS

Extracellular and intracellular compartments as calcium sources in crustacean muscle

In crustacean skeletal muscle, both twitches and contractures elicited by a sudden elevation of $[K^+]_o$ depend on the presence of extracellular Ca^{2+} (Atwater, Rojas & Vergara, 1974; Hidalgo, Luxoro & Rojas, 1979). However, caffeine contractures can be elicited in the absence of extracellular Ca^{2+} (Gainer, 1968). Furthermore, peak tension during a caffeine (20 mM) contracture decreases dramatically during a second application of the methylxanthine as illustrated in Fig. 1 (upper panel), despite the fact that a rest period (10 min) in the absence of the drug was intercalated between the first and second contracture. We interpret these results to indicate that the intracellular Ca^{2+} pool accessed by caffeine has been depleted and that the Ca^{2+} released from that pool has been removed from the cytosol and transported out of the fiber or into an intracellular compartment which is insensitive to caffeine and that 10 min at rest was an insufficient condition for replenishing the caffeine-sensitive compartment. If after a series of caffeine-induced contractures the fiber is exposed to high K^+ ASW (< 5 min), development of tension occurs at a slow rate. A second series of consecutive applications of caffeine (lower panel) elicited a series of contractures similar to those observed in the upper panel. If the muscle fiber is exposed to high K^+ ASW again but this time in presence of caffeine, a vigorous contracture is observed (lower panel, right side). Since high K^+ ASW evokes Ca^{2+} entry through voltage gated Ca^{2+}-channels known to be present in the membrane of T-tubules and clefts (Keynes _et al._, 1974), and, Ca^{2+} uptake by the caffeine-sensitive pool is blocked, it is likely that most of the calcium entering through L-type Ca^{2+}-channels would be available to activate the contractile system.

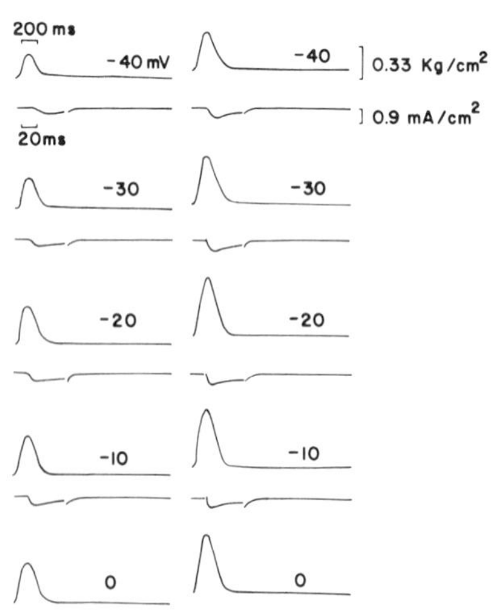

Fig. 2 Caffeine induced potentiation of peak tension during a twitch in barnacle muscle.

Brief (50 ms) depolarizations of the barnacle muscle membrane under voltage-clamp conditions can induce twitches only if Ca^{2+} is present in the external medium (Fig. 2). Two columns of five pairs of records (twitch and inward current records) are shown. First, the fiber was bathed in ASW in the absence of caffeine (left side) and then caffeine (5 mM) was applied (right side). Mechanical activation and area under each inward current record (a measure of the Ca^{2+} influx during and after the depolarizing pulse) as a function of membrane potential in both conditions remained unchanged (not shown). However, the maximum tension during each twitch was markedly augmented in the presence of caffeine.

Taken together, these results suggest that the caffeine-sensitive pool can be depleted by successive applications of the methylxanthine. It is also clear that an exhausted, caffeine-sensitive component of the SR will actually compete with the contractile system for Ca^{2+} entering the fiber, thereby preventing the development of a vigorous contracture in response to either high external $[K^+]$ (Fig. 1) or depolarizing voltage clamp pulses (Fig. 2).

Kinetics of inositol 1,4,5-trisphosphate-induced calcium liberation

Intracellular injection of exogenous $InsP_3$ can activate both Ca^{2+} release and contraction in barnacle muscle (Rojas et al., 1987) despite the unfavorable conditions to access the relevant targets before undergoing hydrolysis by $InsP_3$-ases. Successive applications of Ins_3 (10-160 uM) to minced lobster muscle fibers that had been previously incubated (3-5 min) in a medium containing MgATP ($[Ca^{2+}]$ adjusted to <1 uM) always elicited a multiphasic, transient increase in $[Ca^{2+}]$. Furthermore, the Ca^{2+} transient during the first application of $InsP_3$, regardless of the $[InsP_3]$ used (10-160 uM), decayed monotonically towards the resting level of $[Ca^{2+}]_i$ (Fig. 3, left side). In contrast, after a few applications, $InsP_3$ often generated Ca^{2+} transients which exhibited a clear hump (Fig. 3, right side). The second raise in $[Ca^{2+}]_i$ could be explained by assuming either, a second $InsP_3$-sensitive Ca^{2+} release

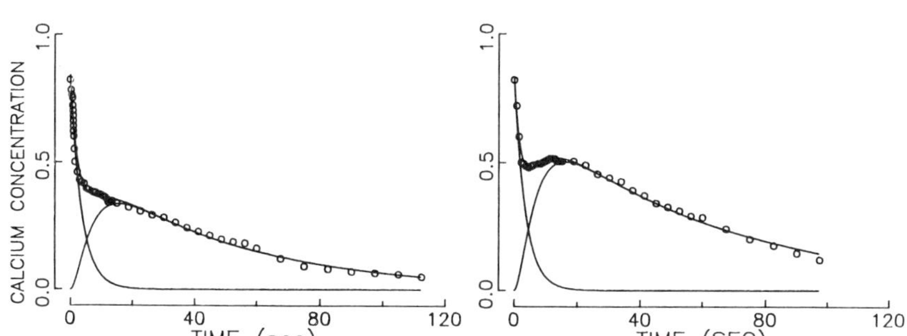

Fig. 3 Analysis of the calcium transient elicited from minced muscle by a rapid application of $InsP_3$ (60 uM) in terms of two components. MgATP was present in the medium; $[Ca^{2+}]_{t=0}$: 110 nM (left side) and 170 nM (right side).

process with different activation rate, or a second Ca^{2+} release process triggered by elevation of $[Ca^{2+}]_i$ caused by $InsP_3$.

Indeed, Ca^{2+} transients elicited by $InsP_3$ can be described by assuming that the Ca^{2+} transient is made up of the sum of two components with different kinetics (Fig.3). The raise to the initial peak of the calcium transient was fast (< 1 s) and could not be resolved with the system used in these measurements. For this reason we fitted the time course of the Ca^{2+}-transient using a minimal empirical model in which the rapid component reaches the maximum value instantaneously and the second component reaches the peak value with a finite activation time constant. The empirical equation is

$$[Ca^{2+}]_i = Ca_{f,max}\, \exp(-t/T_{D-f}) + Ca_{s,max}\, (1-\exp(-t/T_{A-s}))^2\, \exp(-t/T_{D-s}) \qquad (1)$$

where T_{D-f} and T_{D-s} represent the time constants of decay of the fast and slow component respectively and T_{A-s} represents the activation time constant for the slow component. Fig. 3 illustrates the goodness of the fit for the transients obtained during the first and fourth application of $InsP_3$ (60 uM). It may be seen that while the initial value $Ca_{f,max}$ remained unchanged at 0.83 uM the peak value $Ca_{s,max}$ augmented from 0.51 to 0.71 uM. $Ca_{f,max}$ did not increase with $InsP_3$ concentrations greater than 10 uM (not shown). The meaning of this analysis is that the $[InsP_3]$ for 50% $Ca_{f,max}$ response is less than 10 uM. The time constant of decay of the fast component T_{D-f} also remained unchanged at 3.6 ± 0.2 s (n=4) but the time constant of inactivation of the second Ca^{2+} release process seems to increase with $[InsP_3]$. Similar Ca^{2+} release transients were obtained with inositol 1,3,4,5-tetraphosphate ($InsP_4$) as illustrated in Figures 5 and 6.

We have previously shown that intracellular application of a fixed amount (0.2 to 8 nmol) of $InsP_3$ over a brief period (2 sec) into a barnacle muscle fiber induced a vigorous and transitory contracture. While peak tension attained during the first application depended on $[InsP_3]$, the relaxation following the application of $InsP_3$ depended on $[Mg^{2+}]_i$ (Rojas et al., 1987), i.e. low $[Mg^{2+}]_i$ (0.5 mM) slowed the relaxation. This result supports the idea that Mg^{2+}-dependent $InsP_3$-ases present in the myoplasm rapidly degrade $InsP_3$ into $InsP_2$ and InsP. For measurement of Ca^{2+} transients using skinned muscle fibers, the reaction medium contained Mg^{2+}. Thus, it is likely that the exogenous $InsP_3$ added is rapidly degraded by $InsP_3$-ases. Thus, the time constant T_{D-f} presumably measures the rate ($1/T_{D-f}$) of $InsP_3$ degradation. If this argument is correct, the hump in the records representing the maximum release for the second component, must be caused by a different mechanism which is nonetheless intimately dependent on the Ca^{2+} release evoked by the first component. One possible explanation is that the increase in $[Ca^{2+}]_i$ caused by the rapid activation of the Ca^{2+}-release channel in low resting $[Ca^{2+}]_i$ increases the sensitivity of the receptor for $InsP_3$. Although increases in $[Ca^{2+}]$ tended to lower the open probability of the $InsP_3$-sensitive Ca^{2+} release channel, it has been shown that "a critical concentration of calcium is required to allow for maximal agonist ($InsP_3$) potency" (Suárez-Isla et al. 1990).

Comparison between caffeine- and ryanodine-induced calcium release

In vertebrate skeletal muscle, ryanodine causes a progressive and irreversible contracture (Edwards et al., 1948). In barnacle (or lobster) muscle, ryanodine induces Ca^{2+} release from the SR and the ensuing development of tension. Fig. 4 illustrates the time course of the $[Ca^{2+}]$ rise induced by consecutive applications of ryanodine (54 uM) and caffeine (9 mM). Provided the skinned fibers are loaded with Ca^{2+} by incubation in the presence of MgATP high $[Ca^{2+}]$ (1 uM), the elevation in $[Ca^{2+}]$ induced by ryanodine is maintained (Fig. 4). Since, a Ca^{2+} loaded caffeine-sensitive pool would be unable to uptake the Ca^{2+} released from the ryanodine-sensitive pool, the maintained high $[Ca^{2+}]$ level in ryanodine treated skinned fibers must be caused by a ryanodine-induced blockade of the Ca^{2+} uptake system of the ryanodine-sensitive pool. By contrast, in the intact fiber, the elevation of $[Ca^{2+}]$ evoked ryanodine is not maintained and, $[Ca^{2+}]_i$ returns to resting levels of ca. 0.1 uM in the continued presence of the drug (not shown). The explanation for this difference is that in the intact fiber intracellular Ca^{2+} is rapidly transported out of the fiber by a Ca^{2+} transport system known to be present in barnacle muscle (Nassar-Gentina, Luxoro & Rojas, in preparation). It is also apparent from Fig. 4 that the responses were additive. Reversing the order in which ryanodine and caffeine were applied generated Ca^{2+} signals which were essentially of identical size (not shown). The additivity of the responses in terms of $[Ca^{2+}]$ reached at the end, was also true at different concentrations of ryanodine (0.1-61 uM) and caffeine (1-20 mM). This result suggests the two drugs accessed different Ca^{2+} pools. Increases in $[Ca^{2+}]$ induced by ryanodine and caffeine were 350 and 360 nM respectively suggesting that the capacity of the two pools is about equal.

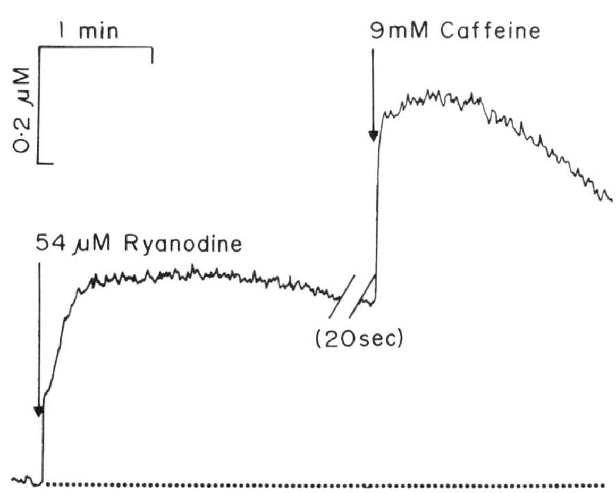

Fig. 4 Ca^{2+} release evoked by ryanodine and caffeine are additive (barnacle muscle).

Application of caffeine (9 mM) to skinned fibers after two consecutive exposures to $InsP_4$ (133 and 123 uM respectively) showed that the size of the caffeine response is not affected by these repetitive treatments with $InsP_4$ (Fig. 5).

In contrast, treatments with $InsP_3$ (or $InsP_4$) prior to the exposure to ryanodine, depending on concentration of $InsP_3$, can exhaust the compartment accessed by ryanodine. As illustrated in Fig. 6, after the consecutive applications of $InsP_4$ (133 uM), while ryanodine (53 uM) elicited a very small response (<u>ca</u>. 200 nM), caffeine (8 mM) gave a large response (<u>ca</u>. 1 uM). Ryanodine is also able to deplete the $InsP_3$-sensitive pool. Exposure of the skinned fibers to saturating concentrations of ryanodine (<u>ca</u>. 50 uM), depending on exposure time and ryanodine concentration, can diminish the release of Ca^{2+} evoked by $InsP_3$ (or $InsP_4$).

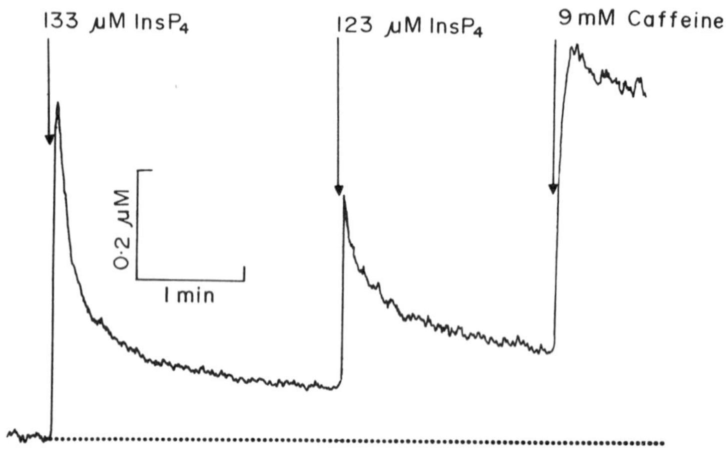

Fig. 5 $InsP_4$- and caffeine-induced Ca^{2+} release responses are independent of each other (lobster).

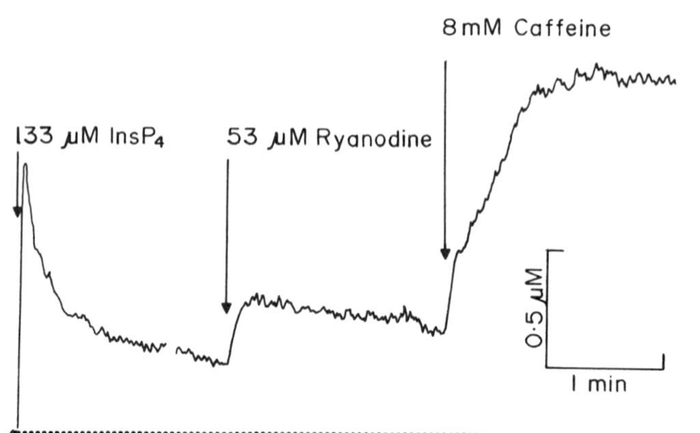

Fig. 6 $InsP_4$ can deplete the ryanodine- but not the caffeine-sensitive pool (barnacle muscle).

Differential effects of tetracaine and procaine on E-C coupling

We have previously shown that tetracaine (2 mM) but not procaine (10 mM) blocks high $[K^+]_o$-induced contractures in barnacle muscle fibers (Luxoro et al., 1990). For the intact fiber under voltage clamp conditions, we further showed that although inward Ca^{2+} currents were unaffected by these anesthetics twitches were also blocked (Fig. 7).

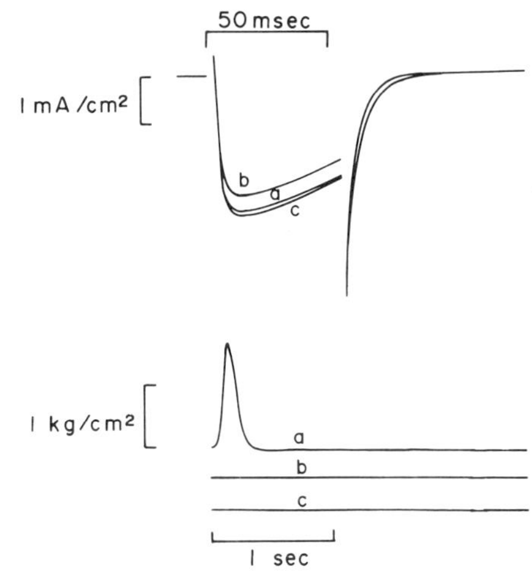

Fig. 7 Tetracaine (2 mM) blocks E-C coupling but not Ca^{2+} entry. A) Ca^{2+} currents: (a) control (10 mM Ca^{2+}); (b,c) in the presence of tetracaine either with 10 mM Ca^{2+} (b) or 15 mM (c). B) Traces of the twitches in response to the Ca^{2+} currents illustrated in A. Traces (b) and (c) were displaced for clarity (barnacle muscle). From Luxoro et al., 1990.

Since direct applications of Ca^{2+} (pCa ca. 5.2) to skinned fibers in presence of either tetracaine or procaine evoked vigorous contractures, we concluded that in the intact fiber tetracaine blocked E-C coupling somewhere after Ca^{2+} entry and before mechanical activation (Luxoro & Nassar-Gentina, 1984). Assuming that $InsP_3$ is indeed the chemical messenger coupling the electrical events at the level of the T-tubule membrane to the Ca^{2+}-release channel known to be present in the SR membranes from crustacean skeletal muscle (Arispe, Olivares, Jaimovich & Rojas, 1991), one could predict that tetracaine would be unable to block $InsP_3$-induced contractures.

As illustrated in Fig. 8, intracellular injection of $InsP_3$ (1 uL of intracellular medium containing 1 mM $InsP_3$) after exposure of the fiber to tetracaine (2 mM) elicited a contracture similar to that obtained during the control application of $InsP_3$ recorded in the absence of the anesthetic. Fig. 8 also shows that the response to caffeine was completely abolished by tetracaine. Taken together these results suggest that tetracaine blocks the coupling between the T-tubule membrane signal and the generation of $InsP_3$. In addition, since the anesthetic effectively blocked caffeine evoked contractures, it can be safely concluded that caffeine acts on the tetracaine blockable Ca^{2+} store which is

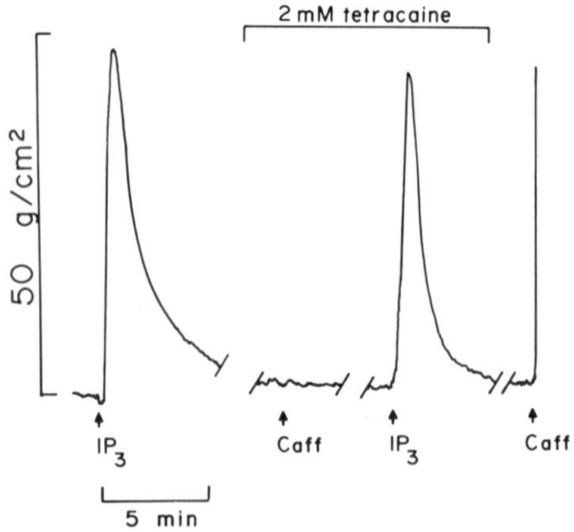

Fig. 8 Differential effects of tetracaine on $InsP_3$- and caffeine-evoked contractures in barnacle muscle fibers. Cut marks: 30-min rest. From Luxoro et al., 1990.

different than that accessed by $InsP_3$ and ryanodine. It follows from these results that, in physiological E-C coupling, the junctional T-tubular membrane is coupled either directly to the caffeine- and tetracaine-sensitive compartment or indirectly, by means of $InsP_3$, to the other Ca^{2+} store. In this second alternative, the anesthetic should block the release of $InsP_3$ from the junctional T-tubular membrane. It is also possible that both modes of E-C coupling coexist crustacean muscle.

DISCUSSION

Our results show clearly that there are at least two, and possible three, intracellular Ca^{2+} pools in crustacean skeletal muscle fibers. This conclusion is based on our observation that ryanodine can deplete the $InsP_3$-sensitive pool which is different than that depleted by caffeine. The presence of pharmacologically distinct Ca^{2+} release compartments in crustacean skeletal muscle is further supported by our observations that in presence of tetracaine (2 mM) application of $InsP_3$ but not caffeine elicited contraction (Fig. 8). The differential effects of tetracaine on caffeine and $InsP_3$-induced contractures indicate that $InsP_3$ and caffeine induce Ca^{2+} release by different mechanisms.

In crustacean skeletal muscle, physiological E-C coupling is totally dependent on the presence of external Ca^{2+} (Hidalgo, Luxoro & Rojas, 1979). Furthermore, peak tension during twitches and Ca^{2+} entry, estimated from the inward Ca^{2+} current, are closely related (Atwater et al., 1974; Hidalgo, et al., 1979). Unlike this situation, vertebrate skeletal muscle can twitch even in the absence of external calcium (Armstrong, Bezanilla & Horowicz, 1972). In addition, tetracaine (0.4 mM) inhibits T-tubule Ca^{2+} currents in vertebrate skeletal muscle (Garcia et al., 1991) but is ineffective (2 mM) in barnacle muscle (Fig. 7). In vertebrate muscle tetracaine also blocks the delayed component of the charge movement (Vergara & Caputo, 1983) thought by some to be a consequence of the Ca^{2+} release process from the SR (Csernoch et al., 1988). In crustacean muscle we found that tetracaine (2 mM) is able to block E-C coupling at a level other than Ca^{2+} entry (Fig. 7). Interestingly, we observed that $InsP_3$, even in presence of tetracaine, was still able to elicit contraction (Fig. 8). This result demonstrates that the $InsP_3$ receptor on the SR membranes is not affected by tetracaine. Thus, assuming that $InsP_3$ is indeed the chemical link in crustacean skeletal muscle, the simplest explanation for the lack of effect of the anesthetic on $InsP_3$-evoked contractures, is that tetracaine might inhibit the liberation of $InsP_3$ from diadic junctional T-tubule membranes. Clearly further work is needed to distinguish between these alternatives.

The inhibition of caffeine-induced contractures by tetracaine was found to be readily reversible (Fig. 8). Thus, the mechanism of blockade cannot assumed to be by Ca^{2+} depletion of the caffeine-sensitive pool caused by tetracaine. Therefore, it is likely that tetracaine acts directly on the Ca^{2+}-release mechanism activated by caffeine. The caffeine-activated Ca^{2+}-channel known to be present in SR membranes from vertebrate skeletal muscle (Suarez-Isla et al., 1986) is, in our opinion, the most probable site of action for tetracaine.

In conclusion, we have provided evidence in support of the idea that under physiological conditions there are at least two distinct Ca^{2+} release mechanisms operating in crustacean skeletal muscle. We propose that the $InsP_3$-sensitive compartment plays a key role in E-C coupling and attribute a less preponderant role to the caffeine-sensitive compartment.

REFERENCES

Arispe, N., Olivares, E., Jaimovich, E. & Rojas, E., 1991, Characteristics of Ca^{2+} release channels from crustacean skeletal muscle, J. Memb. Biol., in press.

Armstrong, C. M., Bezanilla, F. & Horowicz, P., 1972, Twitches in the presence of ethylene glycol bis(ß-aminoethyl ether)-N,N'-tetraacetic acid. Biochim. Biophys. Acta, 267: 605-608.

Atwater, I., Rojas, E. & Vergara, J., 1974, Calcium influxes and tension development in perfused single barnacle muscle fibres under membrane potential control, J. Physiol. (London), 243: 523-551.

Bianchi, C. P. & Bolton, T.C., 1967, Action of local anesthetics on coupling systems in muscle, J. Pharmacol. Exp. Ther. 1957: 388-405.

Csernoch, L., Huang, C. L.-H., Szucs, G. & Kovacs, L., 1988, Differential effects of tetracaine on charge movement and Ca^{2+} signals in frog skeletal muscle, J. Gen. Physiol. 92: 601-612.

Chandler, W. K., Rakowski, R.F. & Schneider, M.F., 1976, A non-linear voltage dependent charge movement in frog skeletal muscle. J. Physiol. (London), 254: 245-283.

Costantin, L.L., 1975, Contractile activation in skeletal muscle, Prog. Biophys. Mol. Biol. 29: 197-224.
Edwards, G.A., Weiant, E.A., Slocombe, A.G. & Roeder, K.D. (1948), The action of ryanodine on the contractile process in striated muscle, Science 108: 330-332.
Franzini-Armstrong, C., Block, B. & Ferguson, D. G., 1990 A molecular architecture of T-SR junctions: Evidence for a junctional complex that directly connects the two membrane systems. In Transduction in Biological Systems, Eds. C. Hidalgo, J. Bacigalupo, E. Jaimovich & J. Vergara, Plenum Press, pp 339-360.
Frank, G.B., 1980, The current view of the source of calcium in excitation-contraction coupling in vertebrate skeletal muscle. Biochem. Pharmocol. 29: 2399-2406.
Garcia, J., Avila-Sakar, A.J. & Stefani, E., 1991, Differential effects of ryanodine and tetracaine on charge movement and calcium transients in frog skeletal muscle, J. Physiol (London), 440: 403-417.
Geiner, H., 1968, The role of calcium in excitation-contraction coupling in lobster muscle. J. Gen. Physiol. 52:88-110.
Heiny, J. & Vergara, J., 1984. Dichroic behavior of the absorbance signal from dyes NK 2367 and WW 375 in skeletal muscle fibers, J. Gen. Physiol. 84: 805-837.
Hidalgo, J., Luxoro, E. & Rojas, E., 1979, On the role of extracellular calcium in triggering contraction in muscle fibres from barnacle under membrane potential control, J. Physiol. (London), 288:313-330.
Hui, C.S. & Gilly, W.F., 1979, Mechanical activation and voltage-dependent charge movement in stretched muscle fibers, Nature, 281: 223-225.
Keynes, R.D., Rojas, E., Taylor, R.E. & Vergara, J., 1973, Calcium and potassium systems of a giant barnacle muscle fibre under membrane potential control. J. Physiol. (London), 229:409-455.
Luxoro, M. & Nassar-Gentina, V., 1984, Potassium-induced depolarizations and generation of tension in barnacle muscle fibers: Effects of external calcium, strontium, and barium, Quart. J. Exp. Physiol. 69: 235-243.
Luxoro, M., Nassar-Gentina, V., Rojas, E., 1990, Excitation-contraction coupling in barnacle muscle fibers: Does calcium entry trigger contraction directly?. In Transduction in Biological Systems, Eds. C. Hidalgo, J. Bacigalupo, E. Jaimovich & J. Vergara, Plenum Press, pp.289-300.
Rojas, E., Forsberg, E. & Pollard, H.B., 1976, Optical detection of calcium dependent ATP release from stimulated medullary chromaffin cells, Adv. Exptl. Med. and Biol., 211: 7-28.
Rojas, E., Nassar-Gentina, V., Luxoro, M., Pollard, M.E. & Carrasco, M.A., 1987, Inositol 1,4,5-trisphosphate-induced Ca^{2+} release from the sarcoplasmic reticulum and contraction in crustacean muscle. Can. J. Physiol. Phar. 65:672-680.

Suárez-Isla, B.A., Marengo, J.J., Irribarra, V. & Bull, R., 1990, Calcium channels in in sarcoplasmic reticulum membranes isolated from skeletal muscle. In Transduction in Biological Systems, Eds. C. Hidalgo, J. Bacigalupo E. Jaimovich & J. Vergara, pp.487-499.
Vergara, J. & Caputo, C., 1983, Effects of tetracaine on charge movements and calcium signals in frog skeletal muscle fibers. Proc. Natl. Acad. Sci. (USA) 80: 1477-1481.
Vergara, J., Tsien, R. & Delay, M., 1985, Inositol 1,4,5-trisphosphate: a possible chemical link in excitation contraction coupling in muscle. Proc. Natl. Acad. Sci. (USA) 82: 760-772.

nACh RECEPTOR-ACTIVITY MODULATING INTRACELLULAR Ca^{2+} (RAMIC); ITS FINDING, PROPERTIES AND ROLES

I. Kimura, H. Tsuneki, T. Kondoh and M. Kimura

Department of Chemical Pharmacology
Faculty of Pharmaceutical Sciences
Toyama Medical and Pharmaceutical University
2630 Sugitani, Toyama 930-01, Japan

Introduction

Noncontractile and contractile Ca^{2+} transients are observed simultaneously by nerve stimulation at neuromuscular synapse of mouse skeletal muscles in the presence of cholinesterase inhibitors [1]. Noncontractile Ca^{2+} transients are more susceptible to n-acetylcholine receptor (nAChR) competitive blockers and external Ca^{2+} than contractile ones. We call the noncontractile ones "RAMIC" (receptor-activity modulating intracellular Ca^{2+}). In this study, we elucidated the mechanisms of RAMIC, compared with those of contractile Ca^{2+} mobilization.

Materials and Methods

We measured simultaneously Ca^{2+}-aequorin luminescence (Ca^{2+} transients) and twitch tension using phrenic nerve-diaphragm muscle preparation isolated from male ddY mice (7-9 weeks old, 28-42 g)[2] treated with neostigmine (0.3 µM). Ca^{2+} transients were led through an acrylic optical fiber to a photomultiplier tube, measured by a photon counter, and analyzed by a FFT analyzer. The signals were averaged 30 times by a signal processor.

Results

1. nAChR competitive blockers suppress RAMIC at much lower concentrations than contractile ones

Pancuronium (10-70 nM) and (+)-tubocurarine (13-130 nM) suppressed RAMIC, but not yet contractile Ca^{2+} transients. Succinylcholine (0.1-1 µM) suppressed both Ca^{2+} mobilization at the same extent, and mainly decreased peak amplitude of RAMIC. Competitive blockers diminished the decay time (T_2), and a depolarizing blocker diminished the time to peak (T_1) in RAMIC signal.

Excitation-Contraction Coupling in Skeletal, Cardiac, and Smooth Muscle
Edited by G.B. Frank *et al.*, Plenum Press, New York, 1992

2. nAChR open channel blockers suppress the activation of RAMIC, and nAChR closed channel blockers accelerate the desensitization of RAMIC

Bupivacaine, an open channel blocker, decreased peak amplitude of RAMIC, whereas chlorpromazine, a closed channel blocker, shortened the duration. Phencyclidine, a mixed type of blocker, decreased both peak amplitude and the duration. All above channel blockers blocked contractile Ca^{2+} transients in the same concentration range [3]. Channel blockers diminished both T_1 and T_2 in RAMIC signal.

3. Calmodulin-related agents may affect RAMIC as a result of nAChR closed channel blocking

Intracellular injection of calmodulin (10,000 U/ml in a pipette) did not change RAMIC and contractile Ca^{2+} transients. Both W-7 (10 µM) and W-5 (10 µM) suppressed RAMIC and contractile Ca^{2+} transients at the same extent. KN-62 (5 µM) did not affect both Ca^{2+} transients.

4. Skeletal muscle Na^+ channel blockers do not affect RAMIC

Geographutoxin II (0.3 µM) and formamide (2 M) blocked completely contractile Ca^{2+} transients, but not RAMIC [3,4].

5. RAMIC is not involved in the activation of muscle Na^+ channel and subsequent Ca^{2+} release from sarcoplasmic reticulum

Caffeine enhanced at 0.25-5 mM, and decreased at 7-20 mM contractile Ca^{2+} transients, but only decreased at 2-10 mM RAMIC [4].

6. G-protein activators increase contractile Ca^{2+} transients

Cholera toxin (CTX, 0.1 µM for 3 hr-treatment), pertussis toxin (PTX, 50 µg/kg, bolus i.v., 3-5 days-elapsed mouse), GTPγS (intracellular injection by a pipette containing 10 mM), and GDPβS (10 mM in an injection pipette) did not affect RAMIC. CTX and GTPγS, however, significantly, and PTX slightly, increased contractile Ca^{2+} transients.

7. Cyclic AMP accumulation increases contractile Ca^{2+} transients

Neither isoproterenol (10 µM) nor theophylline (100 µM) affected RAMIC. Isoproterenol significantly, and theophylline slightly, increased contractile Ca^{2+} transients.

8. Protein kinase A inhibitor decreases noncontractile Ca^{2+} transients

H-89 (0.3 µM), but not its noneffective analogue, H-85 (0.3 µM), decreased noncontractile Ca^{2+} transients and tended to increase contractile ones.

9. Phospholipase A_2 increases contractile Ca^{2+} transients, but not phospholipase C

Phosphatidylinositol (PI)-phospholipase C (PLC)

(intracellular injection by a pipette containing 5 U/ml), phosphatidylcholine (PC)-PLC (5 U/ml in a pipette), anti-PC-PLC (1-10 mg/ml in a pipette) and heparin (10 mg/ml in a pipette) affected neither RAMIC nor contractile Ca^{2+} transients. Intracellular injection of PLA_2 (10,000 U/ml in a pipette) increased contractile Ca^{2+} transients, but not RAMIC.

10. Protein kinase C activator decreases contractile Ca^{2+} transients

Neither staurosporine (10 nM) nor H-7 (10 µM) did not change both types of Ca^{2+} transients. 12-O-tetradecanoylphorbol 13-acetate (TPA, 1 µM) but not its noneffective analogue, 4α-phorbol 12, 13 didecanoate (4α-PDD, 1 µM), however, decreased significantly contractile Ca^{2+} transients.

Conclusion

In the present study, we demonstrate that in neuromuscular synapse of skeletal muscles, RAMIC is more susceptible to nAChR itself than contractile Ca^{2+} mobilization, and is supported increasingly by protein kinase A. Contractile Ca^{2+} mobilization is regulated increasingly by G-protein, cyclic AMP and PLA_2, but decreasingly by protein kinase C. The functional role of RAMIC may be to control contractile Ca^{2+} transients through cross-talk between protein kinase A and C, PLA_2 and nAChR.

References

1. Kimura I, Kondoh T and Kimura M. Postsynaptic nicotinic ACh receptor-operated Ca^{2+} transients with neostigmine in phrenic nerve-diaphragm muscles of mice. Brain Res. 507:309-311 (1989).
2. Kimura I, Kondoh T and Kimura M. Changes in intracellular Ca^{2+} produced in the mouse diaphragm by neuromuscular blocking drugs. J. Pharm. Pharmacol. 42:626-631 (1990).
3. Kimura M, Kimura I, Kondoh T and Tsuneki H. Noncontractile acetylcholine receptor-operated Ca^{++} mobilization: Suppression of activation by open channel blockers and acceleration of desensitization by closed channel blockers in mouse diaphragm muscle. J. Pharmacol. Exp. Ther. 256:18-23 (1991).
4. Kimura I, Kondoh T, Tsuneki H and Kimura M. Reversed effect of caffeine on noncontractile and contractile Ca^{2+} mobilization operated by acetylcholine receptor in mouse diaphragm muscle. Neurosci. Lett. (in press)(1991).

ENERGETICS OF MUSCARINIC STIMULATION OF SMOOTH MUSCLE

Robert M. Levin, Joseph Hypolite, Penelope A. Longhurst, and Alan J. Wein

Division of Urology, 3006 Ravdin Courtyard Building, Hospital of the University of Pennsylvania, 3400 Spruce Street Philadelphia, PA 19104

INTRODUCTION

Urinary bladder contraction is mediated primarily via the stimulation of muscarinic receptors (1). The contractile response to bethanechol stimulation can be dissociated from the metabolic response through the use of diltiazem (calcium channel blockade) (2). KCl stimulates contraction in isolated strips by directly depolarizing the membrane rather than by binding to specific membrane receptors. The current study directly compares the metabolic and contractile activities of bethanechol (muscarinic stimulation) with KCl (direct membrane depolarization).

METHODS

Isolated strips of rabbit urinary bladder body were monitored in-vitro for changes in intracellular free calcium, NADH/NAD ratio, and contraction. Intracellular free calcium was monitored by preincubating of isolated bladder smooth muscle strips with FURA-2 AM and continuously measuring the fluorescence with an MB2 surface spectrofluorometer using excitation wavelengths of 340 and 380 nm, and an emission wavelength of 510 nm. The NADH/NAD ratio was monitored with the MB2 surface spectrophotometer using an excitation wavelength of 366 nm and an emission wavelength of 450 nm. Contraction was monitored using an isometric force transducer connected to a Grass model D polygraph (3).

The effects of bethanechol (1--1000 µM) on contraction, NADH fluorescence, and calcium fluorescence were compared to the effects of KCl (1--140 mM).

RESULTS

Both bethanechol and KCl stimulated a rapid rise in intracellular free calcium and a slower rise in contraction. The calcium response to both agents returned to control levels whereas the smooth muscle tension remained elevated for the duration of stimulation. The maximum contractile response to bethanechol was significantly greater than the maximum response to KCl, however, the maximum increase in intracellular calcium was approximately the same for both.

A comparison of the dose response curves for contraction, calcium fluorescence, and NADH fluorescence for bethanechol and KCL is shown in figures 1 and 2. For Bethanechol: the ED-50 for NADH was significantly lower than the ED-50 for either calcium fluorescence or contraction, which were equal to each other. For KCl: there were no significant differences in the ED-50's among contraction, calcium fluorescence and NADH fluorescence. Both bethanechol and KCl stimulated a decrease in the NADH fluorescence (NADH/NAD ratio) indicating a decrease in metabolic energy.

Diltiazem (200 uM) inhibited both the contractile and calcium responses to bethanechol by approximately 90%, whereas the NADH response was inhibited by only 20%. In contrast, diltiazem (200 uM) inhibited the contractile, calcium, and NADH responses to KCl by approximately 82%.

Excitation-Contraction Coupling in Skeletal, Cardiac, and Smooth Muscle
Edited by G.B. Frank *et al.*, Plenum Press, New York, 1992

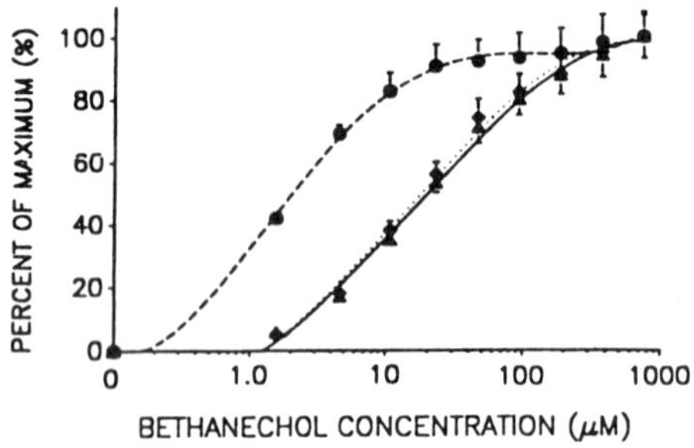

FIGURE 1

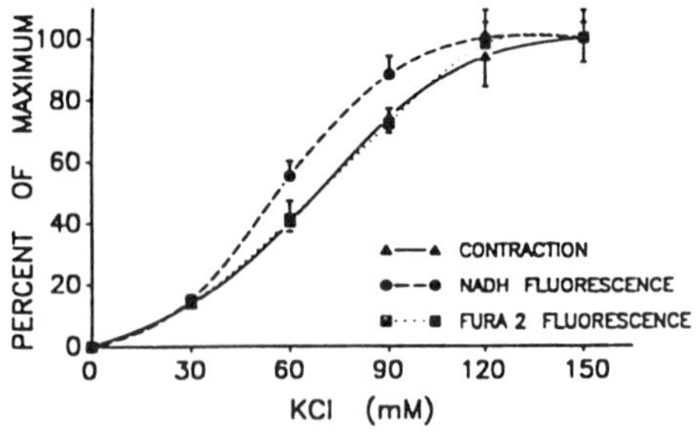

FIGURE 2

DISCUSSION

The results of this study clearly demonstrate that bethanechol stimulates the metabolism of NADH to NAD at concentrations that have virtually no contractile or calcium translocation responses, and that this response is relatively resistant to inhibition by diltiazem. Thus, these results indicate that a major part of this metabolic response to bethanechol (increase in energy utilization) is independent of direct contractile stimulation.

Alternately, contractile stimulation by KCl (direct membrane depolarization) shows the same ED-50 for the increase in intracellular free calcium and NADH fluorescence as it does for contraction. In addition, all three parameters are inhibited to approximately the same degree by the calcium antagonist diltiazem. This would indicate that the metabolic response to KCl is directly related to the contractile response.

It is clear from the current studies and those in the literature, receptor-mediated contraction utilizes a significantly greater amount of energy than contraction mediated by membrane depolarization. (4) It would be reasonable to assume that the energy required for the contraction of the myofilaments would be similar for both cases. The additional energy utilized by receptor-mediated contraction relates to the second messenger systems utilized for translating agonist binding to contraction. In addition to mediating contraction, these intracellular reactions can have significant effects on biochemical events other than contraction. For example, in the present study there was a substantial decrease in the NADH/NAD ratio observed in the absence of the contractile response (in the presence of diltiazem). The mechanisms which mediate this response will be the subject of future studies.

REFERENCES

1) Wein, A.J., Levin, R.M., and Barrett, D.M., Voiding Function: Relevant Anatomy, Physiology, and Pharmacology, in Adult and Pediatric Urology (Gillenwater, J.Y., Grayhack, J.T., Howards, S.S., and Duckett, J.D. eds.) Year Book Medical Publishers, Chicago, 1987, pp 933-1001.
2) Ruggieri, M.R., Wein, A.J., Hypolite, J.A., and Levin, R.M.: Dissociation of the metabolic from contractile response to muscarinic receptor stimulation in the rabbit urinary bladder. Mol. Cell. Biochem., 81: 137, 1988.
3) Franke, H., Barlow, C.H., and Chance, B.: Surface fluorescence of reduced pyridine nucleotide of the perfused rat kidney: interrelation between metabolic and functional states. Contr. Nephrol., 19: 240, 1980.
4) Peterson, J.W.: Effect of histamine on the energy metabolism of K+-depolarized hog carotid artery. Circ. Res., 50: 848, 1982.

NPY-(18-36) AND PERTUSSIS TOXIN DISTINGUISH BETWEEN THE NEGATIVE AND POSITIVE CONTRACTILE EFFECTS OF NEUROPEPTIDE Y (NPY) IN VENTRICULAR CARDIOMYOCYTES

[1]B. Cherie Millar, [1]Barbara J. McDermott, [2]H. Michael Piper, [3]Ambikaipakan Balasubramaniam and [4]Thomas Weis

[1]Department of Therapeutics and Pharmacology, The Queen's University of Belfast, [2]Physiologisches Institut/[4]Pharmacologie Institut, Heinrich-Heine-Universitat, Dusseldorf and [3]Department of Surgery, University of Cincinnati

SUMMARY

The mechanism of action of NPY to produce contractions of adult rat ventricular myocytes was investigated. Positive effects were mediated through an L-type Ca^{2+} channel. Negative effects, linked through an inhibitory G-protein to a transient outward K^+ channel, could be antagonised by NPY-(18-36).

INTRODUCTION

The documented actions of NPY on cardiac contractility are varied, depending on the species and preparation used.[1] In our studies the direct action of NPY on the myocardium has been investigated using as a model the effects of the peptide on the isolated ventricular cardiomyocyte. We previously showed an inhibitory effect of NPY on the contractile response of isoprenaline stimulated ventricular cardiomyocytes from the rat.[2] In this model, it was also shown that NPY inhibits isoprenaline-stimulated adenylate cyclase.[3] Both effects were sensitive to pertussis toxin. In the present study the hypothesis was pursued whether the observed decrease in isoprenaline-stimulated contractility and the lack of direct correlation with attenuation of cAMP accumulation could be a consequence of the binding of NPY to specific receptors linked to ion channels.

METHODOLOGY

Initial electrophysiological studies using voltage-clamp techniques indicated that a transient outward K^+ current (i_{to}) and an L-type Ca^{2+} current (iCa^{2+}) were activated by NPY (10^{-7}M). In pharmacological investigations, 4-aminopyridine (4-AP) and verapamil were used to block the i_{to} and the iCa^{2+}, respectively. Maximum changes in cell length during field-stimulated (0.5Hz, 0.5ms, 60V) contractions were determined in the presence of adenosine deaminase (5U/ml).

RESULTS

NPY had no effect on basal levels of contraction but attenuated the increase in the contractile response induced by 10^{-7}M isoprenaline (IC_{50} = $4x10^{-9}$M). In the

Excitation-Contraction Coupling in Skeletal, Cardiac, and Smooth Muscle
Edited by G.B. Frank *et al.*, Plenum Press, New York, 1992

presence of 4-AP at a concentration of 0.5 mM, which inhibits selectively the i_{to} in adult rat cardiomyocytes,[4] the inhibitory effect of NPY (10^{-7}M) was completely abolished. Also, the basal contractile response to NPY was stimulated (EC_{50} = $2x10^{-9}$M) and this effect could be abolished using verapamil (10^{-8}M). Treatment with pertussis toxin (1μg/ml) completely abolished the negative but not the positive action of NPY on contraction. The negative effect is mediated, therefore, through a receptor linked to an inhibitory G-protein.

The antagonist at the NPY receptor in cardiac tissue, NPY-(18-36) (10^{-6}M),[5] attenuated the negative but not the positive contractile response exerted by NPY.

DISCUSSION

The rat cardiomyocyte , therefore, seems to possess two mechanisms by which NPY can influence the cellular contractile response. One mechanism operates through the transient outward K^+ channel and involves a pertussis toxin-sensitive G protein, and the second is linked to a slow L-type Ca^{2+} channel. The results of this study indicate that these two mechanisms are similar and very sensitive to NPY. On consideration of the present data along with previous results obtained regarding the effect of NPY on cAMP accumulation,[3] it becomes apparent that rat cardiomyocytes may possess a third type of mechanism for NPY, linked to the enzyme, adenylate cyclase, again through a pertussis toxin-sensitive G-protein. The sensitivity of the inhibitory effect of NPY on cytosolic levels of cAMP in cardiomyocytes is lower by one order of magnitude compared to the sensitivity of the negative contractile response to the peptide.[4] Unless it becomes known whether the activity of the i_{to} can be influenced by cAMP or not, it remains possible that the effect of NPY on adenylate cyclase is the basis for the negative contractile response. This is because the activity of the i_{to} could depend on a rapidly saturable subcellular pool of cAMP.

In conclusion, we propose that there may be two subtypes of the NPY receptor on the ventricular cardiomyocyte of the rat, one linked to the i_{to}, possibly through adenylate cyclase, and a second to an L-type Ca^{2+} channel.

ACKNOWLEDGEMENTS

This work was supported by the British Heart Foundation.

REFERENCES

1. J. Wharton, and S. Gulbenkian, Peptides in the mammalian cardiovascular system, Peptides 56: 292-316 (1989).
2. H.M. Piper, B.C. Millar, and B.J. McDermott, The negative inotropic effect of neuropeptide Y on the ventricular cardiomyocyte, Naunyn-Schmiedeberg's Arch. Pharmacol. 340: 333-337 (1989).
3. B.C. Millar, H.M. Piper, and B.J. McDermott, The anti-adrenergic effect of neuropeptide Y on the ventricular cardiomyocyte, Naunyn-Schmiedeberg's Arch. Pharmacol. 338: 426-429 (1988).
4. M.R. Mitchell, T. Powell, D.A. Terrar, and V.W. Twist, Strontium, nifedipine and 4-aminopyridine modify the time course of the action potential in cells from rat ventricular muscle, Br. J. Pharmacol. 81: 551-556 (1984).
5. A. Balasubramaniam and S. Sheriff, Neuropeptide Y (18-36) is a competitive antagonist of neuropeptide Y in rat cardiac ventricular membranes, J. Biol. Chem. 265: 14724-14727 (1990).

EFFECT OF RYANODINE ON GLYCERINATED RABBIT PSOAS MUSCLE FIBERS

Takeshi Tsujimoto

Department of Physiology
Wakayama Medical College
Kyuban-cho, Wakayama 640, Japan

INTRODUCTION

It is well known that contraction and relaxation in skeletal muscle are regulated by Ca^{2+} release from and re-uptake by SR, and the latter to be an ATP-dependent Ca^{2+} pump (Ikemoto,1982). On the other hand, release of Ca^{2+} is the key problem less well understood in E-C coupling, and the process occur in the absence of ATP, but not a simple reversal of Ca ion sequestration. In this study, we utilized a biphasic contracture of glycerinated fiber which showing an interaction of both processes. Thus, to assess the role of ATP in the mechanism of E-C coupling, effects of ryanodine, ruthenium red, nucleoside triphosphate (NTP), and the nonhydrolyzable ATP analog (AMP-PNP) on ATP- and Ca-induced contractures of glycerinated rabbit psoas muscle fiber were studied. Results suggest that ATP affects Ca release in the sarcoplasmic reticulum.

RESULTS AND DISCUSSION

NTP induced biphasic contracture at pCa higher than 9, as described by Reuben et al.(1971). EGTA as high as 10 mM did not reduce the tension, and the contractures likely to occur irrespective with Ca ion regulatory system. When the concentration of ATP exceeded the optimum (pMgATP<4.4), the tension decreased and Ca induced biphasic contracture, however, the fibers are insensitive to Ca in the presence of suboptimal concentration of ATP. Though the other NTP (GTP, ITP, CTP and UTP) produced very slight Ca-tension, but their presence reduced the ATP-tension and really caused relaxation instead of contraction. AMP-PNP itself produced no tension, but its presence reduced ATP-tension, and suggested competition between NTP and ATP on a common site producing tension.

Ryanodine (12 μM) doubly augumented the ATP-induced tension at ATP concentrations higher than the optimum, but had insignificant effect on Ca-tension. However, repeated trial on single fiber reveal the increase. Similar effects were

Excitation-Contraction Coupling in Skeletal, Cardiac, and Smooth Muscle
Edited by G.B. Frank *et al.*, Plenum Press, New York, 1992

confirmed with all NTP. When submaximal dose of Ca was added before ATP, the tension (abbr. as Ca/ATP-tension) was smaller than the sum of ATP- and Ca-induced tensions (abbr. as ATP/Ca-tension). However, the Ca/ATP-tension, in the presence of ryanodine (12 μM), approached that of ATP/Ca. Though the effect of ryanodine was slightly less than that of Brij 58 treatment (0.3%, 30 min), concomitant use of both agents produced an additive increase in the Ca/ATP-tension. Since the effects were observed using glycerinated fiber, it was necessary to assess the effect reside in SR or not (Bianchi & Narayan,1990). Concomitant use of ryanodine and ruthenium red increased the ATP-tension, but the effect disappeared at 10 μM ruthenium red. The competitive increase of Ca-tension with ryanodine against ruthenium red blocking appeared only with higher concentrations of ATP or AMP-PNP. Thus, ryanodine and ATP probably compete with ruthenium red on Ca^{2+} release channel of SR (Meissner, 1986).

Results are compatible with the notion "ryanodine locks the Ca^{2+} release channels in the open state"(Fleischer et al,1985), and "the effect of ATP reside in the calcium release mechanism in the SR" (Endo & Kitazawa, 1976).

REFERENCES

Bianchi, C.P. and Narayan, S., 1990, Bimodal operation of the ryanodine-sensitive transducer calcium channel, Toxicon, 28:1173-1181.

Endo, M. and Kitazawa, T., 1976, The effect of ATP on calcium release mechanisms in the sarcoplasmic reticulum of skinned muscle fibers, Proc. Japan Acad., 52:595-598.

Fleischer,S., Ogunbunmi,E.M., Dixon, M.C. and Fleer, E.A.M., 1985, Localization of Ca^{2+} release channels with ryanodine in junctional terminal cisternae of sarcoplasmic reticulum of fast skeletal muscle, Proc. Natl. Acad. Sci., 82:7256-7259.

Ikemoto, N., 1982, Structure and function of the calcium pump protein of sarcoplasmic reticulum, Ann. Rev. Physiol., 44:297-317.

Meissner, G., 1986, Ryanodine activation and inhibition of the Ca^{2+} release channel of sarcoplasmic reticulum, J. Biol. Chem., 261:6300-6306.

Reuben, J.P., Brandt, P.W., Berman, M. and Grundfest, H., 1971, Regulation of tension in the skinned crayfish muscle fiber I. Contraction and relaxation in the absence of Ca (pCa>9), J. Gen. Physiol.,57:385-407.

MODULATION OF CA^{2+} RELEASE AND NA^{+} CHANNEL FUNCTION IN SKELETAL MUSCLE BY FATTY ACIDS

Jeffrey E. Fletcher,[1] Steven J. Wieland,[1] Jill Beech,[2]
Terry D. Heiman-Patterson,[3] and Henrey Rosenberg[1]

[1]Hahnemann University , [2]University of Pennsylvania, and
[3]Thomas Jefferson Univerisity , Philadelphia, PA USA

SUMMARY

Ca^{2+}- and halothane-induced Ca^{2+} release and Na^{+} currents are modulated by free fatty acids (FAs). FA modulation of ion currents may have important implications for general muscle physiology and skeletal muscle disorders, including malignant hyperthermia (MH).

INTRODUCTION

MH is a potentially fatal metabolic disorder of skeletal muscle that is manifested on exposure to halogenated volatile anesthetics and succinylcholine (humans and swine) or stress (swine). Elevated unsaturated FA production in skeletal muscle of MH-susceptible subjects has been suggested to account for the abnormally low threshold of Ca^{2+}- induced Ca^{2+} release (TCICR) in heavy sarcoplasmic reticulum fractions (HSRFs) isolated from MH susceptible porcine muscle.[1,2] However, the TCICR in HSRFs isolated from human skeletal muscle is only slightly affected by unsaturated FAs.[1] Consequently, the TCICR is normal in HSRFs from MH patients.[1] Application of unsaturated FAs to the inside of primary cultures of normal human skeletal muscle causes an increase in the magnitude of the Na^{+} current, as determined by whole-cell patch clamp.[3] This effect of exogenous FAs on the Na^{+} current is significantly reduced in primary cultures of human MH muscle,[3] possibly due to the presence already of excess endogenous FAs. Sodium currents in primary cultures of human MH muscle also exhibit a greater than normal delayed component of fast inactivation ($Na^{+}{}_{DCFI}$).[4] The present study examined the effects of FAs on the threshold of halothane-induced Ca^{2+} release (THICR) and the $Na^{+}{}_{DCFI}$. Additionally, we examined FA metabolism in primary cultures of normal and MH skeletal muscle.

METHODS

Halothane-Induced Ca^{2+} Release

The HSRFs were prepared, as previously described.[1] Ca^{2+} concentrations in the extravesicular space were monitored with arsenazo III and the TCICR determined using a pyrophosphate method and 10 μM pulses of Ca^{2+}.[1] The HSRFs were then preloaded with Ca^{2+} to 30-35% of the TCICR. Halothane in DMSO was subsequently added at two-fold

Excitation-Contraction Coupling in Skeletal, Cardiac, and Smooth Muscle
Edited by G.B. Frank *et al.*, Plenum Press, New York, 1992

increasing increments starting at 0.25 mM. The final concentration of DMSO did not exceed 0.8% (v/v). The TCICR was that concentration of halothane causing a sustained release of Ca^{2+}. FAs, when used, were added after Mg-ATP.[1]

Whole-Cell Patch Clamp Analysis of Na^+ Currents

Primary cultures of human skeletal muscle were prepared and whole-cell patch clamp studies conducted as previously described.[4] FAs were added to the interior of the cell through the patch clamp microelectrode.[3]

FA Metabolism in Primary Cultures of Human Skeletal Muscle

Primary monolayers of human skeletal muscle were established from collagenase-treated biopsy specimens. The cells were maintained in culture for 3-5 wks before radiolabeling. The lipid fractions were radiolabeled with ^{14}C-linolenic acid (18:3; 10 μM) for 3 days, as previously described for immortalized airway epithelial cells.[5] Three 35 mm culture dishes for each condition were either immediately extracted, or first incubated with or without isoproterenol (±; 10 μM) for 20 min. The lipid extracts were separated into fractions by 1-D TLC and the radioactivity in each spot determined by a scanner.[5]

RESULTS

Halothane-Induced Ca^{2+} Release

The THICR was about 4 mM for HSRFs from both control (n = 4) and MH susceptible (n = 4) swine. Oleic acid (20 μM) reduced the THICR of HSRFs from about 4 mM to about 0.5 mM in preparations from MH porcine muscle. The THICR was similarly reduced by oleic acid (20 μM) in HSRFs from human and equine muscle.

Whole-Cell Patch Clamp Analysis of Na^+ Currents

Oleic acid (15 μM) increased the Na^+_{DCFI} in primary cultures of normal human muscle. Thus the normal cell acquired a Na^+_{DCFI} very similar to that observed previously in MH cultures.

FA Metabolism in Primary Cultures of Human Skeletal Muscle

Exposure to isoproterenol (10 μM; 20 min) to activate hormone-sensitive lipase (HSL) did not alter the ratio of phospholipid to neutral lipid in cultures from normal patients (96 ± 11% of the ratio of cells not exposed to isoproterenol; mean ± SEM, n = 6), suggesting that the HSL is not normally coupled to the ß-adrenergic receptor in these cell cultures. In contrast, cultures from two MH+ patients exhibited changes in the ratio of phospholipid to neutral lipid of 176% and 148% upon challenge with isoproterenol, supporting breakdown of triglyceride and cholesterol esters by the HSL only in the MH cultures.

DISCUSSION

These results provide further evidence that altered FA metabolism plays an important role in MH. Despite the marked effects of FAs on the TCICR of porcine muscle, the TCICR of human skeletal muscle was relatively unaffected by FAs.[1] However, unlike the porcine stress syndrome that is observed in swine in the absence of anesthetics, human MH requires anesthetics to manifest the syndrome. The present study demonstrates that the THICR is affected by FAs regardless of the species (human, horse, pig) from which the HSRFs are isolated. In the present study the addition of FAs to the interior of primary cultures derived from normal human skeletal muscle mimics the abnormal Na^{+}_{DCFI} previously reported in MH cultures.[4] In mature normal human[1] and porcine[2] skeletal muscle isoproterenol enhances the contracture response of normal muscle fibre bundles to halothane, most probably by activation of the HSL and release of FAs. In contrast, we now report that the isoproterenol-stimulated ß-adrenergic receptor is not coupled to the HSL in primary cultures of normal human skeletal muscle. However, the HSL in primary cultures of skeletal muscle from two patients diagnosed as MH susceptible is coupled to the ß-adrenergic receptor, suggesting a possible developmental defect in the coupling mechanism. Several other muscle disorders are also associated with primary or secondary defects in Ca^{2+} and/or Na^{+} regulation. Therefore, a primary or secondary alteration of FA metabolism may also play an important role in the etiology and pathophysiology of skeletal muscle disorders other than MH. It is also possible that catecholamine-enhanced FA metabolism may be involved in normal muscle function by modulating the function of various ion channels.

REFERENCES

1. Fletcher et al. Biochem. Cell Biol. (1990) 68:1195-1201.
2. Fletcher et al. (1991) In: Mechanisms of Anesthetic Action in Muscle (TJJ Blanck and DM Wheeler, eds) Plenum Publishing Corp., New York.
3. Wieland et al. (1991) ibid.
4. Wieland et al. (1989) Am. J. Physiol. 257:C759-C765.
5. Fletcher et al. (1990) Toxicon 28:647-656; Biochem. Int. 21:733-740.

EFFECT OF FREE RADICALS ON EXCITATION-CONTRACTION COUPLING IN ISOLATED RAT AND GUINEA PIG VENTRICULAR MYOCYTES

Magda Horackova and Andrzej Beresewicz

Department of Physiology and Biophysics
Faculty of Medicine, Dalhousie University
Halifax, Nova Scotia, Canada B3H 4H7

SUMMARY: Using the model of isolated myocytes, we have investigated under current- and/or voltage-clamp conditions: (i) the time-course of H_2O_2-induced changes in the action potentials and the contractility. (ii) the effect of pharmacological tools such as ryanodine, TTX and nifedipine on H_2O_2-induced changes; (iii) the effect H_2O_2 on ionic currents underlying the changes in APD; and (iv) the effect of deferoxamine on H_2O_2-induced changes.

INTRODUCTION: It has been proposed that oxygen-derived oxidants, such as the superoxide radical ($\cdot O_2^-$), hydrogen peroxide (H_2O_2), and they hydroxyl radical ($\cdot OH$), could be important mediators of arrhythmias that occur during reperfusion of the ischemic myocardium.[1] Since then, electron paramagnetic resonance and histochemical studies have convincingly demonstrated that $\cdot O_2^-$, H_2O_2 and $\cdot OH$ are generated early during myocardial reperfusion. In addition free radicals have been reported to produce in the membrane potentials of cardiac preparations a variety of changes that are thought to be arrhythmogenic, although the evidence concerning the pattern and time course of these changes is equivocal. Thus, Barrington et al.[2] have described a rapid increase in the action potential duration (APD) followed by the development of arrhythmogenic early afterdepolarizations (EADs) and a shortening of APD in canine myocytes exposed to dihydroxyfumaric acid or to a xanthine/xanthine oxidase system. Similar effect of early increase in APD was also observed in two other studies in which the effects of oxidants generated by Rose Bengal or of H_2O_2 on the action potentials were examined.[3,4] However, several other studies using free radicals reported only delayed (by 10-30 min) shotening of the APD, membrane depolarization and development of arrhythmogenic delayed afterdepolarizations. This uncertainty about the pattern and time-course of free-radical-induced changes hinders efforts to determine which of these changes, if any actually mediate the reperfusion arrhythmias. Furthermore, although reactive metabolites of oxygen ($\cdot O_2^-$, H_2O_2, $\cdot OH$) are known to react with phospholipids and proteins in vitro a definite evidence as to the specific radical(s) important to ischemia/reperfusion injury is lacking. Whereas $\cdot O_2^-$ and H_2O_2 can injure cardiac tissues directly they have relatively short half-lives and are rapidly converted into highly reactive $\cdot OH$ for which there are no known intrinsic cellular scavengers. On the other hand, it has been shown that while the Fe^{3+} chelating agent deferoxamine blocks iron-catalyzed $\cdot OH$ formation from $\cdot O_2^-$ and H_2O_2 in vitro it also reduces reperfusion arrhythmias[5] and the H_2O_2 induced changes in electrical activity of papillary muscles.[4] Thus, it seems possible that ·OH radical formed from H_2O_2 through the iron catalyzed reaction may be one of the important mediators in the development of reperfusion arrhythmias.[5] In order to further investigate this possibility we have used isolated adult rat and guinea pig ventricular myocytes to reevaluate the electrophysiological changes induced by H_2O_2 (in absence or presence of deferoxamine) and to examine the possible ionic mechanisms underlying these changes. Isolated cardiomyocytes represent a better experimental model than intact-tissue preparations for this purpose because their use eliminates problems related to the control of extracellular space and effects of other cell types. Several studies in this laboratory have established that this model is fully compatible functionally with intact-tissue cardiomyocytes from several species.

RESULTS AND CONCLUSIONS: The observed changes are described in three phases: 1: prolongation of action potential duration accompanied by increased contractility. With a prolonged exposure the increased APD was accompanied by early afterdepolarizations (EADs), delayed afterdepolarizations (DADs) and aftercontractions. The changes in APD and EADs were fully and permanently abolished by tetrodotoxin (TTX) but not by nifedipine, while the DADs and aftercontractions were abolished by ryanodine. These changes precede phase 2: which was characterized by APD shortening, a decrease in contractility, membrane depolarization, single or multiple extrasystoles, or steady spontaneous activity; this phase could not be prevented by any of the above pharmacological interventions and resulted in a final phase 3: characterized by full depolarization and inexcitability. All the above changes were indicating that ·OH generated intracellularly in the presence of Fe^{3+} induces the observed changes. The examination of the membrane currents indicated that the increased APD may be due to an increase in the TTX-sensitive Na^+ current as well as to the decreased delayed current, while L-type Ca^{2+} channels appear to be unaffected. The shortening of APD during the second phase was associated with a large increase in the delayed K^+ current. The increased contractility in the first phase appears to be due to increased sarcolemmal Ca^{2+} influx via Na^+-Ca^{2+} exchange (among other possible mechanisms), leading to a loading of sarcoplasmic reticulum that eventually results in Ca^{2+} overload and functional failure. All reported results will be published in near future.[6]

ACKNOWLEDGEMENT: This research was supported by a grant from MRC-Canada to Dr. M. Horackova.

REFERENCES

1. Manning, A.S. and Hearse, D.J., Reperfusion-induced arrhythmias: mechanism and prevention, J. Mol. Cell. Coardiol.,16:497-518 (1984)
2. Barrington, D.L., Meier, Jr. C.F., Weglicki, W.B., Abnormal electrical activity induced by free radical generating systems in isolated cardiomyocytes, J. Mol. Cell. Cardiol., 20:1163-1178 (1988)
3. Tarr, M. and Valenzeno, D.P., Modification of cardiac action potential by photsensitizer-generated reactive oxygen, J. Mol. Cell. Cardiol. 21:539-543 (1989)
4. Firek, L. and Beresewicz, A., Hydrogen peroxide induced changes in membrane potentials in guinea pig ventricular muscle: permissive role of iron, Cardiovasc. Res., 24:483-499 (1990)
5. Bernier, M. Hearse, D.J., Manning, A.S., Reperfusion-induced arrhythmias and oxygen-derived free radicals. Studies with "anti-free radical" interventions and a free-radical-generating system in the isolated perfused rat heart. Circ. Res., 58:331-340 (1986)
6. Beresewicz, A. and Horackova, M., Alterations in electrical and contractile behaviour of isolated cardiomyocytes by hydrogen peroxide: possible ionic mechanisms, J. Mol. Cell. Cardiol. (1991) in press

$[Ca^{2+}]$ AND $[Na^+]$ IN RAT VENTRICULAR MYOCYTES SHOWING NEGATIVE AND POSITIVE FORCE FREQUENCY RELATIONSHIPS

James E. Frampton, Simon M. Harrison and Clive H. Orchard

Department of Physiology, University of Leeds, LS2 9JT. U.K.

SUMMARY

Ventricular cells that show a positive force-frequency relationship show graded changes of intracellular $[Ca^{2+}]$ (Ca_i), $[Na^+]$ (Na_i) and sarcoplasmic reticulum Ca^{2+} content with changes of stimulation frequency. Cells that show a negative force-frequency relationship show smaller changes of Na_i and no change in the Ca^{2+} load of the sarcoplasmic reticulum as stimulation frequency is changed.

INTRODUCTION AND METHODS

Multicellular preparations of cardiac muscle from rat hearts normally show a negative force-frequency relationship. That is, the strength of contraction decreases as stimulation frequency increases (e.g. Orchard & Lakatta, 1985). However, myocytes isolated from the ventricles of rat hearts may show either negative or positive force-frequency relationships (Capogrossi, Kort, Spurgeon & Lakatta, 1986). We have, therefore, investigated possible reasons for these different force-frequency relationships.

Ca_i (monitored using the fluorescent dye Fura-2) or Na_i (determined using the fluorescent dye SBFI) and contraction were monitored in single myocytes isolated from the ventricles of rat hearts. Some cells showed an increase in the size of the twitch as stimulation frequency was increased (positive force-frequency relationship), others showed a decrease in twitch size as the frequency of stimulation was increased (negative force-frequency relationship).

RESULTS

In cells which showed a positive force-frequency relationship, increasing stimulation frequency resulted in increases in Na_i (e.g. increasing stimulation rate from 0.25 to 2 Hz increased Na_i by 1.6 ± 0.2 mM, mean ± s.e.m., n=6), diastolic Ca_i, systolic Ca_i and the amount of Ca^{2+} that could be released from the sarcoplasmic reticulum by caffeine. The rate of decline of the Ca_i transient and the twitch also increased as stimulation frequency

Excitation-Contraction Coupling in Skeletal, Cardiac, and Smooth Muscle
Edited by G.B. Frank *et al.*, Plenum Press, New York, 1992

was increased. In cells which showed a negative force-frequency relationship, increasing stimulation frequency had less effect ($P=0.002$) on Na_i (increasing stimulation rate from 0.25 to 2 Hz increased Na_i by 0.8 ± 0.1 mM, n=6), and had either no effect, or decreased, systolic Ca_i with no change in the amount of Ca^{2+} that could be released from the sarcoplasmic reticulum using caffeine. The rate of relaxation of the Ca_i transient and the twitch again increased as stimulation frequency increased. The pattern and time course of mechanical restitution was the same in both cell types.

DISCUSSION

These results suggest that the final contractile response observed during changes of stimulation frequency may depend on whether the Ca^{2+} load of the cell changes as stimulation frequency is increased. In cells that show a positive force-frequency relationship the increase in the size of the Ca^{2+} transient as stimulation frequency is increased appears to reflect an increase in the Ca^{2+} load of the sarcoplasmic reticulum. In cells which exhibit a negative force-frequency relationship the Ca^{2+} load shows little change as stimulation frequency is increased. In these cells the size of the twitch may then be dominated by other factors (e.g. incomplete restitution).

The reason why some cells show changes in Ca^{2+} load with changes of stimulation frequency while others do not is unclear. However it is possible that the change of Ca^{2+} load reflects the changes of Na_i which will, via Na^+/Ca^{2+} exchange, alter Ca_i and that cells that do not show changes of Ca^{2+} loading cannot do so because they are already fully Ca^{2+} loaded (possibly secondary to being fully Na^+ loaded) at all the stimulation frequencies studied in these experiments.

ACKNOWLEDGEMENTS

This work was supported by the British Heart Foundation, The Wellcome Trust and the Medical Research Council.

REFERENCES

Capogrossi, M.C., Kort, A.A., Spurgeon, H.A. & Lakatta, E.G., 1986, *J. Gen. Physiol.*, 88: 589-613.

Orchard, C.H. & Lakatta, E.G., 1985, *J. Gen. Physiol.* 86: 637-651.

ACTION OF VASOCONSTRICTOR HORMONES AND CALCIUM ENTRY MODULATORS ON ELECTRICAL ACTIVITY AND INTRACELLULAR CALCIUM OF SMOOTH MUSCLE CELLS

U.T. Ruegg, H.J. Knot, M.M. de Ree, and B.H. Gahwiler[1]

Preclinical Research, Sandoz Pharma Ltd, CH-4002 Basle and
[1]Brain Research Institute, University of Zurich, August
Forel-Str. 1, CH-8029 Zurich, Switzerland

The A7r5 smooth muscle cell line, which was originally derived from rat aorta, was studied. These cells show a spontaneous electrical activity. Resting membrane potential was about -45mV. The cells displayed spontaneous overshooting acting potentials at a frequency of 0.1 to 0.5 Hz.

Effects of the vasoconstrictor peptide vasopressin (10^{-9}M to 10^{-7}M) included a fast hyperpolarization and cessation of the spiking activity followed by a slow depolarization and reinitiation of spiking. Intracellular Ca^{2+} concentrations ($[Ca^{2+}]_i$), measured with the Ca^{2+} indicator fura-2, was increased transiently to a level about 10-fold above basal and then gained a new steady state at about twice the basal value which lasted 10-20 minutes. It appears that vasopressin stimulated Ca^{2+} release from intracellular stores (via IP_3), that this triggered a hyperpolarization due to activation of calcium activated K channels. The late and long lasting $[Ca^{2+}]_i$ elevation was probably due to Ca^{2+} influx through non-dihydropyridine sensitive channels.

The modulation of the electrical activity of these cells by inorganic ions known to affect Ca and K channels was also investigated. Addition of $BaCl_2$ (10^{-4}M) to the bathing solution led to an increase in the spiking rate while $CdCl_2$ (10^{-5}M) blocked the spikes. Organic calcium entry blockers such as darodipine (10^{-8}M) stopped the electrical activity while calcium entry activators (e.g. Bay K8644, 10^{-8}M) increased the frequency of discharges but had no effect on the amplitude and the duration of the spikes. $[Ca^{2+}]_i$ transiently increased in correlation with the frequency of electric discharges. Both up- and downstroke of the $[Ca^{2+}]_i$ signal were slower than the corresponding electrical events. These results suggest that the rapid spontaneous depolarization of these smooth muscle cells is at least in part due to an opening of the dihydropyridine-sensitive (L-type) calcium channel while initial repolarization is caused by activation of K channels (inward rectifier) and the following hyperpolarization is probably due to opening of calcium activated K channels.

SOLUBILIZATION, PHYSIOLOGICAL ROLE, AND LOCALIZATION OF A KEY PROTEIN (31.5 KD) TO EXCITATION-CONTRACTION COUPLING PROCESS OF FROG SKELETAL MUSCLE CELLS, USING PHENYLGLYOXAL

S. Fujino, M. Fujino,* K. Satoh, T. Nakai, T. Kado,
T. Arima and M. Fujino**

Department of Pharmacology, Hokkaido Institute of
Pharmaceutical Sciences, Otaru, Japan
*Department of Pharmacology, Sapporo Medical College, Sapporo, Japan
**Department of Physiology, National Defense Medical College
Tokorozawa, Japan

SUMMARY

A new phenylglyoxal-binding protein (31.5 kD) was solubilized from frog skeletal muscle. The monoclonal AB against the 31.5 kD protein inhibited extracellularly E-C coupling process of single-muscle fibers.

We already reported[1,2] that the excitation-contraction (E-C) coupling process in single fibers from toe muscle was specifically inhibited by phenylglyoxal (PGO), a specific guanidyl modifying reagent[3]. To clarify the mechanism in the process of skeletal muscle, we performed a study on solubilization of the PGO-binding protein, which should play a key role in the process, from transverse tubule membrane-junctional SR complexes (TTM-JSR) of the frog skeletal muscles.

1. In single skeletal muscle fibers, E-C coupling was inhibited by the treatment with 3 mM PGO for 3 min. In whole sartorius muscles, PGO-induced inhibition was evoked by the treatment with 5 mM PGO for 10 min. According to these findings, twenty frog sartorius muscles were immersed in Ringer containing 5 mM ^{14}C-PGO for 10 min, and then washed out. Sixty min after the ^{14}C-PGO removal, when E-C coupling process in the muscles were selectively inhibited, TTM-JSR were isolated on the basis of the previous method[4]. ^{14}C-PGO was shown to be predominantly incorporated into the 31.5 kD protein (PGO-protein). This PGO-protein was proved to be a glycoprotein by the Con A peroxidase method. To know the physiological role and cellular localization of it, monoclonal AB against it was developed.

2. The specificity of the purified AB was established by Western Blot analysis. The AB revealed a single band of 31.5 kD protein from CHAPS extracts in the non-reducing condition.

3. Single muscle fibers driven electrically (1 time/min) were immersed in a slightly hypertonic Ringer solution (185 mM NaCl) containing the AB (10 μg/ml) for 60 min, and then washed out. Thirty min after the AB-removal, both twitch and K-contracture tensions were inhibited markedly without recognizable alterations in resting and action potentials and caffeine contracture (Fig. 1).

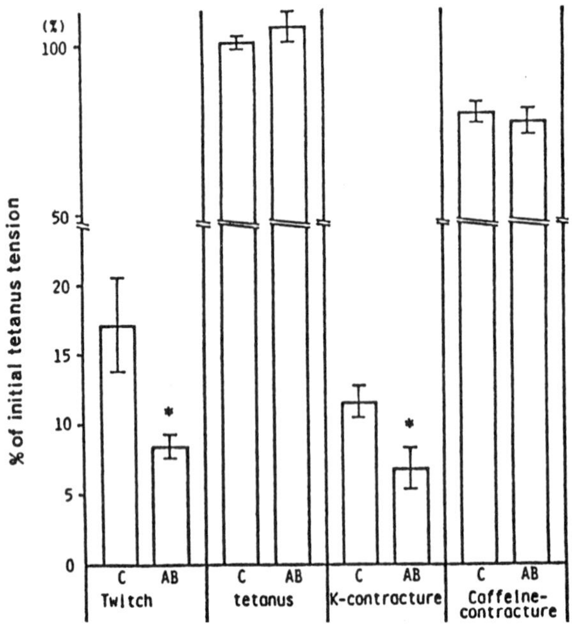

Figure 1

Effect of monoclonal AB against the 31.5 kD protein on E-C coupling of single muscle fibers of frog skeletal muscle.

4. Intact single fibers were immersed in the AB solution under the same condition for immunohistochemical electron microscopic studies. Obtained results showed that the protein localizes on the T-tubule membrane against the foot which occupies the TTM-junctional SR gap.

Conclusively, the protein would be considered to be a significant protein which has not yet been reported in the field of E-C coupling mechanism.

References

1. Fujino, M. et al.; J. Physiol., Soc. Japan., 45: 507 (1983).
2. Fujino, M. et al.; Proc. Int. Union Physiol., 16:207, (1986).
3. Takahashi, K.; J. Biol. Chem., 243:6171 (1968).
4. Fujino, S. et al.; Experientia, 45:467, (1989).

GTPγS ACTIVATES CALCIUM CHANNELS IN NEONATAL RAT VENTRICULAR CELLS

Q.Y. Liu, E. Karpinski, C.G. Benishin, and P.K.T. Pang

Department of Physiology, University of Alberta
Edmonton, Alberta T6C 2H7

INTRODUCTION

It has been reported that G proteins modulate voltage-dependent Ca^{2+} channels in cardiac muscle. G proteins may activate channels directly (acting on Ca^{2+} channel proteins) or indirectly (acting through intracellular second messengers, e.g., cAMP) (Brown, 1990). The effect of the G protein activator, GTPγS, on Ca^{2+} channels in neonatal rat ventricular cells has not been demonstrated. Therefore, the effect of the intracellular application of GTPγS, a non-hydrolyzable analogue of GTP, on Ca^{2+} channel currents in neonatal rat ventricular cells was investigated. Two types of Ca^{2+} channels have been described in the cardiac cells (Bean, 1985). One type of Ca^{2+} channel activates at a low threshold and inactivates very rapidly. This channel is called the low voltage-activated or transient channel (T channel). The other type of Ca^{2+} channel requires larger depolarization for activation and inactivates slowly. This channel is called the high voltage-activated or long-lasting channel (L channel). Therefore, two protocols were used in this study. In one protocol, cells were held at a membrane potential of -80 mV. Test pulses from this holding potential to more positive potentials evoked inward Ca^{2+} currents which consisted of two components (T and L channel components). In the second protocol, the membrane potential was shifted to -40 mV. This inactivated the low voltage (T channel) Ca^{2+} channel current leaving the high voltage-activated (L channel) Ca^{2+} current. Using the above procedures, the effects of GTPγS on the inward Ca^{2+} channel currents were measured in populations of cells. The effects of different intracellular concentrations of GTPγS were also studied. In addition, the effect of GTPγS on the steady-state activation and inactivation of the L channel current was measured. Our results suggest that G proteins are involved in the modulation of voltage-dependent Ca^{2+} channels in ventricular cells.

MATERIALS AND METHODS

Cell Preparation: Single cells were isolated from neonatal rat ventricles by enzymatic dispersion and mechanical disruption (Wang, 1990). The tissue was incubated with HBSS containing 2 mg/ml trypsin and bovine serum at 37°C for 20 min followed by four successive incubations in the same conditions. The cells were used within 12 to 36 hr after seeding.

Recordings: The whole-cell version of the patch clamp technique was used and the current was measured using a List EPC-7 patch clamp. The current was leakage corrected. The recordings of current were made every

Excitation-Contraction Coupling in Skeletal, Cardiac, and Smooth Muscle
Edited by G.B. Frank *et al.*, Plenum Press, New York, 1992

5 min and then compared with the current at time 0. Time 0 was defined as the time at which the inward current was established (initial current, I_0). This was taken as the time at which the outward K^+ current was blocked. The steady-state activation was determined using the magnitudes of the tail currents. The steady-state inactivation was measured using 1 sec conditioning pulses in 5 mV increments from a holding potential of -40 mV followed by a 200 ms test pulse to +20 mV.

Solutions: Internal solution: 130 mM CsCl, 20 mM HEPES, 10 mM EGTA, 5 mM $MgCl_2 \cdot 6 H_2O$, 5 mM ATP. The osmolarity was adjusted to 320 mOsm and the pH to 7.2 using sucrose and $Ba(OH)_2$, respectively. GTPγS (5 to 500 μM) was directly dissolved in the solution. External solution: 150 mM TEA, 0.8 mM $MgCl_2$, 5.4 mM KCl, 20 mM $BaCl_2$, 1 μM TTX, 10 mM HEPES. The osmolarity was adjusted to 340 mOsm and the pH to 7.4 as described above. In all experiments, 20 mM Ba^{2+} replaced Ca^{2+} as the charge carrier.

RESULTS AND DISCUSSION

The ventricular cells used in these experiments were round with diameters of 15 to 20 μm. In experiments where the cells were current clamped, typical ventricular action potentials were obtained. The intracellular application of 500 μM GTPγS produced a large increase in the Ca^{2+} channel current after 2 to 3 min, which was maintained up to 20 min after the initial current (see Methods). GTPγS diffused into the cell gradually after rupture of the membrane and activated G proteins on the cytoplasmic side of the membrane. Population studies were carried out in order to reinforce the preliminary data. Studies on groups of cells confirmed the effect of GTPγS on Ca^{2+} channel currents. The effect of GTPγS on Ca^{2+} channel currents was observed at two holding potentials (-40 and -80 mV). GTPγS increased the current in both situations. However, the effect of GTPγS on Ca^{2+} currents at HP = -80 mV was larger than that at HP = -40 mV, indicating that GTPγS might also increase T currents in our preparations. Three different intracellular concentrations of GTPγS (5, 50 and 500 μM, respectively) were used. The effect on the Ca^{2+} channel current was larger as the concentration of GTPγS was increased. GTPγS not only increased the Ca^{2+} channel current, but also shifted the peak current of the current-voltage relationship toward more negative potentials (about 10 mV). This occurred at both holding potentials. It was found that GTPγS also changed the gating properties of L channels. GTPγS shifted the steady-state inactivation and activation of the L-type Ca^{2+} channel current toward more negative potentials (about 10 mV). In conclusion, GTPγS activated the voltage-dependent Ca^{2+} channels in neonatal rat ventricular cells. GTPγS shifted the peak I-V relationship of the Ca^{2+} channel currents toward more negative potentials. It also shifted the steady-state activation and inactivation curves toward more negative potentials. From our results, it is suggested that G proteins play a regulatory role in the modulation of Ca^{2+} channels in neonatal rat ventricular cells, and thus G proteins may be involved in the modulation of excitation-contraction coupling in the cardiac tissues since Ca^{2+} is essential in this process.

REFERENCES

Brown, A.M., Yatani, A., Van Dongen, A.M.J., Kirsch, G.E., Codina, J. and Birnbaumer, L. Networking ionic channels by G proteins. Soc. Gen. Physiol. Ser. 43, 1-9, 1990.

Bean, B.P. Two kinds of calcium channels in canine atrial cells. Differences in kinetics, selectivity and pharmacology. J. Gen. Physiol. 86, 1-30, 1985.

Wang, R., Karpinski, E. and Pang, P.K.T. Two types of voltage-dependent calcium channel currents and their modulation by parathyroid hormone in neonatal rat ventricular cells. J. Cardiovasc. Pharmacol. 17, 990-998, 1990.

EFFECTS OF CYCLOPIAZONIC ACID, AN INHIBITOR OF THE Ca^{++} ATPase OF SARCOPLASMIC RETICULUM, ON Ca^{++} TRANSPORT, CONTRACTION AND RELAXATION IN CARDIAC MUSCLE

Philippe Lahouratate, Marie Josée Quiniou and Claude Léoty§

SmithKline Beecham Laboratoires Pharmaceutiques, BP 58
F-35762 St. Gregoire Cedex, France
§ UA CNRS 1340, Laboratoire de Physiologie Générale, ENV
F-44087 Nantes Cedex 03, France

SUMMARY

Cyclopiazonic acid is a potent inhibitor of cardiac sarcoplasmic reticulum Ca^{++} ATPase. It scarcely affects inotropism but significantly impairs lusitropism suggesting a greater role for cardiac sarcoplasmic reticulum in the control of cardiac relaxation than in the control of cardiac contraction.

INTRODUCTION

The mycotoxin, cyclopiazonic acid (CPA), has been shown to specifically inhibit the Ca^{++} dependent ATPase and Ca^{++} transport activities of skeletal muscle sarcoplasmic reticulum (SR)[1,2]. This study investigated whether CPA also inhibits Ca^{++} dependent ATPase and Ca^{++} transport activities in cardiac SR vesicles and assessed the role of SR Ca^{++} pump in cardiac contraction and relaxation.

METHODS

SR vesicles prepared from dog left ventricle according to Jones et al.[3] were incubated in a Ca^{++} buffered assay medium containing the ionophore A 23187. Ca^{++} dependent ATPase activity was taken as the difference between ATPase activities with and without Ca^{++} which were assessed by determining colorimetrically the rates of inorganic phosphate liberation. Free Ca^{++} concentrations were calculated using Fabiato's computer program[4]. Oxalate supported Ca^{++} uptake rates were assessed by determining the amount of $^{45}Ca^{++}$ sequestrated by SR vesicles using the Millipore filtration method.

The effect on Ca^{++} sensitivity of myofibrils was assessed in Triton-skinned fibres (TSF), prepared from rabbit papillary muscle by measuring the shift of the pCa-tension curve fitted to the Hill equation. In saponin-skinned fibres (SSF), from rabbit papillary muscle, SR Ca^{++} load was controlled by using a buffered Ca^{++} medium. The amplitude (A) and the half relaxation decay (T1/2), dependent on SR Ca^{++} content and on SR Ca^{++} reuptake, respectively, were determined on transient contractures (TCC) induced by caffein applied to the SSF in low EGTA medium[5].

Excitation-Contraction Coupling in Skeletal, Cardiac, and Smooth Muscle
Edited by G.B. Frank *et al.*, Plenum Press, New York, 1992

In isolated ferret ventricular trabeculae, rapid perfusion of caffein produces a phasic contracture (TCC) mainly related to Ca^{++} release from the SR. The repriming of TCC was studied in this preparation.

Isometric contractile force (CF) and relaxation time (RT) were studied in electrically driven (0.5 Hz) rabbit papillary muscle and dog ventricular trabeculae.

Drugs: Cyclopiazonic acid obtained from Sigma was dissolved in DMSO and added to the assay media at less than 1.8% (V/V) as DMSO, cumulatively for cardiac contractility studies and in single applications for all other experiments.

RESULTS AND DISCUSSION

The Ca^{++} dependent ATPase activity and the Ca^{++} uptake were linear over time during all the assay. CPA produced a dose-dependent inhibition of both activities when compared to those obtained with solvent alone. IC_{50} presented in Table 1 show that CPA is a potent inhibitor of both Ca^{++} dependent ATPase activity and Ca^{++} uptake of dog cardiac SR vesicles as previously reported for rat skeletal muscle SR vesicles[1,2].

Similarly, dose dependent inhibition of the amplitude of TCC produced in rabbit SSF was observed when CPA was added during the Ca^{++} loading step (see IC_{50} in Table 1). That inhibition could be related to the decrease of Ca^{++} uptake by the SR. The fact that inhibition of TCC amplitude occurred at higher concentrations than those able to inhibit Ca^{++} transport by SR vesicles remains unexplained.

<u>Table 1</u> Effects of CPA on Ca^{++} transport (mean value; n=4-8).

	SR vesicles		Saponin Skinned Fibres
	Ca^{++} ATPase	$^{45}Ca^{++}$Uptake	Caffein-contractures (A)
IC_{50} (μM)	0.6	0.4	17

Application of CPA during the releasing step, significantly slowed relaxation of TCC which could be related to the decrease of reuptake of Ca^{++} by the SR. That impairment of TCC relaxation cannot be explained by the sole moderate Ca^{++} sensitising effect of CPA observed in rabbit TSF. On intact ferret ventricular fibres the repriming phenomenon of TCC was delayed with concentrations of CPA higher than 10 μM, in agreement with the observed effects of CPA at the SR level.

After the addition of 100 μM CPA, the amplitude of TCC produced in ferret fibres did not reach the maximal amplitude of control TCC, even with longer repriming time. This cannot result from the impairment of SR Ca^{++} uptake and might be related to the effect of CPA on TCC produced in rabbit SSF when applied before the releasing step: By increasing the time of contact of 100 μM CPA from 2 to 5 min, the amplitude of TCC was reduced from 87% to 70% of the control amplitude. This might be explained either by a leak of Ca^{++} from the SR or by a decrease in the amount of Ca^{++} released by caffein after long application of high concentration of CPA.

Table 2 shows that in rabbit and dog isometrically working cardiac muscle, the negative inotropic effect produced by CPA is negligible except for the highest concentration studied in rabbit. In both species, the negative lusitropic effect of CPA occurred at lower concentrations as shown by the increase in RT. That moderate but significant increase in RT can, partly at least, result from the alteration of SR Ca^{++} transport by CPA. Moreover it occurs at concentrations similar to those required to decrease the amplitude of TCC in SSF.

Table 2 Effects of CPA on cardiac contractility (mean absolute change; n=4-6).

CPA (μM)	Rabbit papillary muscle CF(mg)	Rabbit papillary muscle RT(msec)	Dog ventricular trabeculae CF (mg)	Dog ventricular trabeculae RT (msec)
1	-148 (-84)	0 (0)	-150 (-123)	+3 (-1)
10	-190 (-132)	+10* (+2)	-190 (-180)	+13* (-1)
100	-438*(-196)	+13* (0)	-242 (-230)	+12* (-3)
basal value	1222 (1136)	102 (97)	1358 (1503)	105 (98)

The values in parentheses indicate the corresponding solvent values.
*:Statistically different from corresponding solvent effect (Student's *t*-test)

CONCLUSION

CPA was shown to be a potent inhibitor of Ca^{++} transport in canine cardiac SR vesicles and to strongly affect the TCC, in rabbit as well as in ferret cardiac muscle, suggesting a similar action in several mammalian species. At high concentration of CPA, Ca^{++} release is also affected. The weak negative inotropic effect displayed by CPA, both in rabbit and dog preparations, suggests that when SR is blocked,Ca^{++} required for contraction might be provided, in these species at least, by another source. Relaxation of cardiac muscle seems to be more dependent on the SR activity; however, isometric relaxation remains only moderately impaired even when activity of the SR is severely depressed.

ACKNOWLEDGEMENTS

The authors greatly appreciated the technical collaboration of J.Albadine, N. Brument, J.C. Camelin, J.Guibert, V. Lahouratate and M.C. Le Boulch.

REFERENCES

1. D. E. Goeger, R. T. Riley, J. W. Dorner, and R. J. Cole, Cyclopiazonic acid inhibition of the Ca^{2+}-transport ATPase in rat skeletal muscle sarcoplasmic reticulum vesicles, Biochem. Pharmacol., 37:978 (1988).
2. N. W. Seidler, I. Jona, M. Vegh, and A. Martonosi, Cyclopiazonic acid is a specific inhibitor of the Ca^{2+}-ATPase of sarcoplasmic reticulum, J.Biol.Chem., 264:17816 (1989).
3. L. R. Jones, H. R. Besch, J. W. Fleming, M. McConnaughey, and A. M. Watanabe, Separation of vesicles of cardiac sarcolemma from vesicles of cardiac sarcoplasmic reticulum, J.Biol.Chem., 254:530 (1979).
4. A. Fabiato, Computer programs for calculating total from specified free or free from specified total ionic concentrations in aqueous solutions containing multiple metals and ligands, Meth. Enzym., 157:378 (1988).
5. E. J. Krane, and J. Y. Su, Comparison of the effects of halothane on skinned fibers from newborn and adult rabbits: II. Effects on sarcoplasmic reticulum, Anesthesiology, 71:103 (1989).

ROLE OF EXTERNAL CALCIUM AND SARCOLEMMAL Na/Ca EXCHANGE IN THE DEPOLARISATION-CONTRACTION COUPLING IN MAMMALIAN SKELETAL MUSCLE

C. Huchet, A. Elkhammari, J. Noireaud, and C. Léoty

UA 1340, General Physiology Laboratory, ENV
F-44087 Nantes Cedex 03, France

SUMMARY

In normal or denervated soleus muscle, all manoeuvres (changes in [Na]o or addition of PXB, PLD) which are known to affect the proportion of the rapidly exchangeable Ca bound to the sarcolemma and related to Na-Ca exchange, induce concommitant changes in the d-c coupling.

While in frog skeletal muscle an influx of calcium was found not to be necessary to trigger the contraction (Milcdi et al. 1984), some results have suggested that external calcium may participate in the mechanical activation by maintening tension during long depolarizing steps (Lüttgau and Spiecker 1979) and/or by replenishing intracellular calcium pools (Caputo et al. 1984). However, in heart muscle, calcium ions bound to the sarcolemma are of great importance in e.c. coupling (Langer and Rich 1985) and according to the result found by Léoty and Noireaud (1987), a similar role in mammalian skeletal muscle could be postulated. In rat twitch muscles, Léoty and Léauté (1982) noticed that the relationship between tension and membrane potential became modified by lowering [Na]o. To determine the contribution of external calcium ions to the coupling between depolarization and contraction, a series of experiments was conducted where [Ca]o and [Na]o were varied and substances known to modified the amount of calcium associated to the anionic fixed charge of the membrane were tested.

The experiments have been carried out on fibers of soleus muscles from normal and denervated adult rats and from new born (1-20 days old) ones. The mechanical responses were produced either by means of high-K (constant KCl product) solutions and/or in Na-reduced solutions and/or by using depolarizing pulses under voltage-clamp conditions (double mannitol gap). [Na]i was measured with Na-sensitive microelectrodes filled with an ion-exchange resin Na cocktail.

In long times denervated soleus fibers or in new born rat soleus cells, the contraction developed became strongly dependent on [Ca]o. The study of the time course of the decrease in the contraction amplitude, induced by depolarizing steps of 2.5 sec, following reduction in [Ca]o (3 to 0 mM) has shown that the modifications were correlated to the frequency, to the duration and to the amplitude of the depolarizing pulse. For each [Ca]o tested, the maximum amplitude of the contraction was closely related to [Ca]o and the application of nominally Ca-free solution reversibly eliminated the contractile response.

Excitation-Contraction Coupling in Skeletal, Cardiac, and Smooth Muscle
Edited by G.B. Frank *et al.*, Plenum Press, New York, 1992

Glycerol-treated fibers isolated from normal soleus muscle generated tension which were similar to those recorded in non-treated muscle and were also dependent on [Ca]o.

On sarcolemmal membrane isolated from heart muscle, phospholipase D produces a specific increase of phosphatidic acid; therefore, an increase in the net anionic charge of the membrane. In the presence of phospholipase D (PLD, 1.5 U/ml) the contraction developed by soleus muscle was enhanced by 20% and resembled the effect resulting from an increase in [Ca]o.

The Polymyxin B displaces calcium ions from the membrane in a dose-dependent manner in rat myocardium . In normal Ringer, adding 0.1 mM Polymyxin B (PXB) resulted in a shift to more negative potential of the contractile threshold and for larger depolarization the tension decreased.

The exposure to Na-free solutions (substituted by TEA) in normal soleus muscle induced a decrease about 75-80% of [Na]i and a hyperpolarization (4-7 mV) of the membrane. An increase of [Ca]o from 2 to 10 mM, in the presence of 140 mM [Na]o, induced a reduction in [Na]i equivalent to that obtained by lowering [Na]o to 80 mM ([Ca]o 2mM). These observations strongly suggest that an electrogenic Na/Ca exchange is operative in slow-twitch muscle and is implicated in the contractile response observed following [Na]o changes. The application of [Na]-reduced (<70 mM) or Na-free medium to the muscle resulted in the development of a transient contracture which maximum amplitude was about 40-80% of the twitch. In intact and glycerol-treated preparations, a steep dependence of the Na-withdrawal contracture on [Na]o was found and [Na]o which induced the half maximal response at each [Ca]o, was the same. The amplitude of Na-free contracture was sensitive to changes in [Ca]o and the tension was absent in nominally Ca-free.

The present work have shown that in normal or denervated soleus muscle, all manoeuvres which are known to affect the proportion of the rapidly exchangeable Ca bound to the sarcolemma, induce concommitant changes in the dependency of the contractile responses against membrane potential. The sarcolemmal-bound Ca is one of the aspect of the Na/Ca exchange, which carrier system has also been shown to be present in soleus muscle.The main consequence of these findings is that, in complement to charge movements and calcium channels, the Na/Ca exchange should be integrated in any interpretations concerning d-c coupling of mature and immature mammalian skeletal muscles.

REFERENCES

Caputo C., Bolanos P., Gonzales G.F. (1984) Effect of membrane polarization on contractile threshold and time course of prolonged contractile responses in skeletal muscle fibers. J Gen Physiol 84:927-943

Langer G.A., Rich T.L. (1985) Phospholipase D produced increased contractile force in rabbit ventricular muscle. Cir Res 56:146-149

Léoty , Léauté M. (1982) Membrane potential and contractures in segment cut from rat fast and slow twitch muscles. Pflügers Arch.395 : 42-48

Léoty C., Noireaud J. (1987) Effects of external cations and calcium-channel blockers on depolarization-contraction coupling in denervated rat twitch skeletal muscles. Pflügers Arch 408:146-152

Lüttgau H.C., Spiecker W. (1979) The effect of calcium deprivation upon mechanical and electrophysiological parameters in skeletal muscle fibres of the frog. J Physiol 296:411-429

Miledi R., Parker I., Zhu P.H. (1984) Extracellular ions and excitation-contraction coupling in frog twitch muscle fibres. *J Physiol* 351: 687-710

IP_3 AND GTPγS CAUSE CONTRACTION OF PERMEABILISED GUINEA PIG ILEUM LONGITUDINAL SMOOTH MUSCLE VIA DISTINCT MECHANISMS

M.J. Allen and J.R. Carpenter

Smooth Muscle Research Group, Dept. Physiological Sciences, University of Manchester, M13 9PT

Introduction

IP_3 is well established as being a mediator of calcium release from intracellular calcium stores. GTPγS (a stable analogue of GTP frequently used to activate GTP-binding proteins), however, has also been reported to cause calcium release and contractions in permeabilised rabbit main pulmonary artery (Kobayashi *et al.*, 1988). We have previously reported (Allen & Carpenter, 1989) that inositol 1,4,5-trisphosphate (IP_3) causes a contraction of saponin-permeabilised guinea-pig ileum longitudinal smooth muscle. Here we demonstrate that GTPγS also causes a contraction of permeabilised ileum longitudinal smooth muscle, though via a mechanism distinct from that of IP_3.

Methods

Small strips (aprrox 5mm x 0.3mm) were dissected from the longitudinal smooth muscle layer of guinea-pig ileum. They were mounted vertically in small (0.3 ml) plastic organ baths for isometric tension recording. The tissue strips were permeabilised with saponin (50 μg/ml, 20 min) in standard relaxing solution ("SRS": 130 mM KCl, 20mM Trizma-maleate, 4mM EGTA, 3.3mM Na_2ATP, 1mM $MgCl_2$, pH 6.8). After permeabilisation the tissues were exposed to Ca^{2+} (3μM free ion in SRS, 3 min). The Ca^{2+} served to load intracellular calcium stores and also as a post-saponin control contraction to which later contractions could be compared. The Ca^{2+} was washed out with SRS (30 sec) and the tissue left in EGTA-free SRS (9.5 min) before being exposed to the test substance.

Excitation-Contraction Coupling in Skeletal, Cardiac, and Smooth Muscle
Edited by G.B. Frank *et al.*, Plenum Press, New York, 1992

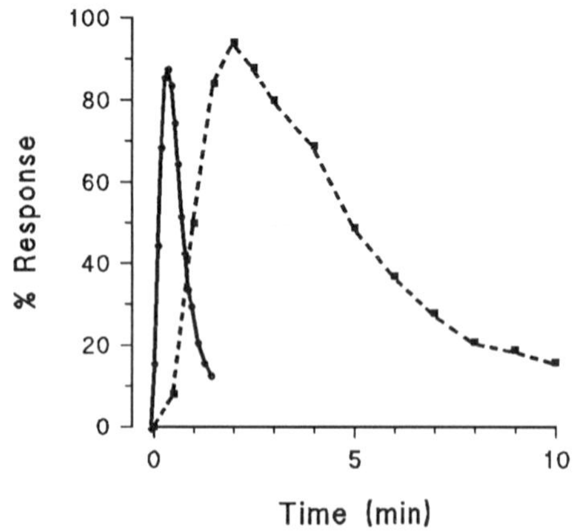

Fig. 1. Time–course of the response of saponin-permeabilised ileum longitudinal smooth muscle to (●——) IP_3 (100 μM) and (■----) GTPγS (100 μM). Results show % maximal tension reached by each agent. Data points represent mean of 5–8 experiments.

Results

IP_3 caused a concentration dependent contraction of the tissue strips with an EC_{50} of 7.2 μM. The maximal contraction (occuring at 100μM) was equivalent to that produced by 0.8μM Ca^{2+} given in time matched controls. The contraction was transient, peaking at 25 seconds and then declining over approximately 60 seconds (Fig 1). GTPγS caused a concentration dependent contraction of the tissue strips with an EC_{50} of 5.5 μM. The maximal contraction (occuring at 100μM) was equivalent to that produced by 1.1μM Ca^{2+} given in time matched controls. The contraction was transient but had a much slower time-course than IP_3, peaking in 2 minutes and declining over 10 minutes (Fig 1).

To examine whether IP_3 and GTPγS were causing contractions by a common mechanism, maximal concentrations of each agent were applied after the response to a maximal concentration of the other had declined, without washing out of the first agent. IP_3 applied directly after the response to an initial dose of IP_3 had declined (i.e without washing out) elicited no response. Likewise GTPγS, applied after the response to an initial dose of GTPγS had declined, elicited no response. Both IP_3 and GTPγS, however, elicited responses after the response to the other had declined, and the magnitude of the responses were not significantly different ($p>0.05$, Mann-Whitney) from time matched controls (Fig 2). This suggests that IP_3 and GTPγS do not cause a contraction via a common mechanism.

In addition to studying the direct contractile effect of IP_3, GTP and GTPγS, the effect of these compounds on the response to calcium was also investigated. Permeabilised tissue strips were first exposed to the standard (3 μM, 3 min) dose of Ca^{2+}.

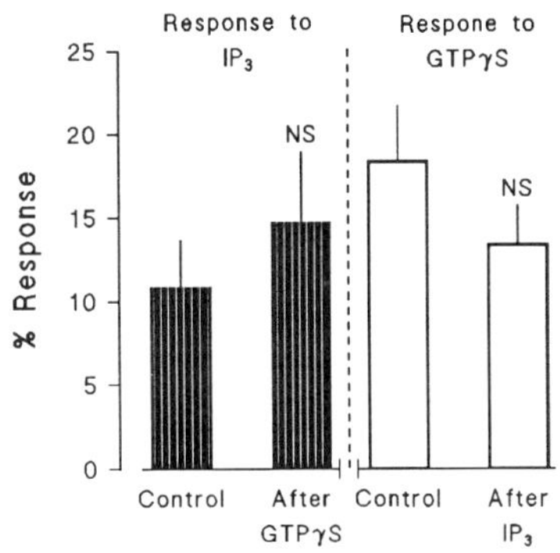

Fig. 2. Response of saponin-permeabilised ileum longitudinal smooth muscle to IP_3 (100 μM) given after GTPγS (100 μM), or vice versa, without washing out of the first substance. Response is shown as % response elicited by the initial 3 μM Ca^{2+}. Bars represent mean ± s.e.m (n=5). NS $p>0.05$ vs time matched control.

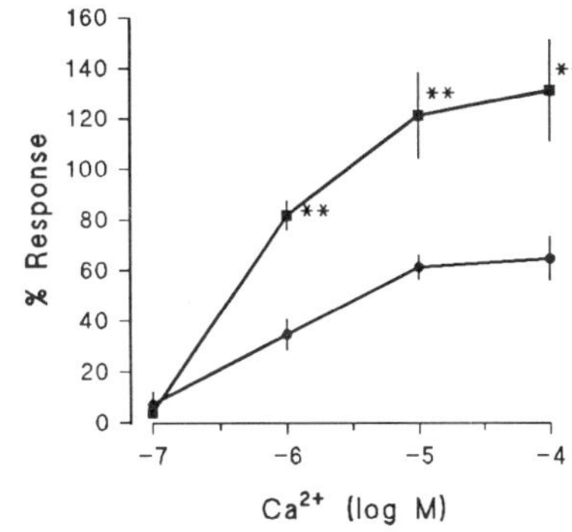

Fig. 3. The effect of GTPγS (100 μM) on the response of saponin-permeabilised ileum longitudinal smooth muscle to Ca^{2+}. (●) control and (■) with GTPγS added 5 min prior to Ca^{2+}. Response is shown as % response elicited by initial 3 μM Ca^{2+}. Data points represent mean ± s.e.m (n=5). ** $p<0.01$ vs control.

This was washed out with SRS and the tissues left for 5 min before IP_3 (100 μM), GTP (100 μM) or GTPγS (100 μM) were applied (since the tissue was being bathed in EGTA-containing SRS, no response to any of these compounds was observed *per se*). After a further 5 minutes a second dose of Ca^{2+} (1 μM, 3 min) was applied. IP_3 and GTP did not have any significant effect on the response to Ca^{2+}. GTPγS, however, caused a doubling of the response to Ca^{2+} that was statistically significant ($p<0.01$). In order to see whether this increase was due to a change in sensitivity to Ca^{2+} (i.e. a decrease in the EC_{50}) or due to enhanced responsiveness (an upward shift of the log concentration-response curve) the effect of GTPγS on a range of Ca^{2+} concentrations was investigated (Fig 3). GTPγS did not cause a marked shift in the EC_{50} for Ca^{2+} (0.6 μM in the presence of GTPγS c.f. 0.9 μM control), but did cause a pronounced upward shift in the response to Ca^{2+}.

In conclusion IP_3 and GTPγS both cause contraction of saponin-permeabilised ileum longitudinal smooth muscle, but via different mechanisms. GTPγS, but not IP_3 causes an enhanced responsiveness to Ca^{2+}.

Acknowledgement

M.J.Allen is in receipt of a Fisons studentship. J.R.Carpenter acknowledges the support of the Wellcome Foundation.

References

Allen, M.J. & Carpenter, J.R. (1989). *Br. J. Pharmacol.* **98**, 792p.
Kobayashi, S., Somlyo, A.P. & Somlyo, A.V. (1988). *J. Physiol.* **403**, 601-619.

Ag^+-INDUCED INWARD CURRENT ON FROG SKELETAL MUSCLE

Takako Aoki and Toshiharu Oba

Dept. Food & Nutr. Sci., Gifu City Women's Coll., Gifu 502 and Dept. Physiol., Nagoya City Univ. Med. Sch., Nagoya 467, Japan

SUMMARY

Ag^+ caused an inward current on voltage-clamped skeletal muscle. The current was carried by Ca^{2+}, Mg^{2+}, and Na^+, and was blocked by Cd^{2+} or Ni^{2+} but not by nifedipine or D600. Channel gating is supposed to be modified by Ag^+.

INTRODUCTION

Ag^+ induces frog intact muscle contraction (Oba and Hotta, 1985). Though it is not physiological contraction, Ag^+-induce contraction has several identical characters to physiological twitch; it occurs without extracellular Ca^{2+}, it disappears after detubulation (glycerol-treatment), pretreatment of the fiber with 20 mM K^+ decreases the contraction, and dantrolene inhibits it. These similarities suggest that Ag^+ triggers contraction through the similar pathway to physiological excitation-contraction coupling. We studied membrane current under voltage-clamped condition to clarify the mechanism of Ag^+-induced contraction.

MATERIALS AND METHODS

Single fibers were dissected from leg muscle (semitendinosus or anterior tibialis) of frog (Rana catesbeiana or Rana japonica) in Ringer's solution. Fiber diameters were from 60 to 140 µm. The fiber was voltage-clamped by a double mannitol gap device (Potreau and Raymond, 1982). The width of the central (test) compartment was 200 µm. Cl^- was replaced by NO_3^- for Ag^+ solution (in mM: 115 $NaNO_3$, 2.5 KNO_3, 10 MOPS, and 1.8 $Ca(NO_3)_2$ or 3 $Mg(NO_3)_2$, pH 7 with NaOH). Contraction was picked up at the central compartment and recorded with current on a paper oscillograph.

RESULT AND DISCUSSION

More than 1 µM Ag^+ induced a large slow inward current in single fibers clamped the membrane potential at -90 mV (Fig. 1a). The delay time from application of 10 µM Ag^+ to the onset of inward current varied widely (2-50 s, but 70 % was less than 15 s). Time from onset to half

Excitation-Contraction Coupling in Skeletal, Cardiac, and Smooth Muscle
Edited by G.B. Frank *et al.*, Plenum Press, New York, 1992

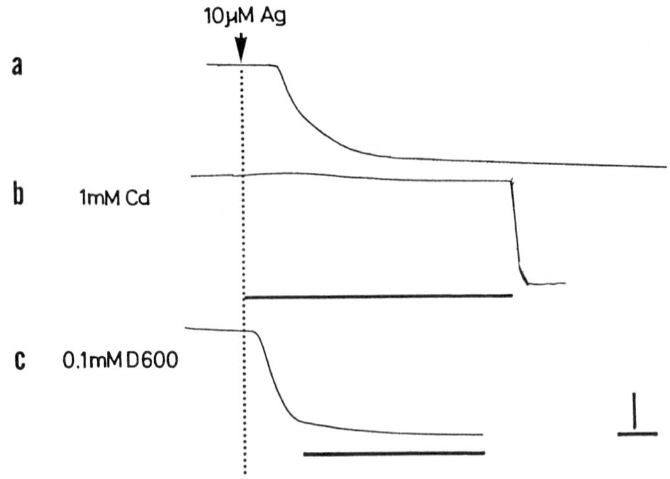

Fig. 1. Ag^+-induced inward current. The fiber was exposed to 10 μM Ag^+ at the arrow. Membrane potential was clamped at -90 mV throughout. Straight lines under the traces in b and c show the existence of Ca^{2+} channel blockers. Calibration: vertical, 4 μA for a, 3 μA for b and c; abscissa, 10 s.

maximal current was 15 ± 4 s (mean ± SEM, n=7) and the maximal current was 9 ± 1 μA in the solution with 3 mM Mg^{2+}. Washout of Ag^+ from the solution slowed or stopped the increase of inward current but did not decrease the current. Dithiothreitol (1 mM) reversed Ag^+-induced current partially.

Depolarization decreased the inward current. The current-voltage relation was linear at the range of -90 - +10 mV. Mean reversal potential was -37 mV.

Removal of divalent cation and replacement of Na^+ by choline ion eliminated the current, meaning that the current is not a nonspecific leak current. The current was insensitive to 1 μM tetrodotoxin, 50 μM nifedipine, 100 μM D600 (Fig. 1c) and 1 mM Ba^{2+}, while it was inhibited by 1 mM Cd^{2+} (Fig. 1b) or 1 mM Ni^{2+}. The inhibition did not seem to the effect of Cd^{2+} or Ni^{2+} on fixed negative charge of the surface membrane because increase of Mg^{2+} to 9 mM did not reduce the current.

These characters do not fit with Na^+ channel nor L-type Ca^{2+} channel. Modification of channel protein by Ag^+ probably changes the ion selectivity and the gating mechanism of the channel, resulting in fixation of the channel at the open state. Ca^{2+} influx through this channel caused contraction. The time course and the concentration dependency on external Ca^{2+} of this contraction are different from Ag^+-induced contraction on an intact muscle fiber (Oba and Hotta, 1985).

REFERENCES

Oba, T. and Hotta, K., 1985, Silver ion-induced tension development and membrane depolarization in frog skeletal muscle fibres, Pflügers Arch., 405: 354.

Potreau, D. and Raymond, G., 1982, Existence of a sodium-induced calcium release mechanism on frog skeletal muscle fibers, J. Physiol., 333: 463.

'REVERSE USE-DEPENDENT' EFFECTS OF 4-AMINOPYRIDINE ON THE TRANSIENT OUTWARD POTASSIUM CURRENT IN FERRET RIGHT VENTRICULAR MYOCYTES

Donald L. Campbell, Yusheng Qu, Randall L. Rasmusson*, and Harold C. Strauss+

Depts.of Pharmacology, *Bioengineering, and +Medicine
Duke University Medical Center
Durham, North Carolina, USA 27710

A transient outward potassium current "I_{to}" has recently been described in many different mammalian single isolated myocyte preparations. Due to the role of I_{to} both in (i) initiating early "phase 1" repolarization and (ii) frequency-dependent modulation of the action potential configuration I_{to} can act as an important modulator of the EC coupling characteristics of cardiac myocytes. It has been hypothesized that the net macroscopic I_{to} in many cardiac myocytes is composed of two separate current components: "$I_{to,1}$", a larger 4-aminopyridine (4-AP) sensitive Ca^{2+}-independent transient outward K^+ current, and "$I_{to,2}$", a smaller Ca^{2+}-activated current that remains after block of I_{to1} by 4-AP. I_{to1} and I_{to2} have been reported to have different characteristics that could be of fundamental importance. For example, recovery of I_{to2} has been reported to be more rapid than I_{to1}, and upon rapid stimulation I_{to2} amplitude increases to a size approaching that of I_{to1}. This has lead to the interesting suggestions that (i) I_{to1} mainly contributes to repolarization at slow heart rates, (ii) I_{to2} plays the more prominent role in repolarization at normal heart rates, and (iii) during tachyarrhythmias I_{to2} would help to protect the myocyte from Ca^{2+} overload.

However, all of these intriguing I_{to1}-I_{to2} hypotheses have been based upon the use of 4-AP to separate the putative properties of I_{to1} from I_{to2}. To more carefully test these hypotheses we have characterized the basic blocking characteristics of 4-AP on the Ca^{2+}-independent I_{to} (= "I_{to1}") in myocytes isolated from the right ventricles of male ferrets using the whole cell configuration of the gigaseal patch clamp technique. Our results strongly suggest that block of the Ca^{2+}-independent I_{to1} by 4-AP occurs through binding to the <u>closed</u> state of the I_{to1} channel.

Our basic conclusions are as follows: (1). Block of Ca^{2+}-independent I_{to1} in isolated ferret right ventricular myocytes by 4-AP is complex, in that 4-AP not only reduces I_{to1} amplitude but also slows activation and apparent inactivation of the macroscopic I_{to1} waveform. This produces a <u>crossover</u> of the I_{to1} currents before and after application of 4-AP (Figure 1). (2). Relative block by 4-AP is voltage dependent, with block being relieved at depolarized potentials. (3). 4-AP appears to <u>associate</u> slowly with a closed state(s) of the I_{to1} channel at hyperpolarized potentials (Figure 2.A) and <u>dissociate</u> relatively slowly (compared to normal I_{to1} gating kinetics) at depolarized potentials (Figure 2.B).

Excitation-Contraction Coupling in Skeletal, Cardiac, and Smooth Muscle
Edited by G.B. Frank *et al.*, Plenum Press, New York, 1992

(4). Both association and dissociation can be reasonably described as exponential processes with time constants that depend upon membrane potential. (5). As a result block by 4-AP is not only voltage-dependent but also frequency-dependent, with block being relieved at more rapid rates of stimulation to a given fixed potential (i.e. "reverse use-dependence"). (6). 4-AP appears not to interact with the inactivated state of unblocked channels.

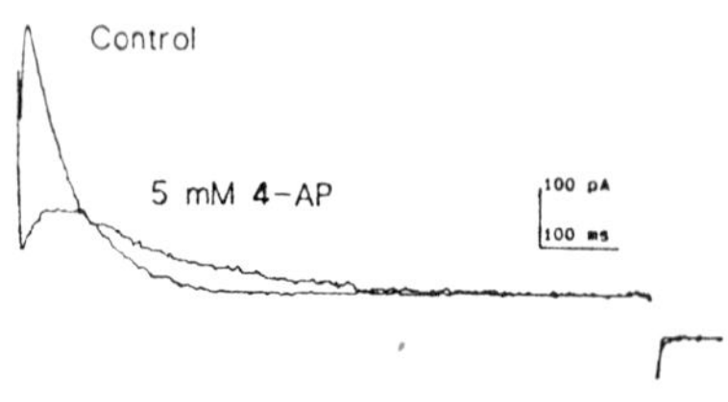

Figure 1. Basic observation: Effect of 4-AP on I_{to1}. 800 ms voltage clamp pulses to +50 mV in control and 5.0 mM 4-AP. Note that 4-AP not only reduces I_{to} but also alters its kinetics in that it slows activation and the time course of current decay. This results in a crossover of the current waveforms.

(A) (B)

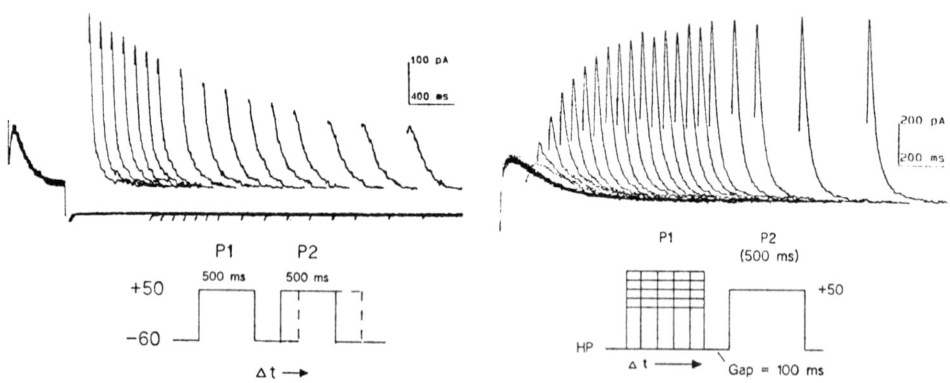

Figure 2. (A). Association of 4-AP at hyperpolarized potentials. 500 ms double pulse recovery protocol (see inset; frequency = 0.125 Hz) conducted at a holding potential of -60 mV in 10 mM 4-AP. Current traces obtained with interpulse intervals of 200 ms and greater illustrated. Note that initially the P2 current is greater than the control P1, but with increasing interpulse interval duration the P2 I_{to1} declines back to the control P1 level. Such biphasic recovery behavior is consistent with 4-AP dissociating during P1 and slowly associating with a closed state(s) during the hyperpolarized interpulse interval. (B). Kinetics of dissociation of 10 mM 4-AP at +50 mV. (protocol illustrated in inset; the gap of 100 ms allowed sufficient recovery of I_{to} to occur with only a minimal amount of 4-AP association). Note that with increasing P1 duration (increments of 50-100 ms) P2 I_{to} progressively increases, indicating that 4-AP dissociates during P1.

The simplest hypothesis which we propose at present is that 4-AP binds to a closed state(s) of the Ca-independent I_{to1} channel, the channel can not conduct until 4-AP dissociates from the closed state(s), and 4-AP dissociation becomes the rate-limiting step in the channel gating process at depolarized potentials. Due to both its frequency-and potential-dependent blocking/unblocking properties results obtained after the application of 4-AP could give the appearance that a second I_{to2} current component has been pharmacologically unmasked in the presence of 4-AP. Our results suggest that conclusions concerning the properties of a putative 4-AP resistant Ca^{2+}-dependent "I_{to2}" current obtained in previous studies on various cardiac myocytes may have to be carefully reinterpreted.

INTERACTION OF I_{Ca} WITH $I_{Na/Ca}$2 $[CA^{++}]_i$ AND CA^{++} BINDING PROTEINS IN FROG ATRIUM: A THEORETICAL STUDY

R.L. Rasmusson[1] and D.L. Campbell[2]

Departments of [1]Biomedical Engineering and [2]Pharmacology
Duke University, Durham, N.C.

The sarcoplasmic reticulum (S-R) of cells from the atrium of the bullfrog (*Rana catesbeiana*) is not involved in the cyclic uptake and release of calcium under normal physiological conditions. Consequently, bullfrog atrial cells provide a system in which the interactions between the "L"-type calcium current (I_{Ca}), sodium-calcium exchange current ($I_{Na/Ca}$) and intracellular Ca^{++} binding proteins (calmodulin and two different classes of Ca^{++} binding sites on troponin: T_{Ca}, a calcium-specific binding site thought to be responsible for phasic tension generation, and T_{Mg}, a Ca^{++}-Mg^{++} competitive binding site) may be examined in isolation from complications arising from S-R uptake and release. A Hodgkin and Huxley type model[1] incorporating $I_{Na/Ca}$ and intracellular Ca^{++} binding sites has been used to examine the potential interactions between these different mechanisms during the cardiac action potential.

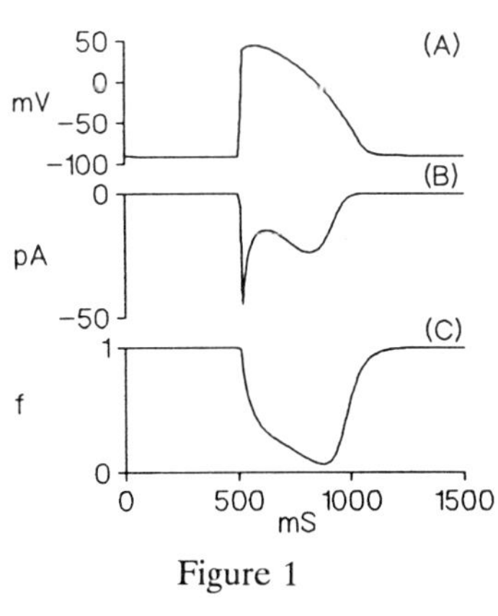

Figure 1

(A) Model action potential.
(B) I_{Ca} during action potential.
(C) Time course of inactivation gating variable f during action potential.

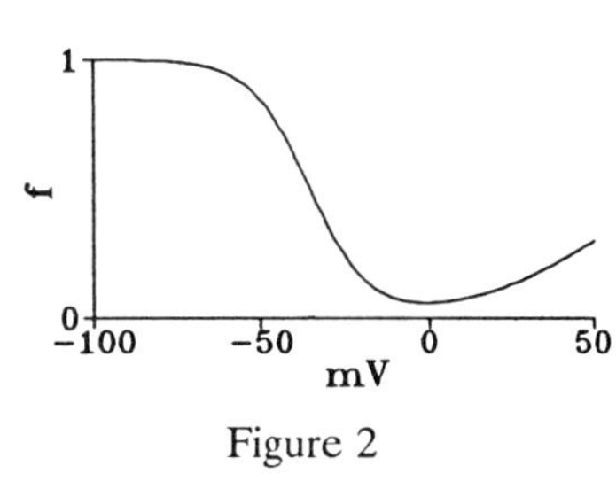

Figure 2

Steady state curve for f, the inactivation gating variable for I_{Ca}. Note that inactivation is never complete.

Excitation-Contraction Coupling in Skeletal, Cardiac, and Smooth Muscle
Edited by G.B. Frank *et al.*, Plenum Press, New York, 1992

Fig. 1a,b shows a simulated model action potential and the underlying I_{Ca}. Note that following the initial "spike" of I_{Ca} there is a sustained inward component. This sustained phase is due to the incomplete inactivation kinetics of the I_{Ca}. As described in Fig. 2, the steady state voltage dependence of the inactivation gating variable, *f*, never completely inactivates, but actually reactivates upon depolarization to potentials greater than 0 mV. During the plateau, the membrane potential is in the range in which the *f* curve "bends up": as a result, I_{Ca} only very slowly and incompletely inactivates (Fig. 1c). As the plateau progresses and the membrane begins to repolarize the driving force for Ca^{++} flow through the calcium channel progressively increases. As a result, I_{Ca} increases again, producing the secondary inward hump. The model therefore predicts that incomplete inactivation produces a maintained component of I_{Ca} during the action potential plateau as has been observed in other cardiac cell types (e.g. [2,3]).

This sustained component of Ca^{++} influx is important in maintaining the action potential plateau and the development of tension. While Ca^{++} binding to the calcium-specific site on T_{Ca} mirrors the $[Ca^{++}]_i$ transient during the action potential (Fig. 3) with occupancy peaking during the approximate midphase of repolarization, the Ca^{++} occupancy on T_{Mg} continues to absorb Ca^{++} as long as Ca^{++} remains elevated. Although occupancy of T_{Mg} by Ca^{++} does not directly mediate contraction, the ability of this site to bind significant amounts of Ca^{++} during repolarization makes it potentially important in delaying the development of tension.

Even thought the occupancy of T_{Mg} by Ca^{++} is relatively high at the end of diastole, it nonetheless binds a significant amount of Ca^{++} during the action potential. During the final phases of repolarization and the subsequent diastolic interval the T_{Mg} slowly releases Ca^{++}. Thus the total occupancy of T^{Mg} by Ca^{++}, and hence the total amount of available Ca^{++} binding sites, is dependent upon the amount of time elapsed since the last action potential. These results suggest that T_{Mg}, due to both its diastolic occupancy level and time course of Ca^{++} binding, can both buffer intracellular Ca^{++} transients and provide some "memory" of past contractile events that can produce rate-dependent changes in twitch tension[1].

The time course of $I_{Na/Ca}$, as well as the magnitude and time-course of the Ca^{++} transient, is predicted to be very different in the presence or absence of buffers (Fig. 4). In the presence of Ca^{++} binding proteins, $I_{Na/Ca}$ is initially outward and becomes inward only during late repolarization due to buffering of $[Ca^{++}]_i$ (i.e.

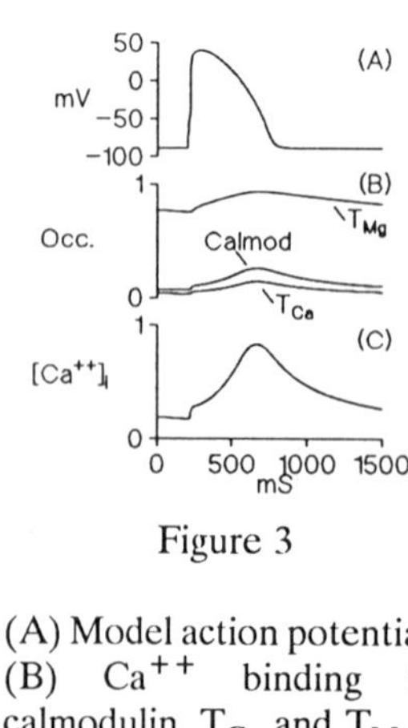

Figure 3

(A) Model action potential.
(B) Ca^{++} binding to calmodulin, T_{Ca} and T_{Mg}.
(C) Free $[Ca^{++}]_i$ during the action potential.

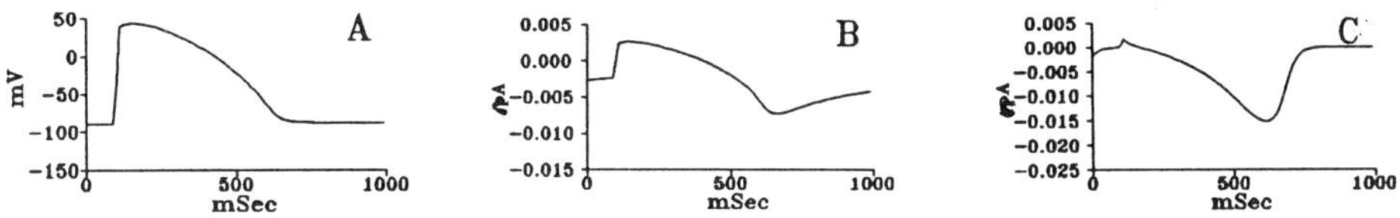

Figure 4

Effect of removing buffers on Inaca.
(A) Control action potential
(B) $I_{Na/Ca}$ in the presence of buffers.
(C) $I_{Na/Ca}$ in the absence of buffers.

$E_{Na/Ca}$ remains in the hyperpolarized range). When intracellular Ca^{++} binding sites are removed from the model, I_{Ca} and action potential configuration are relatively unaltered, but the $[Ca^{++}]_i$ transient is elevated, and $I_{Na/Ca}$ is always inward.

In conclusion, incomplete inactivation of the I_{Ca} may provide a maintained component of extracellular Ca^{++} entry during the action potential plateau. The peak of the $[Ca^{++}]_i$ transient may be delayed and diminished due to the presence of intracellular Ca^{++} binding proteins, which in turn may determine the magnitude and time course of $I_{Na/Ca}$.

References

1. Rasmusson, R.L., Clark, J.W., Giles, W.R., Robinson, K., Clark, R.B., Shibata, E.F., and Campbell, D.L., Am. J. Physiol., 259:H370-H389, (1990).
2. Doerr, T., Denger, R. and Trautwein, W., Pfluegers Arch. 413:599-603, (1989).
3. Mazzanti, M. and DeFelice, L.J., Biophys. J. 58:1059-1065, (1990).

EVIDENCE FOR THE EXISTENCE OF ENDOTHELIAL FACTORS REGULATING CONTRACTILITY IN RAT HEART

S. Winegrad, G. McClellan, and A. Weisberg

Department of Physiology
University of Pennsylvania
Philadelphia, PA

Force developed by isolated papillary muscles decreases as the cross-sectional area increases. The basis for this decline in force is not clear in as much as theoretical considerations and experimental data have indicated that the rate of diffusion of oxygen into thin bundles should not be limiting. Decline of maximum Ca-activated force with increasing cross-sectional area of detergent skinned papillary muscles can be attributed to the accumulation of inorganic phosphate in the center of the bundle. In both cases, the bundle of intact cells with a possible limitation of diffusion of oxygen into the bundle and of skinned cells with a limitation of diffusion of P_i outward, the lowest level of activity should be in the center of the bundle. We have used quantitative histochemistry for measuring Ca- and actin-activated myosin ATPase activity in cryostatic sections of rapidly frozen isolated trabeculae. The technique is very sensitive and has sufficient spatial resolution to resolve individual myofibrils. At different times after dissection, ventricular trabeculae were quickly frozen, transversely sectioned and Ca- and actin-activated myosin ATPase measured in serial sections both without and with 1 μM cAMP in the assay solution. In none of over 40 trabeculae studied was there an inward gradient of actin-activated ATPase activity of myosin. The most superficial cells had very low enzymatic activity. Cyclic AMP decreased the gradient by raising the enzymatic activity of the less active cells more than the more active cells. Ca-activated myosin ATPase was always uniform across the transverse section. These observations are incompatible with a limitation of diffusion of oxygen inward or of an inhibitory factor produced by muscular activity outward. The results are, however, consistent with the production of a regulatory factor within the bundle.

Excitation-Contraction Coupling in Skeletal, Cardiac, and Smooth Muscle
Edited by G.B. Frank *et al.*, Plenum Press, New York, 1992

ACTIVE AND PASSIVE FORCES DURING ISCHEMIA AND REPERFUSION OF RAT FAST AND SLOW TWITCH SKELETAL MUSCLES

Andrew J. Carvalho and Nancy H. McKee

Departments of Surgery and Community Health
University of Toronto, Toronto, Ontario M5S 1A8, Canada

SUMMARY

The present study shows for the first time, the response in active and passive forces of skeletal muscles to tourniquet ischemia. These responses were shown for both fast and slow twitch muscle types in the same rat hindlimb. Our data agrees partially with metabolic studies on ischemic skeletal muscles. The differences may be due to inhibition of excitation of myofilaments.

INTRODUCTION

Skeletal muscles are known to have decreased contractility following ischemia and to recover less with increasing duration of ischemia. However, it is not known how skeletal muscle function is affected during ischemia. Metabolites such as PCr and ATP, pH[1] and muscle membrane potential all decreased during skeletal muscle ischemia while Pi increases[1]. These changes suggest decreased active tension, increase passive tension[3] and decreased myofilament Ca^{2+} sensitivity[4] during ischemia. Depending on the duration of ischemia, these parameters may recover toward preischemic values. The purpose of this study was to investigate the responses in active and passive forces of fast and slow twitch skeletal muscles to ischemia and reperfusion. The effects of increasing duration of ischemia were also determined.

METHODS

Fifteen female Wistar rats (225-250g) were used. Muscle twitch and passive forces were measured during 1 (n=4),2 (n=5) or 3 (n=6) hrs. of ischemia followed by 2 hrs. of reperfusion. The distal tendons of the EDL (fast twitch) and soleus (SOL, slow twitch)

Excitation-Contraction Coupling in Skeletal, Cardiac, and Smooth Muscle
Edited by G.B. Frank *et al.*, Plenum Press, New York, 1992

muscles from the same rat hindlimb were dissected free and connected to two independent force transducers. The knee joint was pinned, and the foot securely taped to the experimental setup. Optimal muscle lengths (Lo) for both muscles were determined by direct muscle stimulation (250Hz, 400ms train, 0.2ms pulse, 70V, 32°C). Initial muscle twitch forces were determined at Lo before the induction of hindlimb tourniquet ischemia (350mmHg). Muscle twitch forces were measured simultaneously once every 15 min. during ischemia and reperfusion. Passive forces were measured before each twitch. Levels of significance were determined using Newmans-Keules procedure. Statistical significance was accepted for $p<0.05$.

RESULTS

Our results indicate an initial fast decay in twitch force from the onset of ischemia to approximately 1.5 hrs. of ischemia (Fig. 1, 2, and 3). However, EDL twitch forces were higher than that of SOL. After this initial phase, twitch forces were essentially constant for each muscle ($p<0.05$) and were not different between muscle types ($p<0.05$) (Fig. 3). Compared to preischemic valued, twitch forces recovered to 94.71 ± 5.02% (mean ± SE) for SOL and to 84.01 ± 4.41% for the EDL after 1 hr. of ischemia and 2 hrs. of reperfusion (Fig. 1). Recovery of twitch forces were 58.60 ± 13.53% for SOL and 41.99 ± 13.59% for EDL following 2 hrs. of ischemia and 2 hrs. of reperfusion (Fig. 2). Compared to preischemic twitch forces, neither muscle demonstrated recovery ($p<0.01$) during reperfusion following 3 hrs. of ischemia, 18.38 ± 4.75% for SOL and 12.30 ± 3.07% for EDL (Fig. 3).

The responses of passive forces during 3 hrs. of ischemia were different between the two muscle types (Figs. 4 and 5). Passive force for the DEL increased to 122% during the 1st hr. of ischemia compared to a 101 increase in the SOL. However, the increase in passive force in the SOL was more dramatic during the 3rd hr. of ischemia (143.88% to 252.19% to 169.98%; Fig. 4). Both muscle types demonstrated a rapid decline in passive force during the first 15 min. of reperfusion. At the end of reperfusion, the passive force for SOL was greatly reduced (77%; $p<0.005$; Fig. 5) while the EDL recovered to preischemic values (111.68%; $p>0.05$; Fig. 4).

DISCUSSION

The general decline in active tension observed during ischemia agrees partially with metabolic studies of ischemic rat muscle[1,5]. Although ATP and PCr decline faster in fast twitch muscle during ischemia[1,5] our results show twitch forces decreased less in the fast EDL than in the slow SOL during the first hour of ischemia. The excitation-contraction (e-c) mechanism of slow twitch muscle may be inhibited[4] sooner during ischemia. However this inhibition may be abolished during reperfusion since the SOL recovered better following 1 or 2 hours of ischemia. Resting membrane potential has been shown to become less negative during ischemia of (white, fast twitch) skeletal muscles[2] and to partially recover during reperfusion and may help explain the changes in twitch forces measured. The higher degree of ATP degradation in fast twitch muscles during ischemia may help explain the faster initial rise in passive force for the EDL. It is possible that myofilament and Z-line integrity were maintained better in the SOL and hence its ability to sustain higher increases in passive tension during ischemia. However, since SOL passive force was well below preischemic levels at the end of reperfusion, subsequent structural damage may have occurred. Thus 1) active and passive forces of fast and slow twitch muscles are qualitatively but not

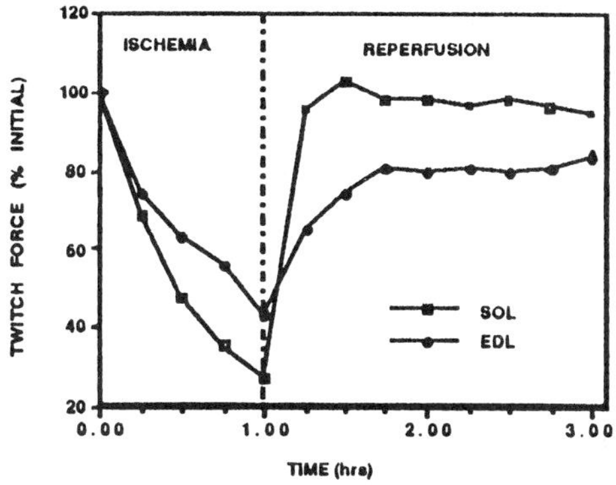

Fig. 1. Twitch forces during 1 hr. ischemia and 2 hr. reperfusion. Twitch forces are expressed as a percentage of initial preischemic twitch force, n=4.

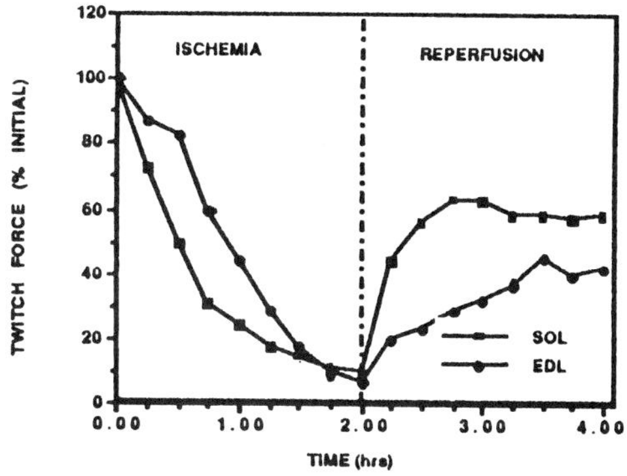

Fig. 2. Twitch forces during 2 hr. ischemia and 2 hr. reperfusion, n=5.

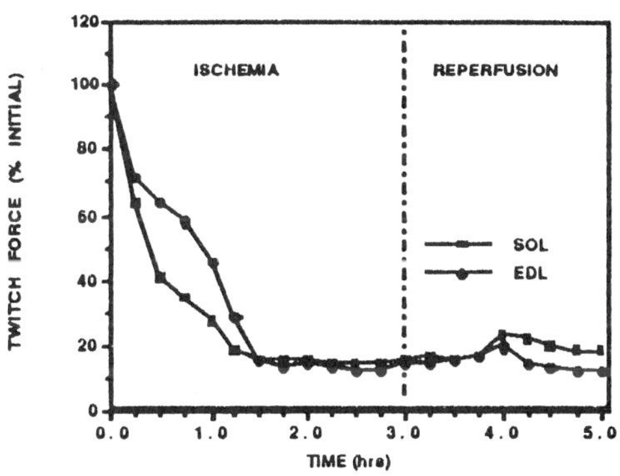

Fig. 3. Twitch forces during 3 hr. ischemia and 2 hr. reperfusion, n=6.

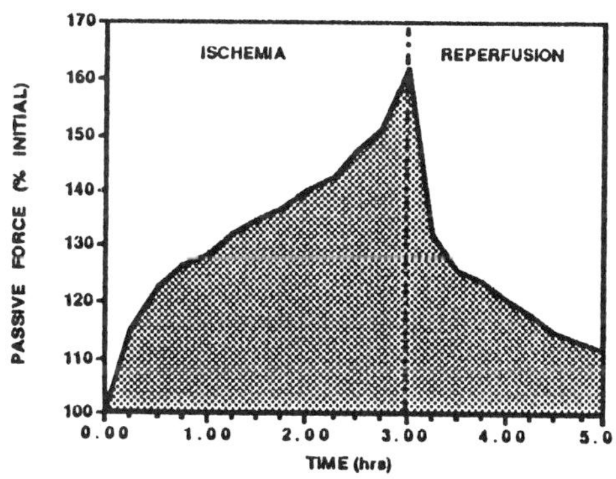

Fig. 4. EDL passive force during 3 hr. ischemia and 2 hr. reperfusion. Passive force was measured before each twitch was made, n=6.

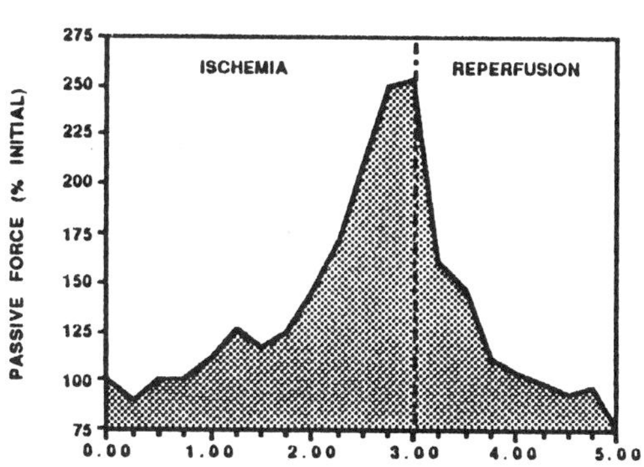

Fig. 5. SOL passive force during 3 hr. ischemia and 2 hr. reperfusion, n=6.

quantitatively similar during ischemia and reperfusion, 2) active forces of slow twitch muscles decline faster than fast twitch muscles during acute ischemia, 3) active forces of slow twitch muscles recover better following acute ischemia, 4) passive forces in fast twitch muscles increase faster during acute ischemia, but is higher in slow twitch fiber during prolonged ischemia, and 5) recovery from ischemia in fast muscles appears to be associated with inhibition of excitation while in slow muscles both inhibition and structural damage may occur.

ACKNOWLEDGEMENTS

Supported by: MRC (MA10536) and PSI (88-83).

REFERENCES

1. B. Soussi, J.P. Idstrom, A.C. Bylund-Fellenius, and T. Schersten. Dynamics of skeletal muscle energetics during ischemia and reperfusion assessed by in vivo ^{31}PNMR. *NMR in Biomedicine* 3:71-77 (1990).
2. E. Jennische, H. Hagberg, and H. Haljamae. Extracellular potassium concentration and membrane potential in rabbit gastrocnemius muscle during tourniquet ischemia. *European J. Physiol.* 392: 335-339 (1982).
3. W.J. Leijendekker, W.D. Gao, and H.E. D.J. ter Keurs. Unstimulated force during hypoxia of rat cardiac muscle: stiffness and calcium dependence. *Am. J. Physiol.* 258: H861-H869 (1990).
4. R.J. Hajjar and J.K. Gwathmey. Direct evidence of changes in myofilament responsiveness of Ca^{2+} during hypoxia and reoxygenation in myocardium. *Am. J. Physiol.* 259: H784-H795 (1990).
5. J.P. Idstrom, B. Soussi, A. Elander, and A. C. Bylund-Fellenius. Purine metabolism after in vivo ischemia and reperfusion in rat skeletal muscle. *Am. J. Physiol.* 258: H1668-H1673, 1990.

CA^{2+} INFLUX THROUGH VOLTAGE- AND PURINOCEPTOR-OPERATED CHANNELS ESTIMATED FROM $[CA^{2+}]_C$ SIGNALS (MYOCYTES FROM GUINEA-PIG URINARY BLADDER)

G. Isenberg, V. Ya. Ganitkevich, and P. Schneider

Department of Physiology, University of Cologne
Germany

SUMMARY

The rise in free cytosolic calcium was studied by a combination of the techniques of microspectrofluorometry and whole-cell patch clamp. By comparing the membrane currents with their effect on $[Ca^{2+}]_c$, the relative importance of Ca^{2+} influx could be quantified for both L-type Ca^{2+} channels and non-selective channels activated by extracellular ATP.

INTRODUCTION

The increase in the concentration of calcium ionized in the cytosol $[Ca^{2+}]_c$ is a crucial step in the activation of contraction. In smooth muscle, the rise in $[Ca^{2+}]_c$ is due to Ca^{2+} influx from the extracellular space and/or Ca^{2+} release from intracellular stores. Ca^{2+} influx can occur via voltage-dependent Ca^{2+} channels or through receptor-operated non-selective cationic channels [1,2]. The introduction of fluorescent Ca^{2+} probes such as Indo-1 [4] has resulted in detailed studies of the processes lining depolarization, Ca^{2+} influx, rise in $[Ca^{2+}]_c$ and activation of contraction in smooth muscle [1,2].

METHODS

Myocytes were continuously superfused with a solution that contained (in mM) 150 NaCl, 3.6 $CaCl_2$, 1.2 $MgCl_2$, 5.4 KCl, 20 glucose, 5 HEPES, pH 5.4, temperature 36°C. The pipet solution contained (in mM) 130 CsCl, 5 creatine, 5 succinic acid, 5 oxalacetic acid, 5 sodium pyruvate, 10 HEPES/CsOH (pH 7.3) and 0.1 Na_5Indo-1. Whole cell currents were not corrected for leakage and recorded on line. Fluorescence measurements were performed with Indo-1 using the single excitation wavelength and dual emission principle [2-4]. At 410 and 470 nm, subtraction of background fluorescence was adjusted when the Indo-1 containing electrode was cell attached. After Indo-1 loading for about 5 min, fluorescence was measured from a 30 μm wide spot of the cell. $[Ca^{2+}]_c$ was evaluated from the ratio 410nM/470nm and intracellular calibration [3].

Ca^{2+} Influx Through Voltage-gated L-type Ca^{2+} Channels

At a holding potential of -60 mV, resting $[Ca^{2+}]_c$ was 114 ± 22 nM (mean ± S.D.).

Excitation-Contraction Coupling in Skeletal, Cardiac, and Smooth Muscle
Edited by G.B. Frank *et al.*, Plenum Press, New York, 1992

During 160 ms depolarization steps to 0 mV, $[Ca^{2+}]_c$ rose to 885 ± 140 nM. With steps of varied duration, peak $[Ca^{2+}]_c$ increased with the time of depolarization up to about 1 s. Upon repolarization $[Ca^{2+}]_c$ recovered to resting levels with half decay time of about 1 s; recovery was not significantly changed with repolarization potentials between -50 and -100 mV as if Ca^{2+} efflux via Na/Ca-exchanger is not the determinant. This conclusion is further supported by the voltage dependence of the above peak $[Ca^{2+}]_c$ transients which was bell shaped similar as peak I_{Ca} is. The curve had a threshold around -40 mV and a maximum at 0 mV. During depolarization steps to potentials to +80 mV, $[Ca^{2+}]_c$ did not significantly rise. During step depolarizations to 0 mV lasting 10 s or longer, $[Ca^{2+}]_c$ peaked within 814 ± 18 ms and then decayed to a sustained level of 250 ± 60 nM. The rate of rise, the amplitude of the $[Ca^{2+}]_c$ peak and the time course its decay depended on the amplitude of I_{Ca}. When I_{Ca} was small at potentials close to the threshold (-40 mV) or when it was reduced by partial inactivation, the depolarizations increased $[Ca^{2+}]_c$ to a sustained level with no clearly defined peak. The sustained level of $[Ca^{2+}]_c$ was studied with 10 s depolarizations, it followed the same bell-shaped voltage-dependence as the peak $[Ca^{2+}]_c$ and I_{Ca}, i.e. the threshold was at -40 mV and the maximum close to 0 mV. A bell-shaped voltage-dependence for $[Ca^{2+}]_c$ was also found when Ca^{2+} influx was activated by ramp-like depolarizations. However, when the rate of depolarization was low (7.5 mVs^{-1}), simultaneous inactivation of I_{Ca} shifted the peak $[Ca^{2+}]_c$ to more negative potentials (-15 mV). The results are compatible with the idea that Ca^{2+} influx through voltage-operated Ca^{2+} channels is the key event in depolarization-mediated changes in $[Ca^{2+}]_c$.

Ca^{2+} influx through ATP-gated channels

In the urinary bladder, activation of P_{2x} purinoceptors couples to the increase in $[Ca^{2+}]_c$ and the contraction by activating ATP gated channels which may conduct influx of some Ca^{2+} ions together with larger influx of monovalent cations. At a constant potential of -60 mV, bath application of ATP induced an inward current $I_{ns,ATP}$ that peaked within 0.4 s and then decayed. $I_{ns,ATP}$ was activated half-maximally with 1.1 μM ATP and it saturated at 50 μM ATP to -1.1 ± 0.2 nA. At 3.6 mM $[Ca^{2+}]_o$, $I_{ns,ATP}$ had a reversal potential (E_{rev}) of -5 ± 2 mV. Reduction of $[Na^+]_o$ shifted E_{rev} to more negative, elevation of $[Ca^{2+}]_o$ to more positive values, suggesting that approx. 7% of $I_{ns,ATP}$ is carried by Ca^{2+} ions.

50 μM ATP increased $[Ca^{2+}]_c$ transiently from resting 130 ± 49 nM to 730 ± 100 nM. At 22°C, $[Ca^{2+}]_c$ rose at a rate proportional to the instantaneous current amplitude of $I_{ns,ATP}$. This relation was lost, however, after warming to 36°C which increased the peak $I_{ns,ATP}$ (Q_{10} 1.25) but reduced the peak of the ATP induced $[Ca^{2+}]_c$ transient (Q_{10} 0.75). We suggest that warming to 36°C stimulated Ca^{2+} sequestration and Ca^{2+} efflux via Ca-ATPases more than Ca^{2+} influx with the result that peak $[Ca^{2+}]_c$ is attenuated significantly. We suggest for these cells that contribution of Ca^{2+} release is insignificant because the ATP-effects were not modified by i.c. heparin and because depletion of i.c. stores by pretreatment with 20 mM caffeine did not attenuate significantly the ATP-gated rise in $[Ca^{2+}]_c$.

The simultaneous measurement of membrane currents and $[Ca^{2+}]_c$ transients provides another possibility for evaluating the Ca^{2+} component of $I_{ns,ATP}$. We assumed that I_{Ca} is a pure Ca^{2+} current and compared how much more charge (time integral of current) had to be transported by $I_{ns,ATP}$ to increment $[Ca^{2+}]_c$ by the same concentration (22°C). For an increment of 450 nM, I_{Ca} had to transport 41 pC but $I_{ns,ATP}$ had to transport 760 pC. Since $I_{ns,ATP}$ had to transport 19 times more charge than I_{Ca} we conclude that 5.8 ± 0.8% of $I_{Ca\ ns,ATP}$ is carried by Ca^{2+}. This estimate is close to the 7% extracted from the reversal potentials. With a simple model, the numbers can be translated into a permeability ratio of pNa:pCa ≈ 1:1.

One could expect that the simultaneous activation of voltage- and ATP-gated channels would result in a $[Ca^{2+}]_c$ signal that peaks to the sum of the individual increments. When $I_{ns,ATP}$ was triggered on top of the I_{Ca} induced $[Ca^{2+}]_c$, the increments were additive indeed.

Additivity was lost, however, when I_{Ca} was evoked during the bath application of ATP which inhibited peak I_{Ca} inhibited by 80 ± 15%. Inhibition of I_{Ca} diminished to 20 ± 8% after cell dialysis with 40 mM EGTA, and it was 19 ± 7% when extracellular Ca^{2+} had been substituted by Ba^{2+}. These results are in agreement with the hypothesis of "I_{Ca} inactivation by Ca^{2+}". The ATP induced $[Ca^{2+}]_c$ transients and the reduction of peak I_{Ca} recovered along a similar time course. For this period, reduction of peak I_{Ca} was analyzed and plotted as a function of elevated $[Ca^{2+}]_c$. The resulting non-equilibrium concentration-effect curve yielded a half maximal efficacy (K_D) of 670 ± 50 nM $[Ca^{2+}]_c$ and a Hill coefficient (H) of 3. The curve significantly deviates from published steady state data (dialysis of Ca-EGTA: K_D=100 nM, H=1). We discuss that the 5-10s of ATP induced rise in $[Ca^{2+}]_c$ is too short that Ca^{2+} binding comes to equilibrium.

For comparison with in vivo, $[Ca^{2+}]_c$ transients and membrane potential was investigated under current clamp (K^+ electrodes). 50 μM ATP depolarized the membrane from -50 to -30 mV for only 0.4 s, then, when $[Ca^{2+}]_c$ rose at maximal rate, it repolarized to -55 ± 5 mV. Repolarization was most likely due to Ca^{2+} activated K^+ conductance. In the absence of ATP, action potentials or sustained depolarizations to -20 mV (by current injection) resulted in $[Ca^{2+}]_c$ transients smaller than those induced by ATP. Thc rcsults suggest the importance of Ca^{2+} influx through ATP gated non-selective channels for $[Ca^{2+}]_c$ also for non-clamped cells and the complexity by which changes in $[Ca^{2+}]_c$ and membrane potential interact with channel-activation and Ca^{2+} influx.

REFERENCES

1. P.L. Becker, J.J. Singer, J.V. Walsh, and F.S. Fay, Science, 244:211-214 (1989).
2. C.D. Benham, J. Physiol., 419:689-701 (1989).
3. V.Ya Ganitkevich, and G. Isenberg, J. Physiol. 435:187-207 (1991).
4. G. Grynkiewicy, M. Ponie, and R.Y. Tsien, J. Biol. Chem. 260:3440-3450.

CALYCULIN A MODULATES VOLTAGE-DEPENDENT OUTWARD CURRENT IN GUINEA-PIG TAENIA COLI SMOOTH MUSCLE CELLS

Kazuo Obara, Toshihiro Usuki, and Hideyo Yabu

Department of Physiology
Sapporo Medical College
Sapporo 060, Japan

INTRODUCTION

Protein kinases modulate the electrical activity of various cells through their action on membrane ion channels. In cardiac and smooth muscle cells, enhancement of a number of K^+ currents by β-adrenergic agonists, such as isoproterenol (ISP), may result from a direct phosphorylation of K^+ channel protein by cAMP-dependent protein kinase (A kinase)(Kolb, 1990). On the other hand, calyculin A (CL-A) which is a toxin isolated from marine sponge (Discodermia calyx) has a potent inhibitory action on protein phosphatase activities (Ishihara et al., 1989).

In this study, we examined the effect of CL-A on the voltage-dependent outward current carried by K^+ in guinea-pig taenia coli smooth muscle cells.

MATERIALS AND METHODS

Single smooth muscle cells were isolated with a combination of 0.3% collagenase and 0.4% trypsin inhibitor from guinea-pig taenia coli (Obara, 1984).

The voltage-dependent outward currents were recorded by using whole-cell patch clamp technique. The pipette solution contained 140 mM KCl, 4 mM ATP, 0.5 mM EGTA and 5 mM HEPES (pH 7.4). Drugs were applied by adding them in different concentrations to the perfusing solution.

RESULTS AND DISCUSSION

In guinea-pig taenia coli smooth muscle cells, the outward current in whole-cell recording had both a Ca^{2+}-activated component and a voltage-dependent component which was insensitive to external Ca^{2+}. When external Ca^{2+} was reduced, the outward current was reduced by ∿20%. This suggests that the voltage-dependent component appears to be predominant current component in whole-cell recording.

CL-A (10 nM) increased the voltage-dependent outward current. CL-A did not appear to affect the threshold in activation of the voltage-dependent outward current. These results suggest that CL-A facilitates the opening of the voltage-dependent K^+ channel. The gating of K^+ channels is suggested to be modulated by phosphorylation of the channel proteins by A kinase. The state of phosphorylation of protein is thought to be determined by the activity balance of protein kinase and protein phosphatase. Ishihara et al. (1989) reported that CL-A inhibited the activity of the catalytic subunits of protein phosphatase type 1 and type 2A. Therefore, it is suggested that CL-A produces its increasing effect on the outward current through suppression of dephosphorylation process resulting in an increased phosphorylation of protein(S) related to the K^+ channel, which in turn facilitates the opening of K^+ channels. In guinea pig taenia coli smooth muscle, ISP causes relaxation by hyperpolarization of membrane (Tomita et al., 1985) and elevation of the level of intracellular cAMP (Yamada and Obara, 1985). The hyperpolarization caused by ISP is likely to be due to an increase in K^+ conductance (Tomita et al., 1985). ISP (1 μM) and the membrane-soluble cAMP analog dibutyryl cAMP (DBcAMP; 200 μM) increased the outward current. The effect of DBcAMP on the outward current was inhibited by A kinase inhibitor H-8 (100 μM). The effect of CL-A on the outward current was also inhibited by H-8. Therefore, the present results suggest that the effect of CL-A on the outward current may involve A kinase-mediated modulation of the voltage-dependent K^+ channel.

In conclusion, in guinea-pig taenia coli smooth muscle cells, CL-A facilitates the opening of the voltage-dependent K^+ channels and the effect of CL-A on the outward current may involve A kinase activity.

REFERENCES

Ishihara, H., Martin, B.L., Brautigan, D.L., Karaki, H., Ozaki, H., Kato, Y., Fusetani, N., Watanabe, S., Hashimoto, K., Uemura, D. and Hartshorne, D.J., 1989, Calyculin A and okadaic acid: Inhibition of protein phosphatase activity, Biochem. Biophys. Res. Commun., 159: 871-877.

Kolb, H.A., 1990, Potassium channels in excitable and non-excitable cells, Rev. Physiol. Biochem. Pharmacol., 115: 51-91.

Obara, K., 1984, Isolation and contractile properties of single smooth muscle cells from guinea pig taenia caeci, Jpn. J. Physiol., 34: 41-54.

Tomita, T., Tokuno, H. and Takai, A., 1985, Action of isoproterenol and epinephrine on the guinea pig taenia coli studied by intracellular microelectrodes, Jpn. J. Smooth Muscle Res., 21 (Suppl): 55-59.

Yamada, T. and Obara, K., 1985, Changes in cyclic AMP and cyclic GMP levels in the contraction-relaxation cycle of isolated smooth muscle cells from guinea pig taenia caeci, Jpn. J. Smooth Muscle Res., 21: 25-37.

EXCITATION-CONTRACTION COUPLING IN CANINE TRACHEAL SMOOTH MUSCLE CELLS

J.-P. Bourreau, C.Y. Kwan and E.E Daniel

Department of Biomedical Sciences, McMaster University
1200 Main Street West
Hamilton, Ontario, Canada L8N 3Z5

INTRODUCTION: In canine tracheal smooth muscle, the significance of depolarization induced by acetylcholine is puzzling because it has no role in contraction. We proposed earlier a model where refilling of ACh-sensitive intracellular Ca^{2+} stores was in part achieved by a mechanism involving a direct influx of extracellular Ca^{2+} into the ACh-sensitive Ca^{2+} store through a dihydropyridine-sensitive channel. Putney proposed that different phases of the calcium response in non excitable cells may be intimately linked with IP_3 first stimulating release of calcium from intracellular stores and the empty stores themselves then providing the signal that stimulates calcium entry. We examined in tracheal smooth muscle the essential feature of Putney's model, that the intracellular stores themselves exert a major control over calcium entry.

METHODS: Tracheal smooth muscle strips were placed in organ chambers filled with Krebs solution at 37° C for isometric tension recording. Agonist-sensitive intracellular calcium stores were evaluated indirectly by recording the transient contraction induced by 3 μM ACh in a medium free of calcium (0 Ca^{2+} added + 0.1 mM EGTA). The maximum amplitude of the transient ACh contraction in these conditions was used as an index of sarcoplasmic reticulum calcium loading.

RESULTS: Loading of empty agonist-sensitive Ca^{2+}stores required extracellular calcium, was partly inhibited by nifedipine (2 μM), was potentiated by BAY K 8644 (10 nM) and also by increased external calcium concentration. Most refilling was prevented by cyclopiazonic acid (CPA, 10 μM), an inhibitor of the sarcoplasmic reticulum calcium-pump. However, BAY K 8644 was able to reverse this inhibitory effect of CPA. This suggests two distinct loading pathways; one pathway involves extracellular Ca^{2+} influx via a dihydropyridine-sensitive calcium channel; the other uptake of cytosolic Ca^{2+} via a CPA-sensitive SR Ca^{2+}-ATPase. Moreover, in Krebs solution, Bay K 8644 itself (up to 1 μM) or in the presence of CPA did not affect muscle resting tension, but increased it substantially in the presence of 30 μM ryanodine. Taken together, these results suggest the existence of a direct pathway between plasmalemma and sarcoplasmic reticulum contributing to internal Ca^{2+} store loading. This direct pathway seems to involve a plasmalemma L-type like calcium channel (dihydropyridine-sensitive) and a sarcoplasmic reticulum calcium release like channel (ryanodine sensitive).
In contrast with the response in normal Krebs solution, serial stimulation of the smooth muscle with 3 μM ACh in the continuous presence of 2 μM nifedipine resulted in a significant decrease in the developed tension.

This effect was associated with a decrease in the content of the agonist-sensitive intracellular calcium stores. Tissue incubation with CPA (10 μM) or ryanodine (30 μM), which both reduce the internal agonist-sensitive Ca^{2+} store, amplified this effect of nifedipine (table).

TABLE 1

	2.5 mM Ca^{2+}		0 mM Ca^{2+}	
Serial Stimulation #	# 1	# 9	# 1	# 10
control	137 ± 17 (20)	138 ± 20 (20)	72 ± 7 (6)	58 ± 12 (20)
nifedipine (2 μM)	120 ± 10 (6)	79 ± 6 (6)	ND	34 ± 5 (6)
CPA (10 μM)	132 ± 18 (6)	129 ± 9 (6)	59 ± 3 (6)	34 ± 5 (6)
ryanodine (30 μM)	148 ± 17 (7)	149 ± 14 (7)	44 ± 2 (4)	26 ± 3 (7)
CPA + nifedipine	88 ± 8 (6)	56 ± 10 (6)	ND	29 ± 3 (6)
ryanodine + nifedipine	58 ± 6 (6)	43 ± 2 (6)	ND	15 ± 2 (6)

Data are expressed as a % of 30 mM KCl contractions.

Although the global ACh contraction in Ca^{2+} containing medium did not seem to depend upon internal Ca^{2+}store loading, the nifedipine resistant part of it was directly related to the amount of store loading. The smaller the transient ACh contraction in Ca^{2+}free medium was, the more potent nifedipine was to inhibit ACh contraction in Ca^{2+}containing medium (cf.Fig 1). The nifedipine sensitivity of ACh contraction did not appear to be modified by membrane potential since it was not influenced by the presence in the medium of either a potassium channel opener (cromakalim, 3 μM) or a potassium channel blocker (TEA, 10 mM)(cf.Fig 2).

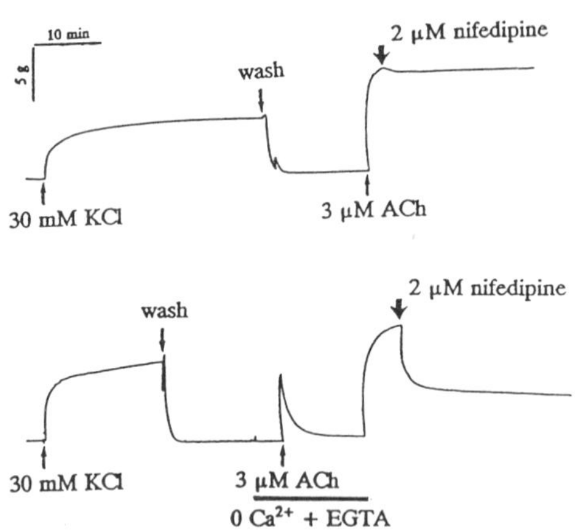

FIGURE 1

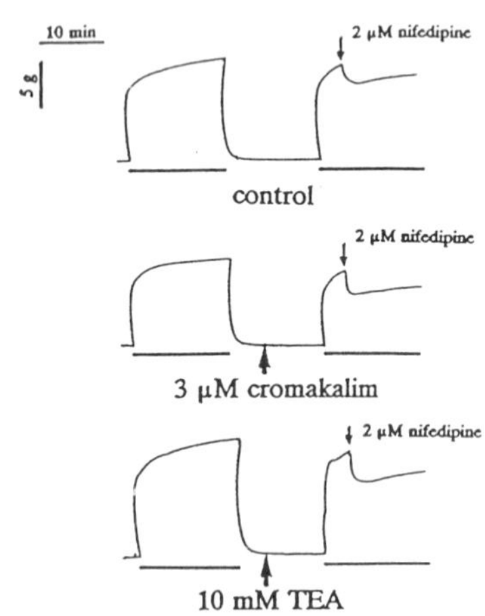

FIGURE 2

CONCLUSION: Dog tracheal smooth muscle cells appear to possess two different dihydropyridine-sensitive Ca^{2+}channels. One corresponds to the classical voltage-operated L-type channel, and the other is implicated in the maintenance of the internal agonist-sensitive Ca^{2+}store. This latter one does not recognize plasma membrane potential, but rather, its opening seems to be controlled by the status of the internal agonist-sensitive Ca^{2+}store. Figure 3 summarizes possible calcium movements triggered by acetylcholine stimulation in canine tracheal smooth muscle.

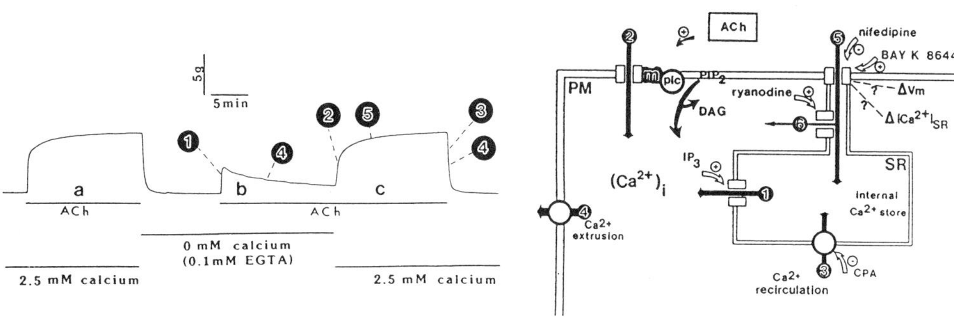

5g
5min
a
b
c
ACh
ACh
0 mM calcium
(0.1mM EGTA)
2.5 mM calcium
2.5 mM calcium
ACh
nifedipine
BAY K 8644
PM
plc
PIP2
DAG
ryanodine
ΔVm
Δ[Ca2+]SR
SR
IP3
(Ca2+)i
internal
Ca2+ store
Ca2+
extrusion
Ca2+
recirculation
CPA

FIGURE 3

NEGATIVE INOTROPIC EFFECTS OF CAFFEINE RELATED TO LEAK OF Ca^{2+} FROM THE SARCOPLASMIC RETICULUM

B.R. MacIntosh

Human Performance Laboratory, and Medical Physiology
University of Calgary, Calgary, Alberta T2N 1N4

Caffeine has been reported to have either a positive or a negative inotropic effect on cardiac muscle, and it is not known what circumstances are necessary for each of these effects. In this study, rat atria were exposed to caffeine (1-5) mM with regular stimulation at 0.1 - 3 Hz. Postrest contractions were also observed to determine if inhibition of contractions was always seen when stimulation was interrupted (Endoh and Iijima, 1981). As expected, addition of caffeine to the superfusate resulted in a transient positive inotropic response, when stimulation was 0.5 or 2.0 Hz. In contrast with this, when stimulation was 0.1 Hz, there was only a negative inotropic response. This decrease in developed tension was greater with 1 mM than with 5 mM caffeine. (see Table 1).

Contraction time (C_t) increased significantly from 31.9 ± .9 to 36.5 ms when stimulation frequency was 0.1 Hz, and caffeine was 1 mM. Note that this coincided with a negative inotropic response. It may be that this prolongation of C_t is related to a prolongation of the duration of Ca^{2+} entry via Ca^{2+} channels. There were no significant changes in C_t when stimulation frequency was 0.5 or 2.0 Hz at this concentration of caffeine. With 5 mM caffeine, there was a significant increase in C_t only for 0.5 and 2.0 Hz stimulation.

Relaxation time was measured from 50% relaxation to 75% relaxation (H_t). H_t was not significantly affected by 5 mM caffeine when stimulation frequency was 0.1 Hz, but was significantly increased when stimulation was at 0.5 or 2.0 Hz. This measure of relaxation was not significantly increased at any frequency of stimulation with 1.0 mM caffeine. It seems unlikely therefore, that caffeine at this concentration slows uptake of Ca^{2+} by the SR.

TABLE 1
DEVELOPED TENSION WITH CAFFEINE

TIME:	2 min	5 min	10 min
A. 0.1 Hz and 1 mM caffeine:	66.3 ± 2.5	44.5 ± 3.1	39.0 ± 3.9
B. 0.1 Hz and 5 mM caffeine:	79.7 ± 4.8	65.5 ± 3.5	59.0 ± 3.6
C. 0.5 Hz and 1 mM caffeine:	144 ± 5.6	131.7 ± 4.3	124.7 ± 2.3
D. 0.5 Hz and 5 mM caffeine:	160 ± 10.1	115.2 ± 7.4	97.2 ± 6.9

Note: values are % of developed tension prior to caffeine exposure.

Excitation-Contraction Coupling in Skeletal, Cardiac, and Smooth Muscle
Edited by G.B. Frank *et al.*, Plenum Press, New York, 1992

It was observed in these experiments that developed tension for contractions with regular stimulation at 0.5, 1 or 2 Hz was not affected by caffeine after 15 minutes of exposure. Furthermore, potentiation was evident with brief rest intervals when stimulation frequency prior to the rest was 1 or 3 Hz. These observations confirm that in the presence of caffeine, the rat atria were able to release Ca^{2+} and reaccumulate it in the SR, with a time course similar to that prior to caffeine exposure. However, when there was a prolonged rest between beats, as with 0.1 Hz stimulation or with an imposed rest interval, there was a negative inotropic effect.

Rat atrial muscle, like rat ventricular muscle, demonstrates potentiation when there is a prolonged interval between beats. It has been proposed that this potentiation is a result of inward transport of Ca^{2+} via Na^+/Ca^{2+} exchange (Shattock and Bers, 1989; Schouten et al.1987), and recent evidence supports this proposal (Lewartowski and Zdanowski, 1990). Caffeine apparently prevents this potentiation, and in fact in the case of regular stimulation at 0.1 Hz, caffeine causes a progressive loss in developed tension. This is consistent with attenuation or possibly a reversal of the proposed direction of Na^+/Ca^{2+} exchange. I would like to propose that a caffeine-induced leak of Ca^{2+} from the SR would be all that is required to effect this response.

The negative inotropic effects of caffeine on rat cardiac muscle appear to be dependent on the interval between beats. With regular stimulation at 0.1 Hz, or with an imposed rest interval , there is a strong negative inotropic response.

If the reversal potential for Na^+/Ca^{2+} exchange is calculated, it can be shown that in rat ventricular muscle, influx of Ca^{2+} would predominate during the diastolic interval (Shattock and Bers, 1989). Assuming this is also the case with rat atria, then the question can be raised, what changes could occur to result in a decrease (less negative) in the reversal potential? If this could happen, then the exchange mechanism would transport Ca^{2+} out of the cell during the diastolic phase, and a negative inotropic response would be observed due to a net loss of Ca^{2+} from the SR. A small increase in intracellular $[Ca^{2+}]$ (ie. from 100 nM to 150 nM) would be enough to effect outward transport of Ca^{2+} via the Na^+/Ca^{2+} exchanger. This could happen with an increased leak of Ca^{2+} from the SR in the presence of caffeine.

REFERENCES

Endoh, M. and Iijima, T. 1981. Twitch potentiation by rest in canine ventricular muscle: effects of theophylline. *Am.J.Physiol.* 241:H583-H590.

Lewartowski, B. and Zdanowski, K. 1990. Net Ca2+ influx and sarcoplasmic reticulum Ca2+ uptake in resting single myocytes of the rat heart: comparison with guinea-pig. *J.Mol.Cell.Cardiol.* 22:1221-1229.

Schouten, V.J.A., van Deen, J.K., de Tombe, P. and Verveen, A.A. 1987. Force-interval relationship in heart muscle of mammals: a calcium compartment model. *Biophys.J.* 51:13-26.

Shattock, M.J. and Bers, D.M. 1989. Rat vs. rabbit ventricle: Ca flux and intracellular Na assessed by ion-selective microelectrodes. *Am.J.Physiol.* 256:C813-C822.

ENDOTHELIUM-DEPENDENT AND-INDEPENDENT RELAXATION OF AORTAS FROM COPPER-DEFICIENT RATS

Jack T. Saari

USDA, ARS, Human Nutrition Research Center
Grand Forks, North Dakota 58202

Summary. Isolated aortas from copper-deficient rats exhibited reduced ability to relax in response to acetylcholine, histamine and sodium nitroprusside, which suggests that release of endothelium-derived relaxing factor or its interaction with the contractile mechanism is impaired.

Dietary copper (Cu) deficiency in experimental animals produces vascular defects which include structural changes in aortas,[1] altered vasoactive responses to norepinephrine in portal veins and pulmonary arteries,[2] enhanced microvascular protein leakage in response to endogenous histamine release[3] and reduced microcirculatory platelet thrombus formation in response to photoactivation of a light-sensitive dye.[3] Structural changes in aortas may affect the endothelium[1] and some of the functional defects noted above (permeability, thrombus formation) necessarily involve endothelium. This knowledge suggested the present study, which was initiated to determine whether endothelium-dependent relaxation of aortas was affected by Cu deficiency.

Male, weanling Sprague-Dawley rats were fed a purified diet that was either deficient (CuD, <0.5 μg/g diet) or adequate (CuA, 5.0-5.5 μg/g diet) in Cu for four weeks. Cu deficiency was verified by measurement of liver Cu concentration; mean values (SEM) were 9.86 (0.21) for CuA and 1.03 (0.05) μg Cu/g wet weight for CuD-fed rats. Aortic rings (3-5 mm long) from paired CuD and CuA-fed rats were isolated from the descending thoracic aorta, placed in tandem tissue baths containing Krebs-Henseleit solution (37°C, aerated with 95% O_2 and 5% CO_2) and attached to force transducers. Over a 90 min equilibration period, tension on the aortas was gradually increased until a resting force of 2 g was reached. Aortas were then precontracted with phenylephrine (PE, $3x10^{-7}$ M) and the degree of force reduction was measured after successively increasing the dose of acetylcholine (ACh, 10^{-8}-10^{-5} M, 19 pairs of rings), histamine (Hist, 10^{-6}-10^{-3} M, 15 pairs of rings) or sodium nitroprusside (NaNP, 10^{-9}-10^{-6} M, 6 pairs of rings). Relaxation of 4 pairs of aortas was also tested with 10^{-4} or 10^{-5} M papaverine.

Relaxation as a percent of maximum force of contraction with PE is shown below for ACh, Hist and NaNP; values are means (SEM). Two-way, repeated measures analysis of variance (ANOVA) indicated that Cu

Excitation-Contraction Coupling in Skeletal, Cardiac, and Smooth Muscle
Edited by G.B. Frank *et al.*, Plenum Press, New York, 1992

deficiency marginally reduced the relaxation response to ACh (0.05<p<0.10) and significantly reduced the relaxation responses to Hist and NaNP (p<0.05). Comparison of means (Bonferroni contrasts) at specific drug doses indicated that differences in response were significant (* = p<0.05) at the three highest doses of each relaxing agent. Relaxation of aortas from both CuD and CuA rats was 100% with either 10^{-5} or 10^{-4} M papaverine.

Table 1

EFFECT OF CU DEFICIENCY ON PERCENT RELAXATION OF PHENYLEPHRINE-CONTRACTED AORTIC RINGS BY VARIOUS AGONISTS

-log [agonist]	*Agonist*					
	ACh		Hist		NaNP	
	Diet					
	CuA	CuD	CuA	CuD	CuA	CuD
9					3(2)	-1(1)
8	2(1)	0(1)			28(5)	12(3)*
7	23(3)	15(3)*			88(4)	69(5)*
6	44(4)	35(5)*	5(2)	1(1)	105(1)	100(1)*
5	52(4)	41(5)*	25(3)	15(2)*		
4			41(4)	28(3)*		
3			47(4)	32(4)*		

Because the relaxation responses to both ACh and Hist in rat aorta have been shown to be dependent on the presence of endothelium,[4] the reduction of these responses suggest that endothelium, or its interaction with smooth muscle, is disrupted in dietary Cu deficiency. Nitrovasodilators bypass the endothelium and relax smooth muscle directly; they are thought to mediate their responses by interaction with smooth muscle guanylate cyclase, a mechanism similar to that of endothelium-derived relaxing factor (EDRF). The reduction in response to NaNP indicates that it is the interaction of EDRF with smooth muscle that is disrupted in Cu deficiency. This suggestion does not rule out the possibility, however, that the endothelium of Cu-deficient aortas releases less EDRF. The ability of papaverine, a phosphodiesterase inhibitor, to cause complete relaxation in both CuD and CuA-fed rats suggests that the responsiveness of the contractile mechanism to elevated c-GMP is intact.

Impairment of endothelium-dependent responses in Cu-deficient rats may be caused by altered activity of any of a variety of enzymes.[5] For example, 1) reduced lysyl oxidase activity could result in a disruption of underlying connective tissue, thus reducing endothelial integrity, 2) reduction of activity of several cellular and sub-cellular antioxidant enzymes in Cu deficiency may make endothelium and/or smooth muscle vulnerable to oxidative damage, 3) reduction of Cu,Zn-superoxide dismutase activity may increase EDRF (i.e., nitric oxide) destruction by superoxide and 4) reduced activity of Cu-dependent guanylate cyclase may impair interaction of EDRF with the contractile mechanism.

Blood pressure is known to be increased in adult Cu-deficient animals; the findings of this study offer a potential mechanism for that alteration. Further study of the effect of dietary Cu deficiency on blood pressure regulation and its possible role in human hypertension appears warranted in light of the findings presented here.

ACKNOWLEDGEMENT

The author gratefully acknowledges the technical assistance of Jackie Keith and Gwen Schelkoph and the editorial comment of Dr. C.B. Allen.

REFERENCES

1. H. A. Hunsacker, M. Morita, and K. G. D. Allen, Marginal copper deficiency in rats. Aortal morphology of elastin and cholesterol values in first-generation adult males, Atherosclerosis 51:1-19 (1984).
2. S. Kitano, Membrane and contractile properties of rat vascular tissue in copper deficient conditions, Circ. Res. 46:681-689 (1980).
3. D. A. Schuschke, J. T. Saari, D. M. Ackermann, and F. N. Miller, Microvascular responses in copper deficient rats, Am. J. Physiol. 257:H1607-H1612 (1989).
4. R. F. Furchgott, and P. M. Vanhoutte, Endothelium-derived relaxing and contracting factors, FASEB J. 3:2007-2018 (1989).
5. J. R. Prohaska, Biochemical functions of copper in animals, in: "Essential and Toxic Trace Elements in Human Health and Disease," A. S. Prasad, ed., Alan R Liss, Inc., New York, pp. 105-124 (1988).

MODULATION OF L-TYPE CALCIUM CURRENT IN MAMMALIAN VENTRICULAR MYOCYTES BY PHOTOLYSIS OF CAGED CALCIUM

Susan E. Bates and Alison M. Gurney

Department of Pharmacology
UMDS, St. Thomas's Hospital
London, England

The caged calcium compound nitr-5 was used to investigate the effects of a rapid increase in intracellular calcium concentration on the L-type calcium current (I_{Ca}) of guinea pig ventricular cells. Short term inactivation and longer lasting potentiation were observed.

Cells were isolated using a modified version of the method of Harding et al (1990) and were usually stored in "KB" medium (Isenberg and Klockner 1982) for 30 min before being transferred to a storage/recording solution containing (in mM) NaCl 120, KCl 2.5, $MgCl_2$ 2.5, NaH_2PO_4 0.5, KH_2PO_4 0.5, $NaHCO_3$ 15, HEPES 5, glucose 10, $CaCl_2$ (or $BaCl_2$) 1 (pH 7.4). The cells were continuously superfused and were voltage-clamped at -70 or -80mV using the whole-cell configuration of the patch-clamp technique. The stimulus protocol consisted of a 100ms prepulse to -40mV to inactivate Na current and any T-type Ca current that might be present, followed by a 300ms step to 0mV to elicit I_{Ca}. Pipette solution contained 130mM CsCl and 10mM HEPES (pH 7.2); 2mM nitr-5 or DM-nitrophen were included, with or without added Ca (usually 1mM). Photolysis was triggered during I_{Ca}, using a xenon flashlamp ($\lambda > 320$nm).

Previous studies have shown that a rapid increase in intracellular Ca causes potentiation of I_{Ca} in frog and guinea pig cardiac cells (Gurney et al 1989). However, in guinea pig the potentiation was preceded by a small (about 10%) inactivation. In the present investigation, photolysis-induced inactivation reached a maximum within 50ms and recovered within 5-10s. I_{Ca} was subsequently potentiated by up to 200%. The time course of the potentiation varied under different experimental conditions. At low rates of stimulation ($\leqslant$ 0.1Hz), I_{Ca} amplitude increased exponentially with a single time constant of about 4 min. At higher rates of stimulation ($\geqslant$ 0.2Hz), a single flash activated a more rapidly developing potentation, with a time constant of about 20s, in addition to the slower component. These differences may be due to increased Ca loading of nitr-5 at higher stimulation frequencies as a result of greater Ca influx. Additionally, varying the stimulation frequency is likely to affect the relative concentrations of Ca in various cellular compartments, since at high frequencies both the sarcolemmal Na/Ca exchanger and the sarcoplasmic reticulum Ca pump spend a greater proportion of time in an activated state. Submaximal responses to low intensity flashes could be repeated in a single cell. However, after a maximal response, apparent recovery of I_{Ca} was followed by a gradual deterioration in cell appearance and the development

Excitation-Contraction Coupling in Skeletal, Cardiac, and Smooth Muscle
Edited by G.B. Frank *et al.*, Plenum Press, New York, 1992

of large, outward currents at the holding and stimulus potentials; a second flash produced only inactivation of I_{Ca}.

To investigate the role of intracellular Ca in these responses, we repeated the experiments with the Ca chelator BAPTA (12mM) added to the recording pipette and Ca omitted. With low frequency stimulation, flashes still induced inactivation and potentiation, but the latter response was accelerated (time constant about 1.5 min) with no reduction in its magnitude. When I_{Ca} was elicited at a higher rate, the size of the inactivation was increased. Similar effects were seen when 25mM EGTA was substituted for BAPTA or when Ca was simply omitted from the pipette. Thus it seems likely that, even in the presence of BAPTA, Ca influx during I_{Ca} provides the nitr-5 with sufficient Ca to produce a response on photolysis. In order to limit the source of Ca we replaced extracellular Ca with Ba. Under these conditions, I_{Ca} potentiation was suppressed. However, the response was completely blocked only if the cells were isolated and stored in a 0 Ca medium, in order to deplete intracellular Ca stores. It seems, therefore, that I_{Ca} potentiation is caused by increased $[Ca]_i$ after photolysis of nitr-5. However, in non-Ca depleted cells, intracellular storage sites can provide the nitr-5 with sufficient Ca to produce potentiation on its release.

It was of interest to determine whether similar effects could be produced with DM-nitrophen, a form of caged Ca with different Ca-binding properties to those of nitr-5. At 2mM, DM-nitrophen would produce Ca jumps in the M range (Kaplan et al 1988), much larger than those produced by nitr-5. The greater magnitude of the jump under these conditions was evidenced by an immediate increase in the Na/Ca exchange current following the flash, which was never observed with nitr-5. Photolysis of DM-nitrophen, bound to varying degrees with Ca, produced inactivation of the current but not potentiation. The time course of the inactivation was similar to that observed with nitr-5 and was compatible with open channel block. Nevertheless, since the Ca jumps were probably large, there may also have been a component due to a reduced driving force for Ca passing into the cell. I_{Ca} usually recovered fully after photolysis of DM-nitrophen, and several responses could be recorded from a single cell.

These results suggest that the effect on I_{Ca} in guinea pig ventricular cells of a rapid increase in intracellular Ca is complex. It consists of an inactivation followed by potentiation, which may comprise at least two components and is highly dependent on the pre-flash Ca level, the size of the increase and the rate at which I_{Ca} is elicited. The inability of BAPTA to suppress the responses additionally suggests that a smaller increase in intracellular Ca is required to produce potentiation in guinea pig cells than in frog (Gurney et al 1989).

REFERENCES

Gurney AM, Charnet P, Pye JM and Nargeot J (1989) Augmentation of cardiac calcium current by flash photolysis of intracellular caged-Ca^{2+} molecules. Nature, 341:65-68

Harding SE, O'Gara P, Jones SM, Brown LA, Vescovo G and Poole-Wilson PA (1990) Species dependence of contraction velocity in single isolated cardiac myocytes. Cardioscience, 1:49-54

Isenberg G and Klockner U (1882) Calcium tolerant ventricular myocytes prepared by preincubation in a "KB medium". Pfluger Arch, 395:6-18

Kaplan JH and Ellis-Davies GCR (1988) Photolabile chelators for the rapid photorelease of divalent cations. Proc Natl Acad Sci USA, 85:6571-6575

EFFECTS OF INORGANIC PHOSPHATE ON CALCIUM AND TENSION OSCILLATIONS IN SAPONIN-TREATED RAT CARDIAC MUSCLE

G.L. Smith, D.S. Steele, and C.A. Crichton

Institute of Physiology, University of Glasgow
Glasgow, Scotland, G12 8QQ

SUMMARY

The effects of inorganic phosphate (P_i) on the sarcoplasmic reticulum (s.r.) were studied in saponin-treated rat cardiac muscle. Ca^{2+}-uptake and release by the s.r. was detected using Indo-1. Oscillations of Ca^{2+} and tension occurred when the [Ca^{2+}] of the bathing solution was around the threshold for force production. Introduction of P_i (20mM) caused a slow transient release of Ca^{2+} from the s.r. The oscillations were abolished in the presence of P_i. Removal of P_i caused a rapid fall in Ca^{2+} followed by a large transient increase in [Ca^{2+}] and tension. Results are consistent with a phosphate-induced decrease Ca^{2+} available for release.

RESULTS

During myocardial ischaemia, there is a marked rise in inorganic phosphate. Experiments on skinned cardiac preparations have shown that P_i inhibits Ca^{2+}-activated force, contributing to contractile failure (Kentish, 1986). In this study, we investigated the effects of P_i on the s.r. in saponin-permeabilised rat ventricular trabeculae (80-150μm dia.). Changes of [Ca^{2+}] within the preparations were detected using Indo-1 which was included in all solutions. The preparation was illuminated at 340nm and [Ca^{2+}] expressed as the ratio of the emitted light at 405 and 495nm. Oscillations of Ca^{2+} and tension occurred in the presence of 5μM Ca^{2+}. Such oscillations result from the cyclic uptake and release of Ca^{2+} from the s.r. Introduction of P_i (20mM) caused a slow transient release of Ca^{2+} from the s.r. This was not accompanied by a change in tension, consistent with a decrease in myofilament Ca^{2+}-sensitivity. Oscillations were abolished in the presence of P_i. Removal of P_i caused a rapid fall of [Ca^{2+}], interpreted as net Ca^{2+}-

Excitation-Contraction Coupling in Skeletal, Cardiac, and Smooth Muscle
Edited by G.B. Frank *et al.*, Plenum Press, New York, 1992

uptake by the s.r. This was followed by a large transient increase in Ca^{2+} and tension, before oscillations returned to control levels. Similar results were obtained with 4 other preparations.

The inhibition of Ca^{2+}-oscillations is consistent with a decrease in Ca^{2+} available for release in the presence of phosphate of P_i. This could occur by (1) activation of s.r. Ca^{2+}-channels or (2) inhibition of Ca^{2+}-uptake by the s.r. This study cannot distinguish between these two possibilities. However, an increase in P_i would be expected to reduce the free energy of ATP hydrolysis (Kammermeier et al, 1982) making the latter possibility more likely. This phenomenon may contribute to the decrease in contractility found during ischaemia in intact cardiac muscle.

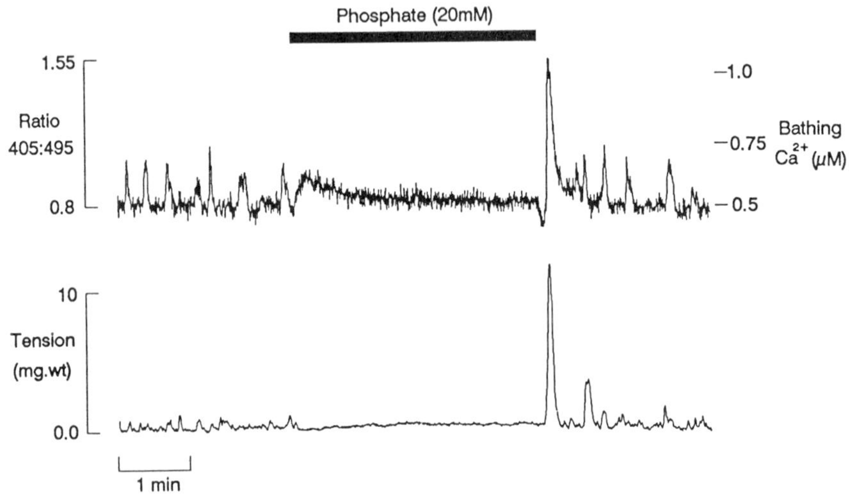

Fig (1) Simultaneous record of [Ca^{2+}] and tension from a rat ventricular trabecula. All solutions contained (mM except where otherwise indicated) 5, Na2ATP; 0.2, EGTA; 7, MgCl2; 130, KCl; 25, HEPES; 4μM Indo-1; pH 7.0. Phosphate was added by exchange with chloride.

REFERENCES

KENTISH, J.C. (1985). Journal of Physiology 370, 585-604.

KAMMERMEIER, H., SCHMIDT, P. & JUNGLING, E. (1982). Journal of Molecular and Cellular Cardiology 14, 267-277.

EFFECTS OF LOW IONIC STRENGTH AND CALMODULIN ON THE DECLINE IN MAXIMAL CALCIUM-ACTIVATED FORCE IN PERMEABILIZED SMOOTH MUSCLE

C.A. Crichton and G.L. Smith

Institute of Physiology
University of Glasgow, Glasgow, Scotland, G12 8QQ

SUMMARY

Maximal calcium-activated tension (Tmax) was measured from saponin-and alpha-toxin permemabilized strips of rat anococcygeus muscle stimulated repetitively. 50 mins after saponin treatment, Tmax had declined to 20% of the initial value. Calmodulin (1μM) only partially reduced this decline. Lowering the ionic strength of the bathing solution from 0.2M to 0.07M markedly reduced the rate of decline of tension. Addition of calmodulin to the bathing medium of low ionic strength prevented the decline of force still further; 50 mins after permeabilization, Tmax declined to 80% of the original level. Loss of endogenous calmodulin should be prevented by permeabilization with alpha-toxin from Staphylococcus aureus. Nevertheless, Tmax still declined to approximately 50% of the original value after 50mins. However, negligible force reduction was observed if the bathing solution had an ionic strength of 0.07M.

INTRODUCTION

A common problem with saponin permeabilized smooth muscle is the loss of cytosolic proteins (Kossman et al, 1987; Gardner et al, 1989) and Ca-activated force (Itoh et al, 1986). A new permeabilization technique using alpha-toxin from Staphylococcus aureus produces smaller pores in the sarcolemma which allow the free diffusion of molecules below a molecular weight of 4000. Hence, the cytosolic proteins should be retained. In the present study we monitored the decline of maximal calcium-activated force (Tmax) in strips of rat anococcygeus muscle after permeabilization with (1) saponin and (2) alpha-toxin.

METHODS

Male Wistar rats (250-270g) were killed by a blow to the back of the head followed by exsanguination. The anococcygeus muscle was then isolated and strips (100μm in diameter and 2-3mm long) were dissected. All solutions had the following basic composition (mM): EGTA, 10; ATP, 5; Creatine phosphate, 15; HEPES, 25; Cl, 120; K^+, 130; Na^+ 40; total Mg^+, 7; pH 7.0; 20°C; total Ca^{2+} was varied. Ionic strength was lowered by decreasing the concentration of K^+ in the mock intracellular solution to 20mM. Strips of smooth muscle were permeabilized by incubation in a mock intracellular solution containing either (a) 50μg/ml saponin for 30 mins or (b) 4mg/ml alpha-toxin from Staphylococcus aureus for 10 minutes.

Excitation Contraction Coupling in Skeletal, Cardiac, and Smooth Muscle
Edited by G.B. Frank *et al.*, Plenum Press, New York, 1992

RESULTS

Figure 1 shows the effect of different bathing solutions on the decline in Tmax in saponin- and alpha-toxin-treated muscle. Four different bathing solutions were examined: (1) ionic strength (I.S.) of 0.2M; (2) I.S. of 0.2M plus 1μM calmodulin; (3) I.S. of 0.07M; and (4) I.S.of 0.07M plus 1μM calmodulin. All four conditions were examined in saponin-treated muscle and (1) and (3) were examined in alpha-toxin-treated muscle. Figure 1A shows that calmodulin had only a small effect on the decline at an ionic strength of 0.2M, but markedly slowed the decline in Tmax at an ionic strength of 0.07M. The decline in the amplitude of Tmax in toxin-treated muscle was less than that in saponin-treated muscle (Figure 1B). The decline in alpha-toxin-treated muscle was prevented at an ionic strength of 0.07M.

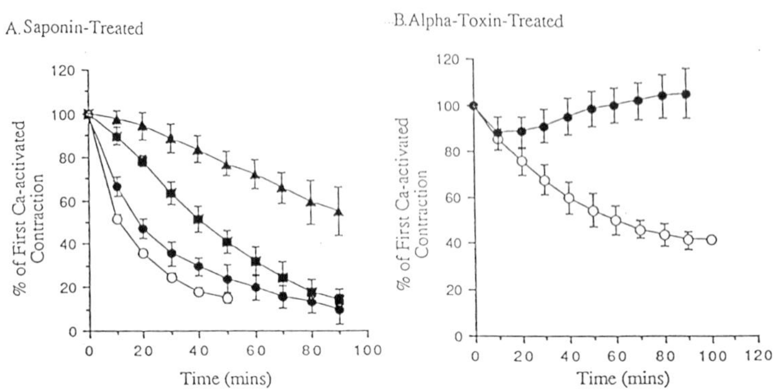

Figure 1 Panel A shows the decline in calcium-activated force in saponin -treated muscle bathing in solutions with: (○) ionic strength (I.S.) 0.2M; (●) I.S. 0.2M, calmodulin 1μM,; (■) I.S. 0.07M; (▲) I.S. 0.07M, calmodulin 1μM. Panel B shows the decline in calcium-activated force in alpha-toxin-treated muscle bathed in solutions with: (○) I.S. 0.2M and (●) I.S. 0.07M.

DISCUSSION

SAPONIN-TREATED MUSCLE: In the present experiments the decline in Tmax could be reduced by lowering the ionic strength of the solution in the presence of calmodulin. The fact Tmax declined nevertheless indicates that other cytoplasmic factors are lost from the muscle. The factors may include myosin and actin as suggested by Kossman, Furst and Small (1987).

ALPHA-TOXIN-TREATED MUSCLE: Tmax declined with time even though calmodulin and other proteins which control force production are not lost. The decline was prevented by lowering the ionic strength of the bathing solution, through the reduction of the KCl concentration.

These results indicate that as well as the loss of cytosolic proteins that the composition of the mock intracellular solution affects the decline in Tmax.

Calmodulin was a gift form Dr K Torok. British Heart Foundation and the Wellcome Trust provided financial Support.

REFERENCES

Itoh, T., Kanumura, Y. & Kuriyama, H.(1986) Inorganic phosphate regulates the contraction-relaxation cycle in skinned muscles of the rabbit mesenteric artery. Journal of Physiology 376, 231-252

Kossman, T., Furst, D. & Small, J.V. (1987) Structural and biochemical analysis of skinned smooth muscle preparations. Journal of Muscle Research and Cell Motility 8, 135-144

Gardner, J.P., Stout, M.A. & Harris, S.R. (1989) Calmodulin loss in vascular smooth muscle following Triton X-100 or saponin skinning. Pflugers Archiv 414, 484-491

CORRELATION BETWEEN THE RESULTS OF FUNCTIONAL AND RECEPTOR BINDING STUDIES IN CARDIAC MUSCLE

E. Fejér, J. Szeredi, J. Gaál, and P. Sz. Körmöczy

CHINOIN Pharmaceutical & Chemical Works Co. Ltd.
Budapest, Hungary

Studies comparing the negative inotropic effect of calcium antagonists and "related" receptor-affinities showed inconsistent results. We tested the possible relationship between the potency in inhibiting $CaCl_2$-induced positive inotropy and receptor affinity. Therefore, we determined the relative potency of Nifedipine, Verapamil, Fendiline, Bepridil and Diltiazem both in isolated rabbit papillary muscle against $CaCl_2$-induced positive inotropy (pA_2) and their relative affinity ($-\log K_D$) in receptor binding experiments in rat cardiac homogenates.

In isolated organ experiments papillary muscles derived from the right ventricle of rabbits, were suspended in a 35 ml organ bath, containing Krebs-Henseleit solution, bubbled through with a mixture of 95 % O_2 + 5 % CO_2. The electrically-driven muscles (1.1 Hz, 3 msec, 1 V) were preloaded with 0.5 g. Grass FT-03 isometric transducer and Experimetria-Sefram amplifying-recording system served for recording of contractions.

The dose range of $CaCl_2$ and calcium antagonist compounds was between 10^{-4} - 3×10^{-2} M and 10^{-10}- 10^{-5} M, respectively. For quantitative comparisons of competitive antagonist potency, pA_2 values were calculated according to Schield.

The receptor binding studies were performed in cardiac crude membrane fractions prepared from male Wistar rats. Trimmed ventricles were homogenized with Teflon pestle in 20 vols of ice-cold buffer (50 mM TRIS/HCl, pH = 7.4, 25 ^{o}C), filtration through cheese-cloth, centrifugation at 12000 g for 10 min twice, at 4 ^{o}C.

Dihydropyridine and Verapamil binding sites were labelled by [^{3}H]PN-200-110[1] and [^{3}H]DMO-Verapamil[2], respectively.

The dissociation constants of the compounds were determined according to Batke and Gaál[3]. The negative logarithm of the dissociation constant was calculated and used for comparison to pA_2 values derived from functional experiments.

To characterize the possible interrelationship between the results of functional (pA_2) and binding studies ($-\log K_D$), correlation coefficient was calculated according to Kwon[4].

The correlation coefficient between $-\log K_D$ and pA_2 values was 0.974. Thus, there is a good correlation between the results of functional and receptor binding studies indicating that calcium is directly involved both in cardiac muscle function and in the mechanism of action of calcium-antagonist compounds.

Excitation-Contraction Coupling in Skeletal, Cardiac, and Smooth Muscle
Edited by G.B. Frank *et al.*, Plenum Press, New York, 1992

TABLE 1

INHIBITORY EFFECT ON $CaCl_2$-INDUCED POSITIVE INOTROPY IN RABBIT PAPILLARY MUSCLE

COMPOUND	n	pA_2 + SEM	Relative potency
NIFEDIPINE	6	8.43 + 0.56	1.00
DILTIAZEM	6	7.36 + 0.78	0.87
VERAPAMIL	6	7.20 + 0.78	0.85
FENDILINE	6	7.04 + 0.83	0.84
BEPRIDIL	6	6.44 + 0.76	0.76

TABLE 2

RECEPTOR BINDING AFFINITY VALUES IN RAT CARDIAC MUSCLE HOMOGENATES

COMPOUND	n	[^{3}H]PN-200-110 K_D(M)	[^{3}H]PN-200-110 -log K_D	[^{3}H]DMO-Verapamil K_D(M)	[^{3}H]DMO-Verapamil -log K_D	Relative* potency
NIFEDIPINE	3	3.6×10^{-9}	8.44	(5.0×10^{-5})	(4.30)	1.00
VERAPAMIL	3	(1.0×10^{-3})	(3.00)	2.0×10^{-8}	7.70	0.91
DILTIAZEM	3	no effect		1.0×10^{-7}	7.00	0.83
FENDILINE	3	(9.0×10^{-6})	(5.05)	1.0×10^{-7}	7.00	0.83
BEPRIDIL	3	(3.0×10^{-6})	(5.52)	3.2×10^{-7}	6.50	0.77

* irrespective of the nature of binding site

TABLE 3

COMPARISON OF RELATIVE POTENCIES DERIVED FROM FUNCTIONAL AND RECEPTOR BINDING STUDIES

COMPOUND	RECEPTOR BINDING -log K_D	RECEPTOR BINDING rel. potency	PHARMACOLOGY pA_2	PHARMACOLOGY rel. potency
NIFEDIPINE	8.44	1.00	8.43	1.00
VERAPAMIL	7.70	0.91	7.20	0.85
DILTIAZEM	7.00	0.83	7.36	0.87
FENDILINE	7.00	0.83	7.04	0.84
BEPRIDIL	6.50	0.77	6.44	0.76

REFERENCES

1. D. R. Ferry, and H. Glossmann, FEBS Lett. 148:331 (1982).
2. M. N. Garcia, M. I. Turmble, J. P. Reuben, G. I. Kaczorowski, J. Biol. Chem. 259:15013 (1984).
3. J. Batke, and J. Gaál, J. Biochem. Biophys. Meth. 12:203 (1986).
4. Y. W. Kwon, G. Franckowiak, D. A. Langs, M. Hawthorn, A. Joslyn, and D. J. Triggle, Naunyn-Schmiedeberg's Arch. Pharm. 339:19 (1989).

PULMONARY BETA RECEPTOR NUMBER CHANGES UNDER THE INFLUENCE OF OVALBUMIN SENSITIZATION AND GLUCOCORTICOID TREATMENT

P. Sz. Körmöczy, M. Szabó, J. Gaál, and E. Fejér

CHINOIN Pharmaceutical & Chemical Works Co. Ltd.
Budapest, Hungary

Characteristic changes have been reported in adrenergic receptor number of asthmatics.

We tested the beta-receptor number before and after ovalbumin sensitization and immediately after ovalbumin (OA) challenge with and without glucocorticoid treatment, using a separate group of animals for each.

Four-membered groups of male guinea pigs either left intact (control) or sensitized with the i.p. application of 1 µg OA + 100 mg $Al(OH)_3$ were used. Forty days later the animals were challenged by the i.v. administration of 5 µg/kg OA, according to the method of Andersson et al[1].

To carry out binding studies, the guinea pigs were killed, their lungs were removed and homogenized in 20 vols of buffer (50 mM Tris/HCl; pH = 7.7, 25 ^{o}C) filtrated through cheesecloth, centrifugated at 12000 g for 10 min, twice. The membranes (0.3-0.6 mg protein) were incubated at 25 ^{o}C with 0.1 - 12 nM $[^3H]$DHA (60 Ci/mmol, Amersham) for 20 min. Propranolol (10^{-4} M) was used to define the ratio of specific binding. The bound radioactivity was trapped on Whatman GF/C filters and measured by liquid scintillation spectrometry[2].

The dissociation constant (K_d) and B_{max} were determined according to Batke and Gaál[3]. The B_{max} of control group was taken as 100 % and the B_{max} values of treated groups were expressed as percentage change from control. For statistical evaluation the unpaired t test was used.

RESULTS

CONTROL GROUP I. (n=4)

K_d = 0.71 + 0.19 nM
B_{max} = 359 + 39 fmol/mg protein

SENSITIZED GROUPS:
Ovalbumin + 40 days
K_d = 0.84 + 0.17 nM
B_{max} = 287 + 61 fmol/mg protein - **20 %** (NS)

Ovalbumin + 40 days + challenge (OA)

K_d = 0.92 + 0.31 nM

B_{max} = 439 + 102 fmol/mg protein **+ 22 %** (NS)

TREATED, NONSENSITIZED GROUP

Dexamethasone (0.5 mg/kg for 10 days)

K_d = 0.95 + 0.11 nM

B_{max} = 403 + 43 fmol/mg protein **+ 12 %** (NS)

Following OA sensitization there is a 20 percent drop in beta-receptor number while immediately after challenge (OA) a 22 percent increase can be detected. Dexamethasone treatment (daily oral dose of 0.5 mg/kg) in itself is without effect.

CONTROL GROUP II. (n=4)

K_d = 1.63 + 0.15 nM

B_{max} = 357 + 45 fmol/mg protein

SENSTITIZED AND TREATED GROUPS

Ovalbumin + 40 days with dexamethasone treatment for the last 10 days

K_d = 1.8 + 0.02 nM

B_{max} = 495 + 66 fmol/mg protein **+ 39 %** ($p < 0.05$)

Ovalbumin + 40 days with dexamethasone treatment for the last 10 days + challenge (OA)

K_d = 2.2 + 0.66 nM

B_{max} = 519 + 58 fmol/mg protein **+ 45 %** ($p < 0.01$)

Dexamethasone (oral daily dose of 0.5 mg/kg) treatment after OA sensitization both with and without OA challenge, elevates beta adrenoreceptor number in the guinea pig lung.

In summary, sensitization of guinea pigs with ovalbumin characteristically alters beta-adrenerg receptor number in the lung, as does dexamethasone treatment. This finding is probably related to the change in responsiveness of respiratory smooth muscle.

REFERENCES

1. P. Andersson, C. Brange, B. Kogerer, B. Sonmark, and G. Stahre, Int. Arch. Allergy Appl. Immunol. 87:32 (1988).
2. D. B. Bylund and S. H. Snyder, Mol. Pharmacol. 12:568 (1976).
3. J. Batke and J. Gaál, J. Biochem. Biophys. Meth. 12:203 (1986).

ROLE OF GUANYLATE CYCLASE ACTIVATION IN THE SMOOTH MUSCLE ACTIONS OF DIHYDROPYRIDINES

Christopher R. Triggle, Sean O'Neill and D. Bieger *

Dept. of Pharmacology and Therapeutics, Faculty of Medicine, The University of Calgary, AB T2N 4N1 and *Division of Basic Medical Sciences, Memorial University of Newfoundland, St. John's, NF A1B 3V6

Certain 1,4-dihydropyridines (DHPs) interact with the 'L'-type voltage-operated Ca^{2+} channel (VOCC) and thereby modulate Ca^{2+} entry into a variety of excitable tissues. Radioligand binding data indicates that the DHP nifedipine binds with high affinity to the VOCC and electrophysiological data indicates that a Ca^{2+} current can be modulated by such compounds[1]. The high affinity that many DHPs show for the VOCC is suggestive that other potential sites of action are of secondary importance with respect to their effects on smooth muscle function. It is thus of interest to note that comparatively low affinity sites have been described for the organic Ca^{2+} channel modulators[2]. Although such sites are of low affinity it must be emphasized that since the organic Ca^{2+} channel antagonists are hydrophobic drugs, they are capable of accumulating within vascular smooth muscle cells at tissue:medium ratios of 200-500:1[3].

We have previously reported[4], confirming an earlier report by Mikkelsen and co-workers[5], that Bay K 8644 can sensitize certain smooth muscle preparations to light. Our data, however, failed to support the hypothesis of Mikkelsen and colleagues[5] that there is a photosensitive protein(s) associated with the VOCC and that light induces a switch in Bay K 8644 binding from an agonist to an antagonist site. We hypothesized[4] that this action of Bay K 8644 was related to an intracellular action of this DHP that was independent of any action modulating the movement of extracellular Ca^{2+} across the plasma membrane. We have replicated these observations with other DHPs and determined that DHPs could be classified into four groups (Fig. IA-D). Comparison of active and inactive compounds suggests that an $-NO_2$ or a benzo- furazan grouping are essential structural features required to support the photo-dependent relaxation response.

We now report that a number of Bay K 8644 analogues, all of which contain at least one $-NO_2$ grouping, support photorelaxation in the rat thoracic aorta (Fig. IE). This data suggests that it is the $-NO_2$ group on the DHP nucleus that confers photosensitivity. The Bay K 8644 analogues are effective photosensitizers in the 5 nmole range reflecting a similar affinity for both the cellular mechanism that is involved in the photorelaxation response and the VOCC. The ability, however, to support photorelaxation is not unique to DHPs and we have shown that structurally diverse compounds including the diabetogenic agent, streptozotocin, the NO-synthase inhibitor L-nitro-arginine methyl ester (L-NAME), but not nitrobenzoic acids or alcohols, also confer similar sensitivity to light. In contrast to the effects of the DHPS the sensitivity to photorelaxation conferred by non-DHPs, but not DHPs, is readily terminated by wash-out from the organ bath; thus suggesting that the cellular mechanisms for DHPs versus non-DHPs are different.

We previously reported that the putative c-GMP lowering quinolinedione, LY83583,

Excitation-Contraction Coupling in Skeletal, Cardiac, and Smooth Muscle
Edited by G.B. Frank *et al.*, Plenum Press, New York, 1992

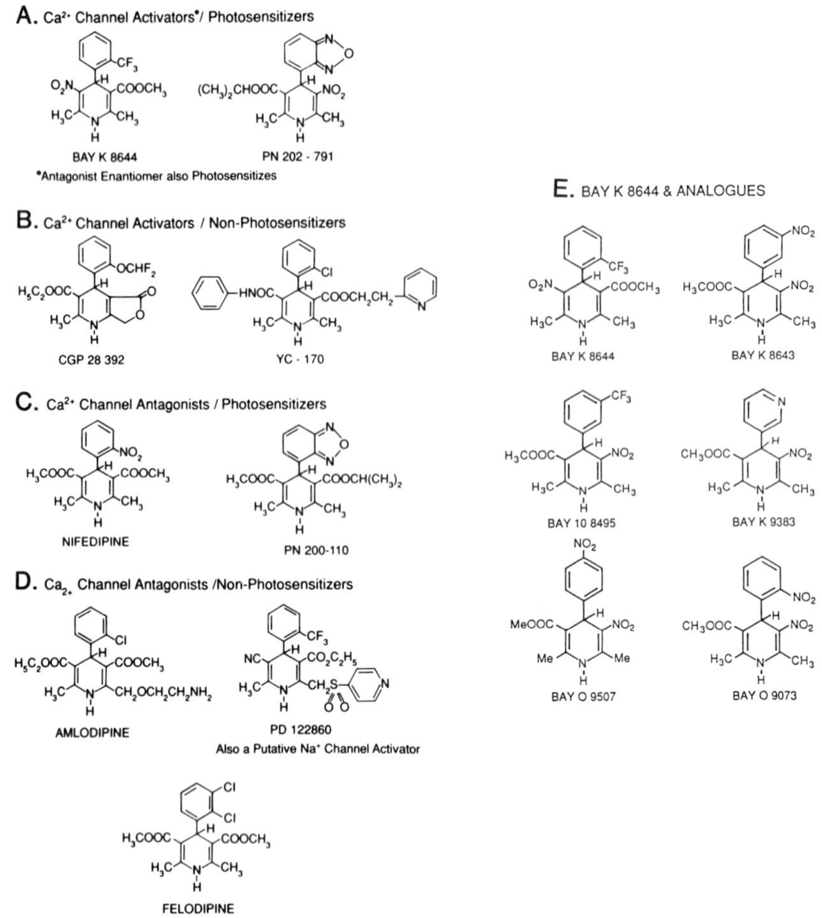

Figure I. Structures of DHPs used in the study.

attenuated the DHP supported photorelaxation response[4]. Further evidence that an increase incellular levels of c-GMP mediates the DHP supported relaxation response is provided by radioimmunoassay data that indicates that 0.1 μmole Bay K 8643 or 8644 elevates c-GMP levels in the rat thoracic aorta by greater than twenty-fold above basal.

These data are consistent with the guanylate cyclase enzyme being the site of action, directly or indirectly, for certain DHPs and that photorelaxation is mediated by an increase in cellular levels of c-GMP. (Supported by the Medical Research Council of Canada.)

REFERENCES

1. Janis, R.A., Silver, P.J. and Triggle, D.J., Adv. Drug Res. 16:309 (1987).
2. Zernig, G., Trends in Pharmacol. Sci. 11: 38 (1990).
3. Bostrom, S.-L, Jung, B., Norlander, M. and Johannson, B., in: "Calmodulin Antagonists and Cellular Physiology", H. Hidaka and D.J. Hartshorne, eds., Academic Press, New York, p. 273 (1985).
4. Triggle, C.R. and Bieger, D., Proc. West. Pharmacol. Soc. 33:227 (1990).
5. Mikkelsen, E., Kazda, S. and Nyborg, N.C.B., Acta Pharmacol. Toxicol. 56:126 (1985).

α-MEDIATED VASOCONSTRICTION: TRANSMEMBRANE SIGNALLING AND RECEPTOR RECLASSIFICATION

C.R. Triggle and R. Tabrizchi

Department of Pharmacology and Therapeutics, Faculty of Medicine
University of Calgary, Calgary, AB, Canada, T2N 4N1

Summary: Apparent differences in the usage of different calcium pools for the processes of excitation-contraction provides support for further sub-classification of the α-adrenoceptors in blood vessels. Presently we have addressed this question by assessing the routes of calcium utilization and the involvement of G-proteins in mediating vasoconstriction *in situ*.

In comparison to the responses that are mediated by activation of α_1-adrenoceptors excitation-contraction mediated via the activation of vascular α_2-adrenoceptors is predominantly dependent on an influx of calcium. However, this generalization *per se* is not applicable to all α_1-adrenoceptor agonists and differences do exist between α_1-adrenoceptor agonists such that, for instance, certain α_1-adrenoceptor agonists are more dependent on extracellular calcium for initiating excitation-contraction coupling. An attempt to address this question has been made by relating the efficacy of an agonist to intracellular versus extracellular calcium utilization and this concept has been linked to the availability of receptor reserve. It has thus been proposed that the availability of a large pool of receptors would reduce the effectiveness of calcium channel antagonists as inhibitors against α_1-mediated contractions in blood vessels (1). However, this hypothesis has its limitations and, for example, we have shown that the alkylating agent benextramine does not increase the effectiveness of calcium channel antagonists once the receptor reserve has been exhausted. In fact, since the concomitant use of benextramine and nifedipine does not enhance the inhibitory actions of nifedipine, we have concluded that benextramine and nifedipine may affect a common site on the α_1-adrenoceptors in the vasculature (2). The key observation that forms the basis for the hypothesis that receptor reserve is responsible for the lack of effectiveness of calcium channel antagonists has come from numerous studies that have employed the irreversible α-antagonist phenoxybenzamine (POB). Such studies show that as receptor reserve is decreased following POB treatment the sensitivity of the response to inhibition by calcium channel antagonists increases (1). Although, we have obtained similar data based on the acute use of POB we have found that following chronic POB treatment the removal of receptor reserve diminishes the inhibitory effects of nifedipine. Han *et al.*, (3) have proposed that there are, at least, two sub-types of α_1-adrenoceptor with, [for instance, the rat vas deferens], a "high-affinity" α_{1a} coupled to a dihydropyridine calcium channel, and "lower affinity", α_{1b}, coupled to PLC. The basis for these conclusions is the use of the irreversible alkylating agent, chlorethylclonidine (CEC),

Excitation-Contraction Coupling in Skeletal, Cardiac, and Smooth Muscle
Edited by G.B. Frank *et al.*, Plenum Press, New York, 1992

which inactivates only a subpopulation of the α_1-adrenoceptors that are linked to PLC. This CEC sensitive site is the so called "low affinity", or α_{1b}, site. Our finding with POB leads us to suggest that POB may also be able to inactivate this same sub-type of α_1-adrenoceptor, α_{1b}, which is linked via PLC, to phosphoinositol turnover.

The nature of the G-proteins that couple the α-adrenoceptors to cellular events is also a matter of conjecture and we thus have examined the role of pertussis and cholera toxins in affecting pressor responses to both α_1- and α_2-adrenoceptor agonists. It has been proposed that a G_i-like, pertussis toxin-sensitive, protein couples the α_2-, but not the α_1-adrenoceptors to a verapamil-sensitive calcium channel (4), furthermore evidence has been presented which suggests that the pressor effects of α_2-adrenoceptors in the pithed rat preparation are mediated via the activation of a pertussis toxin-sensitive G-protein which is also coupled to a calcium channel (5). Following POB treatment, to reduce receptor reserve, vasoconstriction mediated via α_1-adrenoceptors also becomes highly sensitive to inhibition by pertussis toxin treatment (5). However, we believe that this reduction is not necessarily related to receptor reserve being able to buffer the inhibitory effects of pertussis toxin but rather that POB inactivates a site which is coupled to a pertussis toxin-insensitive G-protein; probably that coupled to PLC. Such a relationship would explain the additive inhibition observed with concomitant use of POB and pertussis toxin.

Recently we carried out studies which confirms, that pre-treatment with pertussis toxin greatly attenuates the pressor response to α_2-adrenoceptor agonists, reduces the effects of partial α_1-adrenoceptor agonists, and has minimal effects on full α_1-adrenoceptor agonists. Pre-treatment with cholera toxin , on the other hand, significantly reduced responses to full and partial α_1-adrenoceptor agonists but had no effect on pressor responses to α_2-adrenoceptor agonists. These data suggest that α_1-adrenoceptors are coupled to at least two G-proteins, a pertussis and a cholera toxin-sensitive G-protein(s), and that α_2-adrenoceptors are coupled to a pertussis toxin-sensitive and cholera toxin-insensitive G-protein. The data presented by Nichols *et al.*, (5) has also been interpreted as indicating that the pertussis sensitive G-protein that is coupled to the α_2-adrenoceptor also directly couples this receptor to the nifedipine-sensitive calcium channel. Data from our laboratory are not in complete support of this latter hypothesis since, in our studies, the antagonistic effects of nifedipine and pertussis toxin, against the pressor effects of α_1-adrenoceptor agonists, were additive. We thus postulate that the α_{1a}-sub-type couples to a pertussis toxin-sensitive G-protein which in turn, most likely indirectly regulates a dihydropyridine-sensitive calcium channel in vascular smooth muscle. Finally we believe that the α_{1b} receptor in the vasculature is coupled to PLC via a POB-sensitive but benextramine-insensitive processes.

(The support of Canadian and Alberta Heart and Stroke Foundation is acknowledged)

References

1. Nichols & Ruffolo, 1988 *Tren Pharmacol Sci* **9:**236-241.
2. Tabrizchi & Triggle, 1989 *Life Sci*, **45:**2577-2585.
3. Han, *et al.*, 1987, *Nature*, **329:**333-335.
4. Haeusler, *et al.*, 1986, *J Cardiovasc Pharmacol* **supp 8:**S107-S110.
5. Nichols, *et al.*, 1989, *J Pharmacol Exp Therap* **249:**203-209, 1989.

THE EFFECTS OF CHANGES IN pO_2 ON ENDOTHELIUM-DEPENDENT AND -INDEPENDENT RELAXATIONS OF THE HUMAN UMBILICAL ARTERY AND RAT AORTA

Hui Xie and Christopher R. Triggle

Department of Pharmacology and Therapeutics, Faculty of Medicine, Health Sciences Centre, The University of Calgary, Calgary, AB T2N 4N1

In 1980, Furchgott and Zawadzki[1] reported that the relaxant effect of acetylcholine (ACh) on a pre-contracted rabbit aorta was mediated, indirectly, by the release from the endothelial cells of a relaxant substance which was later termed the endothelium-derived relaxing factor or EDRF. The role of the endothelial cell in the regulation of vascular smooth muscle tone has now been extensively investigated with current opinion indicating that nitric oxide (NO), probably derived from L-arginine, is one of the EDRFs released and that the cellular effects of NO involve the activation of soluble guanylate cyclase[2]. In addition, it is accepted that the endothelial cell also releases contracting , for instance endothelin, as well as relaxing factors[3].

Despite extensive investigation of many vascular beds animal species comparative studies with human tissue are less complete and very few investigators have examined endothelial cell function in human umbilical artery (HUA) function. At birth, the non-innervated HUA constricts and, since the processes involved in initiating this constriction are unknown, consideration should be given to the role of endothelial-derived mediators. The HUA is normally subjected to a low oxygen environment (pO_2~15 mm Hg) and since changes in pO_2 do occur during parturition, our studies have focused on investigating the effects of ΔpO_2 of the role of endothelial cell-smooth muscle interactions in modulating HUA tone. Since many investigations of endothelial cell-vascular smooth muscle interactions have involved the use of the rat thoracic aorta (RA) we established protocols which allowed a comparison of the effects of ΔO_2 on the regulation of HUA and RA tone.

Relaxations initiated by the calcium ionophore A23187 and acetylcholine (ACh) at either the high pO_2 ($pO_2 > 550$ mm Hg) obtained with the use of 95% O_2/5% CO_2 gas mixtures were compared with those obtained with low oxygen tension ($pO_2 < 50$ mm Hg) in precontracted ring preparation from the HUA of the RA. The HUA and RA were pre-contracted with 5-hydroxytryptamine (5HT). The role of the endothelial cell layer was assessed by comparing preparations in which the endothelial cell layer was intact to those denuded of endothelial cells by gentle rubbing with a cotton swab. The presence or absence of the endothelial cell layer was confirmed in representative samples by scanning electron microscopy. The data reveal the following:

1. At either high or low pO_2, ACh and A-23187 relax pre-contracted tissues in a concentration-dependent manner. Relaxations were endothelium-dependent in the RA, but independent in the HUA.

2. Relaxation responses in the RA, but not in the HUA, were sensitive to Δ in pO_2. Thus: EC_{50} values for ACh and A-23187 at low pO_2 in the RA were $1.6 \pm 0.06 \times 10^{-8}$ and $1.3 \pm 0.11 \times 10^{-9}$ M respectively; but were significantly higher at high pO_2: $3.2 \pm 0.07 \times 10^{-8}$ and $1.2 \pm 0.09 \times 10^{-8}$. In the HUA, changing from high to low pO_2 did not affect the sensitivity to ACh: $1.1 \pm 0.16 \times 10^{-6}$ vs $1.6 \pm 0.13 \times 10^{-6}$.

3. The guanylate cyclase inhibitor methylene blue (50 μM) and the nitric oxide synthase inhibitor L-nitro-arginine methyl ester, L-NAME), at 100 μM inhibited both ACh and A23187 mediated relaxations in the RA but not in the HUA.

4. Superoxide dismultase, at 100 U/ml, and L-arginine, at 1 μM, potentiated the effects of ACh at both high and low pO_2 in the HUA and the RA.

5. Relaxations to L-arginine were inhibited by 100 μM L-NAME at low pO_2 in both the HUA and the RA, but only in the RA at high pO_2.

6. 10 μM Indomethacin, in the HUA, did not significantly affect the response to either ACh or A-231-87 at high or low pO_2.

Our data confirms that in the RA relaxations initiated by ACh and A-23187 are mediated by endothelial- and NO-dependent processes that are enhanced by both low pO_2 and superoxide dismultase. In the HUA, however, such relaxations are independent of the endothelium and not mediated by NO or cycloxygenase products or affected by changes in pO_2. The ability of L-arginine to relax the HUA by an L-NAME insensitive process suggests that relaxations mediated by this amino acid can either result independent of NO generation <u>or</u> that the NO-synthase associated with the HUA is less sensitive to L-NAME than in the RA. Thus, at least in response to the endogenous mediator ACh, the endothelial cell-layer of the HUA may not play an important role in modulating vascular tone. Furthermore, the cellular process(es) mediating ACh-induced relaxation of the HUA does not seem to involve a change in c-GMP levels. It is of interest to note that in a recent study, Chaudhuri et al.[4] report that histamine, bradykinin and A-23187, but not ACh, can release an EDRF from the human umbilical vein (HUV). Furthermore, in the same study it was reported that the sensitivity of the HUA to the EDRF-derived from the HUV was low and this observation supports that of an earlier study by van de Voorde et al.[5] that noted that the HUA may have a poor relaxant response to c-GMP. We conclude that the endothelial cell layer of human umbilical arteries differs from other vascular beds with respect to its ability to respond to stimuli that in most other vessels lead to the synthesis and release of NO. (Supported by the Alberta Heart and Stroke Foundation.)

<u>References</u>

1. Furchgott, R.F. & Zawadzki, J.V., <u>Nature (Lond.)</u> 288:373 (1990).
2. Ignarro, L.J., <u>FASEB J.</u> 3:31 (1989).
3. VanHoutte, P.M., <u>Hypertension</u> 13:658 (1985).
4. Chaudhuri, G., Buga, G.M., Gold, M.E., Wood, K.S. and Ignarro, L.J., <u>Br. J. Pharmacol.</u> 102:331 (1991).
5. Van de Voorde, J., Vanderstichele, H. and Leusen, I., <u>Circ. Res.</u> 60:517 (1987).

CA IONS AND POTASSIUM CONTRACTURE IN THE RETRACTOR UNGUIS MUSCLE OF THE COCKROACH

Toshio Yamaguchi, Hiroshi Washio* and Yusuke Kato

Department of Biology, Division of Natural Sciences, International Christian University, Mitaka-shi, Tokyo, 181, Japan
*Laboratory of Neurophysiology, Mitsubishi-kasei Institute for Life Sciences, Machida-shi, Tokyo, 194, Japan

Potassium contracture consists of a phasic and a sustained component. Extracellular Ca ions are essential for both components. In the phasic component the Ca-induced Ca release mechanism is activated by the entry of the Ca ions.

INTRODUCTION

Although there has been much information on the role which Ca ions play in the activation of the mechanical responses in vertebrate muscle tissues, little is known about the insect muscle of which membrane excitation is solely dependent on the inflow of Ca ions. To clarify the mechanism of the contractile activation of the insect muscle, experiments were performed by examining potassium-induced contracture in Ca-free salines or in the presence of Ca antagonists. Caffeine-induced contracture was also examined for this purpose.

METHODS

The retractor unguis muscle was dissected from the metathoracic leg of the cockroach (*Periplaneta americana*), and the contracture of the muscle was recorded isometrically by flushing test salines into a trough of 0.5ml for several seconds. High K^+ salines were prepared so that the [K][Cl] product was the same as that of standard saline (Hodgkin and Horowicz, 1959). Most of the experiments were performed using 100mM K^+ saline which developed a peak tension of about 60% of that induced by 160mM K^+ saline. Ca-free saline contained 2 mM EGTA and 7.5mM Mg^{2+} instead of 5.0mM Ca^{2+}. A Ca channel blocker, nifedipine, of concentrations from 1×10^{-8}M to 3×10^{-5}M was employed. A 20 minute soak in the nifedipine containing salines was necessary to obtain the maximum effect on the contractures. An inorganic Ca channel blocker, Mn^{2+} and an inhibitor of Ca-induced Ca release mechanism, procaine, were also employed. Caffeine contractures were obtained by applying 30 mM of the drug, at and above which concentration the contracture tension attained to a maximum value.

RESULTS AND DISCUSSION

The K^+-induced contracture consisted of an initial phasic and a sustained component. These are ascribed to the "white" type and the "red" type of the fibers in the muscle, respectively (Wood and Usherwood, 1979). If the muscles were exposed to Ca-free saline, both components were eliminated. Nevertheless, in agreement with the locust slow muscle (Aidley, 1965), a sustained contracture was induced if Ca ions were added to this depolarized muscle. At the order of 10^{-8}M, nifedipine had different effects on two components, and had a tendency to increase the sustained component. But, in general, the decreasing effect of the drug on both components was concentration dependent and finally, both were eliminated. However, this effect was less potent to the sustained component than to the phasic one. Mn^{2+} depressed the phasic component, but this ion had a tendency to enhance the sustained component. Procaine also depressed the phasic component prominently. Caffeine-induced contracture was of the phasic type, relaxing rapidly during exposure to the drug. Even if the muscle was soaked in Ca-free saline, the contracture was elicited though decreased to some degree, probably due to a partial loss of cellular Ca. In the K^+-depolarized muscle, caffeine induced an enhanced phasic contracture which was followed by a long-lasting irregular contracture. The phasic component of K^+ contracture was selectively eliminated when 160 mM K^+ saline was applied immediately after the termination of caffeine contracture. These results suggest that the extracellular Ca ions are essential for the activation of the contractile responses of both the phasic and the sustained components, but their underlying mechanisms are different. It is also suggested that in the phasic component, the Ca-induced Ca release mechanism is triggered by the entry of extracellular Ca ions upon membrane depolarization. This is in accordance with the proposal of Hidalgo et al. (1979) and Fabiato (1983).

REFERENCES

Aidley, D. J. (1965) *J. Physiol.* 177: 94-102
Fabiato, A. (1983) *Am. J. Physiol.* 245: C1-C14
Hidalgo, J., Luxoro, M. and Rojas, E. (1979) *J. Physiol.* 288: 313-330
Hodgkin, A. L. and Horowicz, P. (1959) *J. Physiol.* 153:404-412
Wood, M. R. and Usherwood, P. N. R. (1979) *J. Ultrastruct. Res.* 68; 265-280

SPHINGOSINE INHIBITS CALCIUM RELEASE FROM SARCOPLASMIC RETICULUM MEMBRANES

R. Betto#, A. Teresi#, G. Fachechi-Cassano#, G. Salviati# and R.A. Sabbadini*

#CNR Unit for muscle Biology and Physiopathology, Institute of General Pathology, University of Padova, Padova, Italy and *Dept. Biology and Rees-Stealy Research Foundation, San Diego State University, San Diego, CA. 92182

Summary: Sphingosine displayed a very potent, dose-dependent inhibitory effect on calcium release from purified junctional sarcoplasmic reticulum vesicles and from skinned skeletal muscle fibers of the rabbit.

Sphingosine (SPH) is a metabolically active lipid second messenger derived from sphingomyelin (SM) with well-known inhibitory effects on protein kinase C and related cellular processes (Hannun and Bell, 1989). Recently, direct effects of SPH have been reported (Zhang et al., 1990), including a report that SPH and its conjurers induce calcium release from smooth muscle cells in culture (Ghosh et al., 1990). Considering that skeletal muscle transverse tubular membranes are rich in the SPH precursor, SM, (Sumnicht and Sabbadini, 1982), we postulated that SPH could have similar effects on skeletal sarcoplasmic reticulum (SR) calcium release. SR vesicle preparations derived from the terminal cisternae (junctional SR membranes) were isolated according to the method of Saito et al. (1984). SR calcium release studied by following the absorbance of the metallochromic indicator dye, antipyrylazo III, (Palade, 1987) as calcium was released by caffeine (usually 5 mM), AMP-PNP (1 mM), or doxorubicine (50 μM). Calcium release from skinned muscle fibers was performed as previously described (Volpe et al., 1986). As seen in Fig. 1, the rate of 5 mM caffeine-induced calcium release from SR vesicles was significantly inhibited by SPH in a dose-dependent manner with near 100% inhibition at 3 μM SPH and displaying IC_{50} of 0.8 μM. Table I shows that neither SM nor SPC were capable of inhibiting calcium release at the 3 μM concentration level. 3 μM SPH was capable blocking by 63% calcium release in response to 20 mM caffeine. Similar inhibitory effects of SPH on SR calcium release were obtained when examining caffeine-induced calcium release from chemically skinned skeletal fibers from the rabbit. In both the isolated junctional SR and in the skinned fibers, doxorubicine and AMP-PNP induced release was also substantially inhibited. For example, the rate of doxorubicine-induced calcium release from SR was 278 +/-11.6(n=4) nmol/min/mg, but only 35.0 +/-8.5 nmol/min/mg in the presence of 3 μM SPH.

Table I

Condition	nmol/min/mg
5 mM caffeine	136 +/- 7.5 (6)
+ 3μM SPH	6.1 +/- 3.5 (3)
+ 3μM SM	149 +/- 8.2 (5)
+ 3μM SPC	157 +/- 13.6 (4)

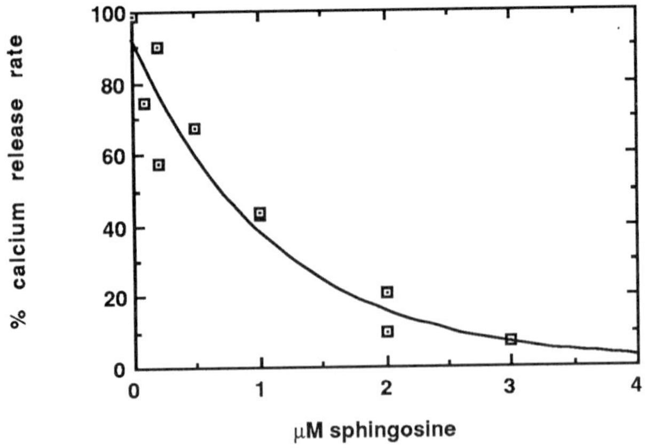

Figure 1. Dose dependence of SPH effect on calcium release

In order to determine if the SPH effect was a direct one or if it was mediated by protein kinase C, we tested the effects of the protein kinase C inhibitor, H7, on its ability to prevent the SPH effect. As seen in Table II, 25 μM H7, was unable to prevent the SPH blockage of calcium release from an isolated membrane preparation which displayed a very fast calcium release rate. These data indicate that the SPH effect is independent of protein kinase C and that SPH may directly interact with the ryanodine receptor.

Table II

Condition	nmol/min/mg R4
5 mM caffeine	954.0
+ 3μM SPH	0.0
+ SPH,+25μM H7	0.0
+ 25μM H7	515.0

Considering that sphingosine is a soluble lipid with concentrations in the cell reaching μM levels, our data indicate that sphingosine generated by the transverse tubule membranes may be a physiologically-relevant mechanism for modulating SR calcium release.

Acknowledgements

This work was supported by grants from the California Metabolic Research Foundation and institutional funds from the Consiglio Nazionale delle Ricerche and the Ministero della Pubblica Istruzione.

References

Ghosh et al., 1990. Science 248, 1653-1656.
Hannun and Bell, 1989. Science 243, 500-507.
Palade, 1987. J. Biol. Chem. 262, 6142-6148.
Saito et al., 1984. J. Cell Biol. 99, 875-885.
Sumnicht and Sabbadini, 1982. Arch. Biochem. Biophys. 215, 628-637.
Volpe et al., 1986. J. Gen. Physiol. 87, 289-303.
Zhang et al., 1990. J. Biol. Chem. 265, 76-81.

PURIFICATION OF A RYANODINE-SENSITIVE CHANNEL PROTEIN FROM HUMAN NEUTROPHILS

David P. Rardon and Philip C. Krause

Krannert Institute of Cardiology
Indiana University School of Medicine
Indianapolis, Indiana

In cardiac and skeletal muscle excitation-contraction coupling is dependent upon release of calcium (Ca^{2+}) from junctional sarcoplasmic reticulum via a ryanodine-sensitive Ca^{2+}-release channel. Similar oscillatory changes in intracellular Ca^{2+} concentration are important for the normal function of neutrophils. Therefore, we postulated the existence of a ryanodine-sensitive Ca^{2+} channel in human neutrophils. This hypothesis is supported by our prior observation that ryanodine (10 nM) increased the intracellular Ca^{2+} content of intact human neutrophils[1].

To purify the neutrophil ryanodine-sensitive channel protein, human neutrophils were isolated[2]. Membrane vesicles were prepared from disrupted neutrophils by centrifugation over a 40% wt/vol sucrose cushion at 150,000g. High molecular weight proteins were isolated from membrane vesicles solubilized in 20 mM MOPS, 250 mM sucrose, 0.5 M KCl, 2 mM DTT, 0.06 mM $CaCl_2$ and 1% digitonin. The supernatant was fractionated on a Fractogel TSK (HW) 65F size exclusion column. Fractions enriched in a 280,000 Dalton protein were identified by SDS-PAGE followed by silver staining. Aliquots of fractions enriched in this protein were examined for channel activity.

Channel activity was recorded after incorporation of proteins in lipid bilayers (phosphatidylethanolamine and phosphatidylserine, 4:5 weight ratio in decane) formed on the tip of a patch clamp pipette. For all experiments the experimental bath contained 1 mM $CaCl_2$, 1 mM EGTA, 125 mM TRIS, 250 mM HEPES, pH 7.4 and the pipette contained 50 mM $Ba(OH)_2$, 250 HEPES, pH 7.4.

Channel activity was observed after addition of 1 μl of fractions enriched in the 280,000 Dalton protein. Under control conditions, channel events were brief, but open channel probability was increased by addition of ryanodine (100 nM) to the experimental chamber. The mean channel slope conductance was 286 ± 42 pS (n = 6). The estimated E_{rev} was 19.4 mV and $P_{Ba}/P_{TRIS} = 9.7$. Channel activity was inhibited by ruthenium red (1 μM). Channel open probability was increased by ATP (1 mM) and decreased by EGTA (5-10 mM). Ryanodine (100-300 μM) did not decrease channel open probability.

The slope conductance of the 280,000 Dalton neutrophil channel was greater than that of the cardiac and skeletal muscle Ca^{2+}-release channels (86 and 110 pS, respectively).

Excitation-Contraction Coupling in Skeletal, Cardiac, and Smooth Muscle
Edited by G.B. Frank *et al.*, Plenum Press, New York, 1992

However, the electropharmacologic effects of ryanodine, ruthenium red, ATP, and EGTA were similar to their effects on the cardiac and skeletal muscle channels. This data suggests that human neutrophils contain a ryanodine-sensitive channel protein with properties similar to the cardiac and skeletal muscle Ca^{2+}-release channel/ryanodine receptor complex.

REFERENCES

1. D.P. Rardon, and L.P. Akard, Ryanodine regulation of intracellular calcium and divalent cation channel activity in human neutrophils, Circulation, 80:II-519, (1989).
2. T.G. Gabig, D. English, L.P. Akard, and M.J. Schell, Regulation of neutrophil NADPH oxidase activation in a cell-free system by guanine nucleotides and fluoride, J. Biol. Chem., 262:1685-1690, (1987).

REGULATION OF SMOOTH MUSCLE CONTRACTION BY THE MYOSIN LIGHT CHAIN PHOSPHORYLATION-DEPENDENT AND -INDEPENDENT MECHANISMS

H. Karaki, M. Hori, K. Sato, H. Ozaki, H. Takano-Ohmuro[1]
T. Tsuchiya and H. Sugi[2]

Department of Veterinary Pharmacology, Faculty of Agriculture
The University of Tokyo, Bunkyo-ku, Tokyo 113, and [2]Department of Physiology, Teikyo University School of Medicine, Itabashi-ku
Tokyo 173, Japan

SUMMARY

Smooth muscle contraction is suggested to be due to increments in cytosolic Ca^{++} level ($[Ca^{++}]_i$), myosin light chain (MLC) phosphorylation and Ca^{++} sensitivity of MLC phosphorylation, and also to a mechanism which is highly sensitive to Ca^{++} but not dependent on MLC phosphorylation.

Smooth muscle contraction has been explained by phosphorylation of MLC resulted from Ca^{++}- and calmodulin-induced activation of MLC kinase. After a prolonged exposure to a stimulant, MLC phosphorylation gradually decreases although muscle tension stays constant. This time-dependent dissociation has been explained by the interaction between unphosphorylated myosin and thin filaments (1). Recently, we have reported that receptor agonists and phorbol esters induce greater contraction than high-K^+ depolarization for a given increase in $[Ca^{++}]_i$ (2-5). The purpose of the present experiments was to clarify the mechanism of the stimulus-dependent Ca^{++} sensitization.

Thoracic aorta was isolated from male Wistar rats and cut into strips. Isometric contractile tension and $[Ca^{++}]_i$ were measured simultaneously in a muscle strip loaded with a fluorescent Ca^{++} indicator, fura-2 (2). $[Ca^{++}]_i$ was monitored with a fluorimeter. MLC phosphorylation was determined by pyrophosphate polyacrylamide gel electrophoresis (PAGE) followed by ureaglycerol PAGE. MLC phosphorylation was measured 20 min after the addition of a stimulant. Force recovery following quick release was determined by imposing a 1.5% shortening step using a servo-controlled length step generator. Stiffness was measured by imposing the sinusoidal vibrations (500 Hz, 0.15-0.18% Lo peak-to-peak) and relative stiffness was calculated from the amplitude of the perturbation of muscle tension by vibration.

Stimulation with 72.7mM KCl, 10μM prostaglandin F2α (PG), 30nM endothelin-1 (ET), 1μM norepinephrine (NE) and 1μM 12-deoxyphorbol 13-isobutyrate (DPB), but not an inactive phorbol ester, 4α-phorbol 12,13-dibutyrate, increased muscle tension, $[Ca^{++}]_i$ and MLC phosphorylation. The levels of contractile tension and MLC phosphorylation at a given $[Ca^{++}]_i$ were greatest in the presence of DPB followed by ET>PG≥NE>high K^+. There was a curve-linear correlation between MLC phosphorylation and muscle contraction, as

Excitation-Contraction Coupling in Skeletal, Cardiac, and Smooth Muscle
Edited by G.B. Frank *et al.*, Plenum Press, New York, 1992

predicted by the Hai and Murphy Model (1). These results suggest that smooth muscle contraction is due to the increments in $[Ca^{++}]_i$ and MLC phosphorylation and that the stimulant-dependent increase in Ca^{++} sensitivity is due to Ca^{++} sensitization of MLC phosphorylation.

CALCIUM RELEASE CHANNEL OF SARCOPLASMIC RETICULUM: AN IMPORTANT TARGET FOR DOXORUBICIN-MEDIATED CARDIOTOXICITY

Isaac N. Pessah

Department of Veterinary Pharmacology and Toxicology
University of California, Davis, CA 95616

Sarcoplasmic reticulum (SR) of cardiac muscle makes essential contributions to excitation-contraction (E-C) coupling by regulating the free calcium (Ca^{2+}) levels in the myoplasm. The ryanodine receptor constitutes the ligand-gated Ca^{2+} channel localized at junctional regions of SR through which Ca^{2+} is released in response to the action potential at the T tubule membrane. The ryanodine receptor complex appears to be co-associated with the voltage-dependent Ca^{2+} channels of the sarcolemma at the triad junction and the interaction of these proteins is critical to the process of E-C coupling. The role of triadic receptors in the etiology of drug-induced cardiomyopathy is explored. Doxorubicin and its anthraquinone (AQ) congeners are clinically valuable chemotherapeutic agents effective against a broad spectrum of human cancers. The repetitive doses of AQ required for control of malignancies coincides with induction of a progressive and irreversible cardiomyopathy which results in arrhythmias, tachycardia, declining ventricular output, and eventually congestive heart failure. Greater than 65% of patients receiving the maximal acceptable cumulative dosages of doxorubicin develop debilitating ventricular dysfunction. The etiology of AQ cardiomyopathy is unknown, but like genetic dilated cardiomyopathy, the drug-induced disease exhibits a similar progression of abnormal diastolic filling and impaired rates of ventricular relaxation.

The integrity of Ca^{2+} transport across sarcolemmal and SR membranes are severely compromised early in the progression of AQ-induced cardiomyopathy and results in pronounced elevation in the level of total cellular calcium. Recent findings (1-7) demonstrate that the acute and chronic effects observed with the clinical use of AQs may be the result of a receptor-mediated shift in the redox equilibrium of critical thiols at the ryanodine receptor complex which leads to 1) cumulative sensitization of the Ca^{2+} release channel to activation by endogenous signal molecules such as Ca^{2+}, ATP, and cAMP followed by (2) long-lived desensitization of the channel.

Anthraquinones are shown to enhance the binding of [^{3}H]ryanodine to SR membranes isolated from rat ventricles with a strong structural requirement. Doxorubicin enhances the binding of [^{3}H]ryanodine to receptor preparations and induces Ca^{2+} release from SR vesicles by increasing the affinity of the Ca^{2+} activator site for Ca^{2+}. Doxorubicin reduces the K_d of the binding of [^{3}H]ryanodine by enhancing the rate of association. Doxorubicin is found to alter the release of Ca^{2+} from actively loaded vesicles monitored with

Excitation-Contraction Coupling in Skeletal, Cardiac, and Smooth Muscle
Edited by G.B. Frank *et al.*, Plenum Press, New York, 1992

antipyrylazo III in a biphasic manner. Doxorubicin (1-30 μM) enhances the initial rate of Ca^{2+} release, whereas higher concentrations are inhibitory. High (>30 μM) doxorubicin induces rapid desensitization of the release channel. Doxorubicin-enhanced rates of Ca^{2+} release from SR vesicles are found to be inhibited by caffeine in a dose-dependent manner. Caffeine competes for the same site with AQ since the caffeine-activated binding of [^{3}H]ryanodine is inhibited by doxorubicin and vice versa, and the effect of caffeine and doxorubicin on the K_d of ryanodine is not additive. Although acute exposure of SR to doxorubicin (5 μM) alters channel function in a fully reversible manner, long-term treatment (up to 24 hr) of SR with doxorubicin increases the sensitivity of the preparation to subsequent acute challenge with doxorubicin. These results reveal a cumulative yet latent alteration of SR channel function consistent with the etiology of AQ-induced cardiomyopathy.

Acute or chronic administration of doxorubicin to rats in vivo alters the native function of the SR Ca^{2+} release channels in isolated cardiac membranes (5). Compared to control SR, actively loaded SR from animals exposed to a single intravenous dose of doxorubicin (2 mg/kg) exhibit faster rates of doxorubicin-induced Ca^{2+} release and the occupancy of [^{3}H]ryanodine is significantly enhanced with repeated exposure to doxorubicin in vitro. The doxorubicin-sensitized channels appear to be persistent, since SR prepared from animals 1 week after a single acute dose exhibit significantly enhances [^{3}H]ryanodine receptor occupancy and Ca^{2+} release rates in response to a second exposure to doxorubicin in vitro. Ryanodine receptors from rats chronically exposed to doxorubicin (total of 9 mg/kg over 4 weeks) consistently exhibit a higher sensitivity to activation by Ca^{2+} which persist 6 weeks following the last injection of drug. Chronic exposure produces a concomitant reduction in the capacity of [^{3}H]ryanodine binding-sites. The marked decrease in receptor density observed with SR from doxorubicin-treated rats appears to be the result of progressive and permanent channel desensitization. Consistent with desensitized receptors, membrane vesicles prepared from rats chronically exposed to doxorubicin take up significantly more Ca^{2+} and exhibit significantly reduced rates of doxorubicin- or Ca^{2+}/ryanodine- induced Ca^{2+} release. These findings demonstrate 1) doxorubicin inflicts cumulative SR channel lesions, 2) a persistent sensitization of the SR channel to activation by Ca^{2+}, and 3) a significant and apparently irreversible desensitization of a large fraction of channel complexes. This mechanism appears to play a significant role in the etiology of doxorubicin-induced cardiomyopathy.

Acknowledgements

This work is supported by grant ES 05002 from the National Institute of Health, and Grant-In-Aid #90-167 from the American Heart Association, California Affiliate with funds contributed by the Riverside County Chapter.

References

1. Zorzato F, Salviati G, Facchinetti T, Volpe P: J. Biol. Chem. 260: 7349-7355 (1985).
2. Abramson JJ, Buck E, Salama G, Casida JE, Pessah IN: J. Biol. Chem. 263: 18750-18758 (1988).
3. Kim DH, Landry AB III, Lee YS, Katz, AM: J. Mol. Cell. Cardiol. 21: 433-436 (1989).
4. Pessah IN, Durie EL, Schiedt MJ, & Zimanyi I: Mol. Pharmacol. 37: 503-514 (1990).
5. Holemberg SRM, Williams AJ: Circ. Res. 67: 272-283 (1990).
6. Ondrias K, Borgatta L, Kim DH, Ehrlich BE: Circ. Res. 67: 1167-1174 (1990).
7. Pessah IN, Schiedt MJ, Shalaby MA, Mack, M, Giri SN: Circ. Res. 1991; submitted.

CALCIUM OR SODIUM LEAKAGE CHANNELS

IN CULTURED MAMMALIAN MYOTUBES

C.G. Carlson and Y. Feng

Dept. of Physiology, University of North Dakota
School of Medicine
Grand Forks, ND 58202

Summary. Calcium or sodium leakage currents (1.2 to 3.1 msec, 6 to 16 pS) were recorded in mammalian myotubes at hyperpolarized membrane potentials.

During the course of studies examining acetylcholine receptor development in cultured mammalian myotubes, an infrequent class of small single channel events have been observed on cell-attached patches that also exhibit acetylcholine receptor (AChR) single channel activity. These events were observed in association with embryonic, low conductance-long open time (LL-AChRs) and adult high conductance-short open time (HS-AChRs) receptors.

Cultured mammalian myotubes are prepared from embryonic (Day 16 to 19) mouse (C57Bl 6J or C57Bl 10 SnJ) skeletal muscle tissue and are maintained in DMEM/HAMS F 12 media plus 10% fetal bovine serum (5% CO_2). Patch clamp recordings are made in a 0 Mg^{2+} - Ringers bathing solution (125 mM NaCl, 5 mM KCl, 24 mM $NaHCO_3$, 2 mM $CaCl_2$, 1 mM NaH_2PO_4, 11 mM glucose 5 X 10^{-8} to 5 X 10^{-7} gm/ml tetrodotoxin, pH 7.3, T=23°C) bubbled with 95% O_2 - 5% CO_2. The pipette (fire polished soft glass) solution contains 0 Mg^{2+} - Ringers plus 5 X 10^{-7} M AChCl.

An amplitude histogram from a patch (pipette potential = $^{+}$75 mV) exhibiting 80% small events and 20% LL-AChRs (Fig 1) indicates two distinct populations with mean amplitudes of 3.03 (small events) and 6.73 pA (LL-AChRs). The open time distribution (5 μsec minimum gap duration) of the small events (Fig 2) was best fit by a double exponential distribution with fast and slow time constants of 1.1 and 2.93 msec, respectively. The fast component of the theoretical distribution accounted for approximately one third of the total openings and the arithmetic mean open time was 3.09 msec at this holding potential. This value declined to 2.66 and 2.34 msec at holding potentials of $^{+}$50 and 0 mV, respectively. The open time distribution for LL-AChRs showed proportionately more long duration openings and was best fit with a single exponential (time constant = 4.03 msec; arithmetic mean open time = 4.83 msec).

The reversal potentials/conductances for the small events and for LL-AChRs were -111mV/16pS and -73.6 mV/45 pS, respectively. Since the reversal potential for AChRs is at a membrane potential of 0 mV, the small events in this patch would reverse at approximately $^{+}$37 mV. Analyses of small events obtained from two other patches (2 to 6.0% small events) were

Excitation-Contraction Coupling in Skeletal, Cardiac, and Smooth Muscle
Edited by G.B. Frank *et al.*, Plenum Press, New York, 1992

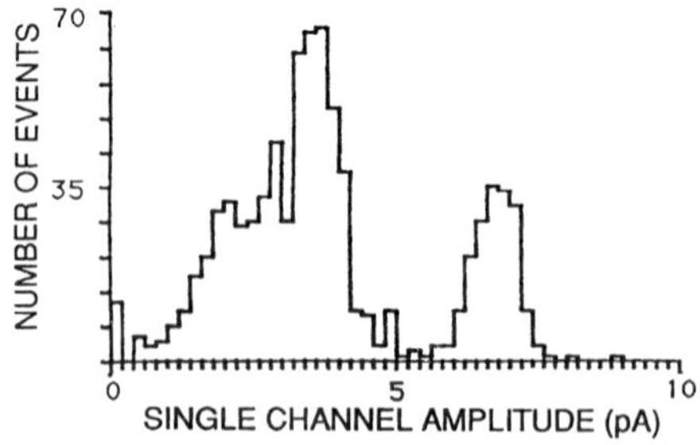

Fig 1. Amplitude histogram of single channel events at $^{+}$75 mV pipette potential.

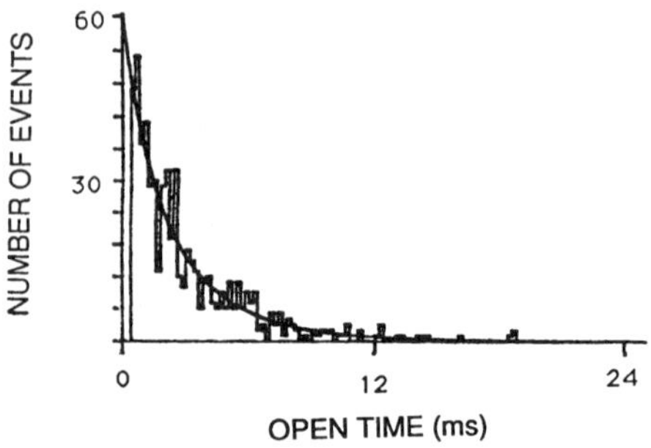

Fig 2. Open time distribution for small events.

roughly consistent with these results; and, in each case, the mean open time of small events increased slightly with hyperpolarization (5 to 10 μsec/mV). The electrophysiological characteristics of the leakage channels observed in this culture system are similar to those previously reported by Fong et al. (1990), and are consistent with a selectivity for calcium and sodium. Calcium/sodium leakage channels may partly regulate passive tension development by providing a constant inward calcium leak at resting (and hyperpolarized) membrane potentials.

Reference

1. Fong, P., Turner, P. R. Denetclaw, W. F., Steinhardt, R. A., 1990, Increased activity of calcium leak channels in myotubes of Duchenne human and mdx mouse origin, Science, 250:673.

CITRATE DECREASES CA CURRENT AND CHANGES SELECTIVITY OF CARDIAC CA CHANNELS

Larry V. Hryshko and Donald M. Bers

Divison of Biomedical Sciences
University of California
Riverside, CA 92521

SUMMARY

Previous studies using citrate as a low affinity Ca buffer have shown that this agent inhibits myocardial contractions[1-3] and neurotransmitter release[4]. Using several different low affinity Ca buffers we have demonstrated that these effects are unlikely to be attributable to Ca buffering *per se*[3]. Here we demonstrate that citrate can directly inhibit Ca current (I_{Ca}). The decrease in I_{Ca} appears to be due to shifts in channel gating (due to surface charge effects) and also to a decrease in channel selectivity for Ca over monovalent ions.

The effects of several low affinity Ca buffers were examined on contractions in rabbit ventricular muscle preparations. These included citrate, nitrilotriacetic acid (NTA), dipicolinic acid (DPA), and acetamido-iminodiacetic acid (ADA). In all cases, free extracellular Ca was tritated to 2.0 mM using murexide absorbance in the presence or absence of 10 mM buffer. Three of these Ca buffers (citrate, DPA, and ADA) were found to inhibit twitch tension

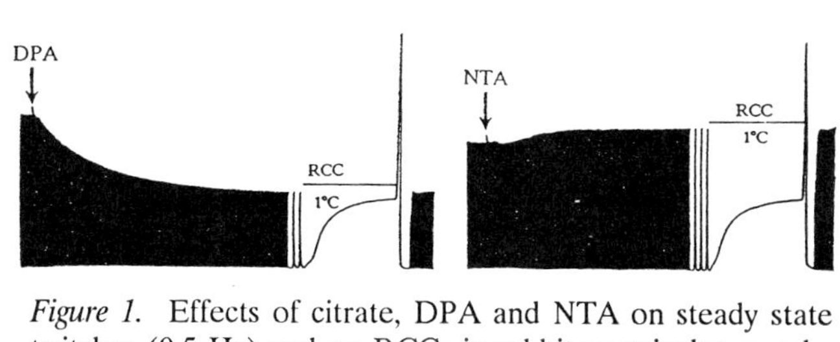

Figure 1. Effects of citrate, DPA and NTA on steady state twitches (0.5 Hz) and on RCCs in rabbit ventricular muscle. The horizontal bars (RCC, 1°C) indicate the duration of the cold superfusion at 1°C.

Excitation-Contraction Coupling in Skeletal, Cardiac, and Smooth Muscle
Edited by G.B. Frank *et al.*, Plenum Press, New York, 1992

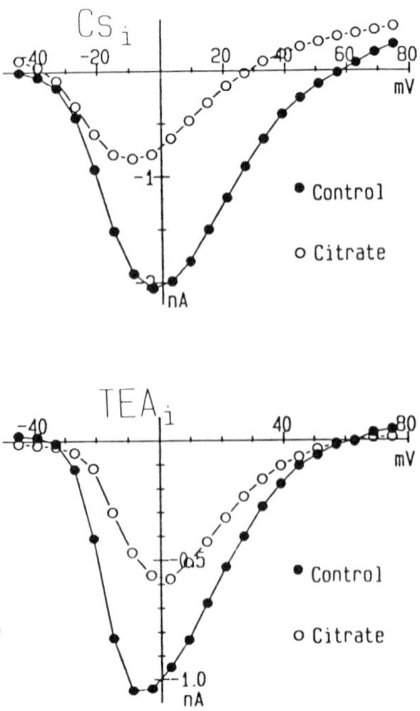

Figure 2. The effects of citrate on whole cell I_{Ca} in isolated rabbit ventricular myocytes at constant 2 mM free [Ca]. Cells were dialyzed with either 140 mM Cs_i (top) where outward I is seen or with 140 mM TEA_i (bottom) to prevent outward I.

whereas a slight increase was observed with NTA (see Figure 1). There was no correlation between Ca affinity of the buffers and their effects on twitch tension. Rapid cooling contractures (RCCs) were used as an index of sarcoplasmic reticulum (SR) Ca content. The depressant effects do not appear to be mediated by changes in SR Ca content as RCC magnitude was unaffected by these buffers. We also found that reducing the Ca buffering abilities of citrate and ADA by acidification *increased* their ability to reduce twitch tension. Thus it does not seem likely that Ca buffering *per se* is responsible for the observed effects on twitch tension.

We examined the effects of citrate and NTA on I_{Ca} in isolated rabbit ventricular myocytes. I_{Ca} was isolated by suppression of competing currents with impermeant ions (Cs, TEA) in the pipette and superfusate. Citrate was found to reduce peak I_{Ca} ~ 50% whereas NTA had no effect. Therefore we will focus on the characteristics and mechanism of I_{Ca} inhibition by citrate. As shown in Figure 2, buffering extracellular Ca with citrate leads to a reduction in peak I_{Ca} and both peak I_{Ca} and reversal potential (E_{rev}) were shifted towards negative potentials. These effects could be greatly exaggerated by lowering free Ca_o to 0.5 mM. In contrast, increasing free Ca_o to 10 mM abolished the effect of citrate on peak I_{Ca}, but the leftward shifts along the voltage axis remained for both peak I_{Ca} and E_{rev}.

The observed shifts in E_{rev} suggested that citrate may have altered the selectivity (P_{Ca}/P_{Cs}) of the Ca channel. To examine this possibility, we conducted experiments were intracellular Cs was replaced by either tetraethylammonium (TEA) or N-methylglucamine (NMG), as these cations are essentially impermeant through the Ca channel. As shown in Figure 2 (for TEA), citrate still produced a depression of I_{Ca}, but no longer caused shifts in E_{rev}. Thus while citrate appears to alter the selectivity of the Ca channel, this result suggests that selectivity is altered by a reduction in Ca permeation, rather than by an augmentation of monovalent permeation.

To further examine the effects of citrate on monovalent permeation, we studied current through Ca channels carried by Na (when extracellular divalent cations are submicromolar). This current has been termed I_{NS} for nonselective cation current. Citrate failed to alter the magnitude of peak I_{NS}, but the voltage dependence of both peak I_{NS} and E_{rev} were still shifted towards negative potentials. This result indicates that citrate can exert direct effects on the Ca channel, independent of Ca buffering. In addition, the I_{NS} shifts along the voltage axis may be mediated via alterations in surface charge[5]. We examined the effects of citrate on Na permeation through the Ca channel at the single channel level (I_{NS}). Citrate did not affect the conductance of the Ca channel for the monovalent cation, supporting the idea that this effect is mediated by a reduction of Ca permeation rather than augmented monovalent permeation.

Finally, we examined the gating properties of the Ca channel by evaluating activation and inactivation parameters for both I_{Ca} and I_{NS}. With both charge carriers, citrate produces shifts towards negative potentials by 5-10 mV for both activation and inactivation. This effect is very similar to that observed for several anions of the lyotropic series (e.g. SCN, perchlorate)[5]. Thus, alteration of surface charge may account for some of the effects of citrate. Recently, certain lyotropic anions have also been shown to reduce the charge movement associated with the cardiac Ca channel[6]. On the other hand the large change of E_{rev} for I_{Ca} implies that channel selectivity is also altered. Thus we conclude that changes in cardiac Ca channel permeation and voltage dependence (and possibly channel number) are sufficient to explain the observed depressant effects of citrate on contractions. These data also suggest that previous studies using citrate as a simple Ca buffer may require re-interpretation.

References

1. Y. Shimoni, and S. Ginsburg, Negative inotropic effect of extracellular calcium buffering in cardiac muscle, Amer. J. Physiol. 252:C248-C252, (1987).
2. S. Ginsburg, and Y. Shimoni, Modulation of effect of extracellular calcium buffering in cardiac muscle, Amer. J. Physiol. 257:H1843-H1850, (1989).
3. D.M. Bers, L.V. Hryshko, S.M. Harrison, and D.D. Dawson. Citrate decreases contraction and Ca current in cardiac muscle independent of its buffering action, Amer. J. Physiol. 260:(May), (1991).
4. S. Ginsburg, and R. Rahamimoff. Is extracellular calcium buffering involved in regulation of transmitter release at the neuromuscular junction? Nature. 306:62-64, (1983).
5. M. Delay, D.E. Garcia, and J.A. Sanchez. The effect of lyotropic anions on charge movement, calcium currents and calcium signals in frog skeletal muscle fibres. J. Physiol. 425:449-469, (1990).
6. E. Rios, R. Shirokov, R. Levis, A. Gonzalez, J. Ma, C. Mundina-Weilenmann and M.M. Hosey. Different effects of perchlorate on skeletal muscle EC coupling, cardiac Ca gating currents and gating of DHP receptors in bilayers. Biophys. J. 59:201a, (1991).

EFFECTS ON CA^{2+} UPTAKE OF MONOCLONAL ANTIBODIES RAISED AGAINST CANINE CARDIAC PHOSPHOLAMBAN AND THEIR CORRELATION WITH EPITOPIC LOCATION

Gary L. Morris and J.H. Wang

University of Calgary
Calgary, Alberta, Canada, T2N 4N1

Monoclonal antibodies raised against canine cardiac sarcoplasmic reticulum (SR) phospholamban were used to study the structure function relationships of phospholamban in terms of the regulation of the cardiac SR (Ca^{2+} - Mg^{2+})-ATPase. Five monoclonal antibodies were assessed for their ability to affect SR Ca^{2+} uptake, 3 of these antibodies had no effect on SR Ca^{2+} uptake, whereas the other 2 monoclonals were able to stimulate SR Ca^{2+} uptake to levels similar to those achieved by cAMP dependent protein kinase phosphorylation of phospholamban. With the use of synthetic peptides corresponding to various portions of phospholamban it was possible to map the epitope sites the monoclonal antibodies bound to. It was found that the monoclonal antibodies which do not stimulate SR Ca^{2+} uptake do not bind to the cytosolic domain of phospholamban, whereas the stimulatory antibodies have been shown to bind to phospholamban residues 7 to 16. These results may help to define the residues of phospholamban which interact with the (Ca^{2+} - Mg^{2+})-ATPase.

Excitation-Contraction Coupling in Skeletal, Cardiac, and Smooth Muscle
Edited by G.B. Frank *et al.*, Plenum Press, New York, 1992

$^{45}Ca^{++}$ EFFLUX STUDIES IN RABBIT T-TUBULE MEMBRANE PREPARATIONS IN THE RANGE OF LATE AFTER POTENTIALS DURING TETANIC CONTRACTIONS

A.M. Öz, S.M.J. Dunn, and G.B. Frank

Department of Pharmacology
University of Alberta
Edmonton, Canada, T6G 2H7

Skeletal muscle t-tubules are the richest known source of Voltage-Dependent Slow Ca^{++} Channels (VDSCCs)[1]. Despite their highly concentrated presence in t-tubules, the functional role of VDSCCs has not been established unequivocally[2]. During twitches, due to the short duration of an action potential, VDSCCs do not have time to open[3,4]. On the other hand, long lasting depolarizing voltage pulses during voltage clamp experiments[5] and high K^+-induced depolarizations can cause the opening of these channels. Under long lasting depolarizing conditions, the opening of these channels can also be modulated by inorganic and organic Ca^{++} channel blockers and activators[6,7]. During tetanic stimuli, the presence of such long-lasting depolarizations has long been known[8,9]. These depolarizations that are due to the accumulation of K^+ ions in t-tubules was called as late after depolarizations (LAPs). We have suggested previously that this LAPs may cause the opening of VDSCCs and permit the entrance of extracellular Ca^{++} during tetanic contractions[5].

Isolated t-tubule membrane preparations have been used to measure the functional responses of VDSCCs. VDSCCs in these t-tubules membrane preparations display the expected voltage dependence of activation and are modulated by Ca^{++} channel blockers and activators[11]. Since the amplitude of LAPs in mammalian muscle fibres[12] are quite similar to the range of LAPs in frog skeletal muscle fibres, in the present study we investigated whether the VDSCCs in isolated t-tubule membrane preparations can function to permit Ca^{++} influx into the muscle fibres at the range of LAPs.

Methods: $^{45}Ca^{++}$ efflux experiments

In all flux experiments t-tubule membrane vesicles from New Zealand White Rabbits were prepared by the method described previously[11]. Vesicles were first equilibrated in low K^+ buffer (10 mM Hepes-Tris, pH: 7.4, 145 mM choline chloride, 5 mM potassium gluconate). $^{45}CaCl_2$ solution in the same buffer was added to give a final concentration of 5 mM total Ca^{++} containing approximately 50 μCi/ml $^{45}Ca^{++}$ (ICN). In order to load Ca^{++} inside the membrane vesicles, the mixture was subjected to two cycles of freezing and thawing using liquid nitrogen. 25 μl of $^{45}Ca^{++}$-loaded membranes were first diluted into 975 μl of high K^+ buffer (10 mM Hepes-Tris, pH: 7.4, 120 mM potassium gluconate, 30 mM choline chloride, 0.133 mM EGTA) containing 0.1 μM valinomycin and, where appropriate, the desired drug. After 5 min incubation at room temperature, 0.9 ml was removed and applied to a GF/C filter and dried under vacuum. Excess buffer was removed from the sample under vacuum, 1 ml

Excitation-Contraction Coupling in Skeletal, Cardiac, and Smooth Muscle
Edited by G.B. Frank *et al.*, Plenum Press, New York, 1992

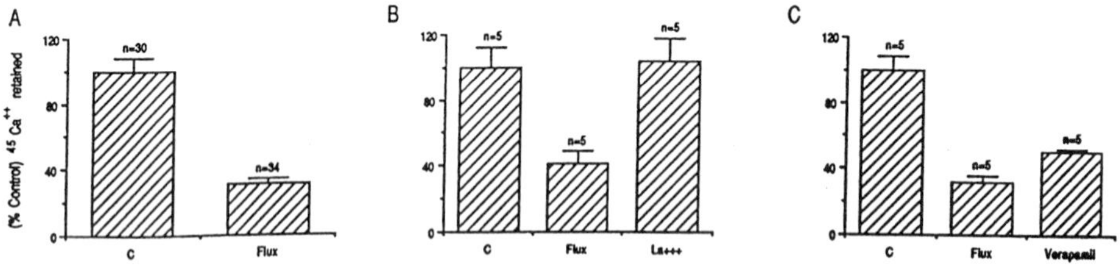

Fig. 1. The effects of partial membrane depolarization, La^{+++} and verapamil on $^{45}Ca^{++}$ efflux in transverse tubule preparations. The amount of efflux was normalized and presented as the percentage of retained $^{45}Ca^{++}$ in control vesicles that had not undergone the preliminary repolarization. A, nonrepolarized control measurements of $^{45}Ca^{++}$ retained in vesicles (c) and the effect of partial membrane depolarization on retained $^{45}Ca^{++}$ (Flux). B, the effect of La^{+++} (1 mM) on $^{45}Ca^{++}$ efflux. C, the effect of verapamil (100 μM).

of high K^+ buffer (10 mM Hepes-Tris, pH: 7.4, 96 mM choline chloride, 54 mM potassium gluconate, 0.133 mM EGTA, 0.1 μM Valinomycin) and the same concentration of appropriate drug was added. Efflux was allowed to continue on the filter for 20s prior to removal of extravesicular solution under vacuum and rapid washing with 2 x 7.5 ml volumes of "stop" solution (10 mM Hepes-Tris, pH: 7.4, 96 mM choline chloride, 54 mM potassium gluconate, 0.1 mM La Cl_3, 30 mM sucrose). The filters were dried, extracted with 5 ml of Aqueous scintillation fluid (Amersham), and counted for $^{45}Ca^{++}$.

Drugs: (±) Verapamil hydrochloride was from Sigma Chemical Co.

Results: Effect of membrane potential on $^{45}Ca^{++}$ efflux

Efflux was stimulated by removing the inactivation of channels with first repolarizing the membranes to -80 mV by dilution into high K^+ buffer and then by partially depolarizing to a potential of -60 mV which is similar to the range of LAPs in mammalian muscle. This treatment led to a rapid efflux of $^{45}Ca^{++}$ from t-tubule membrane vesicles. The amount of this efflux, under partially depolarized conditions, compared to control preparations has been shown in Fig. Ia.

Effect of drugs on $^{45}Ca^{++}$ efflux

In order to characterize the functional responses of VDSCCs, La^{+++} (1 mM) and verapamil (100 μM) were employed. In the presence of La^{+++} and verapamil, $^{45}Ca^{++}$ efflux upon partial depolarization was inhibited to 100% and to 50% of the control values. Results indicating the increase in retained intravesicular $^{45}Ca^{++}$ have been demonstrated in Fig. Ib and c.

Discussion

The present study indicates that VDSCCs can have functional roles during LAPs. The importance of these results comes from the fact that the potential range that was used to activate the VDSCCs in t-tubule membrane vesicles, has been confined to the physiological range of LAPs. Since these LAPs are generated in intact muscle fibres during tetanic muscle contractions, entrance of extra cellular Ca^{++} through VDSCCs during tetanic activity can support the mechanical activity of muscle fibres. Results from $^{45}Ca^{++}$ efflux studies in t-tubule membrane vesicles are also consistent with functional studies[10] indicating that, tetanic contractions require the presence of extracellular Ca^{++} ions.

References

1. M. Fosset, E. Jaimovich, et al., J. Biol. Chem., 258:6086-6092 (1983).
2. W.A. Catterall, Cell Vol., 64:871-874 (1991).
3. C.M. Armstrong, F.M. Bezanilla, et al., Biochim. Biophys. Acta, 267:605-608 (1972).
4. G.B. Frank, Jpn. J. Physiol., 32:589-608 (1982).
5. J.A. Sanchez, E. Stefani, et al., J. Physiol., (London) 337:1-17 (1983).
6. G.N. Beaty, R.W. Tsien, et al., pages 123-140, Alan R. Liss, Inc. (1987).
7. G.B. Frank, Jpn. J. Physiol., 40:205-224 (1990).
8. W.H.J.R. Freygang et al., J. Gen. Physiol., 47:929-952 (1964).
9. P.W. Gage, R.S. Eisenberg, et al., J. Gen. Physiol., 53:298-310 (1969).
10. A.M. Oz, G.B. Frank, et al., J. Pharmacol. Exp. Ther., (1991).
11. S.M.J. Dunn, J. Biol. Chem., 264:11053-11060 (1989).
12. R.H. Adrian, S.H. Bryant, et al., J. Physiol., 240:505-515 (1974).

BIOCHEMICAL PROPERTIES OF OR-1259 - A POSITIVE INOTROPIC AND VASODILATORY COMPOUND WITH AN ANTIARRHYTHMIC EFFECT

Atso Raasmaja, Antti Talo*, Heimo Haikala,
Erkki Nissinen, Inge-Britt Lindén, and Pentti Pohto

Orion Pharmaceutica, P.O. Box 65, Espoo, Finland
*University of Turku, SF 20510 Turko, Finland

OR-1259 is a positive inotropic and vasodilatory compound with antiarrhythmic properties. Biochemically, OR-1259 can act as a troponin modifier, as an phosphodiesterase III inhibitor and as an inhibitor of voltage-sensitive Ca^{2+} current.

OR-1259 ((4-(1,4,5,6-tetrahydro-4-methyl-6-oxo-3-pyridazinyl)phenyl) hydrazono) propanedinitrile is a positive inotropic, vasodilatory and antiarrhythmic compound. However, in spite of the potent and selective inhibition of the cardiac PDE III (IC_{50}= 25 nM) purified from guinea-pig hearts, OR-1259 had also an antiarrhythmic effect. The finding of an antiarrhythmic PDE III inhibitor was surprising, since arrhythmic drawbacks have been observed in clinical use of inotropic PDE III inhibitors. Since Ca^{2+} has a key role in the regulation of contraction, it was important to investigate the effect of OR-1259 on the Ca^{2+} metabolism. Interestingly, OR-1259 (1 μM) decreased the voltage-sensitive Ca^{2+} current in guinea pig cardiomyocytes. NaCa-exchange and NaK-ATPase activities were not changed by OR-1259 (10 and 100 μM, respectively) in sarcolemmal membranes prepared from guinea-pig heart. Further studies of e.g. intracellular Ca^{2+}, voltage-sensitive Ca^{2+} channels and cyclic AMP are in progress with OR-1259. Although the present findings showed no increase in intracellular Ca^{2+}, OR-1259 enhanced the twitch tension (EC_{50}= 0.2 μM) in guinea-pig papillary muscle. In conclusion, OR-1259 has a positive inotropic effect with a parallel decrease in the voltage-sensitive Ca^{2+} current. It is suggested that the decreased Ca^{2+} current may play some role in the observed antiarrhythmic effect of OR-1259.

Excitation-Contraction Coupling in Skeletal, Cardiac, and Smooth Muscle
Edited by G.B. Frank *et al.*, Plenum Press, New York, 1992

CHARACTERIZATION OF A MUSCLE MEMBRANE ATPase GLYCOPROTEIN

P. J. Yazaki, H. B. Cunningham, J. J. Kang, C. Jachec, S. A. Hunt, R.C. Domingo, R. A. Sabbadini, and A. S. Dahms

Departments of Chemistry and Biology, Molecular Biology Institute and Rees-Stealy Research Foundation, San Diego State University San Diego, CA (USA)

In straited muscle (SM), the contraction-relaxation cycle is controlled by unique interfacing elements of the sarcotubular membrane system consisting of the transverse tubules (TT) and the sarcoplasmic reticulum (SR). The TT membranes are continuous invaginations of the sarcolemma (SL), and comprise over 80% of the surface membrane of SM. The topic of TT membranes has been reviewed recently.[1]

Transverse tubules are enriched in a number of metabolically-significant proteins, including the dihydropyridine receptor, insulin receptor, adrenergic receptors, dystrophin, protein kinases C and A, and glucose transporter[1]; a number of TT proteins have been found to serve as substrates for endogenous and exogenous protein kinases.[1] This cholesterol-rich membrane also possesses a complex organization of ion pumps, channels and assorted regulatory enzymes.

Chicken skeletal muscle TT membrane vesicles contain an active MgATPase (~250 umol/hr/mg) that has many unique distinguishing features[1], including: 1) unusual kinetic and thermodynamic properties[2]; 2) insensitivity to FITC and vanadate; 3) modulation by phorbol esters, diacylglycerols, and sphingolipids[3,4]; 4) extreme sensitivity to detergents; and 5) ability to be stimulated by selective lectins. The properties of the enzyme easily distinguish it from other divalent cation-dependent SM ATPases present in the SL, SR or in intracellular organelles; the TT-localized ATPase appears to be designed to catalyze the hydrolytic turnover of ATP under specifically-controlled conditions and mediates unidirectional catalysis (hydrolysis) without any 'partial reaction' characteristic of other types of ATPases[1].

In purified skeletal muscle TT preparations, a glycoprotein of Mr 85 kD has been found to correlate with reactivity towards stimulatory lectins on Western blots, and, to date it is the only polypeptide that can be related to MgATPase activity. Rabbit anti-chicken 85 kD glycoprotein FAB fragments specifically inhibit the TT Mg-ATPase as illustrated in Figure 1.

FPLC size exclusion studies in CHAPS show that the native MW of the enzyme is ~300 kD. Partial purification of the lectin-protected ATPase has been achieved using the non-ionic detergent Triton X-114, whereby the 85kD glycoprotein is highly enriched in the detergent-rich phase (as well as in the insoluble phase) indicating it is an integral membrane protein. The MgATPase is sensitive to deglycosylation with Endo H, N-Glycanase, or TFMSA, whereas O-Glycanase treatment has no effect.

Lectin and Western blots analyses, as well as enzymatic activator/inhibitor studies, of gizzard smooth muscle membranes show both the presence of the MgATPase and the corresponding 85kD glycoprotein.

Excitation-Contraction Coupling in Skeletal, Cardiac, and Smooth Muscle
Edited by G.B. Frank *et al.*, Plenum Press, New York, 1992

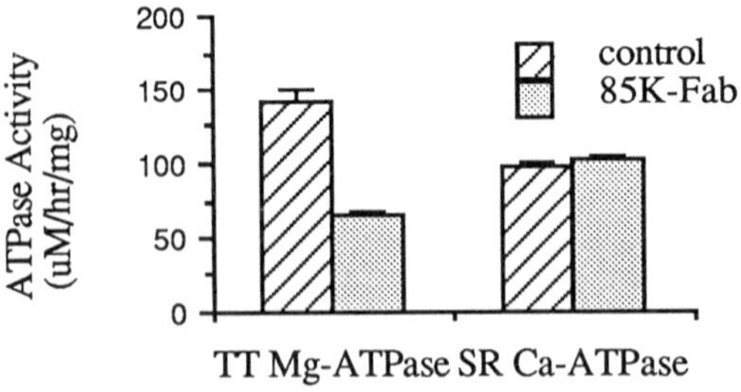

Figure 1. Incubation with the anti-85kD Fab fragment is able to inhibit the Mg-ATPase of TT membrane vesicles whereas there is no effect upon Ca-ATPase activity in sarcoplasmic reticulum membrane vesicles.

As shown in Table 1, chemical/enzymatic characterization and degradation studies show that the glycoprotein from gizzard smooth muscle and skeletal muscle membranes share a number of common features, strongly implicating a common identity. Our laboratory is currently investigating the potential role of the ATPase in intracellular Ca^{2+} mobilization and Ca^{2+} channel gating, as well as their modulation by phospholipid cycle intermediates in skeletal muscle and gizzard smooth muscle. Supported in part by NSF DMB 8613881, NSF INT 8515846, and the California Metabolic Research Foundation.

Table 1. Comparison of stimulatory lectins, deglycosylation and inhibitors showing a common identity between skeletal and smooth muscle Mg-ATPase.

	skeletal		smooth
Lectin		Fold-stimulation	
Con A	4.5X		4.3X
Lentil	4.0X		3.7X
PHA-E	3.5X		3.0X
WGA	2.5X		2.5X
Results of Deglycosylation		Core molecular weight	
N-Glycanase	65kD		65kD
TFMSA	65kD		
Inhibitors		% Control	
Azide (5 mM)	98.0		98.5
Vanadate (100 mM)	99.0		99.0
Cyclopiazonic Acid (10 uM)	97.5		96.5
NBD-F (100 uM)	31.0		23.6

References

1. Sabbadini, R.A. and Dahms, A.S. (1989) J. Bioenerget. Biomembrane 21: 163-213.
2. Moulton, M.M., Sabbadini, R.A., Norton, K.C., and Dahms, A.S. (1986) J. Biol. Chem. 261: 12,244-12,251.
3. Kang, J.J., Cunningham, H.B., Jachec-Schmidt, C., Priest, A.M., Dahms, A.S., and Sabbadini, R.A. (in press) Arch. Biochem. Biophys.
4. Kang, J.J., Cunningham, H.B., Jachec-Schmidt, C. Norton, K.C., Priest, A.M., Sabbadini, R.A., and Dahms, A.S., (1990) Annals NY Academy Sciences, 603: 530-533.

ROLE OF PROTEIN KINASE C (PKC) IN MEDIATING CA^{2+} ENTRY IN VASCULAR SMOOTH MUSCLE CELLS

Y.T. Xuan, O.L. Wang, and A.R. Whorton

Department of Pharmacology
Duke University Medical Center
Durham, North Carolina, 27710

The role of PKC in regulation of Ca^{2+} mobilization is not clear. For a number of years, PKC has been shown to regulate smooth muscle contraction. Although considerable controversy exists, PKC may induce contraction by elevating $[Ca^{2+}]_i$. Thus we begin to investigate the effects of PKC on Ca^{2+} mobilization in vascular smooth muscle cells. When cells were exposed to 25 nM PMA (phorbol myristate acetate) for 2 minutes, $[Ca^{2+}]_i$ rose slowly to a peak value followed by a sustained (>20 minutes) elevation of $[Ca^{2+}]_i$. Both the amplitude and the time to peak response were dose-dependent (EC_{50}=10 nM) with the time to reach peak $[Ca^{2+}]_i$ being inversely related to the concentration used. Although the exact mechanism for PMA elevation of $[Ca^{2+}]_i$ is not clear, several possibilities exist. First, PKC activation may induce release of Ca^{2+} from intracellular pools leading to an increase in $[Ca^{2+}]_i$. This is unlikely since in our hands PMA did not mobilize Ca^{2+} from intracellular pools in cells bathed in Ca^{2+} free buffer. Second, PMA, through activation of PKC, may directly or indirectly activate Ca^{2+} channels including L-type Ca^{2+} channels in these smooth muscle cells. In fact, PKC has been shown to regulate CA^{2+} channels and $[Ca^{2+}]_i$ in neural cells. This appears to be the case in vascular smooth muscle cells, since 1 μM nicardipine reversed the sustained elevation of $[Ca^{2+}]_i$ initiated by PMA. In addition, treatment of cells with nicardipine prior to PMA prevented Ca^{2+} mobilization. Taken together, these data strongly suggest that PMA leads to elevation of $[Ca^{2+}]_i$ by activating the L-type Ca^{2+} channel.

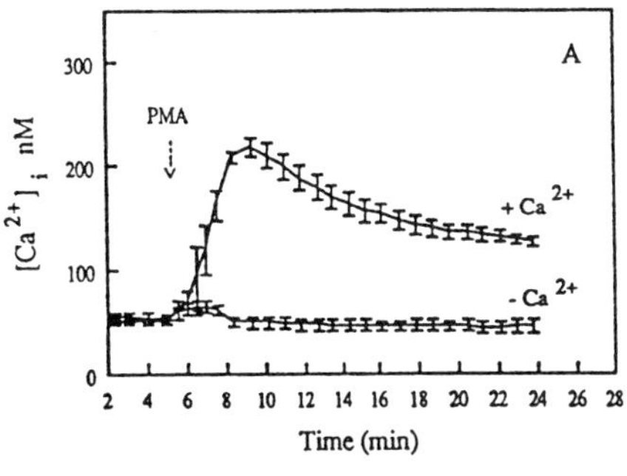

Figure 1

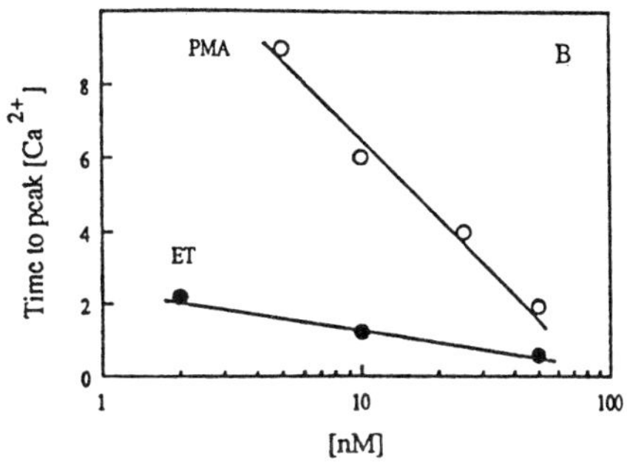

Figure 2

Excitation-Contraction Coupling in Skeletal, Cardiac, and Smooth Muscle
Edited by G.B. Frank *et al.*, Plenum Press, New York, 1992

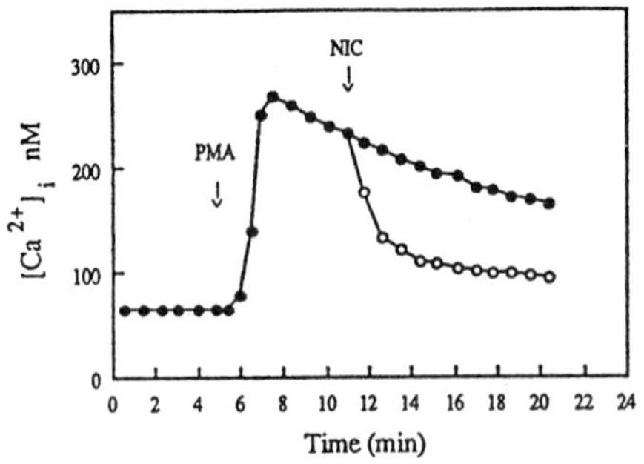

Figure 3

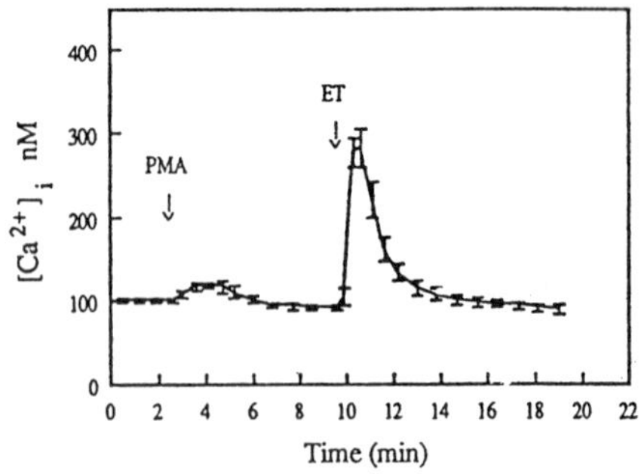

Figure 4

To investigate whether PKC is involved in the agonist activation of the L-type Ca^{2+} channel, we have explored Ca^{2+} influx initiated by endothelin (ET). In our initial experiments we chose to deplete PKC to investigate its activity in mechanisms of ET-induced Ca^{2+} entry. In these experiments, following the preincubation with PMA to deplete PKC, cells were washed and subsequently exposed to either PMA or ET. Pretreatment abolishes the response to 50 nM PMA, reduced the magnitude of the ET-induced Ca^{2+} response and prevented the sustained Ca^{2+} response. These data suggest that PKC is involved in agonist activation of the L-type Ca^{2+} channel and that PKC is particularly important in prolonged Ca^{2+} gating required for sustained elevation of $[Ca^{2+}]_i$. In other experiments, we have investigated the effects of the PKC inhibitor, staurosporine, on ET-induced Ca^{2+} entry using Mn^{2+} as an indicator of activated Ca^{2+} influx. Again, since we had previously shown that the sustained portion of the ET response was due solely to continued activity of the L-type Ca^{2+} channel, we investigated the effects of staurosporine on Ca^{2+} entry 10 minutes following stimulation with ET. Pretreatment of cells with staurosporine completely prevented ET-induced Mn^{2+} entry. In similar experiments, we have examined the effects of staurosporine on ET-stimulated $^{45}Ca^{2+}$ influx. Again we chose to concentrate on the sustained phase of the response. As above, staurosporine significantly inhibited $^{45}Ca^{2+}$ entry stimulated by ET. These data strongly suggest that ET activates PKC and that some regulatory component of the L-type Ca^{2+} channel is phosphorylated as a consequence of the activation process. Even though it is not clear from these data, the maintained activity of the channel may be due to phosphorylation and slow dephosphorylation or to a continuous cycling of phosphorylation/dephosphorylation as long as DAG levels are elevated. In fact, ET has been shown by others to stimulate a prolonged elevation of DAG in vascular smooth muscle cells. It should be pointed out that in these preliminary experiments, we have examined the effects of PKC on sustained Ca^{2+} entry brought about by continued activity of the L-type channels. It is possible that the early activation of the channel may occur through other mechanisms including inositol phosphate activation.

POSTGANGLIONIC NICOTINIC ACTION OF ANATOXIN-A ON GUINEA PIG ILEUM ANTAGONIZED BY THYMOPENTIN

Peter K. Chiang, J. Katzhendler, Richard K. Gordon, B. P. Doctor, and Nesbitt D. Brown

Division of Biochemistry
Walter Reed Army Institute of Research
Washington, DC 20307-5100

Anatoxin-a (ANTX) is a bicyclic amine exotoxin produced by the blue green algae *Anabaena flos-aquae*, and has been found to cause death to livestock and waterfowl via a blockade of neuromuscular transmission and subsequent respiratory paralysis (1, 2). ANTX has been shown to be a potent nicotinic cholinergic agonist in skeletal muscle (3), *Torpedo* electric organ (4) and mammalian central nervous system (5), and can also induce cardiovascular effects by activating nicotinic receptors in the adrenal medulla and sympathetic ganglia (6). We studied the effect of ANTX on guinea pig ileum.

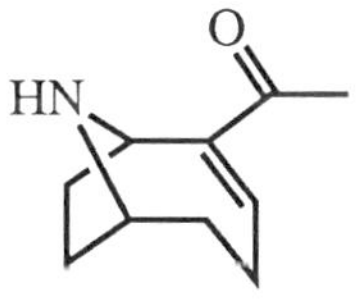

Anatoxin-a

RESULTS

ANTX stimulated the contraction of guinea pig ileum with an ED_{50} of $8.0 \pm 0.1 \times 10^{-7}$ M, which is similar to that of ACh with an ED_{50} of 2.0×10^{-7} M reported by us previously (7). The stimulation of ileum contraction by ANTX was blocked by two nicotinic antagonists, tubocurarine with a pK_B of 5.7 ± 0.02 (apparent $K_B = 1.9 \times 10^{-6}$ M; slope = 0.5) and hexamethonium with a pK_B of 7.9 ± 0.4 (apparent $K_B = 2.7 \times 10^{-8}$ M; slope = 0.5). For comparison, the stimulation of contraction by nicotine, which has an ED_{50} of $1.7 \pm 0.2 \times 10^{-6}$ M, was itself blocked by hexamethonium with a pA_2 of 5.7 ($K_B = 2.0 \times 10^{-6}$ M; slope = 1.0).

Because of the known muscarinic involvement in smooth muscle contraction, atropine was examined for its ability to antagonize the effect of ANTX. ANTX-stimulated ileum contraction was inhibited by atropine; the pK_B was 7.3 ± 0.02 (apparent $K_B = 5.4 \times 10^{-8}$ M; slope = 0.7). In spite of the antimuscarinic effect on the ileum, we failed to show any inhibition of the binding of $[^3H]$N-methylscopolamine to ileum membranes in the presence of ANTX. Nor was there a stimulation by ANTX of the secretion of α-amylase from pancreatic acinar cells, a process that is activated by cholinergic agonists at the muscarinic receptors (7- 9). In contrast, 10^{-5} M hemicholinium, which competes with choline for uptake into nerve endings, inhibited the ANTX-stimulated contraction of guinea pig ileum in a time-dependent manner. The level of inhibition achieved was about 50% after 10 min.

Excitation Contraction Coupling in Skeletal, Cardiac, and Smooth Muscle
Edited by G.B. Frank *et al.*, Plenum Press, New York, 1992

Thymopentin (TP-5), a synthetic peptide corresponding to position 32-36 of thymopoietin (**Arg-Lys-Asp-Val-Tyr**), has been shown to be a specific antagonist of nicotinic synapses (10-14). It was thus of interest to test the effect of thymopentin on the stimulation of ileum contraction by ANTX. The pK_B was 5.7 ± 0.03 (apparent $K_B = 1.9 \times 10^{-6}$ M; slope = 0.6).

DISCUSSION

The stimulation of ileum contraction by ANTX was inhibited by hemicholinium , which inhibits choline uptake and hence ACh synthesis, thus suggesting that the action of ANTX was dependent to a large extent on the *de novo* synthesis of ACh. ACh by itself normally stimulates ileum contraction, but its stimulation unlike that of ANTX, was not subject to an inhibition by hexamethonium or tubocurarine. These results suggest that ANTX was most likely acting as a specific nicotinic agonist at the nicotinic postganglionic synapses of the guinea pig ileum, causing a release of endogenous ACh, which then acted on the postsynaptic muscarinic receptors. This latter effect explained the blocking of ANTX-stimulated ileum contraction by atropine. However, it cannot be determined for now whether the action of ANTX is on the postsynaptic nicotinic receptors of the cell bodies of the postganlionic neurons and/or on presynaptic receptors on the terminals of the postganglionic nerves.

Thymopentin, a derivative of thymopoietin, has been shown to be a potent antagonist at nicotinic receptors in C2 muscle cell cultures (11), to inhibit ligand binding to nicotinic receptors at neuromuscular junctions (12), and to accelerate cholinergic-induced inactivation of nicotinic receptor reconstituted into lipid vesicles (13). The present investigation showed that thymopentin blocked the ileum contraction stimulated by ANTX, and confirmed the specific nicotinic action of ANTX.

Two reasons might explain the observation that the slopes obtained by the Schild plot for all the antagonists, with the exception of hexamethonium on nicotine, were less than unity. First, it could be due to the postganglionic synaptic effect of ANTX and the postsynaptic effect of ACh released by ANTX. Second, ANTX might interact at a noncompetitive site different from the α-bungarotoxin binding site, which has been shown to be a site of action for the parent compound, thymopoietin (14). A slope of unity (1.0) obtained by hexamethonium on nicotine-stimulated ileum contraction seemed to support the last hypothesis.

REFERENCES

1. W. W. Carmichael, D. F. Biggs and P. R. Gorham, 1975, Science 187, 542.
2. J. P. Devlin, O. E. Edwards, P. R. Gorham, N. R. Hunter, R. K. Pike and B. Stavic, 1977, Can. J. Chem. 55, 1367.
3. C. E. Spivak, B. Witkop and E. X. Albuquerque, 1980, Mol. Pharmacol. 18, 384.
4. R. S. Aronstam and B. Witkop, 1981, Proc. Natl. Acad. USA 78, 4639.
5. X. Zhang, P. Stjernlof, A. Adem and A. Nordberg, 1987, Eur. J. Pharmacol. 135, 457.
6. A.-L. Siren and G. Feuerstein, 1990, Toxicol. Appl. Pharmacol. 102, 91.
7. G. A. Miura, P. K. Chiang, R. K. Gordon, B. P. Doctor, C. E. Twine, A. Philip, J. A. Kepler and F. I. Carroll, 1986, Gen. Pharmacol. 17, 477).
8. H. Leader, R. M. Smejkal, C. S. Payne, F. N. Padilla B. P. Doctor, R. K. Gordon and P. K. Chiang, 1989, J. Med. Chem. 32, 1522.
9. R. K. Gordon, E. Breuer, F. N. Padilla, R. M. Smejkal and P. K.Chiang, 1989 , Mol. Pharmacol. 36, 766.
10. A. M. P. Koskinen and H. Rapoport, 1985, J. Med. Chem. 28, 1301.
11. M. Quik, H. E.-Bizri, T. Audhya and G. Goldstein, 1991, Mol. Pharmacol. 39, 324.
12. M. Quik, B. Collier, T. Audhya and G. Goldstein, 1990, J. Pharmacol. Exp. Ther. 254, 1113.
13. E. L. M. Ochoa, S. Medrano, M. C. Ll. de Carlin and A. M. Dilonardo, 1988, Cell. Mol. Neurobiol. 8, 325.
14. M. Quik, U. Babu, T. Audhya and G. Goldstein, 1991, Proc. Natl. Acad. Sci. USA, 88, 2603.

THE SURFACE INPUT SITE RESPONSIBLE FOR EXCITATION-CONTRACTION (E-C) COUPLING MECHANISM IN SINGLE SKELETAL MUSCLE FIBRES OF FROG

M. Fujino, T. Arima, C. Hasegawa, K. Harano,
M. Takahashi, S. Sano, and S. Fujino*

Department of Physiology, National Defense
Medical College, Tokorozawa and *Department of
Pharmacology, Hokkaido Pharmacentical Sciences
Otaru, Japan

The aim of this study was to determine what the feet[1] are in E-C coupling, and to know how input process proceeds upon depolarization. Fortunately, we could extract a protein[2] for E-C coupling using phenylglyoxal, PGO, (PGO-protein); and the protein was a glycoprotein. We have, therefore, used concanavalin A (Con A), a lectin, which inhibits the E-C coupling; while, according to the results so far, phytohemagglutinin-P, a lectin, had no effect on the coupling.

Locations of ferritin attached to Con A in electronmicrographs were measured in all available triads of longitudinal sections from one single fibre and plotted into one figure. From this procedure, we obtained histograms under the conditions of both 4 kinds of K-concentrations and of 5 kinds of developed tension. Results showed that Con A-binding sites are located not at random but center round the localized T-tubular wall attached by foot. In the resting state, the mode of histogram lies over the foot and on central side of T-tubule. With increase of K-depolarization, the mode shifts and begins to appear over the foot and on peripheral side of T-tubules. From these observations, it could be said that sites responsible for binding Con A molecules moves upon depolarization and the grade of movement determines the size of tension developed.

We have already reported[2] an inhibition of E-C coupling due to PGO, of which one molecule has two active sites. Observations have given clear results that the movement aforementioned of the site upon depolarization is completely inhibited by PGO. It would, therefore, be probable that the site responsible for input-mechanism of E-C coupling exists in PGO-protein molecule and has two active groups, which constitute one functional unit, to which PGO binds.

In the input site in question, one positive charge of the two is fixed onto a wall-portion attached by foot; while, the other positive charge exists on the movable arm protrudent from the wall, and Con A binds to the arm. In the resting state, though a strong repulsive force works between the two positive charges, they cannot stand apart from each other

Excitation-Contraction Coupling in Skeletal, Cardiac, and Smooth Muscle
Edited by G.B. Frank *et al.*, Plenum Press, New York, 1992

due to the enormous electric field by resting potential. If membrane depolarizes, electric field is weakened; thus, the positive charge on the arm leaves far from the fixed positive charge. Namely, depolarization is transduced to a mechanical movement and is, therefore, measured as the grade the movement of the arms. Therefore, the set of two active groups and the electric field at membrane constitute together a "potentiometer" or an "electrometer", which measures the change of membrane potential.

If muscle fibres treated with PGO, the set of two positive charges of PGO-protein is considered to interact with two active sites of one PGO molecule. Thus, the "electrometer" cannot be functioned more by the chemical "clamp"; therefore, E-C coupling is uncoupled by PGO.

It would be concluded that (1) the feet form a pathway of signals for E-C coupling and (2) Con A binding portion on T-tubular wall attached by feet is differentiated for E-C coupling process, which seems to be initiated as a mechanical movement of the arm or an indicator of "potentiometer" or an "electrometer", and (3) the movement would probably transmitted as a mechanical signal through feet to lateral sacs.

REFERENCES

1. B. A. Block, T. Imagawa, K. P. Campbell, and C. Franzini-Armstrong, Structural evidence for direct interaction between the molecular components of the transverse tubule/sarcoplasmic reticulum junction in skeletal muscle, J. Cell Biol., 107:2587 (1988).
2. M. Fujino, Y. Sato, T. Arima, and H. Takai, Phenylglyoxal (PGO) and sites for excitation-contraction (E-C) coupling in skeletal muscle cells of frog, J. Physiol. Soc. Japan, 45:507 (1983).

CA-INDUCED CA RELEASE FROM THE SR OF BARNACLE MYOFIBRILLAR BUNDLES, TRIGGERED BY THE PHOTORELEASE OF CALCIUM IONS FROM CAGED CALCIUM

T.J. Lea and C.C. Ashley

University Laboratory of Physiology
Parks Road, Oxford, OX1 3PT
United Kingdom

We have triggered Ca^{2+}-induced Ca^{2+} release (CICR) from the sarcoplasmic reticulum (SR) of isolated myofibrillar bundles of barnacle striated muscle fibres using the rapid photorelease of Ca^{2+} from the caged calcium molecule, nitr-5 (Adams et al, 1988), thus overcoming diffusional delays normally inherent in such experiments.

Myofibrillar bundles of about 100 μm diameter were mounted for isometric tension recording and were first equilibrated in a conventional "skinned" muscle solution containing 0.1 mM nitr-5 as the only Ca^{2+} buffer, at an initial pCa of 6.8 - 6.6. A pulse of UV light of 25 ns duration from a frequency-doubled ruby laser was used to photolyse the nitr-5 in the bundle, thus releasing free Ca^{2+} within 1 ms. This resulted in a phasic contraction which in some experiments reached the maximum Ca-activated tension amplitude (P_0). The mean rise time to half amplitude of the responses was 2.3 s at 12°C (n=12).

The main source of activating Ca^{2+} in these responses was attributed to CICR from the SR since pretreatment with either ryanodine (0.1 mM) or Triton X-100 (1%) abolished a large part of the response. The small tonic response remaining was attributed to direct activation of the myofibrils by the Ca^{2+}, which was released photolytically from the nitr-5. By reference to a pCa-tension curve, obtained with EGTA-based activating solutions, this residual tension could be used to estimate the post-photolysis pCa in these experiments, i.e. the trigger pCa for the CICR. This varied from pCa 6.7 to 6.0 over the range of laser pulse energies used; the CICR response increased in amplitude with increasing photorelease of Ca^{2+}.

The CICR responses, obtained using the photorelease of Ca^{2+} as a trigger, were significantly slower than the fastest tetani which have been recorded from electrically stimulated fibres of the barnacle muscle; these had a mean half rise time of 177 ms at 12°C (Griffiths et al, 1990). The isolated myofibrillar bundle could be made to contract at rates comparable to this, when the photorelease of Ca^{2+} was made sufficiently large so as to fully activate the myofibrils directly, using 2.0 mM nitr-5 at an initial pCa 6.5. These results indicate that either (1) the physiological release of calcium ions from the SR is by some mechanism other than CICR in this crustacean muscle, or (2) if CICR is involved then its efficiency is impaired under the conditions used with the isolated myofibrillar bundles.

Excitation-Contraction Coupling in Skeletal, Cardiac, and Smooth Muscle
Edited by G.B. Frank *et al.*, Plenum Press, New York, 1992

References

S.R. Adams, J.P.Y. Kao, G. Grynkiewicz, A. Minta, and R.Y. Tsien, J. Am. Chem. Soc. 110: 3212-3220 (1988).

P.J. Griffiths, Y.M. Duchateau, Y. Maeda, J.D. Potter, and C.C. Ashley, Pflügers Arch. 415: 554-565 (1990).

CONTRACTILE AND ELECTROPHYSIOLOGIC EFFECTS OF REALKALIZATION IN CARDIAC TISSUES: ROLE OF NA/H EXCHANGE AND INCREASED $[CA]_i$

Jianmin Duan and Margaret P. Moffat

Department of Pharmacology and Toxicology
University of Western Ontario
London, Canada

The effects of lactate acidosis followed by realkalization were studied in fast and slow response ventricular tissues and isolated cardiac myocytes. Both electrophysiological and contractile responses were altered by these conditions in isolated ventricular tissues and $[Ca]_i$ was elevated in isolated myocytes.

Potential mechanisms underlying responses induced by lactate acidosis and realkalization were studied using the isolated right ventricular free wall and isolated cardiac myocytes from the guinea pig. In the isolated right ventricular free wall preparations, transmembrane electrical activity was recorded from both epicardial and endocardial layers using conventional microelectrode techniques. Contractile force was measured by connecting a force transducer to one end of the preparation. In preparations superfused with normal Tyrode's solution (pH 7.4), both electrical and contractile activity remained stable over the test period. Well polarized, fast response tissues were exposed to lactate acidosis (pH 6.80) by reduction of $NaHCO_3$ from 20 to 6 mM, and inclusion of dl lactate (20 mM) in the solution. Lactate acidosis caused a significant decline in contractile force, however, no significant effects on either action potential duration (APD) or resting potential were observed in either the epicardium or endocardium. A small (11%), but significant ($P < 0.05$) increase in transmural conduction time was observed following 20 min. In addition, 2 of 10 preparations developed arrhythmic activity. Realkalization by return to normal Tyrode's solution following 20 min of lactate acidosis induced a transient, but significant increase in contractile force (58% above time control). This change in contractile force was associated with an increase in both APD_{50} and APD_{90} in the epicardium which was statistically significant following 10 and 20 min of realkalization. These electrophysiological responses were much smaller in the endocardium. Five of 10 preparations developed arrhythmic activity, which in 4 experiments occurred within the first 10 min of realkalization. Whether the effects of realkalization on contractility and electrophysiological parameters are related is at present unclear however it is possible that increases in $[Ca]_i$ underlie both the inotropic and arrhythmogenic effects in these tissues.

In order to test whether the effects of lactate acidosis and realkalization were also present in depolarized, slow-response, tissues, similar experiments were carried out in preparations exposed to elevated

KCl (25 mM). In this study, 0.2 mM $BaCl_2$ was added to the solution to reduce the voltage required for stimulation and to minimize decremental conduction. Control preparations remained stable over the test period except for a small decrease in contractile force. Similar to fast response tissues, lactate acidosis induced a significant decrease in contractile force within the first few minutes, especially at lower stimulation frequencies. However, this decline in contractile force was associated with a significant increase in APD_{50} in the endocardium, which was more pronounced at lower stimulation frequencies. In addition, the amplitude of the slow action potentials was increased in the endocardium following 20 min of lactate acidosis. Realkalization caused a significant increase in contractile force at 5 and 10 min at all stimulation frequencies tested, which was associated with an increase in action potential amplitude in the endocardium following 5 min of realkalization. In addition, a small but significant depolarization was observed in the epicardium following 10 and 20 min of realkalization. Thus, both contractile activity and action potential configuration are also modulated by lactate acidosis and realkalization in the depolarized tissues however, the effects differ in several respects from those observed in fast response tissues.

To examine the role of changes in $[Ca]_i$ in the response to lactate acidosis and realkalization we used cell suspensions loaded with indo-1 AM. Ventricular myocytes isolated using a modified procedure of Mitra and Morad were suspended in a HEPES buffer containing 1.5 mM $CaCl_2$ (pH 7.3). Under control conditions, $[Ca]_i$ in the resting cells was 92±15 nM (n=4). Addition of 0.4 mM $BaCl_2$ and 10 nM of isoproterenol to initiate spontaneous cell beating, increased $[Ca]_i$ to 131±11 nM (n=4). A 5 min preincubation of myocytes at $37^{o}C$ in normal HEPES buffer with or without beating agents did not affect $[Ca]_i$ under either resting or beating conditions. However, preincubation for 5 min in HEPES buffer containing 20 mM of dl-lactate (pH 6.8) caused a significant increase (63%) of $[Ca]_i$ in cells in which beating was initiated immediately upon resuspension in normal HEPES solution. This effect was reduced or completely abolished if beating was initiated 1 or 4 min following resuspension, respectively. These results suggest that the cells can recover from an acid load without concommitant increases in $[Ca]_i$ under resting but not beating conditions.

To examine the potential involvement of sodium/hydrogen exchange in the realkalization-induced increase in $[Ca]_i$, hexamethlene amiloride (HMA), a sodium/hydrogen exchange inhibitor, was added to the HEPES buffered solutions. Although HMA (8 μM) did not significantly alter $[Ca]_i$ under control conditions, it did inhibit the realkalization-induced increase in $[Ca]_i$ in beating cells by 34, 45, 67, 89% at concentrations of 1, 2, 4 and 8 μM respectively. In addition, it was found that inclusion of HMA (4 μM) during either lactate acidosis or realkalization alone was also effective, although the effect was less than that observed when HMA was present during both lactate acidosis and realkalization.

These results suggest that lactate-acidosis and realkalization can induce several changes in electrophysiological and mechanical responses in both fast and slow response tissues. Further, the realkalization-induced increase in contractile force is likely to result from a HMA-sensitive increase in $[Ca]_i$. It is possible that Na/H exchange and subsequent Na/Ca exchange may play important roles in the response of cardiac tissues to realkalization.

ACKNOWLEDGEMENT

Supported by grants from the Heart and Stroke Foundation of Ontario and the Medical Research Council of Canada to M.P. Moffat.

LEUKOTRIENE-INDUCED CONTRACTION OF ISOLATED GUINEA PIG GALLBLADDER (GB) STRIPS

Stephen M. Freedman, Debesh K. Basu,
John L. Wallace, and Eldon A. Shaffer

Gastrointestinal Research Unit, Faculty of
Medicine, University of Calgary
Calgary, Alberta, Canada T2N-4N1

Gallstone disease is a significant health problem; an estimated 15 million adults are affected by this biliary tract disease in North America. Cholesterol gallstone formation is accompanied by impaired GB motility. A later stage is cholecystitis- an inflammatory event which may increase smooth muscle tone, resulting in pain[1]. In animal models of cholecystitis, significant increases in the levels of arachidonic acid metabolites in GB perfusates has been demonstrated[2]. These metabolites, comprised of prostanoids (PS) and leukotrienes (LT), have traditionally been associated with many inflammatory processes[3]. The role of PS in the etiology of cholecystitis and gallstone formation has been well defined. However, to date, little is known regarding the role assumed by the LT's in these disease states.

Previous studies have shown that LT's have a stimulatory effect on a number of smooth muscles. Exogenous LT's may either exert their effects directly, or indirectly via the release of intracellular messengers/mediators. LT receptors have been identified, largely through the use of specific receptor antagonists. The use of such specific receptor antagonists permit the characterization of the role of LT's in physiological and pathological processes. LT's contract GB smooth muscle[4], indicating that LT's may be partially responsible for the development of increased GB tone during cholecystitis. The purpose of this study was to examine the mechanism of action of LT-induced contraction in GB smooth muscle with the intention of demonstrating a possible role of LT's in gallbladder inflammation.

In the present investigation, cumulative dose-response curves were constructed with LTC_4 and LTD_4 on isolated guinea pig GB strips. A progressive increase in contraction of the strips in response to increasing concentrations of LTC_4 and LTD_4 (10^{-10} to 10^{-7} M) was noted. Pretreatment of the tissue strips with MK-571™ (L-660,711;1 μM), a specific LTD_4 receptor antagonist on human and guinea pig lung membranes[5], antagonized the contractile response to LTC_4 and LTD_4 at all

Excitation-Contraction Coupling in Skeletal, Cardiac, and Smooth Muscle
Edited by G.B. Frank *et al.*, Plenum Press, New York, 1992

concentrations tested. In both cases, the dose-response curves were shifted to the right. Tissue strips pretreated with the LT biosynthesis inhibitor, MK-886™ (L-663,536; 1 μM), demonstrated no loss of LTD_4-induced contractile activity. Nifedipine (1 μM), a voltage-operated Ca^{2+}-channel antagonist, had no effect on LTD_4-induced contractile responses. In contrast, the same concentration of nifedipine markedly depressed K^+ (70 mM)-stimulated GB contractions. To examine the contribution of Ca^{2+}_i to the contractile process, tissue strips were pretreated with caffeine (5mM) prior to stimulation. Contractions induced by LTD_4 in the presence of caffeine were depressed at all concentrations of LT tested. Caffeine, to a lesser extent, also had an inhibitory effect on K^+-induced GB contractures.

The results from this study confirm earlier findings regarding the stimulatory nature of LT's on GB smooth muscle. The results obtained with MK-571 suggest that the guinea pig GB possesses specific LT receptors. Two explanations may be proposed to account for the inhibition of LTC_4-induced contractions by the LTD_4 receptor antagonist. First, MK-571 is a non-specific antagonist of guinea pig GB LT receptors. Alternatively, LTC_4 may be interconverted to LTD_4 before binding to a LT receptor. The latter explanation proposes the presence of only one population of (MK-571-sensitive) LTD_4 receptor.

Previous studies on lung parenchyma suggest that the contractile effects of exogenous LT's are partially mediated through the release of endogenous arachidonic acid metabolites[6]; this points to an indirect mechanism of LT action. The observation that MK-886 had no effect on the contractile response to LTD_4 indicates that the effects of LT administration on GB smooth muscle are not mediated through the endogenous production of stimulatory LT's. K^+-stimulated GB contraction was highly dependent on transmembrane Ca^{2+} flux via voltage-operated Ca^{2+} channels, while the contractile responses to LTD_4 were not. Ca^{2+}_i may be more important in LTD_4-induced GB contraction, as evidenced by the results obtained with caffeine.

References

1. D.L. Kaminski, Arachidonic acid metabolites in hepatobiliary physiology and disease, Gastroent. 97:781 (1989).
2. D.L. Kaminski, H. Daneshmand, P. Dean, and Y.G. Deshpande, Prostanoids and leukotrienes in experimental feline cholecystitis, Hepatol. 11:1003 (1990).
3. B. Samuelsson, Leukotrienes and lipoxins: Structures, biosynthesis and biological effects, Science 220:568 (1987).
4. P. Yusko, R.A. Hall, and A.W. Ford-Hutchinson, Contraction of guinea pig GB strips by leukotrienes and other agonists, Prostaglandins 25:397 (1983).
5. T.R. Jones, et al., Pharmacology of L-660,711 (MK-571); a novel, potent, and selective LTD_4 receptor antagonist, Can J Physiol Pharmacol. 67:17 (1989).
6. S.E. Dahlen, et al., Mechanisms of leukotriene-induced contractions of guinea pig airways: LTC_4 has a potent direct action whereas LTB_4 acts indirectly, Acta Physiol Scand. 118:393 (1983).

EFFECTS OF RYANODINE ON SR Ca^{2+} RELEASE CHANNELS DEMONSTRATED WITH CAFFEINE-INDUCED TENSION TRANSIENTS IN SKINNED STRIATED AND VASCULAR SMOOTH MUSCLE

Judy Y. Su

Department of Anesthesiology, RN-10
University of Washington School of Medicine
Seattle, WA 98195

Ryanodine has been shown to cause cardiac arrest and contracture of skeletal muscle and to bind to a receptor with high specificity and affinity in isolated sarcoplasmic reticulum (SR).[1] We tested the hypothesis that the ryanodine binding itself, to the receptor is the mechanism of ryanodine's pharmacological effects. We used a skinned muscle fiber preparation[2,3] which is unique in that the ryanodine effect on the SR can be monitored by tension development. This study shows that: (i) ryanodine, in skeletal muscle, enhances the caffeine-induced tension transient, and in striated and vascular smooth muscle, decreases the caffeine-induced tension transient of the subsequent control (C_2 depression); (ii) C_2 depression is dependent on the duration of the open state of the SR Ca^{2+} release channels; and (iii) C_2 depression is a direct function of $[Ca^{2+}]$ and [caffeine], and is an inverse function of $[Mg^{2+}]$.

METHODS

Skinned Muscle Fiber Preparation

Striated muscles and aorta isolated from young adult male rabbits were used. Pieces of striated muscle were homogenized, and the medial layer of aortic strips were treated with saponin. Fiber bundles from myocardium, single fibers from soleus and adductor magnus, and the medial layer from aortic strip were mounted and the isometric tensions were measured.

Basic Bathing Solution

The basic bathing solution contained 70 mM (K^{+}+ Na^{+}) 15 mM creatine phosphate, 2 mM $MgATP^{2-}$, 0.1 mM Mg^{2+}, various free calcium and total EGTA concentrations depending on the specific condition, methanesulfonate (major anion), and 70-80 mM imidazole for striated muscle and 50 mM PIPES for aorta, to achieve an ionic strength of 0.15 with pH of 7.00 ± 0.02 at 23 ± 2°C.

Ca^{2+} Load-Release Cycle in the SR Demonstrated with a Caffeine-Induced Tension Transient

The basic protocol for measuring the caffeine-induced tension transient in skinned muscle fibers has been described previously.[3] Briefly, a load-release cycle consisted of immersing the preparation sequentially in five different solutions to load Ca^{2+} into, and to release Ca^{2+} from the SR with caffeine, resulting in a tension transient. The area of the tension transient was used to estimate the amount of calcium released from the SR (Fig. 1).

Excitation-Contraction Coupling in Skeletal, Cardiac, and Smooth Muscle
Edited by G.B. Frank *et al.*, Plenum Press, New York, 1992

TYPICAL TRACING

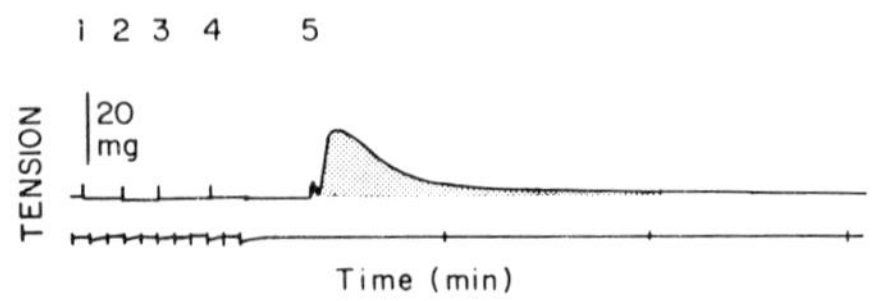

Figure 1.

Figure 1 shows a control caffeine-induced tension transient in skinned myocardial fibers. 1-5 are solution exchange artifacts. The ionic composition of these five solutions includes the basic bathing solution described above and (i) no added Ca^{2+} and 25 mM caffeine for Solution #1, to empty the Ca^{2+} in the SR, (ii) no added Ca^{2+} for Solution #2, to wash away caffeine in the fibers, (iii) pCa 6.5, 7 mM EGTA for Solution #3, to rapidly load Ca^{2+} into the SR, (iv) no added Ca^{2+}, 0.05 mM EGTA for Solution #4, to reduce EGTA in the fibers, and (v) no added Ca^{2+}, 0.05 mM EGTA, 25 mM caffeine for Solution #5, to release Ca^{2+} from the SR, resulting in a tension transient.

In the test cycle, solutions #1, #2, and #4 were the same as the load-release cycle of the controls. The composition of Solution # 5 for all studies, and #3 for the pump activity study, however, varied according to each specific study (Table 1); the duration of immersion was fixed since the ryanodine effect is time-dependent.[4]

Each study consisted sequentially of three cycles: a control (C_1), followed by a test (containing [ryanodine], and ryanodine ± caffeine (for channel/pump studies in solution #3 or #5, and for pCa_i or pMg_i studies in solution #5 only)), and finally a second control (C_2). A $p < 0.05$ was regarded as statistically significant using Student's *t*-test for paired or unpaired data depending on the study.

RESULTS

Sensitivities of striated and vascular smooth muscle to ryanodine were tested in the skinned fiber preparation. We found that ryanodine, present either in the loading or release phase, caused depression of the caffeine-induced tension transient of the subsequent control (C_2).[5,6]

Figure 2 shows that ryanodine, present in the Ca^{2+} release phase, did not significantly change ($p > 0.05$) in papillary muscle and aortic strip (Fig. 2A), and enhanced, in soleus and adductor magnus (Fig. 2B), the immediate (test) caffeine-induced tension transient. However, there was a trend towards a biphasic effect; at 10 nM or lower concentrations this tension transient was enhanced in papillary muscle, but decreased in aortic strip, at 50 nM it was reduced below the control level and at >50 nM it was increased slightly in papillary muscle and aortic strip. Ryandodine depressed, in a dose-dependent manner, the subsequent control (C_2) in all muscle types tested (Fig. 2C & Fig. 2D). Papillary muscle and aortic strip (Fig. 2C, IC_{50} = 50-80 nM) were about 100-fold more sensitive than adductor magnus and soleus (Fig. 2D, IC_{50} = 5-10 μM) to ryanodine for C_2 depression. The dose-range of ryanodine is comparable to that observed in an intact muscle preparation.

TABLE 1. Experimental conditions for the test cycle of five different studies with changes only in the releasing solution #5 (#3 for pump) of the Ca^{2+} load-release cycle.

Studies	Dose	Channel/ Pump	Time- Dependence	[caffeine]	pCa	pMg
pCa (M) (-log [Ca^{2+}])	6.8 - 7.0	6.5	6.8 - 7.0	6.8 - 7.0	>8 - 5.0	6.5
[EGTA] (mM)	0.05	7	0.05	0.05	7	7
pMg (M) (-log [Mg^{2+}])	4	4	4	4	4	4.5 - 3
[Caffeine] (mM)	25	25	25	0.5 - 25	±25	25
Duration	steady state	1 - 10 sec	1 - 130 sec	2 min	2 min	2 min
[Ryanodine]	0.01-10μM	1 or 100μM	0. 5 or 10μM	1 μM	1 μM	1 μM

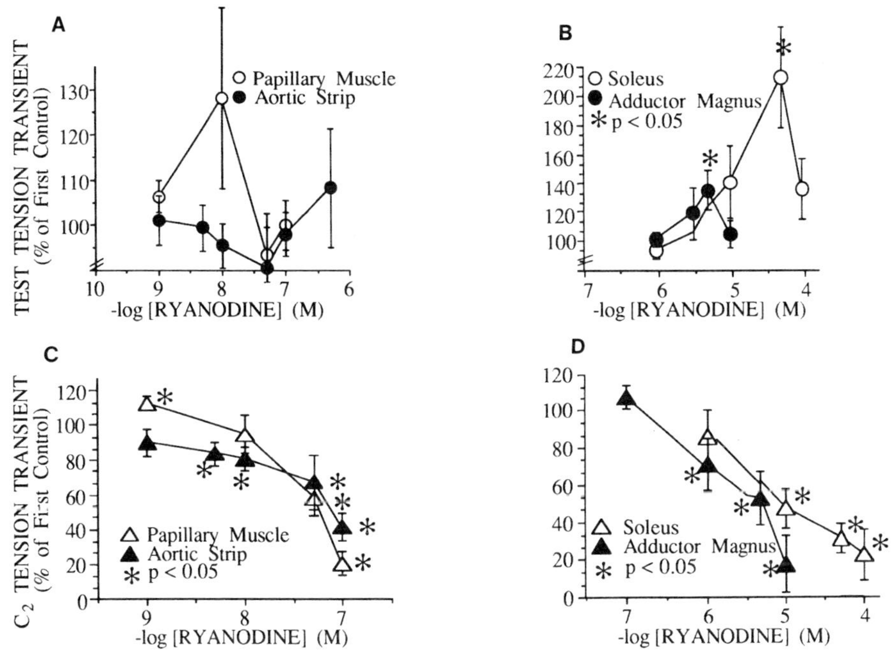

Figure 2.

To test the role of SR-ATPase pump activity in the ryanodine effect, we specially designed a protocol to avoid Ca^{2+} efflux during loading of Ca^{2+}, and re-uptake during Ca^{2+} release; ryanodine was present only for a short period of time (< 3 sec for adductor magnus and soleus with 100 μM ryanodine, and <10 sec for papillary muscle with 1 μM ryanodine).[4] Table 2 shows that ryanodine, in the Ca^{2+} uptake phase, caused no significant effect; however, in the Ca^{2+} release phase, it caused C_2 depression in all three types of striated muscle. This indicates that C_2 depression by ryanodine occurs only when SR Ca^{2+} channels are in an open state.

To test whether C_2 depression by ryanodine was time-dependent, ryanodine (10 μM for striated muscle and 0.5 μM for aortic strip) was present in the releasing solution. Ryanodine was tested for various durations in a caffeine-induced tension transient; i.e., Phase I = time to reach 50% of the peak tension, Phase II = time to reach peak tension, Phase III = time to reach 50% from peak tension to steady-state relaxation, and Phase IV = time from initiation of tension to steady-state relaxation. The duration of a tension transient is different among muscle types (an average of 60 sec for papillary muscle (I & II), 42 sec for adductor magnus (I, II, & III), 22 sec for soleus (I-IV), and 130 sec for aortic strip (I-IV)). Figure 3 shows that there was a time-dependent effect of ryanodine on C_2 depression in skinned fibers from striated muscle (n = 3-5, 10 μM ryanodine, Fig. 3A) and aortic strip (n = 4-5, 0.5 μM ryanodine, Fig 2B).

TABLE 2. C_2 (% of first control), Mean ± SEM (n), * $p < 0.05$ compared to the first control by Student's *t*-test for paired data.

	Initial Uptake Phase	Initial Release Phase
Papillary Muscle	92 ± 5.8(4)	19 ± 11.8(3)*
Soleus	114 ± 16.1(4)	24 ± 12.1(3)*
Adductor Magnus	78 ± 12.5(3)	47 ± 1.3(3)*

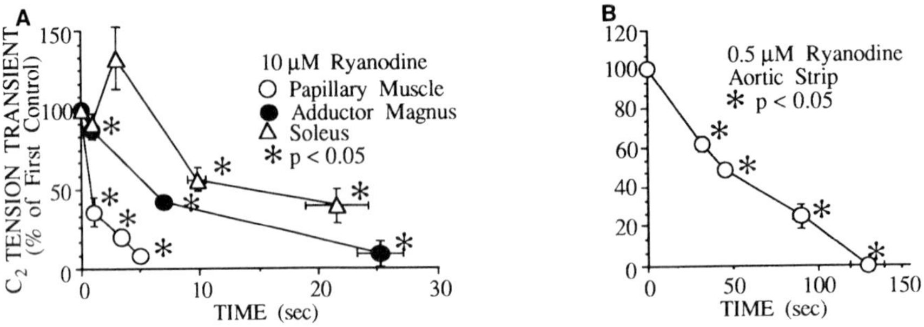

Figure 3.

To test whether activators or inhibitors of ryanodine-receptor binding[1] or single channel activity[7] affect C_2 depression in the skinned fiber preparation, various concentrations of caffeine, Ca^{2+} and Mg^{2+} were tested in <u>skinned myocardial fibers</u>. Figure 4 shows that [caffeine] (Fig. 4A) and pCa_i (Fig. 4B) enhanced, and pMg_i (Fig. 4C) decreased, C_2 depression by ryanodine.

Mean ± SEM (n = 3-9); 0 Rya = no ryanodine, Rya = 1 μM ryanodine, Caff = 25 mM caffeine; * $p < 0.05$ compared to the first control, + = $p < 0.05$ compared between the control in 0 Rya or in the presence of 1 μM ryanodine for the [caffeine] and pMg_i studies, and between 25 mM caffeine and no caffeine for the pCa_i study (see Table 1).

Ryanodine appears to cause the calcium release channel to enter into an open or closed state depending on concentration, but the most stable configuration is the open state which depends on the time of exposure and/or the concentration of ryanodine. The open state induced by caffeine appears to become locked in that open state by ryanodine.

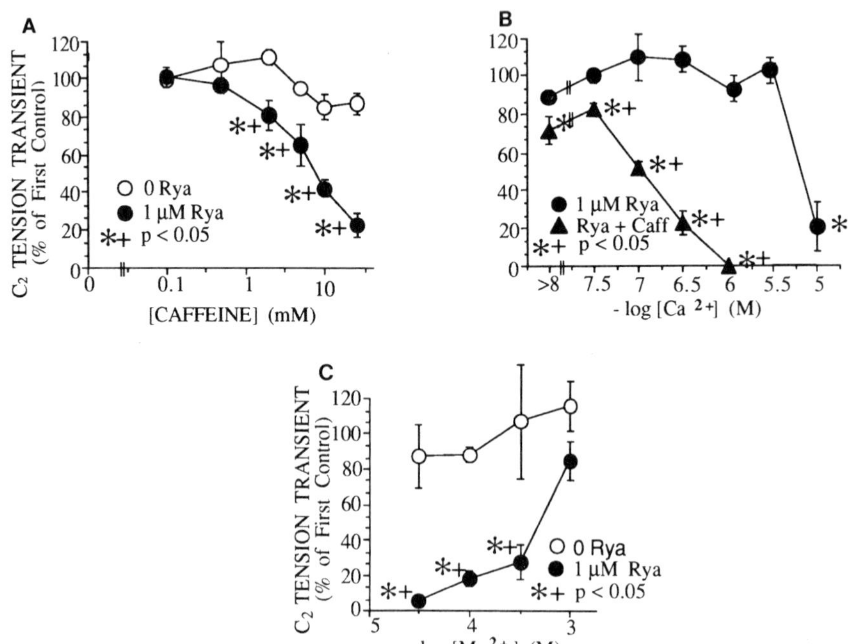

Figure 4.

SUMMARY

In skinned muscle fibers, C_2 depression by ryanodine may account for depression of muscle contraction, and the ryanodine-enhanced caffeine-induced tension transient may account for skeletal muscle contracture. Factors (the open state of the SR Ca^{2+} channels, activators including caffeine and Ca^{2+} and inhibitors including Mg^{2+}) affecting receptor binding and channel activity are consistent with C_2 depression in skinned muscle fibers. (Supported by NIH HL20754)

REFERENCES

1. I. N. Pessah, R. A. Stambuk, and J. E. Casida, Ca^{2+}-activated ryanodine binding: Mechanisms of sensitivity and intensity modulation by Mg^{2+}, caffeine, and adenine nucleotides. Mol. Pharmacol. 31: 232 (1987).
2. J. Y. Su and W. G. L. Kerrick, Effects of halothane on caffeine-induced tension transients in functionally skinned myocardial fibers. Pflügers Arch. 380:29 (1979).
3. J. Y. Su and C. C. Zhang, Intracellular mechanisms of halothane's effect on isolated aortic strips of the rabbit, Anesthesiology 71:409 (1989).
4. J. Y. Su, Mechanisms of ryanodine-induced depression of caffeine-induced tension transients in skinned striated rabbit muscle fibers, Pflügers Arch. 411:371 (1988).
5. J. Y. Su, Effects of ryanodine on skinned skeletal muscle fibers of the rabbit, Pflügers Arch. 410:510 (1987).
6. J. Y. Su, Effects of ryanodine on skinned myocardial fibers of the rabbit, Pflügers Arch. 411:132 (1988).
7. E. Rosseau and G. Meissner, Single cardiac sarcoplasmic reticulum Ca^{2+}-release channel: activation by caffeine. Am. J. Physiol. 256:H328 (1989).

EFFECTS OF DIAPHRAGMATIC FATIGUE ON PHOSPHOINOSITIDE HYDROLYSIS

W.McD. Anderson, J.H. Zavecz, and P. Tso

Departments of Medicine, Pharmacology, Physiology, and Biophysics, Veterans Affairs and Louisiana State University Medical Centers, Shreveport, LA, U.S.A.

SUMMARY

We evaluated the effects of diaphragmatic fatigue on muscle phosphatidylinositol phosphate (PIP) metabolism. Results revealed that the total PIP pool was 76% greater in fatigued rat diaphragms compared to controls, which suggests that fatigue was associated with inhibition of sarcolemmal membrane bound PIP hydrolysis.

The mechanism of excitation-contraction coupling (ECC), as well as fatigue of the specialized skeletal muscle of the diaphragm, is not known. Recent studies have suggested that inositol 1,4,5-trisphosphate ($InsP_3$) may function as a chemical messenger linking t-tubule depolarization to Ca^{2+} release from sarcoplasmic reticulum (SR). Observations which support this proposal include: 1) $InsP_3$ induces Ca^{2+} release and contracture in permeablized skeletal muscle fibers; 2) $InsP_3$ induces Ca^{2+} release in SR vesicles; 3) $InsP_3$ activates SR Ca^{2+} channels in vesicles fused to planar lipid bilayers; 4) the lipid precursors and enzymes responsible for $InsP_3$ production are present in t-tubule membranes; and 5) electric stimulation of muscle fibers increases the intracellular $InsP_3$ concentration milliseconds after stimulation[1]. Other investigators have reported conflicting results which suggest that ECC may result from mechanical rather than chemical coupling[2,3].

We have recently demonstrated that the diaphragm can metabolize phosphatidylinositol 4,5-bisphosphate ($PtdInsP_2$) to $InsP_3$ with subsequent sequential breakdown to inositol 4,5-$InsP_3$ bisphosphate and 4-monophosphate. Furthermore, strips of diaphragm muscle, fatigued in vitro by intermittent direct electrical stimulation have an associated decrease in $InsP_3$ production[4]. The current study was designed to investigate the effect of fatigue on the membrane PIP pool.

METHODS

Diaphragms were removed from Sprague-Dawley rats (200-300g), and divided into four strips, 1-2 cm in length. Tissues were incubated in Krebs-Henseleit (KH) buffer containing 20 uCi of ^3H-inositol (25°C). The costal end of each strip was fastened at the ribs to a rigid support containing a bipolar electrode. All tissues were maintained at 250°C in

Excitation-Contraction Coupling in Skeletal, Cardiac, and Smooth Muscle
Edited by G.B. Frank *et al.*, Plenum Press, New York, 1992

aerated KH buffer (95% oxygen and 5% CO_2) containing 1.5×10^{-6}g/ml d-tubocurarine. Force displacement was measured via a Grass FTO3 displacement transducer attached to the central tendon of each muscle strip. Voltage was adjusted to achieve supramaximal electrical stimulation (Grass, Model S88) followed by optimal fiber length adjustment.

After 25 minutes equilibration in KH with 10 mM LiCl, control tissues were contracted maximally with a 50 Hz tetanus for 10 seconds, to induce IP formation. Experimental tissues were fatigued by direct intermittent stimulation: 20 Hz trains, 2.0 ms pulses, 50% duty cycle, 90 trains/minute for 5 minutes prior to maximal contraction (50 Hz tetanus for 10 seconds). While still in the contracted state, tissues were frozen in liquid N_2 and ground to powder. The PIP and IP fractions were then separated by chloroform/methanol/1M HCl extraction, 1:1:1 (v/v) and after addition of H_2O, quantified by the radioactivity present in the organic and aqueous phases respectively[5].

RESULTS

Determination of the time dependence of ^{3}H-inositol incorporation into the PIP pool in resting muscle strips revealed a significantly greater incorporation at each of the measured time points in dpm/mg wet wt. (1 hour, 257; 2 hour, 1102; 3 hour, 1627). Developed tension decreased 45% in response to the 50 Hz tetanus after fatigue and was significantly decreased (3.6 ± 1g, P 0.05) compared to control tissues (6.5 ± 2g). Inositol polyphosphates were significantly increased (P $\overline{0}$.05) above background in response to a 50 Hz tetanus (13.0 ± 3.4 vs 29.3 ± 4.6 dpm/mg wet wt.), however following fatigue, IP were decreased, 38%, to 19 ± 0.3 in response to a tetanus. Following fatigue, the ^{3}H-PIP pool remaining after the 50 Hz tetanus was significantly greater than control (911 ± 294 vs 518 ± 37 dpm/mg wet wt.) as expected from the decreased levels of IP.

DISCUSSION

The data showed that following fatigue, PIP hydrolysis in response to a maximal contraction was significantly reduced. This was demonstrated by the decrease in IP and the parallel greater amount of PIP remaining in the membrane. Previous studies have shown a parallelism between the generation of $InsP_3$ and the magnitude of a sustained diaphragmatic contraction[4]. In addition other investigators have suggested a role for phosphoinositide metabolism in excitation-contraction coupling as described earlier[1,2,5]. In spite of these findings, hydrolysis of $PtdInsP_2$ to $InsP_3$ may not be the mechanism initiating release of Ca^{2+} from the sarcoplasmic reticulum[3]. $InsP_3$ may be released secondarily following the rise in sarcoplasmic Ca^{2+} and play a role in amplifying the initial Ca^{2+} transient[2]. Whether inhibition of PIP hydrolysis by rapid intermittent stimulation is merely associated with or the cause of diaphragmatic fatigue will require further study.

REFERENCES

1. Sanchez, X., Carrasco, M.A., Vergara, J. and Hidalgo, C. Inositol 1,4,5-trisphosphate activity in membranes isolated from amphibian skeletal muscle. FEBS 279:58 (1991).

2. Volpe, P., DiVirgillo, F., Bruschi, G., Regolisti, G., Pozzan, T. Phosphoinositiol metabolism and excitation-contraction coupling in smooth, cardiac and skeletal muscle. In: "Inositol Lipids in Cell Signaling." Michell, R.H., Drummond, A.H., Downes, C.P., eds. . Academic Press, San Diego, CA. (1989).
3. Walker, J.W., Somlyo, A.V., Goldman, Y.E., Somlyo, A.P., Trentham, D.R. Kinetics of smooth and skeletal muscle activation by laser pulse photolysis of caged inositol 1,4,5-trisphosphate. Nature 327:249, (1987).
4. Zavecz, J.H., Anderson, W.McD., Phillips, K. and McNatt, L. Diaphragmatic fatigue is associated with decreased inositol 1,4,5-trisphosphate formation. Am Rev Respir Dis 141:792 (1990).
5. Lagos, N., Vergara, J. Phosphoinositides in frog skeletal muscle: A quantitative analysis. Biochem Biophys Acta 1043:235, (1990).

PHYSIOLOGICAL ROLE AND LOCALIZATION OF THE NEW OUABAIN RECEPTOR PROTEIN (31.5 KD) FROM CAT CARDIAC MUSCLES, USING THE MONOCLONAL AB AGAINST THE PROTEIN

Masako Fujino*, Sumiko Fujino, Kumi Satoh,
Tohru Nakai, and Tetsuo Kado

*Department of Pharmacology, Sapporo Medical College
Sapporo, Japan
Department of Pharmacology
Hokkaido Institute of Pharmaceutical Sciences
Otaru, Japan

SUMMARY

Role and localization of the new ouabain receptor protein (31.5 kD) were studied by using its monoclonal AB. The protein may be a key protein in the E-C coupling process in the cat cardiac muscle.

We recently reported that a new glycoprotein of 31.5 kD, which has a high affinity for ouabain and is independent of (Na^+-K^+) ATPase, was solubilized from transverse tubule membrane-junctional SR complexes (TTM-JSR) in cat cardiac ventricle muscles[1]. To clarify the mechanism of ouabain potentiation, physiological role and cellular localization of the protein were studied by using the monoclonal AB.

1. The specificity of the purified AB was established by Western blot analysis. The AB revealed a single band of 31.5 kD from CHAPS extracts in the non-reducing condition.

2. Kitten papillary muscles driven electrically (0.1 Hz) were immersed in a slightly hypertonic Tyrode solution (200 mM NaCl) containing the AB for 60 min, and then washed out. Thirty min after the AB-removal, both twitch tension and K-contracture tension were inhibited without affecting resting- and action-potentials and caffeine contracture (Fig. 1).

3. Recently, phenylglyoxal (PGO), a specific guanidyl modifying reagent, has shown to inhibit twitch and K-contracture in frog single skeletal muscles, without affecting resting membrane potential, action potentials, or caffeine-contracture[2,3]. The kitten papillary muscles were immersed in a Tyrode solution containing 5 mM ^{14}C-PGO for 10 min and then washed out. Sixty min after the PGO removal, the twitch was markedly inhibited without affecting the action potential. At this point, ^{14}C-PGO-binding protein from TTM-JSR in the muscles was solubilized by means of Concanavalin A affinity chromatography. The SDS PAGE profile showed that the molecular size of the protein is 31.5 kD in the non-reducing condition. Antagonistic effects of ouabain and PGO on the E-C coupling process in cardiac muscles were observed (Fig. 2).

Excitation Contraction Coupling in Skeletal, Cardiac, and Smooth Muscle
Edited by G.B. Frank *et al.*, Plenum Press, New York, 1992

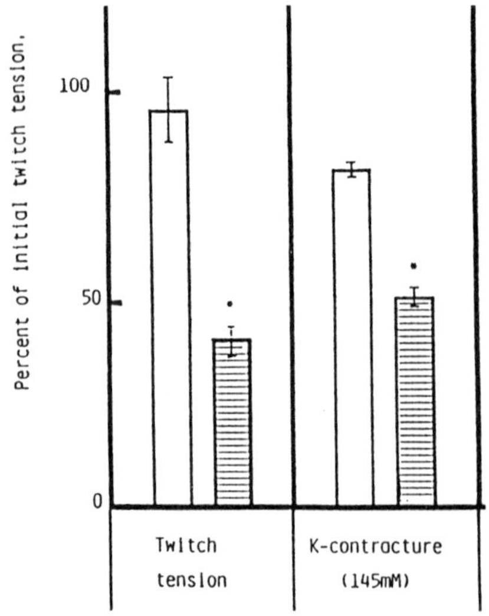

Figure 1

Effect of monoclonal AB against the 31.5 kD protein on contractile responses in cat papillary muscles.

4. Intracellular localization of the protein was studied by immunohistochemical electron microscopy. It seems that the protein localizes on the T-tubule membrane against the bead which occupies the TTM-junctional SR gap. These results suggest that ouabain potentiation takes place through the influence of ouabain on the E-C coupling process[4].

Conclusively, the new ouabain receptor protein may be a key protein in the E-C coupling process of the cardiac muscle.

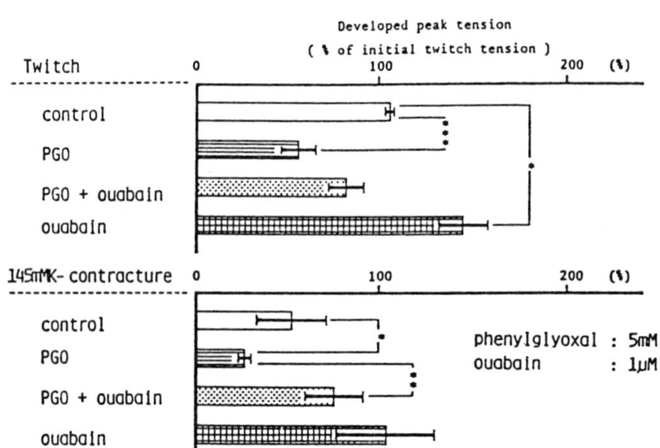

Figure 2

Effects of phenylglyoxal and ouabain on contractile responses in papillary muscle of cat.

References

1. Fujino, S. et al.; Experientia, 45: 467 (1989).
2. Fujino, M. et al., J. Physiol. Soc. Japan, 45: 507 (1983).
3. Fujino, M. et al., Proc. Int. Union Physiol. Sci., 16:207 (1989).
4. Fujino, S. et al., Can. J. Physiol. Pharmacol., 60:542 (1982).

NA-CA EXCHANGE CURRENT DURING THE CARDIAC ACTION POTENTIAL

Glenna Bett, Denis Noble, Susan Noble, Yung Earm, Won-Kyung Ho, and In-Suk So

University Laboratory of Physiology, Oxford, England, and Physiology Department, Seoul National Laboratory, Seoul, Korea

Atrial Cells

Action potentials recorded from rabbit atrial cells have two distinct phases: an initial spike of depolarization followed by a late low plateau. This late plateau is abolished in the presence of ryanodine (which blocks calcium release from the SR) or when external sodium is replaced by lithium ions. The fully activated current-voltage relationship of the plateau current displays a near exponential curve, characteristic of the sodium-calcium exchanger (Earm, Ho & So 1990). The ionic currents that flow during the action potential of an isolated rabbit atrial cell were reconstructed in a model developed by Earm and Noble (1990), based on the multicellular Hilgemann & Noble (1987) model. The separation between the calcium current, I_{Ca}, and the exchange current, I_{NaCa} is almost complete. The calcium current is rapidly deactivated once repolarization (initiated by the outward transient current) begins. The amount of calcium entering the cell via the voltage gated calcium channels is approximately equal to that which is extruded by the exchanger, and thus the intracellular calcium concentration returns to the resting level by the end of the action potential.

Ventricular Cells

Though some ventricular cells (e.g., adult rat) have a similar shape of action potential to those observed in atrial cells, most species (e.g., man, guinea-pig, rabbit, cat, and dog) have a plateau at a greater depolarization than that seen in atrial cells, thus forming a 'squarer' action potential. Determining the relative contributions of the currents to the shape of the action potential in guinea pig is more difficult than in atrial cells because the inactivation of the calcium current occurs more slowly. Calcium still enters the cells through the voltage gated calcium channels when the sodium calcium exchange begins extruding calcium. Experiments performed on isolated guinea-pig ventricular cells by Egan et al., (1989) separated out the sodium calcium exchange contribution by voltage clamping the cell to the resting potential (-90mV) at various times after the start of an action potential. This produces a slowly decaying inward tail current.

Excitation-Contraction Coupling in Skeletal, Cardiac, and Smooth Muscle
Edited by G.B. Frank *et al.*, Plenum Press, New York, 1992

Ventricular Model

A model of the isolated guinea-pig ventricular cell has recently been developed. It is based on the single atrial cell model, with the following major modifications:
(1) the size of the transient outward current, I_{to}, has been reduced;
(2) the rate of inactivation of the voltage gated calcium channel current has been slowed;
(3) the steepness of the voltage dependence of activation of the inward rectifier current, I_{K1}, and the K_m for potassium activation have been altered to fit the experimental results of Sakmann and Trube (1984);
(4) the uptake and release of calcium by the SR has been fitted to data from fura-2 and indo-1 studies.

The model successfully reproduces the tail currents seen by Egan et al. At concentrations above 100nM the relationship between the internal calcium concentration and the exchange current is approximately linear, so the envelope of the tail current peaks reflects the shape of the intracellular calcium transient. This model can also be used to look at the size of the tail currents in the first 50 ms of the action potential, i.e., in the rising phase of the calcium transient, where the tail currents rise slowly to a peak before decaying. Although there is a large efflux of calcium through the sodium calcium exchange at this time, calcium dependent calcium release is still in progress, which produces tail currents with a complex time course. This suggests that the release is at least partly regenerative, and the envelope of the early tail current peaks is not a good indicator of the internal calcium concentration.

The model assumes that the exchanger can be activated at low internal calcium concentrations, but experimental results are not conclusive and the kinetics of activation have yet to be determined. Varying the internal sodium concentration over the range of physiological concentrations has a large effect on the exchange current, varying its amplitude and even changing the direction in which the exchanger works. Thus the sodium calcium exchange is a good regulator of cardiac activity: its strong calcium and sodium dependency form a sensitive feedback control system which, following a sudden change in cellular conditions, initially works to bring the cell back to its steady state.

References

Earm, Y.E., Ho, W.K., and So, I.S., 1990, *Proc. Roy. Soc.* **B 240**: 61-81.
Earm, Y.E., & Noble, D., 1990, *Proc. Roy. Soc.* **B 240**: 83-96.
Egan, T.M., Noble, D., Noble, S.J., Powell, T., Spindler, A.J., & Twist, V.W., 1989, *J. Physiol.* **411**: 639-661.
Hilgemann, D.W., & Noble, D., 1987, *Proc. Roy. Soc.* **B 230**: 163-205.
Sakmann, B & Trube, G., 1984, *J. Physiol.* **347**: 641-657.

All computations were carried out using HEART version 3.3 which is available from OXSOFT Ltd., 49 Old Road, Oxford, OX3 7JZ, UK.

AUTHOR INDEX

INDEX